Studies in Theoretical and Applied Statistics

Selected Papers of the Statistical Societies

For further volumes:
http://www.springer.com/series/10104

Series Editors

Spanish Society of Statistics and Operations Research (SEIO)

Société Française de Statistique (SFdS)

Società Italiana di Statistica (SIS)

Sociedade Portuguesa de Estatística (SPE)

João Lita da Silva • Frederico Caeiro •
Isabel Natário • Carlos A. Braumann
Editors

Manuel L. Esquível • João Tiago Mexia
Associate Editors

Advances in Regression, Survival Analysis, Extreme Values, Markov Processes and Other Statistical Applications

Springer

Editors
João Lita da Silva
Frederico Caeiro
CMA and Faculdade de Ciências
e Tecnologia
Universidade Nova de Lisboa
2829-516 Caparica
Portugal

Isabel Natário
CEAUL and Faculdade de Ciências
e Tecnologia
Universidade Nova de Lisboa
2829-516 Caparica
Portugal

Carlos A. Braumann
Centro de Investigação em Matemática e Aplicações
Universidade de Évora
7000-803 Évora
Portugal

Associate Editors
Manuel L. Esquível
João Tiago Mexia
CMA and Faculdade de Ciências
e Tecnologia
Universidade Nova de Lisboa
2829-516 Caparica
Portugal

ISSN 2194-7767 ISSN 2194-7775 (electronic)
ISBN 978-3-642-44629-0 ISBN 978-3-642-34904-1 (eBook)
DOI 10.1007/978-3-642-34904-1
Springer Heidelberg New York Dordrecht London

Printed on acid-free paper

Springer is part of Springer Science+Business Media (www.springer.com)

To João Tiago Mexia, on his 70th birthday, as a token of gratitude for a lifelong dedication to Statistics and for his unyielding and everlasting support of younger colleagues

João Lita da Silva
Frederico Caeiro
Isabel Natário
Carlos A. Braumann
Manuel L. Esquível

Foreword

Dear reader, On behalf of the four Scientific Statistical Societies – the SEIO, Sociedad de Estadística e Investigación Operativa (Spanish Statistical Society and Operation Research); SFdS, Société Française de Statistique (French Statistical Society); SIS, Società Italiana di Statistica (Italian Statistical Society); and the SPE, Sociedade Portuguesa de Estatística (Portuguese Statistical Society) – we would like to inform you that this is a new book series of Springer entitled "Studies in Theoretical and Applied Statistics," with two lines of books published in the series: "Advanced Studies" and "Selected Papers of the Statistical Societies."

The first line of books offers constant up-to-date information on the most recent developments and methods in the fields of theoretical statistics, applied statistics, and demography. Books in this series are solicited in constant cooperation between the statistical societies and need to show a high-level authorship formed by a team preferably from different groups so as to integrate different research perspectives.

The second line of books presents a fully peer-reviewed selection of papers on specific relevant topics organized by the editors, also on the occasion of conferences, to show their research directions and developments in important topics, quickly and informally, but with a high level of quality. The explicit aim is to summarize and communicate current knowledge in an accessible way. This line of books will not include conference proceedings and will strive to become a premier communication medium in the scientific statistical community by receiving an Impact Factor, as have other book series such as "Lecture Notes in Mathematics."

The volumes of selected papers from the statistical societies will cover a broad range of theoretical, methodological as well as application-oriented articles, surveys and discussions. A major goal is to show the intensive interplay between various, seemingly unrelated domains and to foster the cooperation between scientists in different fields by offering well-founded and innovative solutions to urgent practice-related problems.

On behalf of the founding statistical societies I wish to thank Springer, Heidelberg and in particular Dr. Martina Bihn for the help and constant cooperation in the organization of this new and innovative book series.

Rome, Italy Maurizio Vichi

Preface

This volume of the selected papers from Portugal is a product of the Seventeenth Congress of the Portuguese Statistical Society. The meeting took place at the beautiful resort seaside city of Sesimbra in 2009. At the meeting there were 5 invited 1-hour conferences, 118 papers presented in 20-min talks, and 68 in poster sessions. Following all these presentations, 72 papers were submitted by March 2010. More than 200 participants had fruitful opportunities for learning about the latest ideas and methods being developed in a broad variety of statistical domains such as linear models and regression, survival analysis, extreme value theory, and statistics of diffusions and other Markov processes. Many of the papers and posters presented were applied studies in areas where statistics has a powerful and recognized role as an intelligence-gathering and management tool. The works presented at the congress that were submitted for publication in this volume were thoroughly refereed by a rigorous team of scientific experts, which the editorial team would like to express its heartfelt thanks to. The papers selected were carefully edited in order to provide readers with an extensive, reliable reference work on the subjects treated, as well as a thorough account of the best contributions presented at this major scientific event.

Lisbon, Portugal
June 9, 2012

João Lita da Silva
Frederico Caeiro
Isabel Natário
Carlos A. Braumann
Manuel L. Esquível
João Tiago Mexia

Acknowledgements

The editorial team hereby expresses, to the experts named below, its most heartfelt gratitude for the invaluable assistance in the peer-review process of this selected paper's volume.

Alessandra Luati,
Alessio Fercomeni,
Aline Araújo Nobre,
Ana Bela Cruzeiro,
Ana Cristina Braga,
Ana Gomes,
Ana Luísa Papoila,
Ana Maria Pires,
Ana Paula Martins,
Andreia Hal,
António Pacheco,
Arminda Manuela Gonçalves,
Cinzia Carota,
Cláudia Nunes,
Cláudio Lupi,
Corrado Provasi,
Cristina Rocha,
Dewi Rahardja,
Dinis Pestana,
Dora Gomes,
Dulce Pereira,
Elizabeth Reis,
Elsa Moreira,
Esmeralda Gonçalves,
Fernanda Cipriano,
Fernanda Figueiredo,
Fernanda Sousa,
Fernando Rosado,
Filipe Marques,
Francesco Lagona,
Giovani Silva,
Gracinda Guerreiro,
Helena Bacelar-Nicolau,
Helena Ferreira,
Helena Iglésias,
Inês Sequeira,
Inês Sousa,
Irene Oliveira,
Isabel Barão,
Isabel Fraga Alves,
Isabel Pereira,
João A. Bastos,
João Andrade e Silva,
João Branco,
José Afonso Faias,
Júlia Teles,
Ka Chun Cheung,
Kamil F. Turkman,
Laurens de Haan,
Luís Machado,
Luís Ramos,
Luísa Canto e Castro,
Luísa Pereira,
Luzia Gonalves,
Maria do Rosário de Oliveira,
Manuel Alberto Ferreira,
Manuel Scotto,
Manuela Hill,
Manuela Neves,
Marcel Vieira,
Marcello Chiodi,
Margarida Cardoso,
Maria Antónia Turkman,
Maria Conceição Serra,
Maria Cristina Miranda,
Maria de Fátima Salgueiro,
Nazaré Mendes-Lopes,
Maria do Rosário Grossinho,
Maria Eduarda Silva,
Maria Eugénia Ferrão,
Maria Eugénia Graça Martins,
Maria Graça Temido,
Maria Ivette Gomes,
Maria João Martins,
Maria Lucília Carvalho,
Marília Antunes,
Marília Salomé Cabral,
Mário Figueiredo,
Marta Ferreira,
Mathieu Kessler,
Miguel Fonseca,
Paula Brito,
Paula Milheiro,
Paulo Canas Rodrigues,

Paulo Eduardo de Oliveira,
Paulo Soares,
Pedro Corte Real,
Pedro Duarte Silva,
Pedro Palhinhas Mota,
Ricardo Covas,
Russell Alpizar-Jara,
Sílvia Emiko Shimakura,
Sílvio Velosa,
Sofia M. de Azeredo Lopes,
Stanislaw Mejza,
Suely Giolo,
Tiago Marques,
Valeska Andreozzi,
Wolfgang Schmid.

Contents

Part I

Invited Sessions

Youden Square with Split Units

Stanisław Franciszek Mejza and Shinji Kuriki

Abstract
In this chapter we present the most important problems connected with the design of experiments using Youden squares with split units. In fact we consider two types of designs. The first is connected with different arrangements of subplot treatments on the units of Youden squares. The second is connected with the design of experiments when one or more treatments arranged in Youden squares are control or standard treatments. We characterize some of these designs with respect to general balance property and with respect to design efficiency factors.

1 Introduction

In performing experiments we quite often use a row–column design in order to eliminate real or potential orthogonal disposed heterogeneity of experimental material. In this case the Latin square is the appropriate design. This design possesses many desirable and optimal statistical properties. In the Latin square every treatment occurs once in each row and once in each column. It means that this design uses many experimental units. We can reduce the number of experimental units by using a design in which every treatment occurs once in each row (and not in each column) or vice versa. Then the Youden square is the proper design, with many

S.F. Mejza (✉)
Department of Mathematical and Statistical Methods, Poznan University of Life Sciences, Wojska Polskiego 28, PL-60-637 Poznań, Poland
e-mail: smejza@up.poznan.pl

S. Kuriki
Department of Mathematical Sciences, Graduate School of Engineering, Osaka Prefecture University, 1-1 Gakuen-cho, Naka-ku, Sakai, Osaka 599-8531, Japan
e-mail: kuriki@ms.osakafu-u.ac.jp

J. Lita da Silva et al. (eds.), *Advances in Regression, Survival Analysis, Extreme Values, Markov Processes and Other Statistical Applications*, Studies in Theoretical and Applied Statistics, DOI 10.1007/978-3-642-34904-1_1,

desirable statistical properties (see, e.g., [3]). In the Youden square the treatments occur in completely randomized blocks with respect to rows (columns), while with respect to columns (rows) they occur in a balanced incomplete block design (BIBD). In the experiments considered here the units of a Youden square are subdivided into the same number of subunits. The structure of the experimental material is described formally below.

Let us assume that the experimental material is divided into k_0 superblocks. Each superblock constitutes a row–column design with k_1 rows and k_2 columns. On each unit of the row–column design that is treated as a whole plot, the levels of a factor A ($A_1, A_2, \cdots, A_a$) are arranged. These levels will be called whole-plot treatments. Additionally, each whole plot is divided into k_3 small plots called subplots; on each subplot the levels of the second factor B ($B_1, B_2, \cdots, B_b$) are arranged. These levels are called subplot treatments.

In this chapter we will examine the statistical properties of a design in which each superblock has a Youden square structure with q rows and a columns. It is assumed that a subdesign of the Youden square with respect to columns is a BIBD. More important is the problem of arranging the subplot treatments. We will consider some statistical properties of designs in which subplot treatments will be arranged in BIBD or in a group divisible (GD) partially balanced incomplete block design with two association classes (GDPBIBD(2)) in such a way that the contents of whole plots within a superblock are the same with respect to subplot treatments.

There exist designs for experiments which serve to compare existing treatments (also called test treatments) with a set of control treatments or standards. In this chapter we consider a case in which we wish to compare only whole-plot treatments with certain controls (whole-plot controls).

2 Modelling

In this chapter the statistical properties of the above designs are examined under a mixed linear model of observations. The dispersion structure of a linear model results from the scheme of randomization applied. This scheme includes randomization of superblocks, rows (columns), columns (rows) and subplots. As a result of such randomizations and some additional assumptions, we can describe the observations by a linear mixed model with random superblock, row and column effects and fixed treatment combination effects.

The randomization scheme leads to a linear mixed model having an orthogonal block structure (cf. [10]). Details concerning modelling and analysis of observation obtained in experiments carried out in row–column designs with split units are given in [6]. The design considered here is a particular case of the above-mentioned designs.

3 Analysis

As mentioned above, the linear model of observations being considered has an orthogonal block structure. Thus the overall analysis can be split into so-called strata, as in multistratum experiments (cf. [4]). In our case we have five strata, namely the inter-superblock stratum, inter-column stratum, inter-row stratum, inter-whole-plot stratum and finally inter-subplot stratum (cf. [6]).

In this chapter the treatment combinations will be considered as treatments with the actually used lexicographical order of combinations $A_t B_s$ $(t = 1, 2, \cdots, a; s = 1, 2, \cdots, b)$ and the usual expression of the treatment effect as the sum of the factor effects and the interaction effects. Let $v = ab$ denote the number of treatments.

The statistical properties of the design are connected with the algebraic properties of the so-called information matrices $\mathbf{A}_i$, $i = 1, 2, 3, 4, 5$, which in the considered design have the forms:

$$\mathbf{A}_1 = (k_0 k_1 k_2 k_3)^{-1} \left(k_0 \mathbf{N}_0 \mathbf{N}_0' - \mathbf{r}\mathbf{r}'\right) = \frac{q}{a b_B k_B} \mathbf{J}_a \otimes \left(b_B \mathbf{N}_B \mathbf{N}_B' - r_B^2 \mathbf{J}_b\right)$$

$$\mathbf{A}_2 = (k_1 k_2 k_3)^{-1} \left(k_1 \mathbf{N}_1 \mathbf{N}_1' - \mathbf{N}_0 \mathbf{N}_0'\right) = \mathbf{O},$$

$$\mathbf{A}_3 = (k_1 k_2 k_3)^{-1} \left(k_2 \mathbf{N}_2 \mathbf{N}_2' - \mathbf{N}_0 \mathbf{N}_0'\right) = \frac{a-q}{(a-1)k_B} \left(\mathbf{I}_a - \frac{1}{a}\mathbf{J}_a\right) \otimes \mathbf{N}_B \mathbf{N}_B',$$

$$\mathbf{A}_4 = (k_1 k_2 k_3)^{-1} \left(\mathbf{N}_0 \mathbf{N}_0' + k_1 k_2 \mathbf{N}_3 \mathbf{N}_3' - k_2 \mathbf{N}_2 \mathbf{N}_2' - k_1 \mathbf{N}_1 \mathbf{N}_1'\right)$$
$$= \frac{a(q-1)}{(a-1)k_B} \left(\mathbf{I}_a - \frac{1}{a}\mathbf{J}_a\right) \otimes \mathbf{N}_B \mathbf{N}_B',$$

$$\mathbf{A}_5 = \mathbf{r}^{\delta} - k_3^{-1} \mathbf{N}_3 \mathbf{N}_3' = q \mathbf{I}_a \otimes \left(r_B \mathbf{I}_b - \frac{1}{k_B} \mathbf{N}_B \mathbf{N}_B'\right),$$

where

$$\mathbf{r}\mathbf{r}' = q^2 r_B^2 \mathbf{J}_a \otimes \mathbf{J}_b, \quad \mathbf{r}^{\delta} = q r_B \mathbf{I}_a \otimes \mathbf{I}_b, \quad \mathbf{N}_0 \mathbf{N}_0' = q^2 \mathbf{J}_a \otimes \mathbf{N}_B \mathbf{N}_B',$$
$$\mathbf{N}_1 \mathbf{N}_1' = q \mathbf{J}_a \otimes \mathbf{N}_B \mathbf{N}_B', \quad \mathbf{N}_2 \mathbf{N}_2' = \mathbf{N}_A \mathbf{N}_A' \otimes \mathbf{N}_B \mathbf{N}_B', \quad \mathbf{N}_3 \mathbf{N}_3' = q \mathbf{I}_a \otimes \mathbf{N}_B \mathbf{N}_B',$$
$$\mathbf{N}_A \mathbf{N}_A' = (r_A - \lambda_A)\mathbf{I}_a + \lambda_A \mathbf{J}_a = \frac{q(a-q)}{a-1}\mathbf{I}_a + \frac{q(q-1)}{a-1}\mathbf{J}_a.$$

The matrix $\mathbf{J}_t$ denotes the $t \times t$ matrix of ones; $\mathbf{N}_0, \mathbf{N}_1, \mathbf{N}_2, \mathbf{N}_3$ are the incidence matrices: treatments vs. superblocks, treatments vs. columns, treatments vs. rows and treatments vs. whole plots, respectively; $\mathbf{r}$ denotes the vector of treatment replicates; and $\mathbf{r}^{\delta}$ stands for the diagonal matrix with diagonal elements equal to the numbers of treatment replicates. The other parameters are defined in the next sections.

The statistical properties of a design are related to the eigenvectors and eigenvalues of these matrices. It can be checked that $\mathbf{A}_f \mathbf{1}_v = \mathbf{0}$, i.e., one of the eigenvectors (corresponding to the eigenvalue 0) is proportional to $\mathbf{1}_v$. This means that the other eigenvectors define (basic) contrasts of treatment parameters. These eigenvectors can be the same or different in the strata. The property of a design guaranteeing that all information matrices have the same set of eigenvectors is called general balance (cf. [4, 10]). This property simplifies the analysis of variance and further statistical inference. For details, readers are referred to, for example, [1, 7]

The design is generally balanced iff (cf. [7])

$$\mathbf{A}_f \mathbf{r}^{-\delta} \mathbf{A}_{f'} = \mathbf{A}_{f'} \mathbf{r}^{-\delta} \mathbf{A}_f, \quad f \neq f', \; f, f' = 1, 2, 3, 4, 5.$$

In this chapter we will focus our investigations on the general balance property and the efficiency factors of the designs proposed.

4 Constructions

4.1 Characterization of Youden Square

In this chapter we will examine the statistical properties of a design in which each superblock has a Youden square structure with q rows and a columns. Moreover, let us assume that the subdesign of the Youden square with respect to columns is a symmetrical BIBD with parameters as follows: BIBD($v_A, b_A, r_A, k_A, \lambda_A$).

Then the following relationships hold:

$$k_1 = k_A = r_A = q \quad \text{and} \quad k_2 = b_A = v_A = a, \quad \lambda_A = \frac{q(q-1)}{a-1}.$$

Let $\mathbf{N}_A$ be the treatment $\times$ column incidence matrix in a Youden square. Then the so-called C matrix for the Youden square subdesign with respect to columns is equal to $\mathbf{C}_A = r_A\mathbf{I} - k_A^{-1}\mathbf{N}_A\mathbf{N}'_A$. With this matrix is connected the so-called efficiency factor that is equal to $\varepsilon_A = \frac{a(q-1)}{q(a-1)}$ with multiplicity $\rho_A = a - 1$.

The design with respect to rows is a complete randomized block design.

4.2 Subplot Designs

Case 1. Subplot Treatments in BIBD

Let the subplot treatments be arranged in the BIBD with the parameters $v_B = b$, $b_B, r_B, k_B, \lambda_B, k_0 = b_B, k_3 = k_B$.

Let $\mathbf{C}_B$ denote the C matrix of BIBD for subplot treatments and let $\varepsilon_B = \frac{b(k_3-1)}{k_3(b-1)}$ denote the efficiency factor with multiplicity $\rho_B = b - 1$.

The final design is generally balanced and the efficiency factors are presented in Table 1 (cf. [5]).

Case 2. Subplot Treatments in GDPBIBD(2)

Let the subplot treatments be arranged in a group divisible (GD) partially balanced incomplete block design with two association classes (GDPBIBD(2)) (cf. [2]). In the

Table 1 Stratum efficiency factors

Type of contrasts	Number of contrasts	Strata I	II	III	IV	V
A	$a-1$	0	0	$1-\varepsilon_A$	ε_A	0
B	$b-1$	$1-\varepsilon_B$	0	0	0	ε_B
A×B	$(a-1)(b-1)$	0	0	$(1-\varepsilon_A)(1-\varepsilon_B)$	$\varepsilon_A(1-\varepsilon_B)$	ε_B

Table 2 Stratum efficiency factors

Type of contrasts	Number of contrasts	Strata I	II	III	IV	V
A	$a-1$	0	0	$1-\varepsilon_A$	ε_A	0
$B^{(1)}$	ρ_{B1}	$1-\varepsilon_{B1}$	0	0	0	ε_{B1}
$B^{(2)}$	ρ_{B2}	$1-\varepsilon_{B2}$	0	0	0	ε_{B2}
$A\times B^{(1)}$	$(a-1)\rho_{B1}$	0	0	$(1-\varepsilon_A)(1-\varepsilon_{B1})$	$\varepsilon_A(1-\varepsilon_{B1})$	ε_{B1}
$A\times B^{(2)}$	$(a-1)\rho_{B2}$	0	0	$(1-\varepsilon_A)(1-\varepsilon_{B2})$	$\varepsilon_A(1-\varepsilon_{B2})$	ε_{B2}

GDPBIBD(2) the number of treatments is equal to mn, where m denotes the number of groups each of n treatments. Generally, the parameters of the GDPBIBD(2) for subplot treatments are as follows: $v_B = b = mn$, b_B, r_B, k_B, λ_{B1}, λ_{B2}. The parameters λ_{B1}, λ_{B2} denote the numbers of occurring pairs of treatments from the same group and different groups in the blocks, respectively. The statistical properties of the GDPBIBD(2) are connected with the algebraic properties of the concurrence matrix $\mathbf{N}_B\mathbf{N}'_B$, where $\mathbf{N}_B$ denotes the incidence matrix for subplot treatments. The concurrence matrix $\mathbf{N}_B\mathbf{N}'_B$ has three eigenvalues ω_i with multiplicities ρ_i, where

$$\omega_0 = r_B k_B, \quad \omega_1 = r_B - \lambda_{B1}, \quad \omega_2 = r_B k_B - v_B \lambda_{B2},$$
$$\rho_0 = 1, \quad \rho_1 = m(n-1), \quad \rho_2 = m-1.$$

Let us note that in the incomplete case, as we have for subplot treatments, only a few of the treatments occur on whole plots. In this chapter we assume that the contents of whole plots within each superblock are all the same with respect to subplot treatments. Hence the following relationships hold:

$$k_0 = b_B, \quad k_1 = q, \quad k_2 = a, \quad k_3 = k_B.$$

Let $\mathbf{C}_B$ denote the C matrix of the GDPBIBD(2) for subplot treatments and let $\varepsilon_{Bi} = 1 - \frac{\omega_i}{r_B k_B}$ denote the efficiency factors with multiplicities ρ_{Bi}, $i = 0, 1, 2$, where $\sum_{i=0}^{2} \rho_{Bi} = v_B$. The overall statistical properties of the final design (Table 2) are connected with those efficiency factors (cf. [9]). The ranks of the information matrices $\mathbf{A}_i$, $i = 1, 2, 3, 4, 5$, depend on the type of the GDPBIBD(2).

Table 3 Stratum efficiency factors

Type of contrasts	Number of contrasts	Strata I	II	III	IV	V
$A^{(1)}$	$a-1$	0	0	$1-\varepsilon^*_{A1}$	ε^*_{A1}	0
$A^{(2)}$	$s-1$	0	1	0	0	0
$A^{(3)}$	1	0	1	0	0	0
B	$b-1$	$1-\varepsilon_B$	0	0	0	ε_B
$A^{(1)}\times$B	$(a-1)(b-1)$	0	0	$(1-\varepsilon^*_{A1})(1-\varepsilon_B)$	$\varepsilon^*_{A1}(1-\varepsilon_B)$	ε_B
$A^{(2)}\times$B	$(s-1)(b-1)$	0	$1-\varepsilon_B$	0	0	ε_B
$A^{(3)}\times$B	$b-1$	0	$1-\varepsilon_B$	0	0	ε_B

5 Control Treatments

Experiments are performed in order to compare existing treatments (also called test treatments) with a set of control treatments or standards. In the chapter we consider a case in which we wish to compare only whole-plot treatments with certain controls (whole-plot controls). Then, the supplementation of the Youden square we can express by the following incidence matrix:

$$\mathbf{N}^*_A = \begin{bmatrix} \mathbf{N}_A \\ \mathbf{J}_{s\times b_A} \end{bmatrix}.$$

More exactly, a test treatments of the factor A are assigned in the $(q\times a)$ Youden square, and additionally s control treatments are added.

Using the characterization of the Youden square from Sect. 4.1 with the whole-plot test treatments × columns incidence matrix $\mathbf{N}_A$, we have $k_A = r_A = q$, $k_2 = b_A = v_A = a$, $\lambda_A = \frac{q(q-1)}{a-1}$, $\varepsilon_A = \frac{a(q-1)}{q(a-1)}$, with multiplicity $\rho_A = a-1$.

The final design with respect to the whole-plot treatments has the following parameters:

$$v^*_A = a+s, \quad k^*_A = q+s, \quad b^*_A = b_A, \quad \mathbf{r}^*_A = [q\mathbf{1}'_a \vdots a\mathbf{1}'_s]'$$

$$\varepsilon^*_{A0} = 1, \quad \rho^*_{A0} = s, \quad \varepsilon^*_{A1} = 1 - \frac{k_A(1-\varepsilon_A)}{k_A+s}, \quad \rho^*_{A1} = a-1.$$

Similarly, using characterization of subplot treatments occurring in the BIBD with the parameters: $v_B = b, b_B, r_B, k_B, \lambda_B$ and incidence matrix $\mathbf{N}_B$, we have the parameters of the design with respect to subplot treatments $k_0 = b_B$, $k_1 = a+s$, $k_3 = k_B$ and $\lambda_B = \frac{r_B(k_3-1)}{b-1}$, $\varepsilon_B = \frac{b(k_3-1)}{k_3(b-1)}$, with multiplicity $\rho_B = b-1$.

Finally, the stratum efficiency factors of that design are presented in Table 3.

In the above table, $A^{(1)}$ represents the set of contrasts among effects of whole-plot test treatments only; $A^{(2)}$ represents the set of contrasts among the effects of the whole-plot control treatments only; and $A^{(3)}$ represents the set of contrasts among the effects of the whole-plot test and control treatments only.

Table 4 Stratum efficiency factors

Type of contrasts	Number of contrasts	Strata I	II	III	IV	V
$A^{(1)}$	3	0	0	1/15	14/15	0
$A^{(2)}$	1	0	1	0	0	0
$A^{(3)}$	1	0	1	0	0	0
B	2	1/4	0	0	0	3/4
$A^{(1)} \times B$	6	0	0	1/60	7/30	3/4
$A^{(2)} \times B$	2	0	1/4	0	0	3/4
$A^{(3)} \times B$	2	0	1/4	0	0	3/4

6 Example

Let us consider a two factor experiment, in which there are $a = 4$ whole-plot test treatments as well as $s = 2$ whole-plot control treatments and $b = 3$ subplot treatments. Moreover, the experiment is set up in $k_0 = 3$ superblocks divided into $k_1 = 5$ rows and $k_2 = 4$ columns. This means that in each of the superblock we have 20 whole-plots that are additionally divided into 2 subplots. Then on the three rows we arrange the whole-plot test treatments according to the Youden square scheme.

A_2	A_4	A_3	A_1
A_3	A_1	A_4	A_2
A_4	A_2	A_1	A_3
C	C	C	C
C	C	C	C

The subplot treatments occur in all whole plots in the BIB design according to the scheme:

B_1	B_1	B_2
B_2	B_3	B_3

The final stratum efficiency factors are presented in Table 4.

Finally, the most general case of the considered designs is considered by [8]. In particular, the whole-plot treatments occur in a repeated Youden square, while the subplot treatments occur on subplots in a proper incomplete block design. The statistical properties of the final design are examined.

Acknowledgements The authors would like to thank to JSPS and PAN for giving them the opportunity to carry out this joint research in Japan and in Poland. The research was also partially supported by JSPS Grant-in-Aid for Scientific Research (C) 21540142.

References

1. Bailey, R.A.: General balance: artificial theory or practical relevance? In: Caliński, T., Kala, R. (eds.) Proceedings of the International Conference on Linear Statistical Inference LINSTAT '93, pp. 171–184. Kluwer, Amsterdam (1994)
2. Clatworthy, W.H.: Tables of two-associate-class partially balanced designs. In: NBS Applied Math. Ser. 63, Washington, D.C. (1973)
3. Cox, D.R.: Planning of Experiments. Wiley, New York (1958)
4. Houtman, A.M., Speed, T.P.: Balance in designed experiments with orthogonal block structure. Ann. Stat. **11**, 1069–1085 (1983)
5. Kachlicka, D., Hering, F., Mejza, S.: Control treatments in Youden square with split units. Folia Fac. Sci. Nat. Univ. Masaryk. Brunensis, Mathematica **15**, 137–144 (2004)
6. Kachlicka, D., Mejza, S.: Repeated row-column designs with split units. Comput. Stat. Data Anal. **21**, 293–305 (1996)
7. Mejza, S.: On some aspects of general balance in designed experiments. Statistica, anno LII, **2**, 263–278 (1992)
8. Mejza, S., Kachlicka, D., Mejza, I., Kuriki, S.: Repeated Youden squares with subplot treatments in a proper incomplete block design. Biometrical Lett. **45**(2), 50–61 (2009)
9. Mejza, S., Kuriki, S., Kachlicka, D.: Repeated Youden squares with subplot treatments in a group-divisible design. J. Stat. Appl. **4**(2–3), 201–209 (2009)
10. Nelder, J.N.: The analysis of experiments with orthogonal block structure. R. Soc. Lond. **A283**, 147–178 (1965)

Likelihood and PLS Estimators for Structural Equation Modeling: An Assessment of Sample Size, Skewness and Model Misspecification Effects

Manuel J. Vilares and Pedro S. Coelho

Abstract
This chapter aims to contribute to a better understanding of partial least squares (PLS) and maximum likelihood (ML) estimators' properties, through the comparison and evaluation of these estimation methods for structural equation models with latent variables based on customer satisfaction data. Although PLS is a well-established tool to estimate structural equation models, more work is still needed in order to better understand its properties and relative merits when compared to likelihood methods. Despite the controversy over these two estimation techniques, their complexity makes any analytical comparison very difficult to be made. Therefore, it constitutes a fertile ground for conducting simulation studies. This chapter continues the research of Vilares et al. [Comparison of likelihood and PLS estimators for structural equation modelling: a simulation with customer satisfaction data. In: Vinzi, W.E., Chin, W.W., Henseler, J., Wang, H. (eds.) Handbook of Partial Least Squares. Concepts, Methods and Applications, pp. 289–307. Springer Handbooks of Computational Statistics, Springer (2010)], which has compared PLS and ML estimators using Monte Carlo simulation within three different frameworks (symmetric data, skewed data and formative blocks). It also continues to generate the data according to the ECSI (European Customer Satisfaction Index) model with the assumption that the coefficients of the structural and measurement models are known. This new chapter introduces the effect of sample size and includes two different

M.J. Vilares (✉)
ISEGI, Universidade Nova de Lisboa and Bank of Portugal, Lisbon, Portugal
e-mail: mjv@isegi.unl.pt

P.S. Coelho
ISEGI, Universidade Nova de Lisboa, Lisbon, Portugal

Faculty of Economics, Ljubljana University, Ljubljana, Slovenia
e-mail: psc@isegi.unl.pt

J. Lita da Silva et al. (eds.), *Advances in Regression, Survival Analysis, Extreme Values, Markov Processes and Other Statistical Applications*, Studies in Theoretical and Applied Statistics, DOI 10.1007/978-3-642-34904-1_2,

simulations. The first one is conducted in a context of both symmetric data and skewed response data. This simulation is conducted for the sample sizes $n = 50, 100, 150, 250, 500, 1{,}000$ and $2{,}000$ and uses reflective blocks. A second simulation includes the presence of model misspecifications (omissions of an existent path) for a sample size of 250 observations and symmetric data. In all simulations the ability of each method to adequately estimate the inner model coefficients and indicator loadings is evaluated. The estimators are analysed in terms of bias and dispersion (standard deviation). Results have shown that overall PLS estimates are generally better than covariance-based estimates. This is particularly true when the data is asymmetric, when estimating the model for smaller sample sizes and for the inner model structure.

1 Introduction

Structural equation modeling (SEM) inspires enthusiastic praise as well as persistent rejection. In an Internet survey carried on 2003 [12] one can see quotes emphasizing different points of view, like "the technique of Structural Equation Modeling represents the future of data analysis" and "Nobody really understands SEM." Nevertheless SEM is getting more and more popular. Indeed, the citation frequency in psychological literature has steadily increased since the 1970s, reaching the popularity of ANOVA as it can be checked in the citation frequencies of SEM and (M) ANOVA in the APA PsyncINFO [12]. In terms of estimation methods for SEM, covariance-based methods are undoubtedly the most well-known methods with the result that many social researchers use the terms SEM and covariance-based methods synonymously. Partial least squares (PLS) methods constitute one alternative to estimating SEM. However, in spite of the growing usage of PLS methods in several fields (for instance in customer satisfaction studies), these methods are still often seen *as ad hoc algorithms that have generally not been formally analysed* [11]. Several authors (e.g. [3, 7]) argue that PLS methods present several advantages when compared to covariance-based methods. In fact, it is argued that in order to these later methods produce consistent parameter estimates, the empirical conditions of the data require multivariate normal distribution and independence of observations. Moreover, indicators are typically required to be reflective and unique case values for latent variables cannot be obtained. On the contrary, beyond being based on simpler algorithms, PLS methods don't require any assumptions regarding the joint distribution of indicators or the independence of observations. On the other hand, unique case values for the latent variables are estimated. Also indicators can be modelled in either direction (i.e. formative or reflective). However, there is not neither a formal proof nor a simulation study in the framework of a realistic model that sow these advantages of PLS techniques over covariance-based methods. In fact, Fornell and Bookstein [7] conducted a simulation study for the two kinds of estimation methods, but they used an extremely simple model[1] and they only consider the framework of colinearity between indicators. Cassel et al. [2] also conducted a simulation study to access the performance of PLS estimates. The authors have also used a simplified version of ECSI (European Customer Satisfaction Index) model and access the robustness of PLS estimators in the presence of multicollinearity between manifest or latent variables, in the presence of skewness of manifest variables and in the presence of misspecification (erroneous omission of manifest and latent variables). Nevertheless covariance-based estimates are not obtained in this study and therefore the relative merits of PLS cannot be accessed. Chin and Newsted [4] organize an experiment allowing to access both sample size and block size effects on PLS estimators. However, they do not compare the two estimation methods.

In a very recent chapter, Vilares et al. [16] proceed to a simulation study comparing PLS and ML estimators in the context of two assumptions: the symmetric distribution and the reflective modelling of the indicators. They compare how the two kinds of methods (PLS and covariance-based methods) perform both when these assumptions hold and when they are violated, i.e. when the distribution of the observations is skewed and the indicators are modelled according to a formative scheme. They do this analysis in the framework of the ECSI model and for a fixed sample size of 250 observations. The main goal of this research is to contribute to a deeper comparison of the two estimation methods, through the simulation with a realistic model and on conditions of different sample sizes and model misspecification. We release some of the assumptions carried on [16] in order to overcome some of its shortcomings. More specifically we use the ECSI model and the present simulation will have two goals. The first one is to access the sample size effect on the comparison of PLS and ML estimators in the context of symmetric and skewed response data. It is particularly interesting to evaluate the robustness of the two estimation methods to the presence of small sample sizes and skewed response data. These are usual situations in the typical areas of application for SEM (marketing, psychology, information systems, etc.). The second goal of this research is to compare the robustness of the PLS and ML estimators in the presence of a model misspecification (omission of an existent path in the structural model). In this case, we will use a sample size of 250 observations.

The structure of the remaining part of this chapter is as follows. Section 2 presents ECSI model. Two different estimation procedures for structural equation models are presented in Sect. 3: Sects. 3.1 and 3.2 synthesize the covariance-based methods and PLS, respectively. The organization of the simulation study, including the data-generating process, is shown in Sect. 4. Section 5 presents and analyses the main results obtained in this simulation. The main conclusions are presented in Sect. 6. The chapter concludes with the references (Sect. 6) and an appendix with detailed results of the simulations.

---

[1] The model has only three latent variables (two endogenous and one exogenous). Moreover no relation is assumed between the two endogenous variables.

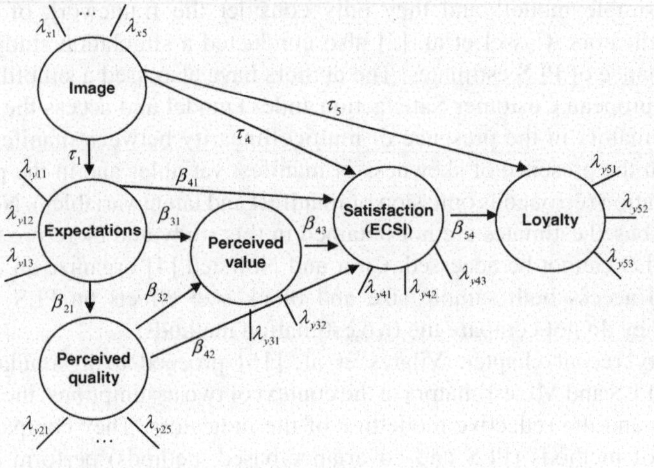

Fig. 1 The ECSI model

2 The ECSI Model

The ECSI model is a framework that aims to harmonize the national customer satisfaction indices in Europe. It is an adaptation of the Swedish Customer Satisfaction Barometer [6] and of the ACSI-American Customer Satisfaction Index [9]. The ECSI model is presented in detail in ECSI (1998) and some of the more relevant issues discussed there are briefly presented in this section.

The ECSI model is composed of two sub-models: the structural model and the measurement model. The structural model includes the relations between the latent or non-observable variables and is represented in Fig. 1. Customer satisfaction is the central variable of this model, having as antecedents or drivers the corporate image of the company, customer expectations, perceived quality of products and services and perceived value. The main consequent of customer satisfaction as specified by the model is customer loyalty. The model is therefore constituted by one exogenous latent variable *(image)* and five endogenous variables. The measurement model defines the relations between the latent variables and the observed indicators or manifest variables. One may have two kinds of measurement models:

- A reflective model when the observed indicators are assumed to be the reflex of the latent variables (the arrow is directed to the observed indicator from its latent variable)
- A cause or formative model when the observed indicators are assumed to cause or form the latent variables (the arrows are directed to the latent variables from their indicators)

The ECSI structural and measurement models may be described by the following equations:

2.1 Structural Model

We have five equations (i.e. the same number of endogenous variables) that can be written in a compact form as

$$\eta = \beta\eta + \tau\xi + \upsilon \tag{1}$$

where η is a vector (5×1) of endogenous latent variables (all except image), ξ is *th* exogenous latent variable (*image*), β and τ are impact matrices and υ is a vector (5×1) of specification errors. We will assume the usual properties about these errors (zero mean, homocedasticity and zero covariance between the errors). More specifically the matrices of structural coefficients β and τ are the following:

$$\beta = \begin{bmatrix} 0 & 0 & 0 & 0 & 0 \\ \beta_{21} & 0 & 0 & 0 & 0 \\ \beta_{31} & \beta_{32} & 0 & 0 & 0 \\ \beta_{41} & \beta_{42} & \beta_{43} & 0 & 0 \\ 0 & 0 & 0 & \beta_{54} & 0 \end{bmatrix}, \qquad \tau = \begin{bmatrix} \tau_1 \\ 0 \\ 0 \\ \tau_4 \\ \tau_5 \end{bmatrix} \tag{2}$$

with ξ: *image*; η_1: *customer expectations*; η_2: *perceived quality of products and services*; η_3: *perceived value*; η_4: *customer satisfaction*; η_5: *customer loyalty*.

2.2 Measurement Model

The ECSI measurement model uses a reflective scheme, described by the equations:

$$\mathbf{y} = \Lambda_y\eta + \epsilon \tag{3}$$

$$\mathbf{x} = \Lambda_x\xi + \delta \tag{4}$$

$$E(\epsilon) = E(\delta) = E(\epsilon|\eta) = E(\delta|\xi) = \mathbf{0}$$

where $\mathbf{y}' = (y_1, y_2, \ldots, y_p)$ and $\mathbf{x}' = (x_1, x_2, \ldots, x_q)$ are the manifest endogenous and exogenous variables, respectively. Λ_y and Λ_x are the corresponding parameters matrices (*loadings*) and ϵ and δ are specification errors.

Representing by $\mathbf{y}'_i = (y_{i1}, y_2, \ldots, y_{iH_i})$ the vector of manifest variables related to the latent endogenous variable η_i and by $\mathbf{x}' = (x_1, \ldots, x_G)$ the vector of manifest variables related to the latent exogenous variable ξ, we can also write the model in the form

$$y_{ij} = \lambda_{yij}\eta_i + \epsilon_{ij}, \quad i = 1, \ldots, 5; j = 1, \ldots, H_i$$

$$x_j = \lambda_{xj}\xi + \delta_j, \quad j = 1, \ldots, G$$

where H_i is the number of manifest variables associated with variable η_i and G is the number of manifest variables associated with variable ξ.

All the indicators are obtained through the administration of a questionnaire to customers, using a scale from 1 to 10 where the value 1 expresses a very negative opinion from the customer and the vale 10 a very positive opinion.

3 Estimation Procedures

The ECSI model to estimate is composed by Eqs. (1)–(4). The estimation of ECSI model as well as of other structural equation models (SEM) faces several difficulties, from which we emphasize:

- The latent variables are not observed.
- The measurement indicators that correspond to the answers to a customer satisfaction questionnaire may not follow a normal distribution. The distribution of the frequencies of these indicators is in general not symmetric and typically presents skewness to the right.
- The measurement variables often present some level of multicollinearity.
- Some blocks hardly can be seen as reflective. This is the case of the exogenous latent variable (*image*), where theory behind the measurement model suggests that the latent variable may be of a formative nature, i.e. the indicators may be viewed as the cause of the latent variable.

Two families of methods have been used to estimate this type of models: the PLS methods and the covariance-based methods. We will present in this section a brief introduction of these two groups of methods.

3.1 Covariance-Based Methods

This group of methods is the most widely adopted. According to [13], the different covariance-based methods are variations on the minimization of a common general discrepancy function:

$$F = (S - \Sigma)' w^{-1} (S - \Sigma) \qquad (5)$$

where S is a vector of the unique (no redundant) elements of the sample covariance or correlation matrix, Σ is a parallel vector of elements from the model-implied matrix and w is a matrix of weights. The different methods correspond to different matrices w. The two most widely used estimation methods are the generalized least squares (GLS) (with w as the variance and covariance matrix of the residuals)[2]

[2]This GLS used in SEM estimation methods is analogous to Generalized Least Square methods used in regression context with an important difference, however. The residuals considered in the discrepancy function F correspond to differences between two types of covariances (of the sample and of the model). On the other hand the residuals in the regression context usually mean differences between observed and estimated values for endogenous variables.

and the maximum likelihood (ML) method (that uses the fitting function $ln|\Sigma| + trace(S/\Sigma) - ln|S| - p$, with p being the number of indicators).

There are different softwares to minimize Eq. (5) or to maximize the likelihood corresponding function. Among these softwares the most known is LISREL (linear structural relations) that is a ML method and it is so much associated to the estimation of structural equation models that it is often confused with the SEM itself. ML methods produce asymptotically unbiased, consistent and efficient estimators under the empirical conditions that the indicators follow a multivariate normal distribution; the sample is large and independence of observations [1]. When these assumptions are not verified, these methods may produce, according to several authors (e.g. [3, 7]), improper solutions such as negative variance estimates. Moreover, these methods do not provide unique values for the scores or case values of latent variables since there is an infinite set of possible scores that are consistent with the parameter estimates. Finally all the indicators must be treated in a reflective manner because to model otherwise would create a situation where we are unable to explain the covariances of all indicators, which is the rationale for this approach [3].

3.2 PLS Methods

PLS for structural equation modelling may be seen as a distribution-free method, since no assumption is made about the distribution of measurement variables or even about the independence of observations. The PLS approach has two stages: in the first one estimates the observations of the latent variables (case values) with an iterative scheme. In the second stage one estimates the parameters of the structural equations and measurement model. In opposition to covariance-based methods, PLS aims to minimize the variance of the dependent variables (both latent and measurement variables).

PLS is supported by an iterative process that iterates between two approximations to the latent variables: the inner approximation and the outer approximation. In each iteration the outer approximation produces an estimate for each latent variable as a linear combinations of their manifests. The inner approximation produces another estimate for the latent variables. Here each variable is obtained as a combination of the external approximation of the other latent variables directly connected to it. Various weighting schemes have been used in this context, being the best well-known: the centroid, the factor and the path weighting scheme. The two estimations are iterated until convergence is reached, i.e. when the weights obtained in outer approximation stabilize. Finally, the structural model (1) is estimated, using Ordinary Least Squares for each equation and each latent variable replaced by its estimate. The adoption of OLS is possible since model (1) is recursive and as a consequence the matrix of the parameters of endogenous variables is triangular.

The case values of the latent variables are inconsistent due to the fact that they are estimated as aggregates of the observed or manifest variables (cf. outside approximation), which include a measurement error. This bias that is the differences between the estimated and the "true" latent variable scores will tend to zero as the number of indicators per both block and sample size increases. This limiting case is termed "consistency" at large and this property has been argued as a justification for using PLS as an estimation method to estimate LISREL parameters in case where the number of manifest variables is large [14].

PLS is the estimation method adopted for estimating ECSI model. There are several presentations of the PLS methodology in this framework (e.g. [2, 5]). More general descriptions of the PLS methodology may be found in [8, 10, 15, 17].

4 The Simulation Study

In reason of the complexity of either SEM models and PLS and covariance-based methods, the analysis of their relative merits and their robustness when some of their assumptions are violated can hardly be assessed in analytical form, particularly in the framework of a realistic model. This is a fertile ground to the use of simulation studies.

4.1 Sample Size and Skewed Response Data

In order to access the sample size effect on the comparison of PLS and ML estimators in the context of symmetric and skewed response data, we have settled a Monte Carlo simulation using two models: one, named symmetric data model but also referred as base model, where all blocks are reflective and the measurement variables show a symmetric distribution; the other, named skewed data model, where all blocks are reflective but the measurement variables show an asymmetric right skewed distribution. This violation was motivated for being typical of customer satisfaction data that frames the study. We used sample sizes of $n = 50, 150, 250, 500, 1,000$ and 2,000 observations.

The simulation aims to analyse the quality of PLS and ML estimates of structural model coefficients (matrices β and τ) and of measurement model coefficients (Λ_y, Λ_x and λ_ξ) in the context of the two variants: symmetric or base model and skewed data model. PLS and ML estimators of model coefficients are analysed in terms of bias and dispersion (as measured by the standard deviation). The bias of an estimator of a generic coefficient β_{ij} is obtained as $B_{\beta_{ij}} = K^{-1}\sum_{k=1}^{K}(\hat{\beta}_{ij,k} - \beta_{ij})$ and the standard deviation by $SD_{\beta_{ij}} = \sqrt{K^{-1}\sum_{k=1}^{K}(\hat{\beta}_{ij,k} - \bar{\hat{\beta}}_{ij})^2}$, where K represents the number of replicates in the simulation and $\hat{\beta}_{ij,k}$ the estimate of β_{ij} obtained with replicate k by one estimation method (PLS or ML). To access the validity of simulation results we have additionally computed the simulation error regarding the estimator biases as $ES_{\beta_{ij}} = 1.96\sqrt{(K(K-1))^{-1}\sum_{k=1}^{K}(\hat{\beta}_{ij,k} - \bar{\hat{\beta}}_{ij})^2}$.

The simulation was run using the SAS system. The PLS approach was implemented through a SAS macro and the ML estimation using CALIS procedure.

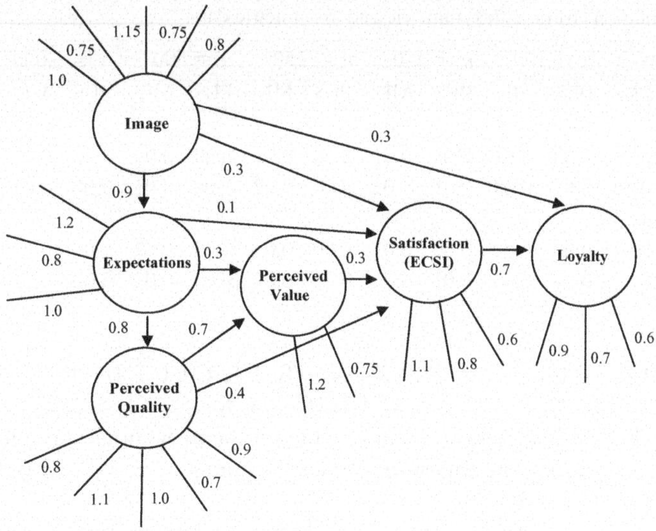

Fig. 2 The postulated ECSI model

4.2 The Data-Generating Process

The starting point of our simulation is the ECSI model (cf. Fig. 1). The data are generated according to the ECSI model, where we have assumed that the coefficients of both models (structural and measurement models) were known.

The values for inner and outer model coefficients were chosen in order to be as similar as possible to the ones that would be obtained with real-world data. For that we have observed typical estimates of model coefficients obtained through the estimation of ECSI model applied to different companies and industries and postulated a model structure consistent with those estimates (Fig. 2).

Thus, the postulated structural model is:

$$\eta_1 = 0.9\xi_1 + \upsilon_1$$

$$\eta_2 = 0.8\eta_1 + \upsilon_2$$

$$\eta_3 = 0.3\eta_1 + 0.7\eta_2 + \upsilon_3$$

$$\eta_4 = 0.3\xi_1 + 0.4\eta_2 + 0.3\eta_3 + \upsilon_4$$

$$\eta_5 = 0.3\xi_1 + 0.7\eta_4 + \upsilon_5$$

where ξ_1 is the exogenous variable image and $\eta_1 - \eta_5$ are endogenous variables that represent customer expectations, perceived quality, perceived value, customer satisfaction and customer loyalty. The measurement models for the endogenous variables are of reflective kind, assuming the following values for the parameters:

Table 1 Simulation errors with symmetric and asymmetric data

	$n = 50$		$n = 100$		$n = 150$		$n = 250$		$n = 500$		$n = 1{,}000$		$n = 2{,}000$	
	PLS	ML	PLS	ML	PLS	ML	PLS	ML	PLS	ML	PLS	ML	PLS	ML
Symmetric data														
Outer	0.00	0.01	0.00	0.01	0.00	0.01	0.00	0.01	0.00	0.01	0.00	0.01	0.00	0.00
Inner	0.01	0.10	0.01	0.05	0.01	0.02	0.00	0.02	0.00	0.01	0.00	0.01	0.00	0.01
Asymmetric data														
Outer	0.00	0.02	0.00	0.02	0.00	0.02	0.00	0.02	0.00	0.02	0.00	0.02	0.00	0.00
Inner	0.01	0.10	0.01	0.05	0.01	0.02	0.00	0.02	0.00	0.02	0.00	0.02	0.00	0.02

$\lambda_{1j} = 1.2, 0.8, 1.0$ for $j = 1,2,3$, $\lambda_{2j} = 0.8, 1.1, 1.0, 0.7, 0.9$ for $j = 1,\ldots,5$, $\lambda_{3j} = 1.2, 0.75$ for $j = 1,2$, $\lambda_{4j} = 1.1, 0.8, 0.6$ for $j = 1,2,3$, $\lambda_{5j} = 0.9; 0.7, 0.6$, for $j = 1,2,3$. The measurement scheme for the exogenous variable, image, is also reflective assuming the following values for parameters: $\lambda_j = 1.0, 0.75, 1.15, 0.9, 0.8$ for $j = 1,\ldots,5$.

For the base model, the cases of the exogenous latent variable were generated using a $\beta(4;4)$ symmetric distribution in the interval [1; 10] and all the errors both in the inner and outer models were generated using a $\beta(3;3)$ symmetric distribution in the interval $[-1.5; 1.5]$. For the skewed data model, we used both for the cases of the exogenous latent variable and the errors a right skewed distribution $\beta(10;4)$. In the two models, the values of the measurement variables were converted into scores in the ten-point scale 1–10, which is the scale used in ECSI and ACSI questionnaires.

For the simulation, we have generated 1,500 data sets of 50 observations and 1,000 data sets of 100; 150; 250; 500; 1,000 and 2,000 observations. A total of 75,000; 100,000; 150,000; 250,000; 1,000,000 and 2,000,000 observations for the 21 measurement variables were obtained.

We have computed the simulation errors regarding the estimator biases for the inner and outer model and using both estimation methods (PLS and ML) in case of symmetric data and of asymmetric data. The results are shown in Table 1. These results show that the simulation errors are very small. They decrease as the sample size increases and they are always smaller when we adopt PLS (instead of ML) estimation procedure.

4.2.1 The Model Misspecification

To compare the performance of PLS and ML estimators in the presence of a model misspecification, we have omitted, in the structural model see Fig. 2, the image-satisfaction path that is of medium size (it is estimated in 0, 3) and also the perceived quality–perceived value path that is of large size (it is estimated in 0, 7). We have estimated by the two methods the symmetric data model with a sample size of 250 observations. We have compared the new biases with the ones obtained with the base symmetric data model to access the robustness of PLS and ML estimators to these model misspecifications.

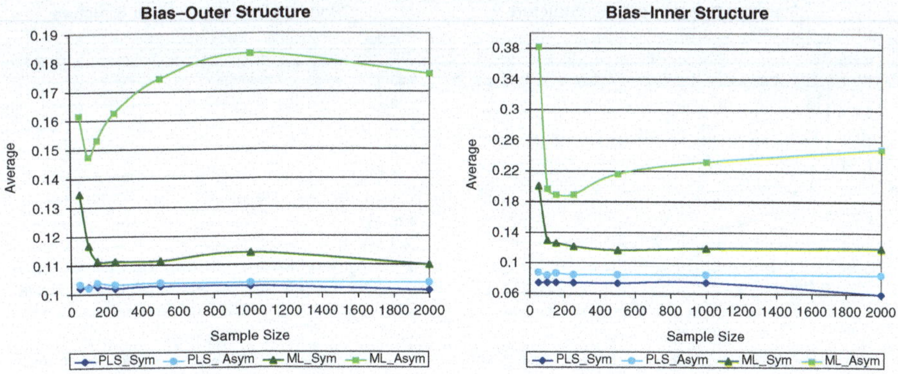

Fig. 3 Bias—outer structure (*left*) and Bias—inner structure (*right*)

5 Simulation Results

The goal of our simulation study is to analyse and compare Maximum Likelihood and PLS estimators properties for structural equation models based on customer satisfaction data, both in terms of bias and precision.

5.1 Sample Size and Skewed Response Data

Simulation results are illustrated in Figs. 3 and 4. Figure 3 illustrates the bias of model parameters (loadings and inner model coefficients) both for the PLS and ML techniques and for the two model formulations (symmetric response data model and skewed response data). The precision of the estimates is assessed through the standard deviations of these parameter estimates and they are presented in Fig. 4. In both figures, the averages of the absolute bias and the averages of standard deviations for the inner and outer structures are also shown[3].

In terms of the biases of the parameters, Fig. 3 illustrates that:

- PLS estimators always show smaller biases than ML estimators (particularly with asymmetric data).
- The bias of PLS estimators is approximately independent of the sample size and it is very similar when using symmetric and asymmetric data.

[3]Detailed simulation results regarding the symmetric data model are shown in Tables 2 and 3 and results referring to asymmetric data appear in Tables 4 and 5 in appendix. Tables 2 and 4 show the bias of model parameters (loadings and inner model coefficients) both for the PLS and ML techniques. The dispersion of the estimators is accessed through the standard deviations which are presented in Tables 3 and 5.

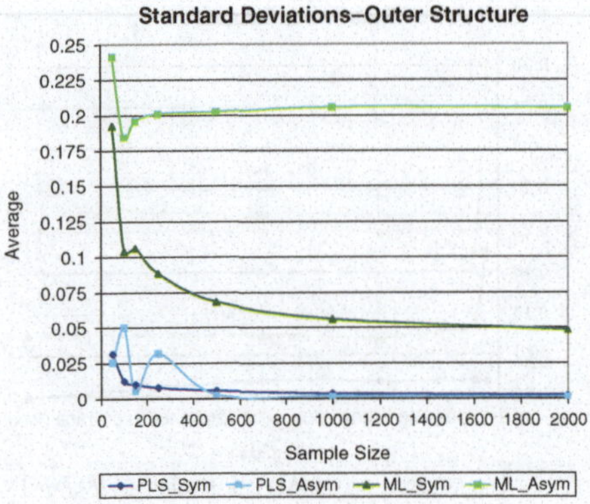

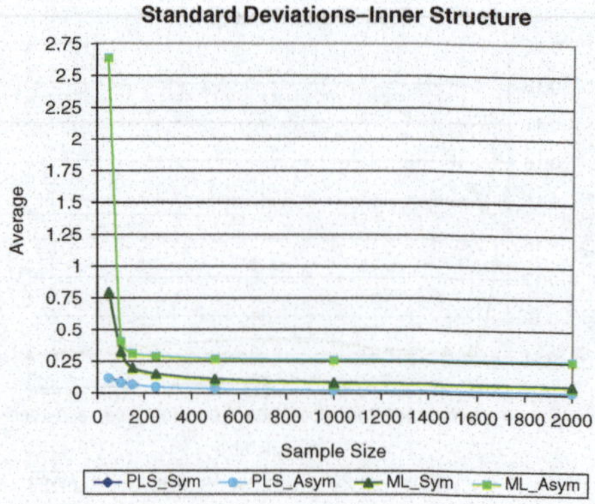

Fig. 4 Standard deviations—outer structure (*left*) and standard deviations—inner structure (*right*)

- ML estimators have a significantly higher biases with asymmetric data; for symmetric data bias is relatively constant after $n = 150$, but with asymmetric data it tends to grow with the sample size, at least until $n = 950$.
- The patterns shown for the outer structure remain in the inner structure.
- The relative advantages of PLS performance in terms of bias are more noticeable in this structure, especially when we deal with small sample sizes.
- With asymmetric data, the differences between biases of both estimators tend to increase, as the bias of ML estimators increase with n (for $n > 150$).

On the other hand, in terms of the precision of the estimators, the following points should be emphasized (see Fig. 4):

- PLS estimators show smaller dispersion than ML estimators for any sample size with sym /asym data.
- With symmetric data both for PLS and ML standard deviation always decrease as n increases.
- PLS tends to show similar dispersion both with symmetric and asymmetric data for $n > 500$. However ML estimators always show higher dispersion in the presence of asymmetries.
- With asymmetric data ML shows a bad property of increase standard deviation with n (after $n = 100$).
- PLS estimators show again smaller dispersion for any sample size.
- PLS shows similar performance both with symmetric and asymmetric data, but ML estimators always show higher standard deviations with asymmetric data.
- The dispersion of the ML estimators reduces as n increases, but the difference with PLS estimators in terms of dispersion never decreases.

5.2 Model Misspecification

Figures 5–8 summarize the results regarding the model misspecifications. Figures 5 and 6 show the results for the outer model in case of the omission of the path

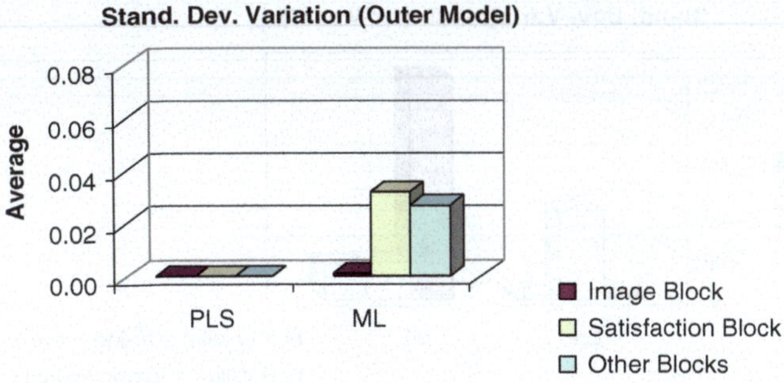

Fig. 5 Bias variation (outer model) in case of image-satisfaction omission

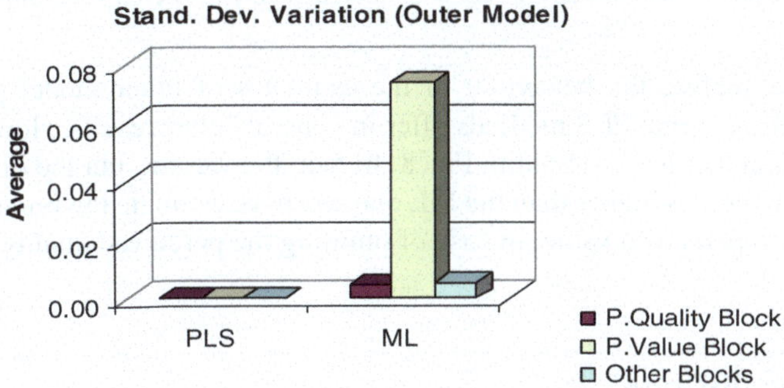

Fig. 6 Bias variation (outer model) in case of perceived quality-perceived value

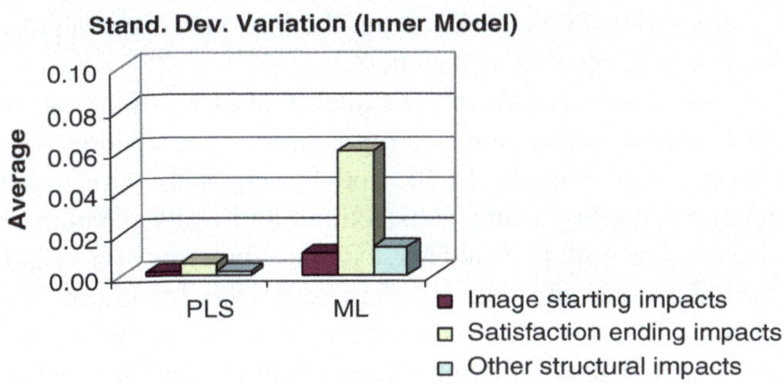

Fig. 7 Bias variation (inner model) in case of image-satisfaction omission

between image and satisfaction and in case of omission of the path between perceived quality and perceived value, respectively. Figures 7 and 8 deal with the inner model for the same omissions.

The first two figures show that in the PLS context, the bias of the loadings is very small in both cases and their estimates remain almost unchanged. On the contrary, when using ML estimators, the bias increases, especially when estimating the loadings included in the explained latent variable in the omitted relationship (satisfaction and perceived value, respectively).

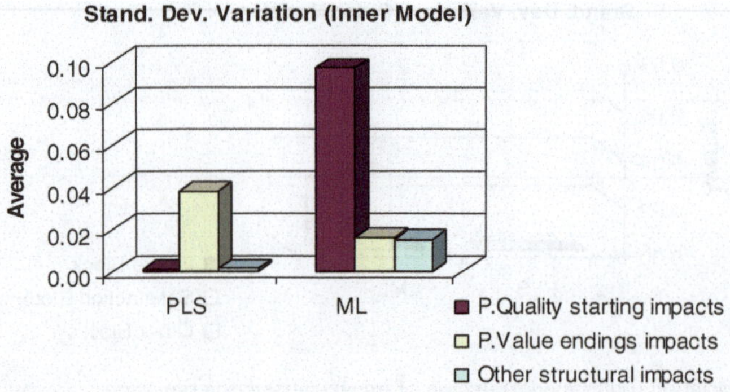

Fig. 8 Bias variation (inner model) in case of perceived quality-perceived value omission

When we analyse the behaviour of the estimates of inner model parameters, Figs. 7 and 8 show that PLS methods offer in general better results. However, there is an exception that is illustrated in Fig. 8. In fact, the variation in the dispersion of the PLS estimators is higher than the ML one when we estimate the impacts that are antecedents to perceived value, in case of omitting the perceived quality perceived value link.

6 Conclusions

Although the covariance-based procedures are by far the most well-known techniques among structural equation modelling, the PLS approach can also be a very useful tool that can be applied by researchers.

Our chapter gives some insights into the quality of PLS estimation when applied to a structural equation model representing customer satisfaction data. We have postulated a model similar to the ECSI model composed by six latent variables (image, expectations, quality, value, satisfaction and loyalty). Within a simulation study we have evaluated both PLS and ML estimates in terms of bias and dispersion when estimating the inner and outer model coefficients for sample sizes ranging from $n = 50$ to $n = 2,000$. We have used two models: one with symmetric data and one variant where data are obtained with a right-skewed distribution. This is a situation that is typical of customer satisfaction data.

Overall, results have shown that PLS estimates are generally better than ML estimates both in terms of bias and dispersion. This is particularly true when using small sample sizes, asymmetric data and when estimating the inner model structure. ML estimators seem to be much more sensitive to the introduction of asymmetries and to the use of small sample sizes, usually producing very poor results for $n < 250$. This means that in the symmetric context, the use of PLS estimators is crucial when using small sample sizes, especially when estimating the inner coefficients. For large sample sizes both methods tend to converge. Nevertheless, when estimating the structural model, the PLS estimators still tend to show slightly better properties.

PLS estimators seem also to be more robust, to the introduction of asymmetric data, for all sample sizes (bias and standard deviations are almost unchanged). The quality of ML estimators is usually very poor (particularly in the inner model and for small sample sizes) and bias shows a bad property of increasing with sample size.

The consideration of PLS and ML simulation errors does not change these conclusions. The PLS estimators are never outperformed by the ML ones, even with large sample sizes. Another interesting result is that when data is symmetric, there is a tendency of PLS estimators to overestimate measurement model coefficients and to underestimate structural model coefficients that seem to be independent of the sample size. The ML method shows exactly the opposite tendency. But with asymmetric data, the ML estimators no longer show a systematic underestimation of the outer model from $n = 50$ to $n = 2,000$.

Finally, concerning sample size effects, the simulation results show that PLS performance, in terms of bias, is superior when estimating the inner model coefficients than when estimating the measurement model coefficients. ML methodology has the opposite tendency, generating smaller biases when estimating the outer structure.

Now, when we incur in model misspecifications, the PLS estimators seem to be more robust to such situation. When a structural path is omitted, the bias and the dispersion of the PLS estimators always show smaller increases than the ML results (with one exception). Moreover, the omission of a path in the structural model influences less the estimation of the outer structure of blocks involved in the omission than the inner structure of the model. For PLS, loadings estimation remain almost unchanged. On the other hand, the paths associated to the latent variables (LV) involved in the omitted link (especially to the explained LV) are the ones whose estimation quality is more affected by the model misspecification. The estimation quality of the other structural coefficients is less affected, especially if PLS is used (ML shows more contamination to the estimation of other coefficients).

The simulations also show that the size of the omitted link in the structural model seems to be important in the estimation of the inner coefficients. The degradation of the estimators properties is higher when omitting the P.Quality–P.Value link than when omitting the image-satisfaction link. A major limitation of this simulation is the fact that we have not considered different levels of skewness in data. Further research should be done in order to understand how different levels of skewness in the measurement variables affect the properties of the two estimators (PLS and ML). Also, further research is needed to validate the sensitivity of both methods to the estimation of blocks with different number of indicators. In our simulation we have accessed the properties of PLS and ML estimators (bias and dispersion) but not the performance of statistical tests based on these estimates. In particular, the ability of each method to detect significant coefficients (of different sizes) should be accessed in future work. Finally, future work should also access the performance of both methods in the presence of multicollinearity and or model misspecification. In fact, with real-world applications, erroneous omissions of model coefficients or manifest and latent variables are common. Also erroneous inclusions of nonexistent relationships between variables may arise. This is fertile ground to a more in-dept study of ML and PLS performance.

Appendix

Table 2 Estimators biases—base model (symmetric data)

	$n = 50$		$n = 100$		$n = 150$		$n = 250$		$n = 500$		$n = 1{,}000$		$n = 2{,}000$	
Parameter	PLS	ML	PLS	ML	PLS	ML	PLS	ML	PLS	ML	PLS	ML	PLS	ML
Indicator loadings														
λ_{x1}	0.08	0.06	0.08	0.04	0.08	0.04	0.08	0.04	0.08	0.04	0.08	0.04	0.08	0.04
λ_{x2}	0.06	0.05	0.06	0.03	0.06	0.03	0.06	0.03	0.06	0.03	0.06	0.03	0.06	0.03
λ_{x3}	0.09	0.07	0.09	0.05	0.09	0.05	0.09	0.05	0.09	0.05	0.09	0.05	0.09	0.04
λ_{x4}	0.07	0.06	0.07	0.04	0.07	0.04	0.07	0.04	0.07	0.04	0.07	0.04	0.07	0.04
λ_{x5}	0.07	0.05	0.07	0.03	0.07	0.03	0.07	0.03	0.07	0.04	0.07	0.03	0.07	0.03
λ_{y11}	−0.01	−0.10	−0.01	−0.14	−0.01	−0.14	−0.01	−0.16	−0.01	−0.17	−0.01	−0.18	−0.01	−0.19
λ_{y12}	0.00	−0.07	0.00	−0.09	0.00	−0.10	0.00	−0.11	0.00	−0.11	0.00	−0.12	0.00	−0.13
λ_{y13}	−0.01	−0.08	0.00	−0.12	−0.01	−0.12	−0.01	−0.13	−0.01	−0.14	−0.01	−0.15	−0.01	−0.16
λ_{y21}	0.09	0.14	0.09	0.10	0.09	0.09	0.09	0.09	0.09	0.08	0.09	0.09	0.09	0.10
λ_{y22}	0.12	0.22	0.12	0.15	0.12	0.13	0.12	0.12	0.12	0.11	0.12	0.13	0.12	0.14
λ_{y23}	0.11	0.20	0.11	0.14	0.11	0.12	0.11	0.11	0.11	0.10	0.11	0.11	0.11	0.12
λ_{y24}	0.08	0.14	0.08	0.10	0.08	0.08	0.08	0.08	0.08	0.07	0.08	0.08	0.08	0.09
λ_{y25}	0.10	−0.18	0.10	−0.20	0.10	−0.19	0.10	−0.19	0.10	−0.20	0.10	−0.20	0.10	−0.20
λ_{y31}	0.02	−0.19	0.02	−0.22	0.02	−0.22	0.02	−0.23	0.02	−0.27	0.02	−0.27	0.02	−0.27
λ_{y32}	0.01	−0.12	0.01	−0.14	0.01	−0.14	0.01	−0.15	0.01	−0.17	0.01	−0.17	0.01	−0.17
λ_{y41}	0.20	−0.36	0.20	−0.28	0.20	−0.27	0.20	−0.26	0.20	−0.23	0.20	−0.23	0.20	−0.22
λ_{y42}	0.15	−0.26	0.15	−0.21	0.15	−0.20	0.15	−0.19	0.15	−0.17	0.15	−0.17	0.15	−0.16
λ_{y43}	0.11	−0.20	0.11	−0.16	0.11	−0.15	0.11	−0.14	0.11	−0.13	0.11	−0.13	0.11	−0.12
λ_{y51}	0.32	−0.12	0.32	−0.08	0.32	−0.08	0.32	−0.08	0.32	−0.08	0.32	−0.08	0.32	−0.08
λ_{y52}	0.25	−0.09	0.25	−0.06	0.25	−0.06	0.25	−0.06	0.25	−0.06	0.25	−0.06	0.25	−0.06
λ_{y53}	0.21	−0.08	0.21	−0.05	0.21	−0.05	0.21	−0.05	0.21	−0.05	0.21	−0.06	0.21	−0.05
Average (abs)	0.10	0.13	0.10	0.12	0.10	0.11	0.10	0.11	0.10	0.11	0.10	0.11	0.10	0.11
Inner model coefficients														
τ_1	−0.03	0.19	−0.03	0.19	−0.03	0.20	−0.03	0.21	−0.03	0.21	−0.03	0.22	−0.03	0.23
β_{21}	0.05	−0.16	0.05	−0.17	0.05	−0.16	0.05	−0.17	0.05	−0.17	0.05	−0.18	0.05	−0.19
β_{31}	0.05	0.06	0.04	0.04	0.04	0.03	0.04	0.04	0.04	0.04	0.04	0.04	0.04	0.03
β_{32}	−0.16	0.31	−0.16	0.29	−0.15	0.29	−0.15	0.29	−0.15	0.31	−0.15	0.30	−0.15	0.31
τ_4	−0.03	0.24	−0.03	0.14	−0.03	0.15	−0.04	0.14	−0.03	0.13	−0.03	0.14	−0.03	0.13
β_{41}	0.06	0.24	0.05	0.06	0.06	0.04	0.06	0.03	0.06	0.01	0.06	0.00	0.06	−0.01
β_{42}	−0.09	0.40	−0.09	0.18	−0.09	0.17	−0.09	0.16	−0.09	0.13	−0.09	0.14	−0.09	0.13
β_{43}	−0.09	0.13	−0.08	0.03	−0.08	0.03	−0.08	0.02	−0.08	0.00	−0.08	−0.01	−0.08	−0.01
τ_5	0.06	0.13	0.07	0.06	0.07	0.06	0.07	0.06	0.07	0.07	0.07	0.07	0.07	0.07
β_{54}	−0.15	−0.15	−0.15	−0.14	−0.15	−0.14	−0.15	−0.12	−0.15	−0.11	−0.15	−0.10	−0.15	−0.09
Average (abs)	0.07	0.20	0.08	0.13	0.08	0.13	0.07	0.12	0.07	0.12	0.08	0.12	0.06	0.12

Table 3 Estimators standard deviations—base model (symmetric data)

	$n = 50$		$n = 100$		$n = 150$		$n = 250$		$n = 500$		$n = 1{,}000$		$n = 2{,}000$	
Parameter	PLS	ML	PLS	ML	PLS	ML	PLS	ML	PLS	ML	PLS	ML	PLS	ML
Indicator loadings														
λ_{x1}	0.03	0.17	0.01	0.03	0.01	0.03	0.01	0.02	0.00	0.02	0.00	0.01	0.00	0.01
λ_{x2}	0.03	0.15	0.01	0.06	0.01	0.05	0.01	0.04	0.01	0.03	0.00	0.02	0.00	0.01
λ_{x3}	0.04	0.22	0.01	0.08	0.01	0.07	0.01	0.05	0.00	0.04	0.00	0.03	0.00	0.02
λ_{x4}	0.03	0.18	0.01	0.07	0.01	0.05	0.01	0.04	0.00	0.03	0.00	0.02	0.00	0.02
λ_{x5}	0.03	0.17	0.01	0.06	0.01	0.05	0.01	0.04	0.00	0.03	0.00	0.02	0.00	0.02
λ_{y11}	0.03	0.22	0.01	0.13	0.01	0.16	0.01	0.08	0.00	0.08	0.00	0.07	0.00	0.07
λ_{y12}	0.03	0.16	0.01	0.10	0.01	0.12	0.01	0.06	0.00	0.06	0.00	0.05	0.00	0.05
λ_{y13}	0.03	0.20	0.01	0.12	0.01	0.14	0.01	0.08	0.00	0.07	0.00	0.06	0.00	0.06
λ_{y21}	0.03	0.19	0.02	0.10	0.01	0.10	0.01	0.09	0.01	0.07	0.00	0.06	0.00	0.06
λ_{y22}	0.04	0.33	0.02	0.17	0.01	0.17	0.01	0.14	0.01	0.11	0.00	0.09	0.00	0.08
λ_{y23}	0.04	0.31	0.02	0.16	0.01	0.15	0.01	0.13	0.01	0.10	0.00	0.08	0.00	0.08
λ_{y24}	0.03	0.22	0.01	0.11	0.01	0.11	0.01	0.09	0.01	0.07	0.00	0.06	0.00	0.05
λ_{y25}	0.03	0.15	0.01	0.08	0.01	0.06	0.01	0.05	0.01	0.03	0.00	0.02	0.00	0.02
λ_{y31}	0.04	0.29	0.01	0.21	0.01	0.25	0.01	0.27	0.00	0.16	0.00	0.12	0.00	0.06
λ_{y32}	0.03	0.19	0.01	0.14	0.01	0.16	0.01	0.17	0.01	0.10	0.00	0.07	0.00	0.04
λ_{y41}	0.04	0.20	0.01	0.14	0.01	0.13	0.01	0.12	0.01	0.10	0.00	0.08	0.00	0.08
λ_{y42}	0.03	0.15	0.01	0.10	0.01	0.10	0.01	0.09	0.01	0.07	0.00	0.06	0.00	0.06
λ_{y43}	0.03	0.12	0.01	0.08	0.01	0.08	0.01	0.07	0.01	0.05	0.00	0.05	0.00	0.04
λ_{y51}	0.04	0.16	0.01	0.09	0.01	0.09	0.01	0.09	0.01	0.08	0.00	0.08	0.00	0.08
λ_{y52}	0.03	0.14	0.01	0.09	0.01	0.08	0.01	0.07	0.01	0.07	0.00	0.06	0.00	0.06
λ_{y53}	0.03	0.12	0.01	0.08	0.01	0.07	0.01	0.06	0.01	0.06	0.00	0.06	0.00	0.05
Average (abs)	0.03	0.19	0.01	0.10	0.01	0.11	0.01	0.09	0.01	0.07	0.00	0.06	0.00	0.05

(continued)

Table 3 (continued)

	$n = 50$		$n = 100$		$n = 150$		$n = 250$		$n = 500$		$n = 1{,}000$		$n = 2{,}000$	
Parameter	PLS	ML	PLS	ML	PLS	ML	PLS	ML	PLS	ML	PLS	ML	PLS	ML
Inner model coefficients														
τ_1	0.03	0.23	0.02	0.13	0.02	0.12	0.01	0.09	0.01	0.08	0.01	0.07	0.00	0.07
β_{21}	0.04	0.20	0.03	0.10	0.02	0.12	0.02	0.08	0.01	0.07	0.01	0.07	0.01	0.06
β_{31}	0.14	0.28	0.09	0.17	0.08	0.16	0.06	0.12	0.04	0.09	0.03	0.07	0.02	0.05
β_{32}	0.14	0.75	0.09	0.29	0.08	0.27	0.06	0.22	0.04	0.18	0.03	0.12	0.02	0.10
τ_4	0.14	0.69	0.10	0.25	0.08	0.24	0.06	0.18	0.04	0.13	0.03	0.09	0.02	0.06
β_{41}	0.17	1,48	0.12	0.22	0.09	0.18	0.07	0.14	0.05	0.10	0.04	0.08	0.03	0.06
β_{42}	0.15	2,92	0.10	0.41	0.08	0.40	0.06	0.32	0.05	0.20	0.03	0.15	0.02	0.09
β_{43}	0.14	0.66	0.14	0.87	0.09	0.23	0.06	0.16	0.04	0.11	0.03	0.07	0.02	0.05
τ_5	0.12	0.59	0.12	0.62	0.09	0.17	0.05	0.11	0.03	0.08	0.03	0.06	0.02	0.05
β_{54}	0.12	0.20	0.12	0.23	0.08	0.15	0.05	0.11	0.03	0.09	0.03	0.08	0.02	0.08
Average (abs)	0.12	0.80	0.09	0.33	0.07	0.20	0.05	0.16	0.04	0.11	0.03	0.09	0.02	0.07

Table 4 Estimators biases—skewed data model $Beta(10, 4)$

	$n = 50$		$n = 100$		$n = 150$		$n = 250$		$n = 500$		$n = 1{,}000$		$n = 2{,}000$	
Parameter	PLS	ML	PLS	ML	PLS	ML	PLS	ML	PLS	ML	PLS	ML	PLS	ML
Indicator loadings														
λ_{x1}	0.07	0.08	0.07	0.12	0.08	0.13	0.07	0.16	0.08	0.17	0.08	0.18	0.08	0.19
λ_{x2}	0.08	0.07	0.08	0.09	0.08	0.10	0.08	0.12	0.08	0.12	0.08	0.14	0.08	0.14
λ_{x3}	0.07	0.10	0.07	0.14	0.07	0.16	0.07	0.18	0.07	0.19	0.07	0.21	0.07	0.22
λ_{x4}	0.08	0.08	0.07	0.11	0.08	0.12	0.08	0.14	0.08	0.15	0.08	0.17	0.08	0.17
λ_{x5}	0.08	0.07	0.07	0.10	0.08	0.11	0.08	0.13	0.08	0.13	0.08	0.15	0.08	0.15
λ_{y11}	−0.02	−0.03	−0.03	0.05	−0.02	0.10	−0.02	0.17	−0.02	0.25	−0.02	0.31	−0.02	0.33
λ_{y12}	0.01	−0.02	0.01	0.04	0.01	0.07	0.01	0.12	0.01	0.16	0.01	0.21	0.01	0.22
λ_{y13}	−0.01	−0.03	−0.01	0.05	−0.01	0.09	−0.01	0.15	−0.01	0.21	−0.01	0.26	−0.01	0.28
λ_{y21}	0.10	0.13	0.10	0.15	0.10	0.15	0.10	0.16	0.10	0.12	0.10	0.09	0.10	0.03
λ_{y22}	0.10	0.19	0.09	0.20	0.10	0.21	0.10	0.22	0.10	0.17	0.10	0.12	0.10	0.04
λ_{y23}	0.10	0.17	0.09	0.18	0.10	0.19	0.10	0.20	0.10	0.15	0.10	0.11	0.10	0.03
λ_{y24}	0.10	0.13	0.10	0.13	0.10	0.13	0.10	0.14	0.10	0.11	0.10	0.08	0.10	0.02
λ_{y25}	0.10	−0.17	0.10	−0.14	0.10	−0.14	0.10	−0.12	0.10	−0.11	0.10	−0.10	0.10	−0.10
λ_{y31}	0.00	−0.25	0.00	−0.23	0.00	−0.28	0.00	−0.34	0.00	−0.41	0.00	−0.49	0.00	−0.57
λ_{y32}	0.03	−0.15	0.03	−0.14	0.03	−0.18	0.03	−0.21	0.03	−0.25	0.03	−0.30	0.03	−0.36
λ_{y41}	0.18	−0.52	0.17	−0.42	0.18	−0.40	0.18	−0.38	0.18	−0.40	0.18	−0.38	0.18	−0.35
λ_{y42}	0.15	−0.37	0.15	−0.31	0.15	−0.29	0.15	−0.28	0.15	−0.29	0.15	−0.27	0.15	−0.26
λ_{y43}	0.13	−0.28	0.13	−0.23	0.13	−0.22	0.13	−0.21	0.13	−0.22	0.13	−0.21	0.13	−0.19
λ_{y51}	0.30	−0.23	0.30	−0.11	0.30	−0.06	0.30	0.00	0.30	0.02	0.30	0.03	0.30	0.02
λ_{y52}	0.25	−0.18	0.25	−0.08	0.25	−0.05	0.25	0.00	0.25	0.02	0.25	0.02	0.25	0.02
λ_{y53}	0.23	−0.15	0.23	−0.07	0.23	−0.04	0.23	0.00	0.23	0.02	0.23	0.02	0.23	0.01
Average (abs)	0.10	0.16	0.10	0.15	0.10	0.15	0.10	0.16	0.10	0.17	0.10	0.18	0.10	0.18

(continued)

Table 4 (continued)

Parameter	$n = 50$		$n = 100$		$n = 150$		$n = 250$		$n = 500$		$n = 1{,}000$		$n = 2{,}000$	
	PLS	ML	PLS	ML	PLS	ML	PLS	ML	PLS	ML	PLS	ML	PLS	ML
Inner model coefficients														
τ_1	−0.02	0.19	−0.01	0.12	−0.01	0.09	−0.01	0.06	−0.01	0.02	−0.01	0.00	−0.01	−0.01
β_{21}	0.06	−0.08	0.06	−0.06	0.06	−0.03	0.06	0.01	0.06	0.08	0.06	0.17	0.06	0.24
β_{31}	0.06	0.15	0.06	0.12	0.06	0.17	0.06	0.25	0.06	0.33	0.06	0.44	0.06	0.54
β_{32}	−0.17	0.78	−0.17	0.38	−0.17	0.45	−0.17	0.52	−0.17	0.60	−0.16	0.69	−0.17	0.74
τ_4	−0.02	0.58	−0.02	0.31	−0.02	0.30	−0.03	0.31	−0.02	0.35	−0.02	0.34	−0.02	0.33
β_{41}	0.08	0.44	0.06	0.19	0.07	0.16	0.07	0.12	0.07	0.10	0.07	0.09	0.07	0.07
β_{42}	−0.10	0.64	−0.09	0.31	−0.10	0.28	−0.09	0.32	−0.10	0.30	−0.09	0.27	−0.10	0.21
β_{43}	−0.10	0.54	−0.09	0.16	−0.10	0.11	−0.09	0.07	−0.09	0.05	−0.09	0.00	−0.09	−0.04
τ_5	0.10	0.29	0.10	0.13	0.10	0.11	0.10	0.09	0.10	0.10	0.10	0.12	0.10	0.14
β_{54}	−0.18	−0.14	−0.18	−0.19	−0.18	−0.20	−0.18	−0.22	−0.18	−0.24	−0.18	−0.21	−0.18	−0.18
Average (abs)	0.09	0.38	0.08	0.20	0.09	0.19	0.09	0.19	0.09	0.22	0.09	0.23	0.09	0.25

Table 5 Estimators standard deviations—skewed data model $Beta(10, 4)$

	$n = 50$		$n = 100$		$n = 150$		$n = 250$		$n = 500$		$n = 1{,}000$		$n = 2{,}000$	
Parameter	PLS	ML	PLS	ML	PLS	ML	PLS	ML	PLS	ML	PLS	ML	PLS	ML
Indicator loadings														
λ_{x1}	0.03	0.20	0.05	0.12	0.00	0.13	0.03	0.14	0.00	0.14	0.00	0.14	0.00	0.14
λ_{x2}	0.02	0.18	0.04	0.11	0.00	0.11	0.03	0.11	0.00	0.11	0.00	0.11	0.00	0.11
λ_{x3}	0.03	0.25	0.06	0.16	0.00	0.16	0.04	0.16	0.00	0.17	0.00	0.16	0.00	0.16
λ_{x4}	0.02	0.20	0.05	0.13	0.00	0.13	0.03	0.13	0.00	0.13	0.00	0.13	0.00	0.13
λ_{x5}	0.02	0.18	0.04	0.12	0.00	0.12	0.03	0.12	0.00	0.12	0.00	0.11	0.00	0.11
λ_{y11}	0.03	0.34	0.06	0.31	0.00	0.34	0.04	0.37	0.00	0.39	0.00	0.40	0.00	0.42
λ_{y12}	0.02	0.24	0.04	0.22	0.00	0.23	0.03	0.25	0.00	0.26	0.00	0.27	0.00	0.28
λ_{y13}	0.02	0.30	0.05	0.27	0.00	0.29	0.03	0.31	0.00	0.33	0.00	0.34	0.00	0.35
λ_{y21}	0.02	0.22	0.05	0.17	0.00	0.19	0.03	0.21	0.00	0.20	0.00	0.22	0.00	0.22
λ_{y22}	0.03	0.35	0.06	0.26	0.01	0.27	0.04	0.29	0.00	0.28	0.00	0.31	0.00	0.30
λ_{y23}	0.03	0.31	0.05	0.24	0.01	0.25	0.03	0.26	0.00	0.26	0.00	0.28	0.00	0.27
λ_{y24}	0.02	0.23	0.04	0.17	0.00	0.18	0.03	0.19	0.00	0.18	0.00	0.20	0.00	0.19
λ_{y25}	0.03	0.17	0.05	0.11	0.00	0.11	0.03	0.10	0.00	0.10	0.00	0.10	0.00	0.10
λ_{y31}	0.03	0.33	0.06	0.25	0.01	0.25	0.04	0.25	0.00	0.25	0.00	0.24	0.00	0.21
λ_{y32}	0.02	0.22	0.04	0.16	0.00	0.16	0.02	0.16	0.00	0.16	0.00	0.15	0.00	0.13
λ_{41}	0.03	0.27	0.06	0.23	0.01	0.24	0.04	0.22	0.00	0.20	0.00	0.18	0.00	0.18
λ_{y42}	0.02	0.20	0.05	0.17	0.01	0.17	0.03	0.16	0.00	0.14	0.00	0.13	0.00	0.13
λ_{y43}	0.02	0.16	0.04	0.13	0.01	0.13	0.02	0.12	0.00	0.11	0.00	0.10	0.00	0.10
λ_{y51}	0.03	0.29	0.06	0.23	0.01	0.27	0.04	0.27	0.00	0.29	0.00	0.32	0.00	0.33
λ_{y52}	0.02	0.23	0.05	0.19	0.00	0.21	0.03	0.22	0.00	0.23	0.00	0.25	0.00	0.25
λ_{y53}	0.02	0.20	0.04	0.16	0.01	0.18	0.03	0.19	0.00	0.20	0.00	0.21	0.00	0.22
Average (abs)	0.03	0.24	0.05	0.18	0.00	0.20	0.03	0.20	0.00	0.20	0.00	0.21	0.00	0.21

(continued)

Table 5 (continued)

Parameter	$n = 50$		$n = 100$		$n = 150$		$n = 250$		$n = 500$		$n = 1,000$		$n = 2,000$	
	PLS	ML	PLS	ML	PLS	ML	PLS	ML	PLS	ML	PLS	ML	PLS	ML
Inner model coefficients														
τ_1	0.03	0.34	0.02	0.20	0.02	0.20	0.01	0.20	0.01	0.17	0.01	0.17	0.00	0.17
β_{21}	0.04	0.39	0.03	0.26	0.02	0.28	0.02	0.32	0.01	0.33	0.01	0.35	0.01	0.34
β_{31}	0.14	0.46	0.10	0.32	0.08	0.35	0.06	0.36	0.04	0.38	0.03	0.40	0.02	0.42
β_{32}	0.14	13.38	0.10	0.46	0.08	0.43	0.06	0.44	0.04	0.44	0.03	0.42	0.02	0.43
τ_4	0.15	1.65	0.10	0.52	0.08	0.33	0.06	0.30	0.04	0.27	0.03	0.22	0.02	0.22
β_{41}	0.17	1.45	0.12	0.51	0.09	0.36	0.07	0.30	0.05	0.25	0.04	0.19	0.03	0.14
β_{42}	0.15	2.11	0.10	0.53	0.08	0.45	0.06	0.40	0.05	0.32	0.03	0.29	0.02	0.27
β_{43}	0.13	5.38	0.09	0.82	0.08	0.35	0.06	0.27	0.04	0.22	0.03	0.17	0.02	0.14
τ_5	0.13	0.89	0.09	0.27	0.07	0.22	0.05	0.19	0.04	0.19	0.03	0.20	0.02	0.24
β_{54}	0.12	0.38	0.08	0.23	0.07	0.22	0.05	0.20	0.04	0.19	0.03	0.23	0.02	0.25
Average (abs)	0.12	2.64	0.08	0.41	0.07	0.32	0.05	0.30	0.04	0.28	0.03	0.27	0.02	0.26

References

1. Bollen, K.A.: Structural Equations with Latent Variables. Wiley, New York (1989)
2. Cassel, C., Hackl, P., Westlund, A.: On measurement of intangibles assets: a study of robustness of partial least squares. Total Qual. Manag. **7**, 897–907 (2000)
3. Chin, W.W.: The partial least squares approach to structural equation modeling. In: Marcoulides, G.A. (ed.) Modern Methods for Business Research, cap.10. Lawrence Erlbaum Associates, MahWah, New Jersey (1998)
4. Chin W.W., Newsted, P.R.: Structural equation modeling analysis with small sampling using partial least squares. In: Hoyle, R.H. (ed.) Statistical Strategies for Small Sample Research (pp. 307–341). Sage, Newbury Park, California (1998)
5. ECSI: European Customer Satisfaction Index: Report prepared for the ECSI Steering Committee (1998)
6. Fornell, C.: A national customer satisfaction Barometer: the Swedish experience. J. Marketing **56**, 6–21 (1992)
7. Fornell, C., Bookstein, F.L.: The two structural equation models: LISREL and PLS applied to customer exit-voice theory. J. Marketing Res. **19**, 440–452 (1982)
8. Fornell, C., Cha, J.: Partial least squares. In: Bagozzi, R.P. (ed.) Advanced Methods of Marketing Research. Blackwell, Cambridge (1994)
9. Fornell, C., Johnson, M.D., Anderson, E.W., Cha, J., Everitt Bryant, B.: The American customer satisfaction index: nature, purpose and findings. J. Marketing **60**, 7–18 (1998)
10. Lomöller, J. B.: Latent Variable Path Modelling with Partial Least Squares. Physica, Heidelberg (1989)
11. Mcdonald, R.P.: Path analysis with composite variables. Multivariate Behav. Res. **31**(2), 239–270 (1996)
12. Nachtigal, C., Kroelune, K., Funke, F., Steyer, R.: (Why) should we use SEM? pros and cons of structural equation modeling. Methods Psychol. Res., Online **8**(2), 1–22 (2003)
13. Ridgon, R.E.: Structural equation modelling. In: Marcoulides, G.A. (ed.) Modern methods for business research. Lawrence Erlbaum Associates, MahWah, New Jersey (1998)
14. Schneeweiss, H.: Models with latent variables: LISREL versus PLS. Contemp. Math. **112**, 33–40 (1990)
15. Tenenhaus, M.: Comparisson between PLS and LISREL approaches for structural equation models. In: Vilares, Tenenhaus, Coelho Vinzi, Morinau (eds.) PLS and Related Methods-Proceedings of the PLS'03 International Conference (2003)
16. Vilares, M.J., Almeida, M.H, Coelho, P.S.: Comparison of likelihood and PLS estimators for structural equation modelling: a simulation with customer satisfaction data. In: Vinzi, W.E., Chin, W.W., Henseler, J., Wang, H. (eds.) Handbook of Partial Least Squares. Concepts, Methods and Applications, pp. 289–307. Springer Handbooks of Computational Statistics, Springer (2010)
17. Vilares, M.J., Coelho, P.S.: A Satisfação e Lealdade do Cliente - Metodologias de Avaliação, Gestão e Análise, 2ł edição, Escolar Editora (2011)

Part II

Communications

A Parametric Cure Model with Covariates

Ana M. Abreu and Cristina S. Rocha

Abstract
Cure models were developed to deal with situations where it is plausible to assume that there are non-susceptible (or cured) individuals within the study population. Usually, in a cure model, the aim is to estimate the proportion of non-susceptible individuals, the survival function of the susceptible individuals and the effect of the covariates, if they have been included in the model. Therefore, researchers are interested in knowing if the event will occur (which is called incidence) and when it will occur, given that it can occur (which is called latency). For each covariate there are two parameters: one that describes how the covariate affects incidence and the other that describes how it affects latency. In this context, the population under study is heterogeneous not only because there are susceptible and non-susceptible individuals but also due to the different values of their covariates. This chapter follows another one Abreu and Rocha [Um novo modelo de cura paramétrico. In: Castro, L.C., Martins, E.G., Rocha, C., Oliveira, M.F., Leal, M.M., Rosado, F. (eds.) Ciência Estatística, pp. 151–162. Edições SPE, Lisboa (2006)], where we proposed a cure model based on the Chen distribution [Chen, A new two-parameter lifetime distribution with bathtub shape or increasing failure rate function. Stat. Probab. Lett. **49**, 155–161 (2000)]. The good results obtained with this new model when fitted to real data was a motivation for including covariates into the model.

A.M. Abreu (✉)
CCCEE and CCM, University of Madeira, Campus Universitário da Penteada 9020-105 Funchal, Portugal
e-mail: abreu@uma.pt

C.S. Rocha
DEIO and CEAUL, Faculty of Science, University of Lisbon, Lisbon, Portugal
e-mail: cmrocha@fc.ul.pt

J. Lita da Silva et al. (eds.), *Advances in Regression, Survival Analysis, Extreme Values, Markov Processes and Other Statistical Applications*, Studies in Theoretical and Applied Statistics, DOI 10.1007/978-3-642-34904-1_3,

1 Introduction

Survival analysis is strongly stimulated by the constant evolution of medicine. In particular, new models were developed to take into account the possibility of cure of certain diseases. It is in this context that cure models appear, because they allow the analysis of survival data in which some subjects can eventually experience, and others never experience, the event of interest. An important property of cure models (mixture and non-mixture) is the fact that they have an improper survival function, which is equivalent to the cumulative hazard function being limited.

Although, frequently, the cure is not observable, the suspicion is based on some features of the data, namely the existence of many censored observations beyond the last observed survival time. Therefore, a long and stable plateau of the Kaplan–Meier survival curve [5] suggests the applicability of the mixture cure model approach [8].

Usually, in a cure model, we want to estimate the proportion of cured individuals, the survival function of the susceptible individuals and the effect of the covariates, if they have been included in the model. There are several ways of modelling the effect of the covariates, **x**, on the survival of the susceptible individuals. For instance, the accelerated failure time model, that is, $S_d(t|\mathbf{x}) = S_{d_0}(te^{\beta'\mathbf{x}})$, where $S_{d_0}(.)$ is independent of the covariates and can be formulated either parametrically [9] or non-parametrically [7]. Another possibility is the proportional odds model, which is used when the hazard functions of individuals with different values of their covariates converge after some time. The most widely used model is undoubtedly the proportional hazards model $S_d(t|\mathbf{x}) = S_{d_0}(t)^{\exp(\beta'\mathbf{x})}$ where, usually, $S_{d_0}(t)$ is non-parametric [10]. Another alternative is to consider a mixture cure model with more than one survival function for susceptible individuals [4]. The logistic regression model is the most common choice to model the effects of the covariates, **z**, on the cure proportion.

In this chapter, we propose a new mixture cure model with covariates based on the Chen distribution [2]. Section 2 describes the general structure of this model, while in Sect. 3 some parameter estimation details are presented. In Sect. 4 the applicability of our model is illustrated with the analysis of leukaemia data and Sect. 5 is reserved to concluding remarks.

2 A Cure Model with Covariates

In this section we describe the structure of the mixture cure model, some features of the Chen distribution and present our new model.

2.1 The Mixture Cure Model

We denote by T the random variable that represents the survival time in a population where there are susceptible and non-susceptible individuals. Let Y denote a binary random variable indicating that an individual is either susceptible ($Y = 1$) or not ($Y = 0$). The mixture cure model can be formulated through the survival function

$$S(t) = p + (1 - p)S_d(t), \tag{1}$$

where $p = P(Y = 0)$ represents the non-susceptible proportion and $S_d(t) = S(t|Y = 1)$ is the (proper) survival function of the susceptible individuals. As $S(t) \to p$ when $t \to \infty$, then $S(t)$ is an improper survival function. Note that, if an individual has censored survival time, then Y is not observable, so we do not know if that individual is susceptible or not.

If we introduce covariates in model (1), we have

$$S(t_i|\mathbf{x}_i, \mathbf{z}_i) = p(\mathbf{z}_i) + (1 - p(\mathbf{z}_i))S_d(t_i|\mathbf{x}_i), \tag{2}$$

where $\mathbf{x}_i$ and $\mathbf{z}_i$ are the vectors of covariates associated to the ith individual ($i = 1, \ldots, n$), $p(\mathbf{z}_i) = P(Y = 0|\mathbf{z}_i)$ is the probability that the ith individual is non-susceptible given a covariate vector $\mathbf{z}_i$ and $S_d(t_i|\mathbf{x}_i) = P(T_i > t_i|Y_i = 1, \mathbf{x}_i)$ is the probability that an individual survives longer than t_i, given that the individual is susceptible and has a covariate vector $\mathbf{x}_i$. Note that $\mathbf{x}_i$ and $\mathbf{z}_i$ can include the same covariates.

2.2 The Chen Distribution

The distribution function proposed by Chen [2] is

$$F(t) = 1 - \exp[\lambda_1(1 - \exp(t^{\lambda_2}))], \quad t > 0, \quad \lambda_1, \lambda_2 > 0, \tag{3}$$

where λ_1 is the scale parameter and λ_2 is the shape parameter. The corresponding survival and hazard functions are, respectively,

$$\overline{F}(t) = \exp[\lambda_1(1 - \exp(t^{\lambda_2}))], \quad t > 0, \tag{4}$$

$$h^*(t) = \lambda_1\lambda_2 t^{\lambda_2 - 1}\exp(t^{\lambda_2}), \quad t > 0.$$

The author refers that $h^*(t)$ can be bathtub-shaped when $\lambda_2 < 1$ and that it increases when $\lambda_2 \geq 1$, which is unusual in most distributions used in survival analysis. In fact, as

$$h^{*'}(t) = \lambda_1\lambda_2 t^{\lambda_2 - 2}\exp(t^{\lambda_2})((\lambda_2 - 1) + \lambda_2 t^{\lambda_2}),$$

for $\lambda_2 < 1$ we have $h^*(t)$ decreasing for $t \in [0, (\frac{1}{\lambda_2} - 1)^{\frac{1}{\lambda_2}}]$ and, for $t \geq (\frac{1}{\lambda_2} - 1)^{\frac{1}{\lambda_2}}$, $h^*(t)$ is an increasing function. Hence, the range of the interval where $h^*(t)$ is decreasing will increase as λ_2 decreases. Therefore, if λ_2 is near zero, for example,

$\lambda_2 = 0.1$, the interval is so large that, from the practical point of view, it is just like having a decreasing hazard function. Reciprocally, as λ_2 approaches 1, the interval where the hazard function is decreasing is so small that it is almost like if the hazard function was always increasing.

2.3 The Cure Model Based on the Chen Distribution with Covariates

Admit that the survival time of susceptible individuals follows the Chen distribution, given by Eq. (3). As stated by Abreu and Rocha [1], the cure model obtained by substituting in Eq. (1) $S_d(t)$ by the expression (4) is

$$S(t) = p + (1-p)\exp[\lambda_1(1-\exp(t^{\lambda_2}))], \quad t > 0,\ \lambda_1, \lambda_2 > 0. \tag{5}$$

If the model is defined in terms of hazard function, we have

$$h(t) = \frac{(1-p)\lambda_1\lambda_2 t^{\lambda_2 - 1}\exp(t^{\lambda_2})\exp[\lambda_1(1-\exp(t^{\lambda_2}))]}{p + (1-p)\exp[\lambda_1(1-\exp(t^{\lambda_2}))]}.$$

Consider the proportional hazards model for the survival time of susceptible individuals. Then we have

$$S_d(t|\mathbf{x}) = S_d(t|\boldsymbol{\beta}'\boldsymbol{x}, \lambda_1, \lambda_2) = S_{d_0}(t|\lambda_1, \lambda_2)^{\exp(\boldsymbol{\beta}'\boldsymbol{x})},$$

where λ_1 and λ_2 are the parameters of the Chen distribution corresponding to the baseline survival function, that is,

$$S_d(t|\mathbf{x}) = [\exp[\lambda_1(1-\exp(t^{\lambda_2}))]]^{\exp(\boldsymbol{\beta}'\boldsymbol{x})}. \tag{6}$$

Let

$$p(\mathbf{z}) = P(Y = 0|\mathbf{z}) = \frac{1}{1+\exp(\boldsymbol{\gamma}'\mathbf{z})} \tag{7}$$

be the function that models the effect of the covariates on the proportion of non-susceptible individuals. In fact, in this context, the logistic regression model is the most commonly used binary regression model.

The mixture cure model of proportional hazards specified by Eqs. (2), (6) and (7) can be written in the form

$$S(t|\mathbf{x},\mathbf{z}) = \frac{1}{1+\exp(\boldsymbol{\gamma}'\mathbf{z})} + \frac{\exp(\boldsymbol{\gamma}'\mathbf{z})}{1+\exp(\boldsymbol{\gamma}'\mathbf{z})}[\exp[\lambda_1(1-\exp(t^{\lambda_2}))]]^{\exp(\boldsymbol{\beta}'\boldsymbol{x})}. \tag{8}$$

3 Parameters Estimation

In this section, the parameters estimation process for the proposed model is presented. With this purpose, we apply the maximum likelihood method, making use of the EM algorithm [3], since here we are dealing with missing data.

3.1 Maximum Likelihood Function

Let us assume that censoring is noninformative. Denote the observed survival time for the ith individual by t_i, $i = 1, \ldots, n$. Suppose we have data in the form $(t_i, \delta_i, \mathbf{x}_i, \mathbf{z}_i)$, $i = 1, \ldots, n$, where $\delta_i = 1$ if t_i is uncensored and $\delta_i = 0$ otherwise, and $\mathbf{x}_i$ and $\mathbf{z}_i$ are two covariate vectors. Without loss of generality, suppose that the first m $(m < n)$ survival times are censored. Then $\delta_i = 0$ if $1 \le i \le m$ and $\delta_i = 1$ if $m + 1 \le i \le n$.

The contribution to the likelihood of an individual for whom the event of interest was observed at t_i is $(1 - p(\mathbf{z}_i)) f_d(t_i|\mathbf{x}_i)$, where $f_d(t_i|\mathbf{x}_i)$ represents the density function of the susceptible individuals, conditional on the corresponding covariates. If the event of interest is not observed until time t_i, then the contribution of the individual to the likelihood is $p(\mathbf{z}_i) + (1 - p(\mathbf{z}_i))S_d(t_i|\mathbf{x}_i)$.

Then, the observed likelihood function is

$$L_O = \prod_{i=1}^{n} \left\{[1 - p(\mathbf{z}_i)] f_d(t_i|\mathbf{x}_i)\right\}^{\delta_i} \left\{p(\mathbf{z}_i) + [1 - p(\mathbf{z}_i)] S_d(t_i|\mathbf{x}_i)\right\}^{1-\delta_i},$$

which can be written as

$$L_O = \prod_{i=1}^{n} \left\{[1 - p(\mathbf{z}_i)]\lambda_1\lambda_2 t_i^{\lambda_2 - 1} \exp(t_i^{\lambda_2} + \boldsymbol{\beta}'\boldsymbol{x}_i)\{\exp[\lambda_1(1 - \exp(t_i^{\lambda_2}))]\}^{\exp(\boldsymbol{\beta}'\boldsymbol{x}_i)}\right\}^{\delta_i}$$
$$\times \left\{\{\exp[\lambda_1(1 - \exp(t_i^{\lambda_2}))]\}^{\exp(\boldsymbol{\beta}'\boldsymbol{x}_i)}\right\}^{1-\delta_i}$$

when the Chen distribution is used for the survival time of susceptible individuals.

Let $y_1, \ldots, y_n$ be such that $y_i = 1$ if the individual is susceptible and $y_i = 0$ otherwise. If all y_i's were observed, the complete likelihood would be

$$L_C = \prod_{i=1}^{n} \left\{[(1 - p(\mathbf{z}_i)) f_d(t_i|\mathbf{x}_i)]^{y_i}\right\}^{\delta_i} \left\{p(\mathbf{z}_i)^{1-y_i}[(1 - p(\mathbf{z}_i)) S_d(t_i|\mathbf{x}_i)]^{y_i}\right\}^{1-\delta_i}.$$

Considering $q(\mathbf{z}_i) = 1 - p(\mathbf{z}_i)$, after some calculations the previous expression can be rewritten as

$$L_C = \prod_{i=1}^{n} q(\mathbf{z}_i)^{y_i}[1 - q(\mathbf{z}_i)]^{1-y_i} \prod_{i=1}^{n} h_d(t_i|\mathbf{x}_i)^{y_i\delta_i} S_d(t_i|\mathbf{x}_i)^{y_i}. \qquad (9)$$

The logarithm of Eq. (9) is given by

$$\begin{aligned}\log L_C = &\textstyle\sum_{i=1}^{n}[y_i \log q(\mathbf{z}_i) + (1-y_i)\log(1-q(\mathbf{z}_i))+ \\ &\textstyle\sum_{i=1}^{n} y_i \delta_i \log h_d(t_i|\mathbf{x}_i) + y_i \log S_d(t_i|\mathbf{x}_i)].\end{aligned} \tag{10}$$

3.2 EM Algorithm

The fact that in most cases cure is not observable, gives origin to an incomplete data situation. In this context, the EM algorithm is a widely used tool for maximizing the likelihood function. In general terms, the maximization of the likelihood is replaced by maximizing its expectation conditional to the current parameter values and the observed data. Thus, the missing values are identified with the corresponding conditional expected value.

In fact, the E step of the EM algorithm consists in obtaining the expectation of the logarithm of the complete likelihood with respect to the distribution of the unobserved Y_i's, given the current parameter values and the observed data $\mathcal{O}$, where $\mathcal{O} = \{\text{observed } y_i'\text{s}, (t_i, \delta_i, \mathbf{x}_i, \mathbf{z}_i), i = 1, \ldots, n\}$. However, since $\log L_C$ is linear in Y_i, to compute the expected value of $\log L_C$, we only need to replace in Eq. (10) each unobserved Y_i by its expected value, denoted by τ_i. Therefore, we have

$$\tau_i = E(Y_i|\mathcal{O}) = P(Y_i = 1|T_i > t_i, \delta_i = 0, \boldsymbol{\theta}) = \frac{[1-p(\mathbf{z}_i)]S_d(t_i|\mathbf{x}_i)}{S(t_i|\mathbf{x}_i, \mathbf{z}_i)} \tag{11}$$

where $\boldsymbol{\theta} = (\boldsymbol{\beta}, \boldsymbol{\gamma}, \boldsymbol{\lambda})$ is the vector parameter of model (8) and $\boldsymbol{\lambda} = (\lambda_1, \lambda_2)$. Thus, in the logarithm of the complete likelihood, each y_i is replaced by ω_i, the probability of the ith individual being susceptible, where $\omega_i = 1$ if $\delta_i = 1$ and $\omega_i = \tau_i$ if $\delta_i = 0$.

At the M step, we need to maximize the following two components of the expected log-likelihood:

$$\begin{aligned}\log L_{E_1} &= \textstyle\sum_{i=1}^{n}[\omega_i \log q(\mathbf{z}_i) + (1-\omega_i)\log(1-q(\mathbf{z}_i))] \\ &= (n-m)\log q(\mathbf{z}_i) + m\log(1-q(\mathbf{z}_i)) + \textstyle\sum_{i=1}^{m} \tau_i [\log q(\mathbf{z}_i) - \log(1-q(\mathbf{z}_i))],\end{aligned}$$

$$\begin{aligned}\log L_{E_2} &= \textstyle\sum_{i=1}^{n}[\delta_i \omega_i \log h_d(t_i|\mathbf{x}_i) + \omega_i \log S_d(t_i|\mathbf{x}_i)] \\ &= \textstyle\sum_{i=1}^{m} \tau_i \log S_d(t_i|\mathbf{x}_i) + \sum_{i=m+1}^{n}[\log h_d(t_i|\mathbf{x}_i) + \log S_d(t_i|\mathbf{x}_i)].\end{aligned}$$

From $\log L_{E_1}$, after some algebra, we obtain the following explicit expression for the estimate of $q(\mathbf{z}_i)$ at the $(k+1)$th iteration:

$$q(\mathbf{z}_i)^{(k+1)} = \frac{1}{n}\left[(n-m) + \sum_{i=1}^{m} \tau_i^{(k)}\right],$$

but only in the case where the covariates are not included in the cure proportion. Making use of the Chen distribution for the survival time of the susceptible individuals, by Eq. (11), we get

$$\tau_i = \frac{q(\mathbf{z}_i)\{\exp[\lambda_1(1-\exp(t_i^{\lambda_2}))]\}^{\exp(\boldsymbol{\beta}'\boldsymbol{x}_i)}}{1-q(\mathbf{z}_i)+q(\mathbf{z}_i)\{\exp[\lambda_1(1-\exp(t_i^{\lambda_2}))]\}^{\exp(\boldsymbol{\beta}'\boldsymbol{x}_i)}}. \tag{12}$$

In what concerns $\log L_{E_2}$, since it can be written as

$$\log L_{E_2} = \lambda_1 \sum_{i=1}^{m} \tau_i \exp(\boldsymbol{\beta}'\boldsymbol{x}_i)[1-\exp(t_i^{\lambda_2})] + (n-m)(\log\lambda_1 + \log\lambda_2) +$$

$$(\lambda_2 - 1)\sum_{i=m+1}^{n} \log t_i + \sum_{i=m+1}^{n} (\exp(\boldsymbol{\beta}'\boldsymbol{x}_i) + t_i^{\lambda_2}) +$$

$$\lambda_1 \sum_{i=m+1}^{n} \exp(\boldsymbol{\beta}'\boldsymbol{x}_i)[1-\exp(t_i^{\lambda_2})],$$

after some algebra, we obtain an explicit formula for the estimator of λ_1,

$$\hat{\lambda}_1 = \frac{n-m}{\sum_{i=1}^{m} \tau_i \exp(\boldsymbol{\beta}'\boldsymbol{x}_i)\left[\exp(t_i^{\lambda_2})-1\right] + \sum_{i=m+1}^{n} \exp(\boldsymbol{\beta}'\boldsymbol{x}_i)\left[\exp(t_i^{\lambda_2})-1\right]},$$

where τ_i is given by Eq. (12). No explicit formula was obtained for the estimator of λ_2. Therefore, we recommend using simultaneously another maximization procedure, such as the Newton–Raphson method.

4 Application to Leukaemia Data

Kersey et al. [6] reported data on patients with refractary acute lymphoblastic leukaemia. Patients receive either an allogeneic transplant (group 1) or an autologous transplant (group 2) and are followed until a recurrence occurs.

If we fit model (5) for each group separately, the estimated survival functions are

$$\hat{S}_1(t) = 0.2714 + 0.7286 \times \exp(0.76112 \times (1 - \exp(t^{0.61397})))$$

for group 1 and

$$\hat{S}_2(t) = 0.1799 + 0.8201 \times \exp(1.15842 \times (1 - \exp(t^{0.6853})))$$

for group 2. We can consider the data from the two groups jointly and fit the same model. The result is

$$\hat{S}(t) = 0.22739 + 0.77261 \times \exp(0.92261 \times (1 - \exp(t^{0.63706}))).$$

For the moment, we restrict our analysis to the case of one binary covariate. So, defining a covariate, x, as the indicator of the patients group, we obtain

$$\hat{S}(t|x) = 0.22821 + 0.77179 \times (\exp(1.15379 \times (1 - \exp(t^{0.65037}))))^{\exp(-0.42x)}. \quad (13)$$

This covariate had no significant effect on the non-susceptible proportion, something expected given the proximity of the values in the two previous models. Note that the survival time of the susceptible individuals follows a Chen distribution with parameters λ_1 and λ_2 when $x = 0$ and with parameters $\lambda_1 \times e^{\beta}$ and λ_2 when $x = 1$. Due to difficulties in the implementation of the EM algorithm, namely convergence problems, the estimate of β was obtained making use of this characteristic.

5 Concluding Remarks

The aim of this article is to increase the options for survival distributions when the use of cure models is relevant. The Chen distribution is very versatile, resulting in a good fit in many cases where other parametric models were unsatisfactory. We introduced covariates in the model in order to make it more suitable for practical situations. So far, some issues in the estimation process are not completely solved. Nevertheless, we obtained significant correlation coefficients (r=0.9946, p=0.000 for group 1 and r=0.9512, p=0.000 for group 2) between the Kaplan–Meier estimates and the fitted values obtained using model (13), indicating a good fit for both groups.

Acknowledgements Ana Abreu's research was supported by FCT, POCTI-219, FEDER, and Cristina Rocha's research is partially sponsored by national funds through the Fundação Nacional para a Ciência e Tecnologia, Portugal—FCT under the project PEst-OE/MAT/UI0006/2011. We would like to thank the reviewers for their thorough and insightful review of the manuscript.

References

1. Abreu, A.M., Rocha, C.S.: Um novo modelo de cura paramétrico. In: Castro, L.C., Martins, E.G., Rocha, C., Oliveira, M.F., Leal, M.M., Rosado, F. (eds.) Ciência Estatística, pp. 151–162. Edições SPE, Lisboa (2006)
2. Chen, Z. : A new two-parameter lifetime distribution with bathtub shape or increasing failure rate function. Stat. Probab. Lett. **49**, 155–161 (2000)
3. Dempster, A.P., Laird, N.M., Rubin, D.B.: Maximum likelihood estimation from incomplete data via the EM algorithm (with discussion). J. R. Stat. Soc. B **39**, 1–38 (1977)
4. Hunsberger, S., Albert, P.S., London, W.B.: A finite mixture survival model to characterize risk groups of neuroblastoma. Stat. Med. **28**, 1301–1314 (2009)
5. Kaplan, E.L., Meier, P.: Nonparametric estimation from incomplete observations. J. Am. Stat. Assoc. **57**, 457–481 (1958)
6. Kersey, J.H., Weisdorf, D., Nesbit, M.E., LeBien, T.W., Woods, W.G., McGlave, P.B., Kim, T., Vallera, D.A., Goldman, A.I., Bostrom, B., Hurd, D., Ramsay, N.K.C.: Comparison of

autologous and allogeneic bone marrow transplantation for treatment of high-risk refractary acute lymphoblastic leukaemia. N. Engl. J. Med. **317**, 461–467 (1987)
7. Li, C–S., Taylor, J.M.G.: A semi-parametric accelerated failure time cure model. Stat. Med. **21**, 3235–3247 (2002)
8. Maller, R.A., Zhou, S.: Survival Analysis with Long-Term Survivors. Wiley, New York (1996)
9. Peng, Y., Dear, K.B.G., Denham, J.W.: A generalized F mixture model for cure rate estimation. Stat. Med. **17**, 813–830 (1998)
10. Sy, J.P., Taylor, J.M.G.: Estimation in a Cox proportional hazards cure model. Biometrics **56**, 227–236 (2000)

Survival Analysis Applied to the Study of Time from Diagnosis of HIV-1 Infection to AIDS in Portugal

Marta Alves, Cristina S. Rocha, and Maria Teresa Paixão

Abstract
HIV infection is characterized by a progressive destruction of the immune system, allowing for the occurrence of several, severe, opportunistic infections and diseases, leading to the clinical stage of AIDS. In Portugal, the national surveillance system for HIV/AIDS registered 17,825 pre-AIDS cases since January 1993 until 31 March 2008. The national database collects clinical, epidemiological and virological data, including dates of onset of major health events. The objective of this work is to study the time from diagnosis of HIV infection to the development of AIDS. Often, a patient may die prior to the development of AIDS, in a pre-symptomatic stage. This is a situation of competing risks because the individual may experience more than one event. The cumulative incidence function was used to estimate the probability of an event and a competing risks proportional hazards model was used to identify important prognostic factors. In this study, the main determinants towards disease progression to AIDS, as well as death prior to the occurrence of opportunistic infections, are the year of diagnosis, which reflects the availability of antiretroviral drugs (HAART), gender, age and transmission category.

M. Alves
DEIO, Faculty of Sciences, University of Lisbon, Lisbon, Portugal
e-mail: marta.l.alves@gmail.com

C.S. Rocha (✉)
DEIO and CEAUL, Faculty of Sciences, University of Lisbon, Lisbon, Portugal
e-mail: cmrocha@fc.ul.pt

M.T. Paixão
National Health Institute Doutor Ricardo Jorge, Lisbon, Portugal
e-mail: teresa.paixao@insa.min-saude.pt

J. Lita da Silva et al. (eds.), *Advances in Regression, Survival Analysis, Extreme Values, Markov Processes and Other Statistical Applications*, Studies in Theoretical and Applied Statistics, DOI 10.1007/978-3-642-34904-1_4,

1 Introduction

Since the 1980s, HIV infection and AIDS are a major public health problem worldwide. HIV infection is characterized by a progressive destruction of the immune system, allowing for the occurrence of several, severe, opportunistic infections and diseases, leading to the clinical stage of AIDS [1]. The quality of life of HIV and AIDS affected persons is closely related to the different incidences of the AIDS-associated opportunistic diseases and to inequalities in access to health care services, which are jointly responsible for the survival pattern observed globally. AIDS has changed from an acute disease with a very short survival time to what is generally described as a chronic disease, due to the advent of HAART (highly active antiretroviral therapy) and improvement in the prevention and treatment of opportunistic infections. The assessment of factors with direct influence in the survival of affected persons is of the utmost importance for the health services in general, and specifically to the development of appropriate support services, as cases, nowadays, have a better life prospect due to therapeutical advances.

In Portugal, HIV infection and AIDS are one of the leading causes of morbidity. However, persons affected by HIV are living longer disease-free periods, and they expect to benefit from the availability of new antiretroviral drugs to extend their survival time. The national database collects clinical, epidemiological and virological data, including dates of onset of major health events as well as death when notified.

Over the years, many studies have focused on HIV-infected persons and the development of a particular health event, either an AIDS-defining disease or death, in order to compare several health-related interventions. In these studies, survival curves estimated by the Kaplan–Meier method [2], as well as Cox proportional hazards model [2], have been used in order to identify factors that influence the occurrence of the event. These methods are based on the existence of only one event of interest and the censoring mechanism is assumed to be noninformative. Although the end point of the natural history of HIV infection is AIDS, during the natural history of the infection, other events may occur, which will prevent the case of progressing in the course of disease, namely death as asymptomatic HIV-1 or symptomatic non-AIDS (symptoms and illnesses which do not classify the case as AIDS). Consequently, we are in a competing risks situation as the case is exposed to more than one type of adverse health event.

2 Study Population

Since 1985, in Portugal, the National Health Institute Doutor Ricardo Jorge has been responsible for collecting reports of HIV/AIDS cases diagnosed in health services (e.g. hospitals, outpatients medical units and private practices).

In Portugal, the notification of HIV/AIDS cases has been mandatory since 2005 and AIDS is defined on the basis of the clinical condition of the patient. The diagnostic criteria include the three AIDS-defining diseases (extrapulmonary tuberculosis, recurrent bacterial pneumonia and invasive cervical carcinoma), according to the 1993 revised European AIDS case definition [3].

The aim of this work is to study the time between the date of diagnosis of HIV-1 infection and the onset of AIDS. With this purpose, we performed a statistical analysis of data from the national surveillance system, concerning cases of HIV/AIDS that have been diagnosed between January 1993 and March 2008. There are 17,825 asymptomatic HIV-1 or symptomatic non-AIDS cases in the dataset. Information was collected on sociodemographic characteristics (age, gender, origin, vital status) as well as clinical and epidemiological data (transmission category, lymphocyte CD4 cell counts and year of diagnosis). The year of diagnosis was used as a proxy for the availability of antiretroviral agents and we defined 1996 as a period of transition from bitherapy to HAART. Some cases were excluded for the following reasons: individuals aged under 13 years at date of diagnosis; individuals infected with HIV-2, double seropositivity or serological status not reported; individuals infected through mother-to-child transmission, haemophilia, transfusion, blood derivative recipient, nosocomial and other/unknown; individuals with follow-up time less than 1 day.

3 Statistical Methods

In studies where the event of interest is the development of AIDS, a patient can experience an event different from the event of interest. In fact, a patient may die as asymptomatic HIV-1 or symptomatic non-AIDS infected case. The occurrence of this event hinders the development of AIDS and so changes the probability of occurrence of the event of interest. This is a situation of competing risks events because the individual may experience more than one type of event.

In the presence of competing risks, the usual survival statistical methods should be applied with caution and one has to be aware of the consequences of their use. Methods of standard survival analysis such as Kaplan–Meier method for estimation of cumulative incidence lead to incorrect and biased results. In fact, if the Kaplan–Meier method is used to estimate the cumulative incidence of development of AIDS, then patients who died without developing AIDS are censored. However, censoring is informative because those patients will never develop AIDS. So, ignoring the competing risk of death can result in substantial overestimation of the cumulative incidence of progression to AIDS.

In this study we analyse data using the nonparametric estimation of cumulative incidence of the event of interest taking into account the informative nature of censoring due to competing risks. To estimate the probability of occurrence of an event we used the cumulative incidence function (CIF) proposed by Kalbfleisch and Prentice [4]. Time was measured in years from the date of diagnosis to the date of occurrence of the event or to last follow-up when the patient did not experience any event. The outcome variables are time to the development of AIDS and time to death as asymptomatic HIV-1 or symptomatic non-AIDS infection, i.e, death without developing AIDS. The CIF, for an event of type c ($c = 1, 2, \ldots, m$), is defined as

$$F_c(t) = P(T \leq t, C = c) = \int_0^t h_c(u) S(u) du.$$

The CIF at time t is the probability that an event of type c occurs at or before time t, where $h_c(t)$ is the cause-specific hazard function and $S(t) = P(T > t)$ is the overall survival function. The nonparametric estimate of the CIF is given by

$$\hat{F}_c(t) = \sum_{i:t_i \leq t} \hat{h}_{ci} \hat{S}(t_{i-1}),$$

where $\hat{h}_{ci}$ is the estimated cause-specific hazard for an event of type c at t_i and $\hat{S}(t_{i-1})$ is the estimated probability of remaining event free prior to t_i. Thus, $\hat{F}_c(t)$ is the estimate of the joint probability of being event free immediately prior to t_i and experiencing an event of type c at t_i.

To compare the CIF of a particular type of event among different groups in the presence of competing risks, we used a test proposed by Gray based on the hazard of the CIF [5].

After testing for the difference between cumulative incidence curves using Gray's method, we performed a competing risks regression analysis using a model proposed by Fine and Gray [6]. This is an extension of the Cox proportional hazards model to account for competing risks. Fine and Gray method was used to identify important prognostic factors, modelling the effect of covariates on the hazard of the CIF for competing risks data. Thus, the hazard function of the CIF is defined as $\gamma(t; \mathbf{z}) = \gamma_0(t) \exp(\boldsymbol{\beta}'\mathbf{z})$, where $\gamma_0(t)$ is the baseline hazard of the CIF, $\mathbf{z}$ is the vector of the covariates and $\boldsymbol{\beta} = (\beta_1, \ldots, \beta_p)'$ is a vector of p regression coefficients.

We used R software version 2.6.2 to perform the statistical analysis.

4 Results

From a total of 17,825 cases diagnosed as asymptomatic HIV-1 or symptomatic non-AIDS, 890 (5.0 %) cases died while 2,103 (11.8 %) cases have progressed to AIDS. For the remaining 14,832 (83.2 %) none of the events was observed (censored observations).

Of the 17,825 cases with a positive HIV-1 diagnosis, 3,705 (20.8 %) were diagnosed previously to 1997 and 14,120 (79.2 %) were diagnosed between 1997 and 2008.

The analysis of HIV-1 cases by gender and age shows that 70.9 % are males and 21.9 % females and that the age group 13–29 years is the most frequently reported, representing 38.8 % of all notified cases. In males the most frequently notified age group is between 30 and 39 years, while females are younger, aged between 13 and 29 years.

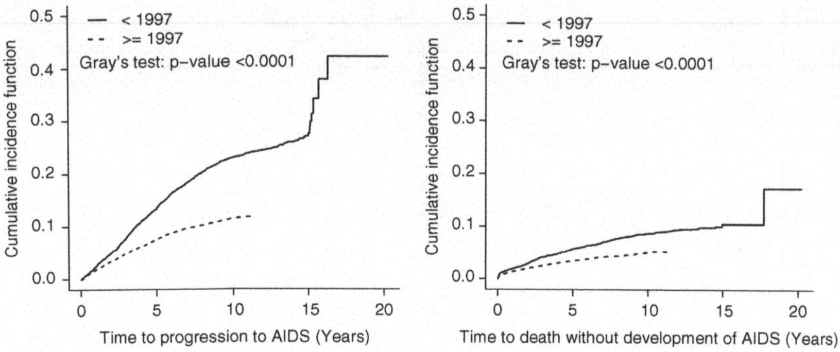

Fig. 1 Cumulative incidence function estimates for the two events by year of diagnosis

The mean age at diagnosis was 34.23 years, age ranging from 13 to 86 years, and a median age of 32 years (31 years for females and 32 for males).

The main routes of transmission of HIV-1 infection are either sexual or related with the intravenous use of drugs (IVDU). In this study, 47 % of HIV-1-infected cases are associated with transmission through IVDU, 40.4 % report sexual (heterosexual) transmission and 11.3 % are male homosexuals (MSM). In males, IVDU are 54.8 % of notified cases, while in females the majority of cases report sexual transmission (71 %) and 28 % of cases are IVDU.

In the natural history of infection, the estimated probabilities of developing AIDS at 5, 10, 15 and 20 years after diagnosis are 0.08, 0.15, 0.20 and 0.37, respectively. The estimated probabilities of dying before developing signs and symptoms, i.e., dying in an asymptomatic HIV-1 or symptomatic non-AIDS stage, up to 5, 10, 15 and 20 years after diagnosis are 0.04, 0.06, 0.08 and 0.16, respectively.

The CIF estimates according to the year of diagnosis (Fig. 1) show that cases diagnosed before 1997 have a greater probability of developing an AIDS indicator disease, as well as dying either in an asymptomatic stage or before developing any opportunistic disease or infection, when compared with those cases diagnosed since 1997.

The CIF estimates (Fig. 2) show that males have a higher probability of developing AIDS and also to die either in an asymptomatic stage or not developing any opportunistic disease or infection, when compared with females. For both males and females, the estimated CIF shows that the probability of developing AIDS at 5, 10 and 15 years is always higher than the corresponding probability of dying in an asymptomatic stage or with a symptomatic non-AIDS infection.

The results of the CIF estimates (Fig. 3) for HIV transmission categories show that cases reporting IVDU are at a greater risk of developing AIDS, as well as dying in an asymptomatic or non-AIDS stage, than cases in any of the other transmission categories. The estimated probabilities of developing an AIDS indicator disease up to 5, 10, 15 and 20 after diagnosis are always higher than the corresponding probabilities of dying asymptomatic or without an opportunistic disease for any of the remaining transmission categories.

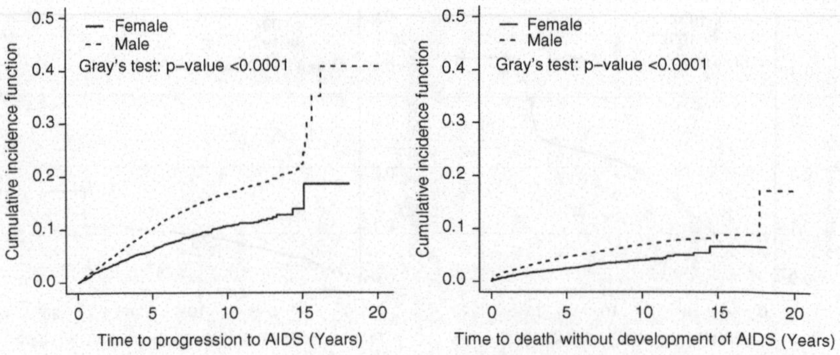

Fig. 2 Cumulative incidence function estimates for the two events by gender

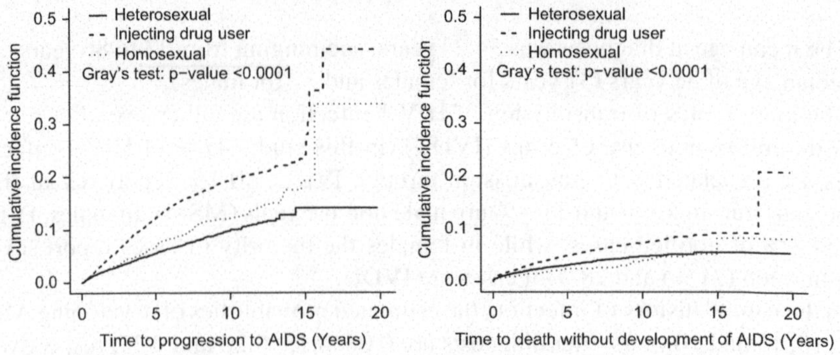

Fig. 3 Cumulative incidence function estimates for the two events by transmission category

For each event, using Gray's test, we found statistically significant differences between the groups defined by year of diagnosis, gender and HIV transmission category.

Table 1 shows the results obtained from fitting Fine and Gray's regression model to data representing time to the onset of AIDS and time to death as asymptomatic HIV-1 or symptomatic non-AIDS. For each variable in turn, considering cases with equal values of the remaining variables, we concluded that:

- Year of diagnosis: cases diagnosed since 1997 have a risk reduction of 54 % of developing AIDS and a risk reduction of 37 % of dying as asymptomatic HIV-1 or symptomatic non-AIDS when compared to the cases diagnosed before 1997.
- Gender: both males and females have the same risk of progression to AIDS, but males have a poorer prognosis than females, as they have 53 % of increased risk of dying as asymptomatic HIV-1 or symptomatic non-AIDS.
- Transmission category: at any time the risk of developing AIDS of injecting drug users is twice the risk associated with cases reporting heterosexual transmission. Injecting drug users have a poorer prognosis as they have a 65 % increased risk of dying before developing AIDS relatively to cases in which HIV infection was acquired by heterosexual transmission.

Table 1 Results of the regression analysis by Fine and Gray model

	Progression to AIDS				Death without developing AIDS			
Classes	$\hat{\beta}$	$\exp(\hat{\beta})$	$se(\hat{\beta})$	p-value	$\hat{\beta}$	$\exp(\hat{\beta})$	$se(\hat{\beta})$	p-value
≥ 1997	−0.78	0.46	0.06	< 0.0001	−0.46	0.63	0.09	3.6×10^{-6}
Male	0.01	1.01	0.07	9.7×10^{-1}	0.43	1.53	0.13	9.7×10^{-4}
30–39 years	0.44	1.56	0.07	7.7×10^{-11}	0.09	1.10	0.11	4.0×10^{-1}
40–49 years	0.62	1.86	0.09	1.1×10^{-12}	0.34	1.41	0.16	3.1×10^{-2}
≥ 49 years	0.56	1.75	0.12	4.1×10^{-6}	1.15	3.14	0.16	2.1×10^{-12}
African origin	0.18	1.20	0.11	8.6×10^{-2}	−0.18	0.83	0.22	3.8×10^{-1}
Other origin	−0.39	0.67	0.26	1.2×10^{-1}	−0.95	0.39	0.58	1.0×10^{-1}
IVDU	0.69	2.01	0.08	< 0.0001	0.50	1.65	0.13	1.3×10^{-4}
Homosexual	0.12	1.13	0.12	3.0×10^{-1}	−0.21	0.81	0.20	3.0×10^{-1}
200–499 CD4 cell	−1.43	0.24	0.07	< 0.0001	−0.27	0.77	0.11	1.7×10^{-2}
> 499 CD4 cell	−2.09	0.12	0.09	< 0.0001	−0.57	0.56	0.13	8.5×10^{-6}

5 Discussion

The study was conducted to assess several clinical, epidemiological and behavioural events, which have an influence in the natural history of HIV infection, namely progression to AIDS and death, in cases diagnosed and notified in Portugal, taking into account the major therapeutic achievements in the past two decades. The national surveillance system has the usual limitations of completeness and timeliness, but results are consistent with those reported at international level.

In this study, on time elapsed from diagnosis of HIV-1 infection (asymptomatic or symptomatic non-AIDS) to AIDS, there were 2,103 cases that developed full-blown AIDS during the study period.

A study carried out in Spain showed a risk reduction of 66 % in disease progression to AIDS, for cases diagnosed between 1997 and 1999, when comparing with cases diagnosed between 1992 and 1996 [7]. In our study, cases diagnosed after 1997 have a reduction in the risk of developing AIDS, of approximately 54 %, when compared with those cases diagnosed prior to 1997 and with equal values of the other variables.

Several studies have shown that there are no differences in gender in the clinical progression to AIDS, while other have opposite results. In the present study, when accounting for the effect of the other variables, gender showed no statistical significance in what concerns progression to AIDS. This result may be related with the unbalanced number of cases diagnosed (and notified) in each transmission category. Also, males showed a 53 % increased risk of dying AIDS-free, when compared to females. Another study carried out in Spain has shown that, for intravenous drug users, progression to AIDS is slower in females than in males [8].

In this study, the main determinants towards disease progression to AIDS as well as death prior to the occurrence of opportunistic infections are year of diagnosis of HIV infection, age, intravenous drug use and the number of CD4 cell count at time of diagnosis. The year of diagnosis is the critical factor associated with disease progression to AIDS, as well as death before developing full-blown AIDS, which reflects the availability of antiretroviral drugs and specifically the general implementation of HAART.

This is one of the first studies performed in this country on survival analysis with this kind of data and further studies, with a more all-inclusive case series, are planned to better clarify issues associated with determinants of living AIDS-free and disease progression to AIDS.

Acknowledgements Cristina Rocha's research is partially sponsored by national funds through the Fundação Nacional para a Ciência e Tecnologia, Portugal—FCT under the project PEst-OE/MAT/UI0006/2011.

References

1. Hoffmann, C., Rockstroh, J.K., Kamps, B.S.: HIV Medicine 2007, 15th edn. Flying Publisher, Paris (2007)
2. Collett, D.: Modelling Survival Data in Medical Research, 2nd edn. Chapman & Hall/CRC, Boca Raton (2003)
3. European Centre for the Epidemiological Monitoring of HIV/AIDS (EuroHIV). 1993 revision of the European AIDS Surveillance case definition. AIDS Surveillance in Europe, Quarterly Report. **37**, 23–28 (1993)
4. Kalbfleisch, J.D., Prentice, R.L.: The Statistical Analysis of Failure Time Data. Wiley, New York (1980)
5. Gray, R.J.: A class of k-sample tests for comparing the cumulative incidence of a competing risk. Ann. Stat. **16**, 1141–1154 (1988)
6. Fine, J.P., Gray, R.J.: A proportional hazards model for the subdistribution of a competing risk. J. Am. Stat. Assoc. **94**, 496–509 (1999)
7. Amo, J., Romero, J., Barrasa, A., Prez-Hoyos, S., Rodrguez, C., Dez, M., Garca, S., Soriano, V., Castilla, J. and the Grupo de Seroconvertores de la Comunidade Madrid: Factors influencing HIV progression in a seroconverter cohort in Madrid from 1985 to 1999. Sex. Transm. Infect. **78**, 255–260 (2002)
8. Garcia de la Hera, M., Ferreros, I., del Amo. J., et al.: Gender differences in progression to AIDS and death from HIV seroconversion in a cohort of injecting drug users from 1986 to 2001. J. Epidemiol. Community. Health. **58**, 944–950 (2004)

A New Independence Test for VaR Violations

P. Araújo Santos and M.I. Fraga Alves

Abstract
Interval forecasts evaluation can be reduced to examining the unconditional coverage and independence properties of the hit sequence. In this work we propose a definition for tendency to clustering of violations and an exact independence test for the hit sequence. This test is suitable to detect models with a tendency to generate clusters of violations and is based on an exact distribution that does not depend on any unknown parameter. Moreover, we provide evidence through a simulation study that the suggested test performs better than other tests presented in the literature.

1 Introduction

We consider a time series of daily log returns, $R_{t+1} = \log(V_{t+1}/V_t)$, where V_t is the value of the portfolio at time t. The corresponding 1-day-ahead VaR forecasts made at time t for time $t+1$, $VaR_{t+1|t}(p)$, are defined by $P[R_{t+1} \leq VaR_{t+1|t}(p)|\Omega_t] = p$, where Ω_t is the information set-up to time t and p is the *coverage rate*. Considering a *violation* in the event that a return on the portfolio is

P.A. Santos (✉)
Departamento de Informática e Métodos Quantitativos, Escola Superior de Gestão e Tecnologia, Instituto Politécnico de Santarém, Santarém, Portugal

CEAUL, Lisbon, Portugal
e-mail: paulo.santos@esg.ipsantarem.pt

M.I. Fraga Alves
Departamento de Estatística e Investigação Operacional, Faculdade de Ciências, Universidade de Lisboa, Lisbon, Portugal

CEAUL, Lisbon, Portugal
e-mail: isabel.alves@fc.ul.pt

J. Lita da Silva et al. (eds.), *Advances in Regression, Survival Analysis, Extreme Values, Markov Processes and Other Statistical Applications*, Studies in Theoretical and Applied Statistics, DOI 10.1007/978-3-642-34904-1_5,

lower than the reported VaR, we define the hit function, also represented by I_t, as

$$I_{t+1}(p) = \begin{cases} 1 \text{ if } R_{t+1} < VaR_{t+1|t}(p) \\ 0 \text{ if } R_{t+1} \geq VaR_{t+1|t}(p). \end{cases} \tag{1}$$

Christoffersen [6] showed that evaluating interval forecasts can be reduced to examining whether the hit sequence, $\{I_t\}_{t=1}^T$, satisfies the unconditional coverage (UC) and independence (IND) properties. UC hypothesis means $P[I_{t+1}(p) = 1] = p, \forall_t$. IND hypothesis means that past violations do not hold information about future violations. A problematic non-verification of the IND hypothesis is the one that leads to clustering of violations, which corresponds to several large losses occurring in a short period of time. As noted by Campbell [4], comparatively with the UC property, the IND property represents a more subtle yet equally important property. When both properties are valid then we write $P[I_{t+1}(p) = 1|\Omega_t] = p, \forall_t$, and we say that forecasts have a correct conditional coverage (CC). In Lemma 1 of Christoffersen [6] it is shown that condition CC is equivalent to $I_{t+1}(p) \overset{iid}{\sim} Bernoulli(p)$, where iid denotes independent and identically distributed. In a recent paper, Berkowitz et al. [1] extend and unify the existing tests by noting that the demeaned hits $\{I_{t+1} - p\}$ form a martingale difference sequence. The hit function and condition CC, imply that $E[(I_{t+1} - p)|\Omega_t] = 0$ and then for any variable Z_t in the time-t information set, $E[(I_{t+1} - p)Z_t] = 0$. This is the motivation for tests based on the martingale property. There are several backtesting procedures for evaluating intervals forecasts; for a detailed review see Campbell [4] and Berkowitz et al. [1]. The Christoffersen [6] Markov IND and CC likelihood ratio tests (Markov), are perhaps the most widely used in the literature. These tests, based on asymptotic distributions, are only sensible to one violation immediately followed by other, ignoring all other patterns of clustering. If we set $Z_t = I_{t-k}$ for any $k \geq 0$, we have $E[(I_{t+1} - p)(I_{t-k} - p)] = 0$. Based on this condition Berkowitz et al. [1] suggested the Ljung-Box statistic (LB), for a joint test of whether the first m autocorrelations of $(I_{t+1} - p)$ and $(I_{t+1-k} - p), k = 1, \ldots, m$, are zero. Considering other data in the information set such as past returns, under CC we have $E[(I_{t+1} - p)g(I_t, I_{t-1}, \ldots, R_t, R_{t-1}, \ldots)] = 0$ for any non-anticipating function $g(.)$. In the same line as Engle and Manganelli [11], Berkowitz et al. [1] consider the autoregression

$$I_t = \alpha + \sum_{k=1}^{n} \beta_{1k} I_{t-k} + \sum_{k=1}^{n} \beta_{2k} g(I_{t-k}, I_{t-k-1}, \ldots, R_{t-k}, R_{t-k-1}) + \varepsilon_t \tag{2}$$

with $n = 1$ and $g(I_{t-k}, I_{t-k-1}, \ldots, R_{t-k}, R_{t-k-1}) = VaR_{t-k+1|t-k}(p)$. These authors proposed the logit model and test the CC hypothesis with a likelihood ratio test considering for the null hypothesis $P(I_t = 1) = 1/(1 + e^{-\alpha}) = p$ and the coefficients β_{11} and β_{21} equal to zero. For the IND hypothesis, the test is adapted considering β_{11} and β_{21} equal to zero. We refer these tests as the CAViaR tests of Engle and Manganelli (CAViaR). A *duration-based* approach emerged in the literature (e.g. [7, 8, 13]). In this set-up, let us define the duration between two consecutive violations as

$$D_i := t_i - t_{i-1} \tag{3}$$

where t_i denotes the day of violation number i. If the IND hypothesis is valid then $I_{t+1}(p) \overset{iid}{\sim} Bernoulli(\pi)$, with $0 < \pi < 1$, and the common distribution of durations Eq. (3) is geometric with probability mass function (pmf)

$$f_D(d;\pi) = (1-\pi)^{(d-1)}\pi, d \in \mathbb{N}, 0 < \pi < 1. \tag{4}$$

The exponential distribution with density function (df)

$$f_D(d;\beta) = \beta \exp(-\beta d), \quad d > 0 \text{ and } \beta > 0, \tag{5}$$

is the continuous analogue of the geometric distribution. Based on the exponential, Christoffersen and Pelletier [7] suggested tests using the duration based approach. Haas [13] showed that tests based on discrete distributions for durations, have higher power. The generalised method of moments test framework suggested by Bontecamps [3] to test for distributional assumptions was extended by Candelon et al. [5] to the case of VaR forecasts accuracy. In the group of duration-based tests it is shown that the proposed GMM tests are the best performers. For the CC and IND hypothesis, the Markov tests are perhaps the most widely used in the literature and this is why we have chosen the Markov independence test for the comparative study. In the group of available duration-based tests we chose the best performers GMM tests. We also selected the CAViaR test, the best performer in the comparative simulation study done by Berkowitz et al. [1]. The rest of the chapter is organized as follows. In Sect. 2 we present the new independence test. Finally, in Sect. 3, we compare its performance with other tests.

2 A New Independence Test

Let $D_{1:N} \leq \ldots \leq D_{N:N}$ be the order statistics (o.s.'s) of durations $D_1, \ldots, D_N$ defined in Eq. (3). The first motivation behind the class proposed is the following: when violations generated by the hit function (1) occur in clusters, the majority of durations are short (the short durations between violations in the clusters) and some durations are very long (the durations between the last violation of one cluster and the first violation of the following cluster). If the majority of durations are short then the median, $D_{[N/2]:N}$, is short (notation: $[x]$ denotes the integer part of x). If some durations are very long, the maximum, $D_{N:N}$, is very long. Finally, with a short median and a very long maximum, the ratio $D_{N:N}/D_{[N/2]:N}$ is large. We illustrate this motivation with an example: we have chosen the returns from the Deutscher Aktien index (DAX) from January 2, 1997 up until December 30, 2008, and we have calculated durations between violations using the popular historical simulation (HS) method for VaR(0.05) with a moving window of size 250. Figure 1 shows the geometric pmf, with $\pi = 0.05$, and the frequency of durations. For short durations,

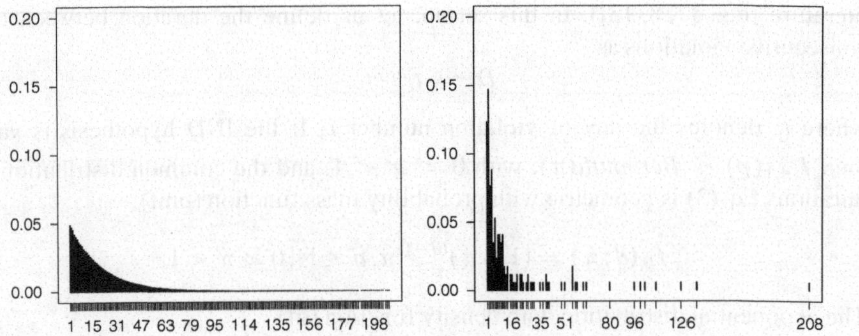

Fig. 1 Geometric ($\pi = 0.05$) pmf (*left*) and frequency of durations (*right*) between violations for DAX index from 2 January 1997 until 30 December 2008

the frequencies in the frequency plot are much higher than the corresponding probability masses in the geometric pmf. The majority of durations are short, either equal or lower than 6 days, and the empirical median is 6, contrasting with the expected value of $D_{85:170}$, under IND, which is close to 14. Moreover, for durations above 60 days, we note higher frequencies in the frequency plot than the probability masses in the geometric pmf. The maximum duration, $d_{170:170}$, is 208 days, almost double the expected value under IND, which is close to 112. The ratio is 34.66, much higher than the median of $D_{170:170}/D_{85:170}$ under IND, which is 8.03 (see the cumulative distribution function (cdf) of Proposition 2.1). In this example, where violations occur in clusters, the majority of durations are short, some durations are very long and, as mentioned before, a high ratio $D_{N:N}/D_{[N/2]:N}$ gives strong evidence against the IND hypothesis. Based on this motivation, we suggest the following definition.

Definition 1 (Tendency to Clustering of Violations). A hit function (1) has a tendency to clustering of violations if the median of $D_{N:N}/D_{[N/2]:N}$ is higher than the median under the IND hypothesis.

For explicitly testing the IND hypothesis versus tendency to clustering of violations, we propose the following test statistic:

$$R_{N,[N/2]} := \frac{D_{N:N} - 1}{D_{[N/2]:N}}. \tag{6}$$

The correction -1 made to $D_{N:N}$ allows us to obtain a pivotal test. Proposition 2.2 allows us to do that as well as to present in Proposition 2.3 a level α test. We will denote Y_i instead of D_i, the durations, when we use the exponential model (5). From now on, we denote

$$a_w = \binom{N - [N/2] - 1}{w}, \quad b_s = \binom{[N/2] - 1}{s}, \quad c_{w,s} = N - [N/2] - w + s$$

$$\gamma_N = \frac{N!}{([N/2]-1)!(N-[N/2]-1)!} \text{ and } R_N^E = Y_{N:N}/Y_{[N/2]:N}.$$

Proposition 2.1. *Let $Y_1, \ldots, Y_N$ be iid exponential random variables (rvs) with common df Eq. (5). The cdf of R_N^E is*

$$1-\gamma_N \sum_{w=0}^{N-[N/2]-1} \sum_{s=0}^{[N/2]-1} (-1)^{w+s} a_w b_s \left([c_{w,s}(w+1)]^{-1} - [c_{w,s}(w+1+c_{w,s}/r]^{-1}\right), \tag{7}$$

with $1 \le r \le \infty$.

Proof. For the Weibull distribution with density function $f_X(x;p;\theta) = \theta p(xp)^{\theta-1}e^{-(px)^\theta}$, $x > 0$, $p > 0$, $\theta > 0$, Malik and Trudel [14] proved that the density of the ratio of the ith and jth o.s.'s with $i < j \le N$, is

$$f_{Z_N}(z;p;\theta) = \frac{\theta C_j}{(i-1)!(j-i-1)!} \sum_{w=0}^{j-i-1} \sum_{s=0}^{i-1} (-1)^{w+s} \binom{j-i-1}{w} \cdot \binom{i-1}{s} z^{\theta-1}[N-j+w+1+(j-i-w+s)z^\theta]^{-2}, \tag{8}$$

with $0 \le w \le 1$ and where $C_j = \prod_{v=1}^{j}(N-v+1)$. To obtain the ratio of the ith and jth o.s.'s, with $i < j \le N$, from the (5) model, in Eq. (8) we substitute θ by 1. We also replace i and j, respectively, by $[N/2]$ and N. Calculating the integral, the cdf for the ratio $Z_N = Y_{[N/2]:N}/Y_{N:N}$ is

$$\gamma_N \sum_{w=0}^{N-[N/2]-1} \sum_{s=0}^{[N/2]-1} (-1)^{w+s} a_w b_s \left([c_{w,s}(w+1)]^{-1} - [c_{w,s}(w+1+c_{w,s}z)]^{-1}\right),$$

with $0 \le z \le 1$. For $R_N^E = 1/Z_N$ the cdf is $1 - F_{Z_N}(1/r)$, and the result follows. □

Proposition 2.2. *Let $D_1, \ldots, D_N$ be iid rv's whose common distribution is geometric with pmf Eq. (4). If we consider $R_{N,[N/2]}$ and R_N^E, then we have*

$$F^{\leftarrow}_{R_{N,[N/2]}}(1-\alpha) < F^{\leftarrow}_{R_N^E}(1-\alpha), \quad \textit{for all } 0 < p < 1, \quad \textit{and } 0 < \alpha < 1.$$

Proof. Let Y be an exponential rv with df Eq. (5) and denote $[Y]$ the integer part of Y and $< Y >$ the fractional part of Y. If we define $X = [Y] + 1$, then

$$f_X(x) = F_Y(x) - F_Y(x-1) = \left(\exp(-\beta)\right)^{(x-1)}\left(1-\exp(-\beta)\right)$$

with $x \in \mathbb{N}$. Note that X is distributed as geometric with $\pi = (1 - \exp(-\beta))$. Now, for $\pi = (1 - \exp(-\beta))$, $D_{i:N} \stackrel{d}{=} X_{i:N} = [Y]_{i:N} + 1 \stackrel{d}{=} [Y_{i:N}] + 1$, and since $Y\beta \stackrel{d}{=} E$, we have

$$\frac{D_{N:N} - 1}{D_{[N/2]:N}} \stackrel{d}{=} \frac{[Y_{N:N}]}{[Y_{[N/2]:N}] + 1} < \frac{[Y_{N:N}] + < Y_{N:N} >}{[Y_{[N/2]:N}] + < Y_{[N/2]:N} >} = \frac{Y_{N:N}}{Y_{[N/2]:N}} \stackrel{d}{=} R_N^E.$$

□

Proposition 2.3. *Let us consider* $\underset{\sim}{D} := \{D_i\}_{i=1}^N$, *the sample of the N durations defined in Eq. (3). Denote by* $Med(R_{N,[N/2]})$ *the median of* $R_{N,[N/2]}$ *and* $r^*_{1/2,N,[N/2]}$ *the particular value under geometric distribution with pmf Eq. (4). At level* α, *for testing the IND hypothesis*

$$H_{0,IND} : D_i \stackrel{iid}{\sim} D \sim Geometric(\pi), \;\; with \;\; 0 < \pi < 1 \;\; and \;\; i = 1, \ldots, N$$

against alternatives expressing tendency to clustering patterns

$$H_1 : Med(R_{N,[N/2]}) > r^*_{1/2,N,[N/2]},$$

the rejection region is defined by $R_{N,[N/2]} > r_{\alpha,N,k}$, *where* $r_{\alpha,N,[N/2]}$ *denotes a quantile* $1 - \alpha$ *of* R_N^E.

Proof. The proof follows straightforward using Propositions 2.1 and 2.2. □

Remark 1. The critical point $r_{\alpha,N,[N/2]}$ implies a conservative approach with a test of level α and not of size α, i.e., we have $P[\text{type I error}] \leq \alpha$.The test is pivotal in the sense that is based on a distribution that does not depend on an unknown parameter.

Remark 2. The test suggested in Proposition 2.3 is based on an exact distribution. The other independence tests, referred in Sect. 1, are based on asymptotic distributions and suffer from small sample bias. To aggravate the problem, the presence of the nuisance parameter p makes it impossible to control the size of the tests using the Monte Carlo testing approach of Dufour [10] as other authors do for the case of joint testing UC and IND (e.g. [1,5,7]); see the paper of Dufour [10] for details.

3 Comparative Simulation Study

In the context of a Monte Carlo study, we compare the power of the test we suggest in Proposition 2.3 with the Markov, the CAViaR and the GMM independence tests, denoted by M_{IND}, $CAViaR$ and $J_{IND}(k)$. We employ the R language [15] and the fGarch package of Wuertz et al. [16] in order to develop the programs. Following other authors (e.g. [1, 5–7, 13]) we consider a GARCH specification for the returns process. Additionally, we use an APARCH model which nests some of the GARCH models with leverage effect.

- Gaussian GARCH(1,1) model [2],

$$r_{t+1} = \sigma_{t+1} z_{t+1} \text{ with } \sigma_{t+1}^2 = w + \alpha r_t^2 + \beta \sigma_t^2, \tag{9}$$

  where the innovations z_{t+1}s are drawn independently from a standard normal distribution. As in Christofferson [6], we chose the parameterization $w = 0.05$, $\alpha = 0.1$ and $\beta = 0.85$.

- APARCH(1,1) model [9],

$$r_{t+1} = \sigma_{t+1} z_{t+1} \text{ with } \sigma_{t+1}^\delta = w + \alpha(|r_t| - \gamma r_t)^\delta + \beta \sigma_t^\delta, \tag{10}$$

  where the innovations z_{t+1}s are drawn independently from a skewed Student's t(ν) distribution with asymmetry coefficient φ, proposed by Fernandez and Steel [12]. We assume a portfolio that replicates the DAX index and we use daily data from beginning of 1997 until the end of 2008, for estimation. The parameterization achieved was $w = 0.03$, $\alpha = 0.086$, $\gamma = 0.64$, $\beta = 0.91$, $\delta = 1.15$, $\varphi = 0.88$ and $\nu = 10$.

As in other power studies with the same purpose, we have chosen the HS method which generates clusters of violations when applied to heteroscedastic processes. We conducted our study with $p = 0.01, 0.05$, $T = 250, 500, 750, 1{,}000$ and set the size of the rolling window equal to 500. For each T and p, we have simulated returns using the models (9) and (10) over 10,000 replications. The empirical power of the tests is obtained by rejection frequencies with 0.1 significance level, excluding the samples with less than 2 violations. The frequencies of excluded samples (FES) are presented in the tables. To explicitly test the IND hypothesis, it is impossible to have a test of size α using a Monte Carlo approach. Therefore, and for all test statistics except Eq. (6), we apply the asymptotic distributions in order to find critical values, conscious of the limitations in the small sample cases. From Table 1, it is clear that the proposed test performs better than the other tests under study. In order to study the empirical type I error rates, we have simulated iid Bernoulli samples. In the CAViaR test we have generated the VaR regressors with a GARCH model that are independent of the Bernoulli samples. Table 2 shows that the Markov and CAViaR tests are undersized for small sample sizes and oversized for large sample sizes. The GMM tests are extremely undersized for small samples. These results confirm that the asymptotic critical values are misleading.

Acknowledgements Research partially supported by national funds through **FCT**—Fundação para a Ciência e a Tecnologia, FCT/ PROTEC, project PEst-OE/MAT/UI0006/2011 and FCT /PTDC/MAT/101736/2008, EXTREMA project. The authors would like to thank the two referees for their comments and suggestions which led to improvements of an earlier version of this chapter.

Table 1 Empirical power of tests ($\alpha = 0.1$)

	$p = 0.01$				$p = 0.05$			
	T = 250	T = 500	T = 750	T = 1,000	T = 250	T = 500	T = 750	T = 1,000
Gaussian GARCH(1,1)								
$R_{N,[N/2]}$	0.295	**0.452**	**0.567**	**0.630**	**0.377**	**0.575**	**0.694**	**0.757**
M_{IND}	0.115	0.156	0.210	0.214	0.144	0.247	0.327	0.374
CAViaR	**0.316**	0.411	0.507	0.566	0.334	0.483	0.596	0.667
$J_{IND(3)}$	0.098	0.182	0.284	0.378	0.205	0.448	0.638	0.748
$J_{IND(5)}$	0.080	0.165	0.262	0.362	0.162	0.374	0.556	0.673
FES	0.292	0.041	0.002	0.000	0.003	0.000	0.000	0.000
Skewed t APARCH(1,1)								
$R_{N,[N/2]}$	0.375	**0.603**	**0.762**	**0.854**	**0.496**	**0.809**	**0.928**	**0.969**
M_{IND}	0.145	0.217	0.287	0.338	0.205	0.384	0.527	0.633
CAViaR	**0.392**	0.505	0.605	0.675	0.436	0.619	0.748	0.829
$J_{IND(3)}$	0.214	0.386	0.559	0.697	0.378	0.745	0.910	0.970
$J_{IND(5)}$	0.166	0.354	0.535	0.676	0.314	0.681	0.876	0.953
FES	0.373	0.093	0.009	0.001	0.041	0.001	0.000	0.000

Table 2 Empirical type I error rates with $\alpha = 0.1$

	$p = 0.01$				$p = 0.05$			
	T = 250	T = 500	T = 750	T = 1,000	T = 250	T = 500	T = 750	T = 1,000
$R_{N,[0.5N]}$	0.088	0.092	0.095	0.093	0.077	0.078	0.084	0.081
M_{IND}	0.023	0.029	0.039	0.037	0.054	0.111	0.158	0.134
CAViaR	0.080	0.056	0.066	0.057	0.083	0.099	0.130	0.124
$J_{IND(3)}$	0.003	0.006	0.011	0.015	0.018	0.032	0.045	0.045
$J_{IND(5)}$	0.001	0.004	0.007	0.012	0.012	0.021	0.028	0.033
FES	0.292	0.038	0.004	0.000	0.000	0.000	0.000	0.000

References

1. Jeremy Berkowitz, J., Christoffersen, P., Pelletier, D.: Evaluating Value-at-Risk models with desk-level data. Management Science, **57**(12), 2213–2227 (2011)
2. Bollerslev, T.: Generalized autoregressive conditional heteroskedasticity. J. Econom. **31**, 307–327 (1986)
3. Bontemps, C.: Testing distributional assumptions: a GMM approach. Working Paper (2006)
4. Campbell, S.D.: A review of backtesting and backtesting procedures. J. Risk **9**(2), 1–18 (2007)
5. Candelon, B., Colletaz, G., Hurlin, C., and Tokpavi, S.: Backtesting value-at-risk: a GMM duration-based test. HAL, Working Paper (2008)
6. Christoffersen, P.: Evaluating intervals forecasts. Int. Econ. Rev. **39**, 841–862 (1998)
7. Christoffersen, P., Pelletier, D.: Backtesting value-at-risk: a duration-based approach. J. Financial Econom. **2**(1),84–108 (2004)
8. Danielsson, J., Morimoto, Y.: Forecasting extreme financial risk: a critical analysis of practical methods for the Japanese market. Monetary Econ. Studies **18**(2), 25–48 (2000)
9. Ding, Z., Engle, R.F, Granger, C.W.J.: A long memory property of stock market return and a new model. J. Empirical Finance **1**, 83–106 (1993)

10. Dufour, J.M.: Monte Carlo tests with nuisance parameters: a general approach to finite sample inference and nonstandard asymptotics. J. Econom. **127**(2), 443–477 (2006)
11. Engel, R.F., Manganelli, S.: CAViaR: conditional autoregressive value-at-risk by regression quantiles. J. Bus. Econ. Stat. **22**, 367–381 (2004)
12. Fernández, C., Steel, M.F.j.: On Bayesian modelling of fat tails and skewness. J.Am. Stat. Assoc. **93**, 359–371 (1998)
13. Haas, M.: Improved duration-based backtesting of value-at-risk. J. Risk **8(2)**, 17–36 (2005)
14. Malik, R.J., Trudel, R.: Probability density function of quotient of order statistics from the pareto, power and weibull distributions. Commun. Stat. Theor. Methods **11**(7), 801–814 (1982)
15. R Development Core Team: R: A Language and Environment for Statistical Computing. R Foundation for Statistical Computing, Vienna, Austria (2008). ISBN 3-900051-07-0, URL http://www.R-project.org
16. Wuertz, D., Chalabi, Y., Miklovic, M.: fGarch: Rmetrics—Autoregressive Conditional Heteroskedastic Modelling (2008). R package version 290.76. http://www.rmetrics.org

Discrimination Between Parametric Survival Models for Removal Times of Bird Carcasses in Scavenger Removal Trials at Wind Turbines Sites

Regina Bispo, Joana Bernardino, Tiago A. Marques, and Dinis Pestana

Abstract

Wind power is one of the most promising energy sources found in nature. Despite being considered a clean energy source, the existence of potential environmental impacts, namely, on flying vertebrates, is broadly recognized. In monitoring studies, estimation of avian (or bats) mortality caused by collision has particular interest and must take into account carcass removal by scavengers. For this purpose, scavenger removal trials are conducted at wind turbine sites. Data from scavenger removal trials refer to time until removal of the carcass and are "classical" examples of survival times.

Parametric survival models based on the exponential, Weibull, log-logistic, and lognormal distributions are among the most repeatedly used throughout

R. Bispo (✉)
ISPA - Instituto Universitário, Lisboa, Portugal

CEAUL - Centro de Aplicações e Estatística da Universidade de Lisboa, Lisbon, Portugal
e-mail: rbispo@ispa.pt

J. Bernardino
Bio3 - Estudos e Projectos em Biologia e Valorização de Recursos Naturais, Almada, Portugal
e-mail: joana.bernardino@bio3.pt

T.A. Marques
Centre for Research into Ecological and Environmental Modeling, The Observatory, Buchanan Gardens, St Andrews, Scotland, UK

CEAUL - Centro de Aplicações e Estatística da Universidade de Lisboa, Lisbon, Portugal
e-mail: tiago@mcs.st-and.ac.uk

D. Pestana
Departamento de Estatística e Investigação Operacional, Faculdade de Ciências da Universidade de Lisboa, Lisboa, Portugal

CEAUL - Centro de Aplicações e Estatística da Universidade de Lisboa, Lisbon, Portugal
e-mail: ddpestana@fc.ul.pt

J. Lita da Silva et al. (eds.), *Advances in Regression, Survival Analysis, Extreme Values, Markov Processes and Other Statistical Applications*, Studies in Theoretical and Applied Statistics, DOI 10.1007/978-3-642-34904-1_6,

literature. In this study we aim to discriminate between these four competing parametric models to analyze removal data from trials conducted in ten Portuguese wind farms. Plotting procedures and model selection strategies are used and discussed.

1 Introduction

Nowadays, wind is considered as one of the most promising energy sources found in nature. Despite being considered a clean energy source, the existence of potential environmental impacts, namely, on flying vertebrates, is broadly recognized [8]. There is a major concern with the mortality caused by collision with wind plant structures [4]. To fully understand the importance of this impact, mortality estimation is necessary.

Mortality assessment is based on counting bird carcasses in the wind farms. However, the observed number of fatalities is different from the true fatality namely because carcasses are removed either by predators/scavengers or decomposition. To account for carcass removal, the observed mortality must be corrected by the probability of permanence of a carcass. To estimate this probability, wind farm monitoring plans include scavenger removal trials. Typically, in these trials, a certain number of carcasses are randomly placed underneath the wind turbines for a a priori fixed period of time. For each placed carcass, time until removal, i.e., time until a carcass is no longer available for detection (corpses absent from the location of placement), is recorded.

Time until removal is typically positively skewed and often includes censored observations. Hence, proper survival analysis should be used to analyze this type of data [2]. Parametric survival methods, by assuming a specific form for the underlying data distribution, have the advantage to enable probability estimation and allow more precise inferences [2]. However, because the parametric survival methods are strictly dependent on the validity of the distributional assumption, the selection of the lifetime distribution has crucial importance.

Several methods are described in the literature to assess the distributional form of the survival times. Plotting procedures based on a linear transformation of the survivor function are often used. Also, empirical and parametrical estimated functions can be drawn together to visually check the model adjustment. Both types of plots may be constructed in strata defined by the components of the regression vector, whenever models include covariates [9]. The comparison of the adjustment between several plausible models can also be made on the basis of statistics such as the Akaike's information criteria (AIC) or the Bayesian's information criteria (BIC). These statistics are suitable for comparisons between non-nested models. Additionally, procedures based on residual analysis are important as they enable to check the models assumptions and assess special features of the data, such as extreme observations [11].

To avoid reporting removal rates exclusively on the grounds of empirical estimates or based on an eventually misspecified lifetime distribution, we propose the use of parametric survival models based on a proper comparative goodness-of-fit analysis regarding diverse plausible models. The focus of this chapter is, therefore, (1) to propose a methodological strategy to discriminate between several plausible parametric survival models suitable for modeling the removal times of bird carcasses in scavenger removal trials and (2) to exemplify the proposed methodology using the data collected in trials conducted at ten Portuguese wind farms.

2 Motivating Data

Carcass removal trials were conducted in ten wind farms located in the north and center of Portugal (for confidentiality reasons sites names are coded from WF1 to WF10). The number of carcasses placed in each trial varied between 20 and 80, according to the size of the farm. Trials were spread over two seasons (May/June and September/October or January/February and July/August) to account for weather conditions influence on removal. Additionally, three bird size classes were considered (small: $\leq$ 15 cm; medium: between 15 and 25 cm; large: $>$ 25 cm). Carcasses were placed in randomly chosen locations beneath the wind turbines, independently of size class. To avoid scavenger swamping, carcasses were placed at a minimum distance of 500 m from each other. The carcasses were checked daily and time until removal was recorded for a maximum period of 20 days. Hence, observations are type I right censored and carcasses not removed until day 20, have censored times of removal all equal to 20 days.

3 Discrimination Between Parametric Survival Models

Time until removal was modeled using the accelerated failure time model as, in this context, covariates can affect the rate at which carcass persistence proceeds along the time axis. This is a general model for survival data that encompasses a wide range of lifetime distributions, in which exploratory variables measured on a subject are assumed to act multiplicatively on the timescale [2]. Plausible expected hazard behaviors include either decreasing or hump-shaped removal hazards. Hence, the Weibull, the log-logistic, and the log-normal distributions seemed to be plausible models as they exhibit monotonic decreasing and asymmetric with positive mode hazard behaviors. Despite its implicit hazard of removal being constant, which is implausible under this context, the exponential distribution was included because it is the most commonly used distribution in wind farm mortality estimation (e.g., [7]).

Plots based on the linearization of the survivor function, through an appropriated transformation, can give information on the underlying lifetime distribution [11]. The linear relationships regarding the exponential, the Weibull, the log-logistic, and the log-normal lifetime distributions are summarized in Table 1. For a given sample, plots of time (or log(time)) versus the appropriate transformation of the estimated (Kaplan–Meier) survivor function should be roughly linear if the assumed model

Table 1 Required linear transformations of survival probability and time (t) scales for different lifetime distributions for graphical inspection of the parametric survival models adequacy

Lifetime distribution	Survivor function	Time scale	Probability scale
Exponential	$S(t) = \exp(-\rho t)$	t	$-\log S(t)$
Weibull	$S(t) = \exp[-(\rho t)^{\gamma}]$	$\log t$	$\log(-\log S(t))$
Log-logistic	$S(t) = [1 + (\rho t)^{\kappa}]^{-1}$	$\log t$	$\log\left(\frac{S(t)}{1-S(t)}\right)$
Log-normal	$S(t) = 1 - \Phi[(\log t - \mu)/\sigma]$	$\log t$	$\Phi^{-1}(1 - S(t))$

is correct. The linear agreement can then be appreciated by eye (which can be misleading) or be measured using the standard coefficient of determination.

Another graphical procedure can be achieved by superimposing graphically the empirical (Kaplan–Meier) and the parametricaly estimated survivor functions to check visually the adjustment between the observed and the fitted functions.

For censored data, the described plotting procedures are probably the most widely useful graphical approaches for comparing competing parametric models [3].

As, in this context, the final goal of inference is to use fitted parametric models to estimate carcass persistence probabilities; model selection procedures are particularly important. To choose among competing models, we used the AIC (defined by $AIC = -2\log\hat{L} + 2k$, where $\hat{L}$ is the maximized likelihood and k is the number of the unknown parameters in the model) and the BIC (defined by $BIC = -2\log\hat{L} + k\ln(n)$, where n denotes the number of observations). The lower these measures, the more parsimonious is the fit.

Additionally, fitted model adequacy can be assessed by residual analysis. In this study both deviance and Cox-Snell residuals were analyzed.

Data were analyzed using R software [12]. In particular, we used the `survival` package [13].

4 Results

The percentage of censored observations varied across the wind farm trials ranging from 0 % (at WF6, median time of removal of 2.5 days) to 35 % (at WF2, median time of removal of 8.5 days), depending on the speed of the carcass removal. On average, as expected, an increase in carcass removal speed was associated with the decrease of the censoring degree.

The data analysis showed consistently that removal times were not affected significantly by season and body size factors in 6 out of the 10 wind farms (WF1 to WF6). In WF7 and WF8 wind farms, season proved to have a significant effect ($p < 0.001$) and in WF9 and WF10 wind farms, both covariates had a significant effect on the removal times ($p < 0.001$). Although the described plotting procedures were used for all the ten analyzed data sets, plots based on the linearization of the survivor function are shown only for WF1 to WF6 wind farm data sets (in which covariates were found not to affect significantly the removal times) and plots superimposing

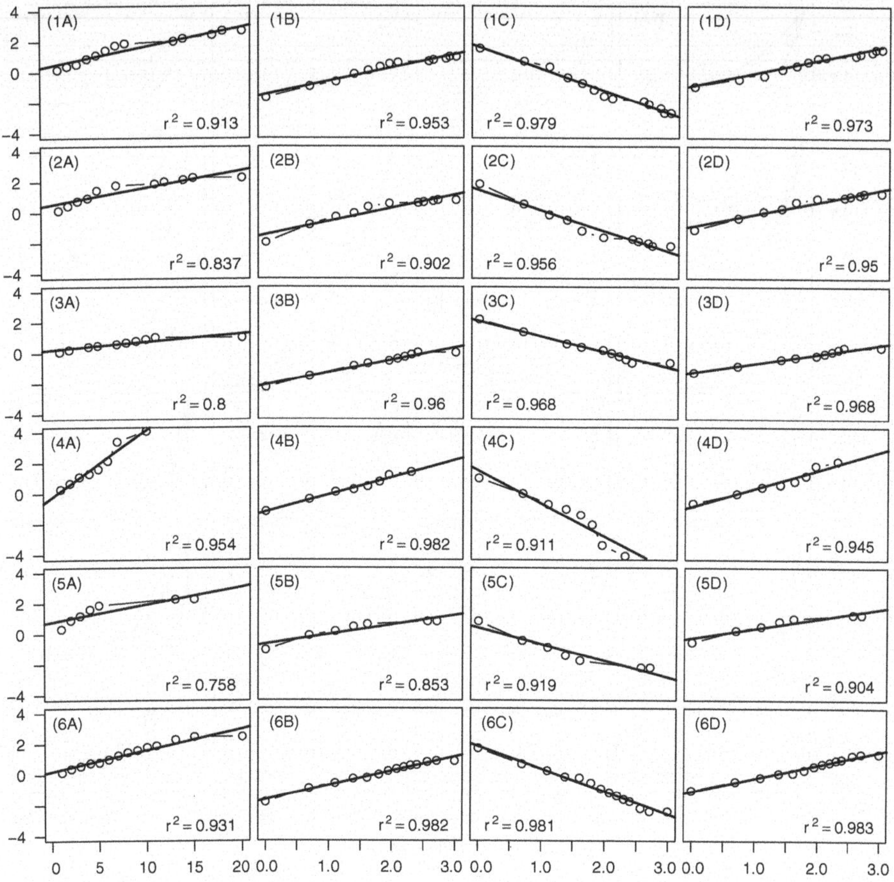

Fig. 1 Plots based on a linear transformation of the survivor function for the inspection of the fitted parametric survival models adequacy in 1-WF1, 2-WF2, 3-WF3, 4-WF4, 5-WF5 and 6-WF6 wind turbine sites, regarding A-exponential, B-Weibull, C-log-logistic and D-log-normal fitted models

the empirical and the adjusted models are used to illustrate the adequacy of the models accounting for dependency on explanatory variables (WF7 to WF10).

AIC and BIC statistics showed a very strong agreement between them, pointing to the same model selection in all the ten analyzed data sets. Hence, we refer here only the results according to the AIC. For the WF1, WF3, WF4, and WF6 wind farms, the AIC was found to be the lowest for the log-normal model, while the best fitting model was the log-logistic for the WF2 and WF5 wind farms. However, differences between AIC values for the log-logistic and log-normal models were minimal, suggesting similar model adjustment, which, in fact, was expected, since these models are very similar. Consequently, inferences based on either model will be, in this case, very similar.

Plots based on a linear transformation of the survivor function (Fig. 1) show that the exponential model has the poorest fit in all six wind farms (smaller coefficients of

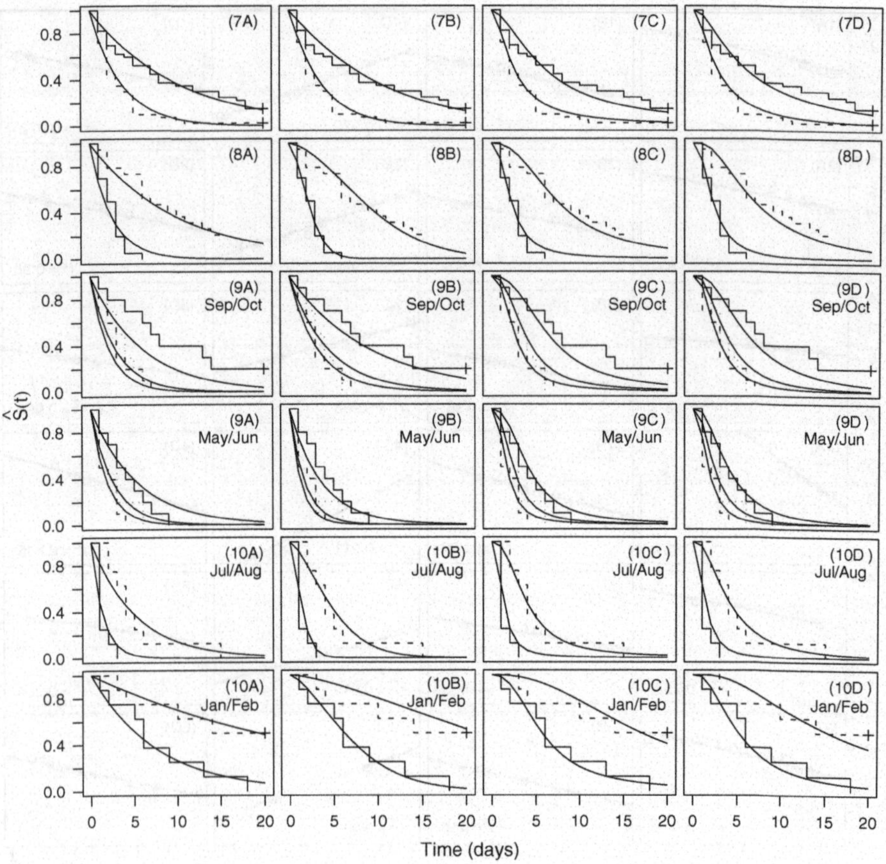

Fig. 2 Empirical (step functions) and fitted parametric survivor functions at 7-WF7 (*step solid line*, Jan/Feb and *step dashed line*, Jul/Aug); 8-WF8 (*step solid line*, May/Jun and *step dashed line*, Sep/Oct); 9-WF9 (*step solid line*, small-size carcasses; *step dashed line*, medium-size carcasses and *step dotted line*, large-size carcasses) and 10-WF10 (*step dashed line*, medium-size carcasses and *step dotted line*, large-size carcasses) wind turbine sites, regarding A-exponential; B-Weibull; C-log-logistic and D-log-normal models

determination), which reflects the relative inadequacy of the exponential distribution to model removal times under this context. The remaining parametric models give fairly good approximated linear relationships, with slight differences between them. The coefficients of determination point to the log-logistic and the log-normal models as the most suitable, matching the results from AIC. For the WF4 wind farm the best linear relationship was found for the Weibull model.

Comparisons between the four fitted models, based on plots shown in Fig. 2, seem hard as differences between the models are almost eye imperceptible. Hence, model selection based on these type of plots is risky and can be misleading. The relative goodness-of-fit measures assume, therefore, a specially important role in this context.

Regarding the WF7 and the WF8 wind farms, lowest AIC values were found for the log-normal and the Weibull models, respectively, suggesting these models as the most suitable to model carcass removal times at these wind farms. For the WF9 and WF10 data, AIC indicates the Weibull and the log-logistic models as the most suitable. However, AIC values were very similar for the Weibull, the log-logistic, and the log-normal models, which, in fact, was expected given the minor differences between corresponding plots displayed in Fig. 2. The analysis of the residuals revealed no major problems with any of the best fitting models.

5 Concluding Remarks

While we focus on wind farm wildlife fatalities, the methodological approach proposed and explored here is, nonetheless, broadly applicable in many other contexts. In particular, we propose the use of the described methods in all the situations in which mortality by collision with anthropogenic structures is a source of concern and, hence, whenever mortality estimation is mandatory. Among these situations we underline wildlife mortality resulting, e.g., from collision with power lines [5, 6], communication towers [1], or cars on roads [14], or from pesticide applications in agricultural systems [10]. In these situations monitoring studies are conducted aiming to estimate the number of fatalities. In all of them, to correctly estimate mortality it is important to consider carcass removal. For that reason it is a standard procedure to conduct carcass removal trials, collecting data regarding carcass removal times. Carcass removal trials results are always site-specific and the permanence probability estimated for a specific wind farm should never be used to correct the observed mortality at another site. As the estimation of this probability through the use of parametric models implies a distributional assumption, procedures used to check model adequacy are particularly important. Lawless (2003) [11] underlines that

> Often data are analysed under a particular model simply because (1) the model has been used before in similar situations, or (2) it fits the data on hand. This does not imply any absolute validity of the model, and we should ask whether inferences change much if another similar "plausible" model is used instead.

So, recognizing that the carcass persistence probability can, in fact, depend heavily on the model selected, this study proposes and applies a methodology to discriminate between competing survival models when analyzing data from carcass removal trials.

We found plotting procedures to be insufficient for model selection. Eye judgment of differences between the statistical models based on plots analysis was difficult. The analysis of the plots based on a linear transformation of the survivor function has the advantage of being interpreted in terms of coefficients of determination, leading to less ambiguous choices. Although plotting procedures do not discriminate sufficiently enough the fitted models, they enable to illustrate model adjustment after model choice. The use of the AIC allowed to choose the best relative fitted model.

The discrimination between the competitive parametric survival models is strictly dependent on sample size and on censoring degree. Small sample sizes and higher censoring degrees lead to, in general, a less efficient estimation process and, therefore, the efficiency in discriminating between alternative competing models may be compromised. These sources of error are still poorly explored. Hence, future work should be considered to evaluate extensively these effects under the context of modeling carcass removal time for wildlife mortality assessment.

Acknowledgments This research was conducted in collaboration with BIO3—Estudos e Projectos em Biologia e Valorização de Recursos Naturais. The research was partially sponsored by national funds through the Fundação Nacional para a Ciência e Tecnologia, Portugal—FCT under the project PEst-OE/MAT/UI0006/2011.

References

1. Ball, L.G., Zyskowski, K., Griselda, E.S.: Recent bird mortality at a Topeka television tower. Kansas Ornithol. Soc. Bull. **46**, 33–35 (1995)
2. Collett, D.: Modelling Survival Data in Medical Research. Chapman & Hall/CRC, Boca Raton, Florida (2003)
3. Cox, D.R., Oakes, D.: Analysis of Survival Data. Chapman & Hall, London (1998)
4. Drewitt, A.L., Langston, R.H.W.: Collision effects of wind-power generators and other obstacles on birds. Ann. N. Y. Acad. Sci. **1134**, 233–266 (2008)
5. Ferrer, M., Riva, M., Castroviejo, J.: Electrocution of raptors on power lines in southwestern Spain. J. Field Ornithol. **62**, 181–190 (1991)
6. Hass, D., Nipkow, M., Fielder, G., Schneider, R., Haas, W., Schrenberg, B.: Protecting birds from powerlines. In: Nature and Environment, vol. 140. Council of Europe Publishing, Strasbourg (2005)
7. Huso, M.M.P.: An estimator of wildlife fatality from observed carcasses. Environmetrics **22**, 318–329 (2010)
8. Johnson, G., Erickson, W., Strickland, M., Shepherd, M., Shepherd, D., Sarappo, S.: Mortality of bats at a large-scale wind power development at Buffalo Ridge, Minnesota. Am. Midl. Nat. **150**, 332–342 (2003)
9. Kalbfleisch, J.D., Prentice, R.L.: The Statistical Analysis of Failure Time Data. Wiley, New York (2002)
10. Kostecke, R.M., Linz, G.M., Bleier, W.J.: Survival of avian carcasses and photographic evidence of predators and scavengers. J. Field Ornithol. **72**, 439–447 (2001)
11. Lawless, J.F.: Statistical Models and Methods for Lifetime Data. Wiley, New York (2003)
12. R Development Core Team: R: A Language and Environment for Statistical Computing. R Foundation for Statistical Computing, Vienna, Austria (2011). URL http://www.R-project.org/. ISBN 3-900051-07-0
13. Therneau, T., original Splus→R port by Thomas Lumley: survival: Survival analysis, including penalised likelihood. (2011). URL http://CRAN.R-project.org/package=survival. R package version 2.36-5
14. Trombulak, S.C., Frissel, C.: Review of ecological effects of roads on terrestrial and aquatic communities. Conserv. Biol. **14**, 18–30 (2000)

Uniformity

M.F. Brilhante, M. Malva, S. Mendonça, D. Pestana, F. Sequeira, and S. Velosa

Abstract

Transformations such as $V = X + Y - \mathrm{I}[X + Y]$ or $W = \min\left(\frac{X}{Y}, \frac{1-X}{1-Y}\right)$ and Sukhatme's transformation can be used to augment uniform random samples and uniform order statistics, respectively. We discuss the bearing of these facts in testing uniformity, an important issue in the field of combining p-values in meta-analytical syntheses.

1 Introduction

Let us assume that the p-values $\{p_k\}_{k=1}^n$ are known from testing H_{0k} vs. H_{Ak}, $k = 1, \ldots, n$, in n independent studies on some common issue, and our aim is to achieve a decision on the overall question H_0^* : all the H_{0k} are true *vs.* H_A^* :

M.F. Brilhante (✉)
Universidade dos Açores (DM) and CEAUL, Rua da Mãe de Deus, Apartado 1422, 9501-801 Ponta Delgada, Portugal
e-mail: fbrilhante@uac.pt

M. Malva
Escola Superior de Tecnologia de Viseu and CEAUL, Campus Politécnico de Viseu de Repeses, 3504-510 Viseu, Portugal
e-mail: malva@estv.ipv.pt

S. Mendonça · S. Velosa
Universidade da Madeira (DME) and CEAUL, Campus Universitário da Penteada, 9000-390 Funchal, Portugal
e-mail: smfm@uma.pt; sfilipe@uma.pt

D. Pestana · F. Sequeira
Universidade de Lisboa, Faculdade de Ciências (DEIO) and CEAUL, Bloco C6, Piso 4, Campo Grande, 1749-016 Lisboa, Portugal
e-mail: dinis.pestana@fc.ul.pt; fjsequeira@fc.ul.pt

J. Lita da Silva et al. (eds.), *Advances in Regression, Survival Analysis, Extreme Values, Markov Processes and Other Statistical Applications*, Studies in Theoretical and Applied Statistics, DOI 10.1007/978-3-642-34904-1_7,

some of the H_{Ak} are true. As there are many different ways in which H_0^* can be false, selecting an appropriate test is in general unfeasible. On the other hand, combining the available p_k's so that $T(p_1, \dots, p_n)$ is the observed value of a random variable whose sampling distribution under H_0^* is known is a simple issue, since under H_0^*, $p = (p_1, \dots, p_n)$ is the observed value of a random sample $P = (P_1, \dots P_n)$ from a standard uniform population. In fact, several different sensible combined testing procedures are often used [6, 11].

Therefore an important issue is to test whether a given sequence $\{p_k\}_{k=1}^n$ is or is not a sample from a standard uniform population. Paul [10] discussed new characterizations of the uniform population, but as they are formulated in terms of expected values, they did not lead directly to new simple tests of uniformity. Gomes et al. [5] exploited the possibility of using computationally augmented samples to test uniformity, with the surprising result that power can decrease with sample augmentation in the class of alternatives they used. Sequeira [12] explains why this is so, and in Sect. 2 below we further discuss the question. In this chapter we use Sukhatme's transformation to suggest new ways of dealing with the matter.

Sukhatme's [13] transformation, from which Rényi's representation of exponential order statistics can easily be derived, appears in David and Nagaraja ([2], p. 17–18) and in Johnson et al. ([8], p. 305), with slightly different presentations, applied to the study of exponential and of uniform order statistics, respectively. Durbin [4] used ordered spacings of the uniform to investigate the construction of exact tests. In Sect. 3 we use a Sukhatme's like transformation to augment the set of order statistics from a uniform parent, and in Sect. 4 we investigate power issues when they are used in testing uniformity.

2 Uniformity Versus Mixtures of Uniform and Beta(1,2)

Gomes et al. [5] introduced the family $\{X_m\}_{m\in[-2,2]}$ of absolutely continuous random variables, with probability density function $f_{X_m}(x) = \left(mx - \frac{m-2}{2}\right)\mathrm{I}_{(0,1)}(x)$ (the uniform density corresponds to $m = 0$; for $m \in (0, 2]$, X_m is a convex mixture of Beta(1,1) and Beta(2,1), and for $m \in [-2, 0]$, X_m is a mixture of Beta(1,1) and Beta(1,2)). Observe that for all $m \in [-2, 0)$, $\mathbb{P}[X_m \le x] - \mathbb{P}[U \le x] = \frac{m}{2}\, x\,(x - 1) > 0$ for all $x \in (0, 1)$, and thus pseudorandom numbers generated by X_m tend to be closer to 0 than pseudorandom numbers generated by a standard uniform random variable U. Thus this family can give important hints on nonuniformity of the set of p-values, cf. the concepts of random p-values in Kulinskaya et al. [9] and of generalized p-values in Hartung et al. [6].

Observe also that for $m \in (0, 2]$, X_m tends to take values closer to 1 than the $X_0 \frown$ Uniform(0, 1) random variable, and hence in that range of values it provides a suitable alternative in the case of right one-tailed alternative tests. Moreover, the inverse of the corresponding distribution function is

$$F_{X_m}^{-1}(u) = \frac{\frac{m}{2} - 1 + \sqrt{\left(\frac{m}{2} - 1\right)^2 + 2mu}}{m}$$

and the generation of pseudo-random numbers from X_m for simulation studies is therefore straightforward.

Let U and X be two independent standard uniform random variables. The random variables $V = U + X - \mathrm{I}[U + X]$, where $\mathrm{I}[x]$ denotes the largest integer not greater than x, and $W = \min\left(\frac{U}{X}, \frac{1-U}{1-X}\right)$ are uniform and independent of X (see Deng and George [3]). This fact was used by Gomes et al. [5] for computationally augmenting samples and to assess the power of detecting non-uniformity when the sample comes in fact from X_m, $m \in [-2, 0]$, with the strange result that power does not improve for increased samples.

The explanation is however simple: if X_m and X_p are two independent random variables, with $m, p \in [-2, 2]$, then $\min\left(\frac{X_m}{X_p}, \frac{1-X_m}{1-X_p}\right) \stackrel{d}{=} X_{\frac{mp}{6}}$ (see Brilhante et al. [1]). Hence, in case the algorithm uses uniform pseudorandom numbers to augment the sample, the augmented slice will in fact be a uniform subsample, and power decreases. Brilhante et al. [1] present better results using left-skewed parent pseudorandom numbers.

Still, the use of the family $\{X_m\}_{m \in [-2,2]}$ has many advantages, and instead of augmenting the sample *externally*, as in the above-mentioned papers, by using $V_m = U + X_m - \mathrm{I}[U + X_m]$ and $W_m = \min\left(\frac{U}{X_m}, \frac{1-U}{1-X_m}\right)$, with the spurious effect of always generating uniform pseudo p-values, we can use an alternative approach when the purpose is to test the null hypothesis of uniformity *vs.* X_m parent:

- Choose at random one $p_j \in \{p_k\}_{k=1}^n$.
- Generate $n - 1$ pseudo p's of the form $\min\left(\frac{p_j}{p_k}, \frac{1-p_j}{1-p_k}\right)$, $k \neq j$.

3 Order Statistics, Spacings and Sukhatme's Transformation

Let $\boldsymbol{X} = (X_1, X_2, \ldots, X_n)$ be a random sample from the absolutely continuous positive random variable X with probability density function f_X and $(X_{1:n}, X_{2:n}, \ldots, X_{n:n})$ the corresponding vector of ascending order statistics. For convenience we assume that left-endpoint $\alpha_X = 0$ and we define $X_{0:n} = \alpha_X = 0$.

The joint probability density function of the spacings $S_k = X_{k:n} - X_{k-1:n}$, $k = 1, \ldots, n$, is

$$f_{(S_1,S_2,\ldots,S_n)}(s_1, s_2, \ldots, s_n) = n!\, f_{(X_1,X_2,\ldots,X_n)}(s_1, s_1 + s_2, \ldots, s_1 + \cdots + s_n)$$

($s_k > 0$, $k = 1, \ldots, n$, and if the right-endpoint ω_X is finite, $\sum_{k=1}^n s_k < \omega_X$; in this case we can consider the rightmost spacing $S_{n+1} = \omega_X - X_{n:n}$, but this can be expressed as a function $\omega_X - \sum_{k=1}^n S_k$). Hence, the joint probability density function of the ascending reordering of those n spacings is

$$f_{(S_{1:n},S_{2:n},\ldots,S_{n:n})}(y_1, y_2, \ldots, y_n) = (n!)^2 f_{(X_1,X_2,\ldots,X_n)}(y_1, y_1 + y_2, \ldots, y_1 + \cdots + y_n)$$

where $0 < y_1 < \ldots < y_n$ and $\sum_{k=1}^n y_k < \omega_X$.

Now define

$$W_k = (n+1-k)(S_{k:n} - S_{k-1:n}), \quad k = 1, \ldots, n,$$

(similar to Sukhatme's transformation, as defined in David and Nagaraja [2], but applied to ascendingly ordered spacings), again with the convention $S_{0:n} = 0$.

The joint probability density function of $(W_1, W_2, \ldots, W_n)$ is

$$f_{(W_1,W_2,\ldots,W_n)}(w_1, w_2, \ldots, w_n) = n!\, f_{(X_1,X_2,\ldots,X_n)}\left(\tfrac{w_1}{n}, \tfrac{2w_1}{n} + \tfrac{w_2}{n-1}, \ldots, w_1 + \cdots + w_n\right)$$

$w_k > 0$, $k = 1, \ldots, n$, (observe that the k-th argument is

$$\frac{kw_1}{n} + \frac{(k-1)w_2}{n-1} + \cdots + \frac{(k+1-j)w_j}{n+1-j} + \cdots + \frac{w_k}{n+1-k}, \quad k = 1, \ldots, n),$$

and the joint probability density function of the vector of partial sums $Y_k = \sum_{j=1}^{k} W_j$, $k = 1, \ldots, n$, is

$$f_{(Y_1,Y_2,\ldots,Y_n)}(y_1, y_2, \ldots, y_n) = n!\, f_{(X_1,X_2,\ldots,X_n)}\left(\tfrac{y_1}{n}, \ldots, \sum_{j=1}^{k} \tfrac{(k+1-j)(y_j - y_{j-1})}{n+1-j}, \ldots, y_n\right)$$

with $0 < y_1 < \ldots < y_n$ and the convention $y_0 = 0$.

If $X \frown \text{Uniform}(0, \omega_X)$, then

$$f_{(X_1,X_2,\ldots,X_n)}\left(\tfrac{y_1}{n}, \ldots, \sum_{j=1}^{k} \tfrac{(k+1-j)(y_j - y_{j-1})}{n+1-j}, \ldots, y_n\right) = \frac{1}{\omega_X^n} = f_{(X_1,X_2,\ldots,X_n)}(y_1, y_2, \ldots, y_n),$$

and hence $(Y_1, Y_2, \ldots, Y_n) \stackrel{d}{=} (X_{1:n}, X_{2:n}, \ldots, X_{n:n})$. [1]

This suggests that uniformity can be investigated testing whether $\{X_{k:n}\}_{k=1}^{n}$ and $\{Y_k\}_{k=1}^{n}$ can be considered samples from the same distribution. Unfortunately, under the null hypothesis that the parent distribution is standard uniform, the two vectors are not independent since we can re-express $Y_k = \sum_{j=1}^{k} S_{j:n} + (n-k)S_{k:n}$, and consequently $Y_n = X_{n:n}$. Thus, the Smirnov two-sample test is of no use in the present situation.

However, the observation of Fig. 1, where we compare the empirical distribution function (edf) corresponding to the order statistics $x_{k:n}$ (black) and the y_k (gray), in

[1] Observe that if $\omega_X < \infty$, we can consider $n+1$ spacings, with $S_{n+1} = \omega_X - X_{n:n}$; of course in this situation S_{n+1}, $S_{n+1:n+1}$ and W_{n+1} (where in this case it is convenient to use the transformation

$$W_k = (n+2-k)(S_{k:n+1} - S_{k-1:n+1}),$$

as in Johnson et al. [8], p. 305) can be expressed as simple functions of the predecessor members of the sequence. We still get the result that $(Y_1, Y_2, \ldots, Y_n) \stackrel{d}{=} (X_{1:n}, X_{2:n}, \ldots, X_{n:n})$ in case of standard uniform parent X.

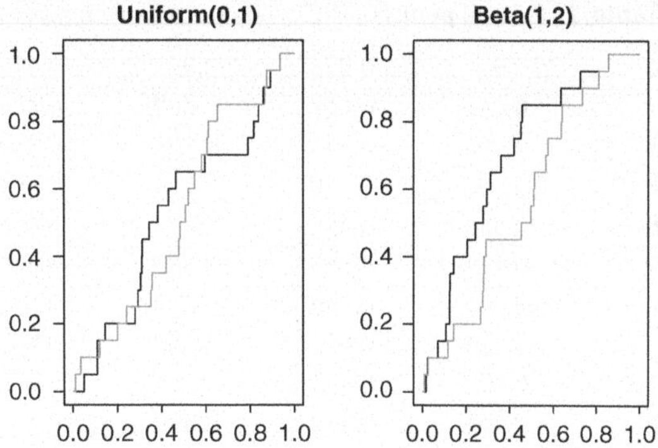

Fig. 1 Empirical distribution functions F_{20}^* and G_{20}^* for Uniform(0,1) and Beta(1,2) parents; this illustrates the general pattern

case of uniform and Beta(1,2) parents, suggests that $D_n^* = \sup_x |F_n^*(x) - G_n^*(x)|$, where F_n^* stands for the order statistics edf and G_n^* for the accumulated y_k edf, will be greater under the alternative $H_A : X$ nonuniform with support (0,1) than under the null hypothesis $H_0 : X \frown \text{Uniform}(0, 1)$.

For uniformity testing purposes we present in Table 1 the upper critical points of D_n^*, $n = 3(1)30(5)100$, when the underlying parent is standard uniform ($U \stackrel{d}{=} X_0$). These points were obtained by generating 10,000 independent replicates of the sample $(D_{n,1}^*, D_{n,2}^*, \ldots, D_{n,50}^*)$ and defining the quantile of order p of D_n^* as the mean of the samples quantiles for $p = 0.9, 0.925, 0.95, 0.975, 0.99, 0.995, 0.999$.

We also performed a simulation study of the proportion of rejections of uniformity when the underlying parent was X_m, $m \in [-2, 0]$ and when making pairwise comparisons of the order statistics $\{x_{k:n}\}$ edf and the $\{y_k\}$ edf (the process of generating $\{y_k\}$ was iteratively repeated 10,000 times). Observe that the rationale for this procedure relies on the fact that if the original observations $\{p_k\}$ are indeed uniform, the "Sukhatme's" $\{y_k\}$ would be order statistics of standard uniform, and hence repeating Sukhatme's algorithm we would obtain again a set of order statistics of standard uniform.

From Fig. 2 we observe that the proportion of rejections of uniformity increases with n. However, the extended Sukhatme's like transformed data performs badly in detecting departures from uniformity when $n < 20$. This situation can obviously constitute a problem when combining p-values in meta-analytical syntheses since the number of available (reported) p-values is usually small.

Another way of assessing the usefulness of this extended Sukhatme's transformation in testing uniformity is by calculating the area limited by the edf's F_n^* and G_n^*, since under the validity of the null hypothesis $X \frown \text{Uniform}(0, 1)$, the area between the two curves should be zero—big area values should indicate a departure from uniformity. In Table 2 we compare the areas obtained by simulation (10,000 runs) for some values of n when the underlying parents are standard uniform and Beta(1,2). Analyzing Table 2 we see that the area is indeed inferior for the standard uniform parent, except for some few cases. However, the differences between the two areas can be very small, which can difficult the task of testing uniformity with this procedure.

Table 1 Critical points of D_n^* when the underlying parent is Uniform (0,1)[a]

n	0.9	0.925	0.95	0.975	0.99	0.995	0.999
3	0.667	0.667	0.667	0.667	0.667	0.667	0.667
4	0.605	0.656	0.703	0.734	0.747	0.747	0.747
5	0.600	0.610	0.634	0.682	0.753	0.753	0.753
6	0.548	0.580	0.62	0.666	0.736	0.736	0.736
7	0.542	0.563	0.589	0.632	0.712	0.712	0.712
8	0.509	0.529	0.558	0.605	0.686	0.686	0.686
9	0.484	0.509	0.540	0.582	0.660	0.660	0.660
10	0.470	0.491	0.518	0.558	0.635	0.635	0.635
11	0.454	0.472	0.498	0.537	0.612	0.612	0.612
12	0.436	0.455	0.482	0.520	0.592	0.592	0.592
13	0.422	0.441	0.466	0.503	0.574	0.574	0.574
14	0.410	0.429	0.452	0.487	0.557	0.557	0.557
15	0.398	0.415	0.438	0.472	0.539	0.539	0.539
16	0.387	0.404	0.427	0.460	0.525	0.525	0.525
17	0.377	0.393	0.416	0.447	0.511	0.511	0.511
18	0.368	0.385	0.406	0.437	0.498	0.498	0.498
19	0.359	0.376	0.396	0.427	0.486	0.486	0.486
20	0.352	0.367	0.387	0.416	0.474	0.474	0.474
21	0.345	0.360	0.379	0.408	0.463	0.463	0.463
22	0.337	0.352	0.371	0.399	0.453	0.453	0.453
23	0.331	0.345	0.363	0.391	0.444	0.444	0.444
24	0.325	0.339	0.357	0.384	0.435	0.435	0.435
25	0.319	0.332	0.350	0.376	0.427	0.427	0.427
26	0.313	0.326	0.344	0.370	0.419	0.419	0.419
27	0.308	0.321	0.338	0.363	0.411	0.411	0.411
28	0.302	0.315	0.332	0.357	0.404	0.404	0.404
29	0.298	0.311	0.327	0.352	0.400	0.400	0.400
30	0.293	0.306	0.322	0.345	0.392	0.392	0.392
35	0.273	0.285	0.300	0.321	0.363	0.363	0.363
40	0.257	0.268	0.282	0.302	0.341	0.341	0.341
45	0.243	0.253	0.267	0.286	0.322	0.322	0.322
50	0.231	0.241	0.254	0.272	0.306	0.306	0.306
55	0.221	0.230	0.242	0.260	0.292	0.292	0.292
60	0.212	0.221	0.232	0.249	0.280	0.280	0.280
65	0.204	0.212	0.224	0.239	0.269	0.269	0.269
70	0.197	0.205	0.216	0.231	0.260	0.260	0.260
75	0.190	0.198	0.209	0.223	0.251	0.251	0.251
80	0.185	0.193	0.202	0.217	0.244	0.244	0.244
85	0.179	0.186	0.196	0.210	0.236	0.236	0.236
90	0.174	0.182	0.191	0.204	0.229	0.229	0.229
95	0.170	0.177	0.186	0.199	0.223	0.223	0.223
100	0.166	0.172	0.181	0.194	0.217	0.217	0.217

[a]The standard errors of the critical points are less than or equal to 0.001

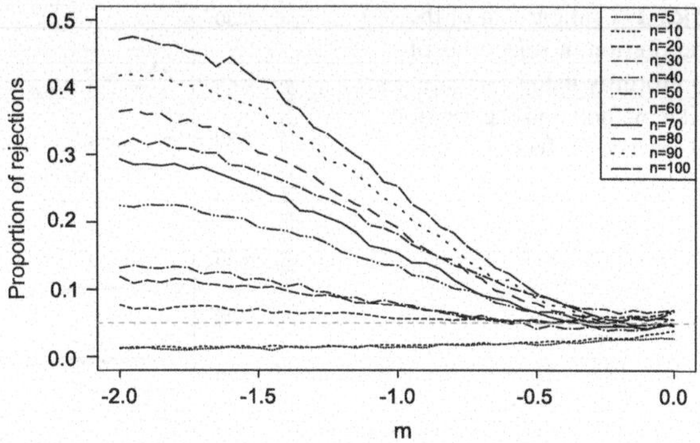

Fig. 2 Proportion of rejections of uniformity at level 0.05 using Sukhatme's like transformation when the underlying parent is X_m, $m \in [-2, 0]$

Table 2 Area limited by the functions F_n^* and G_n^* when the underlying parents are Uniform(0,1) and Beta(1,2)

	Beta(1,2)		Uniform(0,1)	
n	Area	s.e.	Area	s.e.
5	0.0366	0.00188	0.0333	0.00179
10	0.0848	0.00279	0.1027	0.00304
15	0.0794	0.00270	0.1216	0.00327
20	0.0860	0.00280	0.0820	0.00274
25	0.0620	0.00241	0.0608	0.00239
30	0.0823	0.00275	0.0495	0.00217
35	0.0699	0.00255	0.0526	0.00223
40	0.0742	0.00262	0.0411	0.00199
45	0.0665	0.00249	0.0450	0.00207
50	0.1005	0.00301	0.0319	0.00176
55	0.0927	0.00290	0.0370	0.00189
60	0.0774	0.00267	0.0376	0.00190
65	0.0830	0.00276	0.0247	0.00155
70	0.0648	0.00246	0.0425	0.00202
75	0.0371	0.00189	0.1369	0.00344
80	0.0682	0.00252	0.0388	0.00193
85	0.0702	0.00256	0.0403	0.00197
90	0.0901	0.00286	0.0395	0.00195
95	0.0701	0.00255	0.0358	0.00186
100	0.0730	0.00260	0.0498	0.00218

4 Conclusion

It seems worth to point out that the entropy of X_m, $m \in [-2, 2]$, is

$$H(X_m) = -\int_0^1 f_{X_m}(x)\ln(f_{X_m}(x))\mathrm{d}x = 0.5 + \ln(2) + \frac{\ln\left[\left(\frac{2-m}{2+m}\right)^m\right]}{8} - \frac{\ln(4-m^2)}{2} + \frac{\ln\left(\frac{2-m}{2+m}\right)}{2m},$$

(for a detailed study of entropy, cf. [7]), whose graph is concave, and hence the entropy of $\min\left(\frac{X_m}{X_p}, \frac{1-X_m}{1-X_p}\right) \stackrel{d}{=} X_{\frac{mp}{6}}$ is, for $m, p \in [-2, 2]$, nearer to the entropy

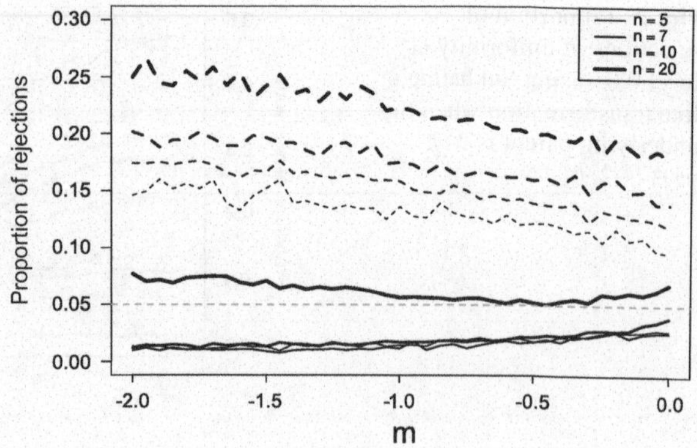

Fig. 3 Comparison of the proportion of rejections of uniformity using Sukhatme's like method and the method described in Sect. 2

of X_0 than to the entropy of X_m and X_p. We would thus expect that Sukhatme's like method of sample augmentation would provide better results than the method explained in Sect. 2. Observe however that further investigation of the matter seems to indicate the reverse, as shown in Fig. 3 (the solid lines correspond to Sukhatme's like method and the dashed lines to the method described in Sect. 2). The general question of comparing analytically edfs of correlated samples remains unsolved, even for simple forms of weak dependence only simulation results in well-defined situations seem feasible.

Acknowledgements This research has been supported by National Funds through FCT—Fundação para a Ciência e a Tecnologia, project PEst-OE/MAT/UI0006/2011. The authors are grateful to the referees for stimulating comments, leading to further results.

References

1. Brilhante, M.F., Pestana, D., Sequeira, F.: Combining p-values and random p-values. In: Luzar-Stiffler, V., et al. (eds.) Proceedings of the 32nd International Conference on Information Technology Interfaces, pp. 515–520 (2010)
2. David, H.A., Nagaraja, H.N.: Order Statistics, 3rd edn. Wiley, New York (2003)
3. Deng, L.-Y., George, E.O.: Some characterizations of the uniform distribution with applications to random number generation. Ann. Inst. Stat. Math. **44**, 379–385 (1992)
4. Durbin, J.: Some methods of constructing exact tests. Biometrika **48**, 4–55 (1961)
5. Gomes, M.I, Pestana, D., Sequeira, F., Mendonça, S., Velosa, S.: Uniformity of offsprings from uniform and non-uniform parents. In: Luzar-Stiffler, V., et al. (eds.) Proceedings of the 31st International Conference on Information Technology Interfaces, pp. 243–248 (2009)
6. Hartung, J., Knapp, G., Sinha, B.K.: Statistical Meta-Analysis with Applications. Wiley, New York (2008)
7. Johnson, O.: Information Theory and the Central Limit Theorem. Imperial College Press, London (2004)
8. Johnson, N.L., Kotz, S., Balakrishnan, N.: Continuous Univariate Distributions, vol. 2, 2nd edn. Wiley, New York (1995)
9. Kulinskaya, E., Morgenthaler, S., Staudte, R.G.: Meta Analysis. A Guide to Calibrating and Combining Statistical Evidence. Wiley, Chichester (2008)

10. Paul, A.: Characterizations of the uniform distribution via sample spacings and nonlinear transformations. J. Math. Anal. Appl. **284**, 397–402 (2003)
11. Pestana, D.: Combining p-values. In: Lovric, M. (ed.) International Encyclopedia of Statistical Science, pp. 1145–1147. Springer, Heidelberg (2011)
12. Sequeira, F.: Meta-Análise: Harmonização de Testes Usando os Valores de Prova. PhD Thesis, DEIO, Faculdade de Ciências da Universidade de Lisboa (2009)
13. Sukhatme, P.V.: On the analysis of k samples from exponential populations with especial reference to the problem of random intervals. Statist. Res. Memoir. **1**, 94–112 (1936)

Asymptotic Comparison at Optimal Levels of Minimum-Variance Reduced-Bias Tail-Index Estimators

Frederico Caeiro and M. Ivette Gomes

Abstract
In this chapter we are interested in the asymptotic comparison of a set of semi-parametric minimum-variance reduced-bias tail-index estimators, at optimal levels and for a wide class of models. Again, as in the classical case, there is not any estimator that can always dominate the alternatives, but interesting clear-cut patterns are found. Consequently, and in practice, a suitable choice of a set of tail-index estimators will jointly enable us to better estimate the tail index, the primary parameter of extreme events.

1 The Estimators Under Study and Scope of the Paper

Let us consider the common set-up of independent, identically distributed (i.i.d.) random variables (r.v.'s) $X_1, X_2, \cdots, X_n$ with a common distribution function (d.f.) F and denote the associated ascending order statistics (o.s.) by $X_{1:n} \le X_{2:n} \le \cdots \le X_{n:n}$. Let us assume a first-order condition, i.e., that there exist sequences of real constants $\{a_n > 0\}$ and $\{b_n \in R\}$ such that $(X_{n:n} - b_n)/a_n$ converges in distribution towards a non-degenerate r.v. Then F belongs to the max domain of attraction of an *extreme value* (EV) d.f.:

F. Caeiro (✉)
Faculdade de Ciências e Tecnologia, Universidade Nova de Lisboa,
2829-516 Caparica, Portugal

CMA, Universidade Nova de Lisboa, Portugal
e-mail: fac@fct.unl.pt

M.I. Gomes
FCUL, Campo Grande, 1749-016 Lisboa, Portugal

CEAUL, Lisbon, Portugal
e-mail: ivette.gomes@fc.ul.pt

J. Lita da Silva et al. (eds.), *Advances in Regression, Survival Analysis, Extreme Values, Markov Processes and Other Statistical Applications*, Studies in Theoretical and Applied Statistics, DOI 10.1007/978-3-642-34904-1_8,

$$EV_\gamma(x) = exp(-(1+\gamma x)^{-1/\gamma}), \quad 1+\gamma x > 0, \quad \gamma \in R, \tag{1}$$

and we write $F \in \mathcal{D}_{\mathcal{M}}(EV_\gamma)$. The parameter γ is the *tail index*, the primary parameter of extreme events, with a low frequency, but a high impact. This index measures the heaviness of the right *tail function* $\overline{F} := 1 - F$, and the heavier the tail, the larger γ is. In this chapter we shall work with heavy-tailed models, i.e., Pareto-type underlying d.f.'s, with a strict positive tail index.

The *second-order parameter*, ρ (≤ 0), rules the rate of convergence in the first-order condition, and is the non-positive parameter appearing in the limiting relation

$$\lim_{t\to\infty} \frac{\ln U(tx) - \ln U(t) - \gamma \ln x}{A(t)} = \frac{x^\rho - 1}{\rho}, \tag{2}$$

which is assumed to hold for every $x > 0$ and where $|A|$ must then be of regular variation with index ρ [7]. To obtain information on the order of the asymptotic bias of second-order reduced-bias tail-index estimators, it is necessary to further assume a third-order condition, ruling the rate of convergence in Eq. (2), and which guarantees that, for all $x > 0$,

$$\lim_{t\to\infty} \frac{\frac{\ln U(tx) - \ln U(t) - \gamma \ln x}{A(t)} - \frac{x^\rho - 1}{\rho}}{B(t)} = \frac{x^{\rho+\rho'} - 1}{\rho + \rho'}, \tag{3}$$

where $|B(t)|$ must then be of regular variation with index $\rho' \le 0$.

In this chapter, similarly to what has been done in Gomes et al. [11], we consider a Pareto-type class of models with a tail function

$$1 - F(x) = Cx^{-1/\gamma}\left(1 + D_1 x^{\rho/\gamma} + D_2 x^{2\rho/\gamma} + o\left(x^{2\rho/\gamma}\right)\right), \quad \text{as} \quad x \to \infty, \tag{4}$$

with $C > 0$, D_1, $D_2 \neq 0$, $\rho < 0$. Note that to assume Eq. (4) is equivalent to say that Eq. (3) holds with $\rho = \rho' < 0$ and that we may there choose

$$A(t) = \alpha\, t^\rho =: \gamma\, \beta\, t^\rho, \qquad B(t) = \beta'\, t^\rho = \beta' A(t)/(\beta\gamma), \qquad \beta,\ \beta' \neq 0, \tag{5}$$

with β and β' "scale" second and third-order parameters, respectively.

For heavy-tailed models, the classical tail-index estimator is Hill's estimator [14], the average of the scaled log-spacings U_i or of the log-excesses V_{ik}:

$$H_n(k) \equiv H(k) := \frac{1}{k}\sum_{i=1}^{k} U_i = \frac{1}{k}\sum_{i=1}^{k} V_{ik}, \tag{6}$$

where

$$U_i := i\left\{\ln \frac{X_{n-i+1:n}}{X_{n-i:n}}\right\} \quad \text{and} \quad V_{ik} := \ln \frac{X_{n-i+1:n}}{X_{n-k:n}}, \quad 1 \le i \le k < n. \tag{7}$$

For intermediate k, i.e., for a sequence of integers $k = k_n$, $1 \le k < n$, such that

$$k = k_n \to \infty \quad \text{and} \quad k_n = o(n), \quad \text{as} \quad n \to \infty, \tag{8}$$

the Hill estimator in Eq. (6) is consistent for $\gamma > 0$ whenever $F \in \mathscr{D}_{\mathscr{M}}(EV_\gamma)_{\gamma>0}$ holds.

The adequate accommodation of the bias of Hill's estimator has been extensively addressed in recent years in the literature. Recently, several authors [4, 10–12] consider, in different ways and under the second-order framework in Eq. (2), minimum-variance reduced-bias (MVRB) tail-index estimators based on the joint external estimation of both the 'scale' and the 'shape' parameters, β and ρ, respectively. These estimators are called MVRB due to the fact that, under adequate restrictions, they are able to reduce the bias without increasing the asymptotic variance, which is shown to be kept at the value γ^2, the asymptotic variance of Hill's estimator, at least for values k such that $\sqrt{k}A(n/k) \to \lambda$, finite, as $n \to \infty$. Gomes et al. [11] consider a tail-index estimator based on a linear combination of the log-excesses V_{ik} in Eq. (7) and given by

$$\overline{WH}_{\hat\beta,\hat\rho}(k) := \frac{1}{k}\sum_{i=1}^{k} e^{-\hat\beta\,(n/k)^{\hat\rho}\,\psi_{ik}(\hat\rho)}\, V_{ik}, \qquad \psi_{ik}(\rho) = -\frac{(i/k)^{-\rho}-1}{\rho\ln(i/k)}, \tag{9}$$

WH standing here for *weighted Hill* estimator. Caeiro et al. [4] consider two estimators of this same type, here denoted:

$$CH_{\hat\beta,\hat\rho}(k) := H(k)\Big(1 - \frac{\hat\beta}{1-\hat\rho}\Big(\frac{n}{k}\Big)^{\hat\rho}\Big), \tag{10}$$

$$\overline{CH}_{\hat\beta,\hat\rho}(k) := H(k)\exp\Big(-\frac{\hat\beta}{1-\hat\rho}\Big(\frac{n}{k}\Big)^{\hat\rho}\Big), \tag{11}$$

where the dominant component of the bias of Hill's estimator in Eq. (6), given by $A(n/k)\ /\ (1-\rho) = \gamma\ \beta\ (n/k)^\rho\ /(1-\rho)$, is thus essentially estimated through $H(k)\ \hat\beta\ (n/k)^{\hat\rho}\ /(1-\hat\rho)$, and directly removed from Hill's classical tail-index estimator. The notation CH stands for *corrected Hill.* A third class has been introduced in Gomes et al. [11], and it has the functional form

$$ML_{\hat\beta,\hat\rho}(k) := H(k) - \hat\beta\ \Big(\frac{n}{k}\Big)^{\hat\rho}\ \Big(\frac{1}{k}\sum_{i=1}^{k}\Big(\frac{i}{k}\Big)^{-\hat\rho}U_i\Big), \tag{12}$$

with U_i given in Eq. (7). These authors consider also

$$\overline{ML}_{\hat\beta,\hat\rho}(k) := \frac{1}{k}\sum_{i=1}^{k}\exp\big(-\hat\beta(n/i)^{\hat\rho}\big)U_i, \tag{13}$$

the estimator directly derived from the likelihood equation for γ with β and ρ fixed and based upon the exponential approximation $U_i \approx \gamma\exp(\beta(n/i)^\rho)E_i$, $1 \le i \le k$, being claimed a better performance of the ML estimator, comparatively to the $\overline{ML}$ estimator, for a large class of models. This is the reason why we shall also work with the bias-corrected Hill estimator

$$WH_{\hat{\beta},\hat{\rho}}(k) := H(k) - \hat{\beta}\left(\frac{n}{k}\right)^{\hat{\rho}}\left(\frac{1}{k}\sum_{i=1}^{k}\psi_{ik}(\hat{\rho})\, V_{ik}\right), \tag{14}$$

with $\psi_{ik}(\hat{\rho})$ given in Eq. (9).

Remark 1. In all the above MVRB tail-index estimators, $\hat{\beta}$ and $\hat{\rho}$ need to be adequate consistent estimators of the second-order parameters β and ρ, respectively. For more details related with the estimation of these parameters, see, for instance, Fraga Alves et al. [6], Gomes and Martins [9] and Caeiro and Gomes [2].

In this chapter, we compare asymptotically, at optimal levels, the above-mentioned MVRB statistics, denoted generically $UH_{\beta,\rho}(k)$ (assuming thus that β and ρ are known or adequately estimated). In Sect. 2, we shall state for the class of models in Eq. (4), the asymptotic properties of $UH_{\beta,\rho}(k)$, and in Sect. 3, we provide a full asymptotic comparison, at optimal levels, of $UH_{\beta,\rho}(k)$ for $UH = CH, ML$ and WH.

2 The Asymptotic Behaviour of the MVRB Tail-Index Estimators

Let $\{E_i\}$ denote a sequence of i.i.d. standard exponential r.v.'s and define

$$Z_k := \frac{1}{k}\sum_{i=1}^{k} E_i \quad \text{and} \quad \overline{Z}_k := \sqrt{k}\left(Z_k - 1\right). \tag{15}$$

Assuming the third-order framework in Eq. (4), we state the following result, a particular case, with a few additions related to the $\overline{UH}$ statistics, of Theorem 3.1 in [5].

Theorem 1. *Under the third-order framework in Eq.* (4), *with $A(t)$ given in Eq.* (5), *$\overline{Z}_k$ given in Eq.* (15), *and for intermediate k, i.e., if Eq.* (8) *holds, we can write*

$$UH_{\beta,\rho}(k) \stackrel{d}{=} \gamma + \frac{\gamma\,\overline{Z}_k}{\sqrt{k}} + \left(b_{UH}A^2(n/k) + O_p\left(\frac{A(n/k)}{\sqrt{k}}\right)\right)(1 + o_p(1)), \tag{16}$$

where, with $\xi = \beta'/\beta$ and $a_2(\rho) := -\frac{1}{\rho^2}\left(\ln(1-2\rho) - 2\ln(1-\rho)\right)$,

$$b_{CH} = \frac{1}{\gamma}\left(\frac{\xi}{1-2\rho} - \frac{1}{(1-\rho)^2}\right), \qquad b_{\overline{CH}} = \frac{1}{\gamma}\left(\frac{\xi}{1-2\rho} - \frac{1}{2(1-\rho)^2}\right),$$

$$b_{ML} = \frac{\xi - 1}{\gamma(1-2\rho)}, \qquad b_{\overline{ML}} = \frac{2\xi - 1}{2\gamma(1-2\rho)},$$

$$b_{WH} = \frac{1}{\gamma}\Big(\frac{\xi}{1-2\rho} - a_2(\rho)\Big), \qquad b_{\overline{WH}} = \frac{1}{\gamma}\Big(\frac{\xi}{1-2\rho} - \frac{a_2(\rho)}{2}\Big).$$

Consequently, even if $\sqrt{k}\ A(n/k) \to \infty$, *with* $\sqrt{k}\ A^2(n/k) \to \lambda_A$, *finite,*

$$\sqrt{k}\ \big(UH_{\beta,\rho}(k) - \gamma\big) \underset{n\to\infty}{\overset{d}{\longrightarrow}} Normal\big(\lambda_A b_{UH}, \sigma^2_{UH} = \gamma^2\big).$$

Remark 2. If $\sqrt{k}\ A^2(n/k) \to \infty$, $\big(UH_{\beta,\rho}(k) - \gamma\big)$ is $O_p(A^2(n/k))$.

Remark 3. Note that $b_{ML} = b_{\overline{ML}} = 0$ whenever $\xi = 1$. This happens for important models like the Burr and the GP, and it is a point in favour of the ML-statistic.

Remark 4. We also add that the results for $\overline{UH}$ follow straightforwardly from the results for UH. Indeed, as $n \to \infty$, $\overline{WH}_{\beta,\rho} - WH_{\beta,\rho} \overset{p}{\sim} a_2(\rho)A^2(n/k)/(2\gamma)$, $\overline{CH}_{\beta,\rho} - CH_{\beta,\rho} \overset{p}{\sim} A^2(n/k)/(2\gamma(1-\rho)^2)$ and $\overline{ML}_{\beta,\rho} - ML_{\beta,\rho} \overset{p}{\sim} A^2(n/k)/(2\gamma(1-2\rho))$.

Remark 5. Note that, as already mentioned in Caeiro et al. [5], since $\lambda_A \geq 0$ and $1/(2a_2(\rho)) > (1-\rho)^2 > 1/a_2(\rho) > 1-2\rho$ for any $\rho < 0$, $b_{\overline{WH}} \geq b_{CH} \geq b_{WH} \geq b_{ML}$. All depends then on the sign of the bias.

3 Asymptotic Comparison of the MVRB Tail-Index Estimators

We shall next proceed to the comparison of the MVRB estimators under study at their optimal levels. This is again done in a way similar to the one used in de Haan and Peng [13] and Gomes and Martins [8] for the classical tail-index estimators. Let us assume that $\widehat{\gamma}^\bullet_{n,k}$ denotes any arbitrary reduced-bias semi-parametric estimator of the tail index γ, for which we have, for any intermediate $k = k_n$,

$$\widehat{\gamma}^\bullet_{n,k} = \gamma + \frac{\sigma_\bullet}{\sqrt{k}} Z^\bullet_k + b_\bullet\ A^2(n/k) + o_p(A^2(n/k)), \tag{17}$$

with $Z^\bullet_k$ an asymptotically standard normal r.v. Then, $\sqrt{k}\big[\widehat{\gamma}^\bullet_{n,k} - \gamma\big] \longrightarrow^d N(\lambda_A b_\bullet, \sigma^2_\bullet)$ provided that k is such that $\sqrt{k}\ A^2(n/k) \to \lambda_A$, finite, as $n \to \infty$. We then write $Bias_\infty\big[\widehat{\gamma}^\bullet_{n,k}\big] := b_\bullet\ A^2(n/k)$ and $Var_\infty\big[\widehat{\gamma}^\bullet_{n,k}\big] := \sigma^2_\bullet/k$. The so-called asymptotic mean square error ($AMSE$) is then given by

$$AMSE\big[\widehat{\gamma}^\bullet_{n,k}\big] := \sigma^2_\bullet/k + b^2_\bullet\ A^4(n/k).$$

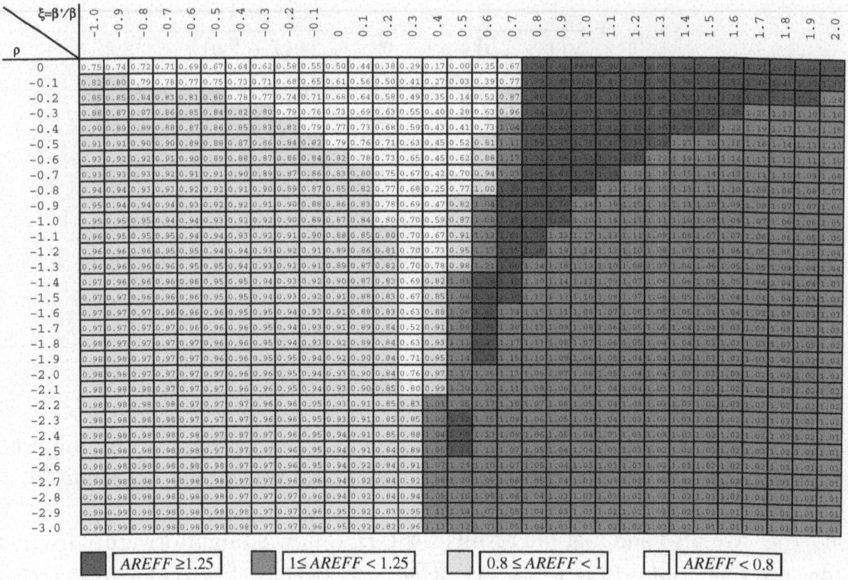

Fig. 1 $AREFF_{CH|\overline{CH}}$, in the (ξ, ρ)-plane

Regular variation theory [1], enables us to show that, whenever $b_\bullet \neq 0$, there exists a function $\varphi(n) = \varphi(n, \gamma, \rho)$, such that

$$\lim_{n\to\infty} \varphi(n)\, AMSE\left[\widehat{\gamma}^\bullet_{n0}\right] = (\sigma_\bullet^2)^{-\frac{4\rho}{1-4\rho}} (b_\bullet^2)^{\frac{1}{1-4\rho}} =: LMSE\left[\widehat{\gamma}^\bullet_{n0}\right],$$

where $\widehat{\gamma}^\bullet_{n0} := \widehat{\gamma}^\bullet_{n,k_0^\bullet(n)}$ and $k_0^\bullet(n) := \arg\inf_k AMSE\left[\widehat{\gamma}^\bullet_{n,k}\right]$. It is then sensible to consider the following:

Definition 1. Given two biased estimators $\widehat{\gamma}^{(1)}_{n,k}$ and $\widehat{\gamma}^{(2)}_{n,k}$, for which a distributional representation of the type of the one in Eq. (17) holds, with constants (σ_1, b_1) and (σ_2, b_2), $b_1, b_2 \neq 0$, respectively, both computed at their optimal levels, the *asymptotic root efficiency* (*AREFF*) of $\widehat{\gamma}^{(1)}_{n0}$ relatively to $\widehat{\gamma}^{(2)}_{n0}$ is

$$AREFF_{1|2} \equiv AREFF_{\widehat{\gamma}^{(1)}_{n0}|\widehat{\gamma}^{(2)}_{n0}} := \sqrt{\frac{LMSE\left[\widehat{\gamma}^{(2)}_{n0}\right]}{LMSE\left[\widehat{\gamma}^{(1)}_{n0}\right]}} = \left(\left(\frac{\sigma_2}{\sigma_1}\right)^{-4\rho}\left|\frac{b_2}{b_1}\right|\right)^{\frac{1}{1-4\rho}}. \quad (18)$$

Remark 6. Note that this *AREFF* indicator has been conceived so that the highest the *AREFF* indicator is, the better is the first estimator.

We first present in Figs. 1–3 the measure $AREFF_{UH|\overline{UH}}$ for $UH = CH, ML, WH$, respectively, in the (ξ, ρ)-plane. Figure 4 shows us the MVRB tail-index estimator with minimum LMSE in the (ξ, ρ)-plane.

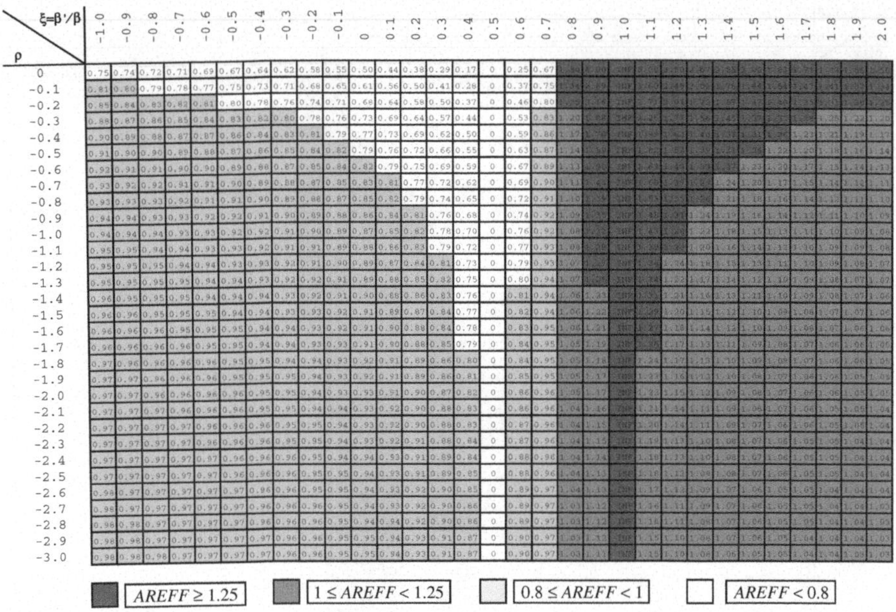

Fig. 2 $AREFF_{ML|\overline{ML}}$, in the (ξ, ρ)-plane

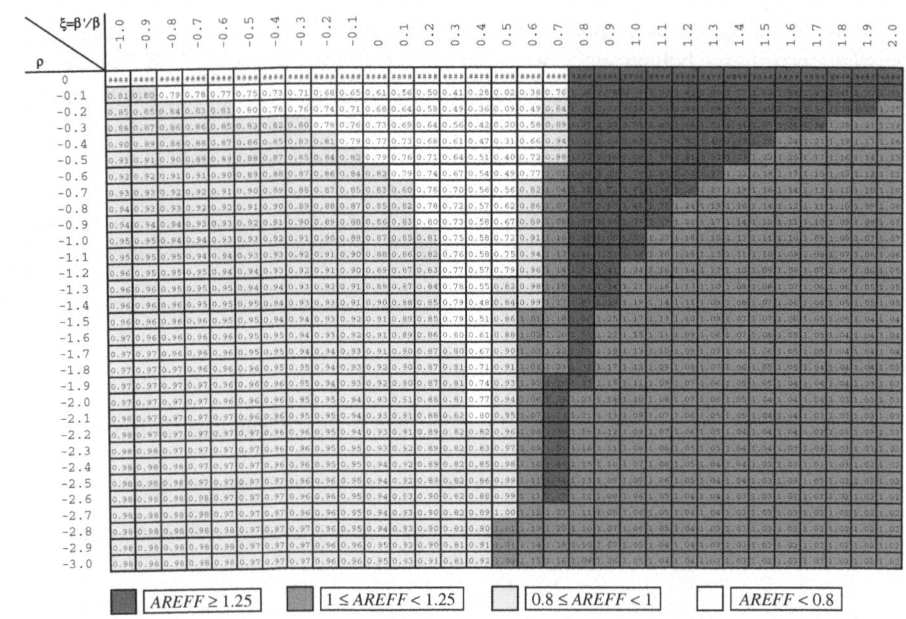

Fig. 3 $AREFF_{WH|\overline{WH}}$, in the (ξ, ρ)-plane

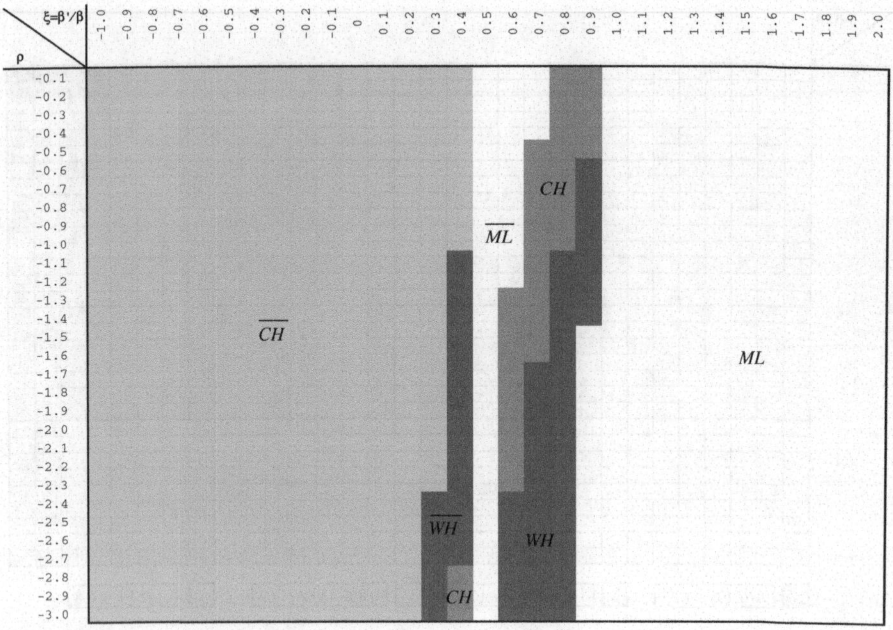

Fig. 4 Minimum LMSE among the CH, ML and WH statistics

From these figures it is possible to see that there is practically no difference between the relative behaviour between UH and $\overline{UH}$ for $UH = CH, ML$ and CH. Figure 4 also shows us that, at optimal levels, none of these estimators outperform the others in all the (ξ, ρ)-plane, but their simultaneous use will for sure enable us to better estimate γ. For a more detailed comparison, see Caeiro and Gomes, [3].

Acknowledgements Research partially supported by FCT—Fundação para a Ciência e a Tecnologia, projects PEst-OE/MAT/UI0006/2011 (CEAUL), PEst-OE/MAT/UI0297/2011 (CMA/UNL) and PTDC/FEDER, EXTREMA.

References

1. Bingham, N.H., Goldie, C.M., Teugels, J.L.: Regular Variation. Cambridge University Press, Cambridge (1987)
2. Caeiro, F., Gomes, M.I.: Minimum-variance reduced-bias tail index and high quantile estimation. Revstat **6**(1), 1–20 (2008)
3. Caeiro, F., Gomes, M.I.: Asymptotic comparison at optimal levels of reduced-bias extreme value index estimators. Stat. Neerl. **65**, 462–488 (2011)
4. Caeiro, F., Gomes, M.I., Pestana, D.D.: Direct reduction of bias of the classical Hill estimator. Revstat **3**(2), 113–136 (2005)
5. Caeiro, F., Gomes, M.I., Henriques Rodrigues, L.: Reduced-bias tail index estimators under a third order framework. Commun. Stat. Theor. Methods **38**(7), 1019–1040 (2009)

6. Fraga Alves, M.I., Gomes, M.I., de Haan, L.: A new class of semi-parametric estimators of the second order parameter. Portugaliae Mathematica **60**(1), 193–213 (2003)
7. Geluk, J., de Haan, L.: Regular Variation, Extensions and Tauberian Theorems. CWI Tract 40, Center for Mathematics and Computer Science, Amsterdam, Netherlands (1987)
8. Gomes, M.I., Martins, M.J.: Generalizations of the Hill estimator - asymptotic versus finite sample behaviour. J. Stat. Plan. Inference **93**, 161–180 (2001)
9. Gomes, M.I., Martins, M.J.: "Asymptotically unbiased" estimators of the tail index based on external estimation of the second order. Extremes **5**(1), 5–31 (2002)
10. Gomes, M.I., Pestana, D.D.: A simple second order reduced-bias tail index estimator. J. Stat. Comput. Simul. **77**(6), 487–504 (2007)
11. Gomes, M.I., Martins, M.J., Neves, M.: Improving second order reduced-bias extreme value index estimation. Revstat **5**(2), 177–207 (2007)
12. Gomes, M.I., de Haan, L., Henriques Rodrigues, L.: Tail Index estimation for heavy-tailed models: accommodation of bias in weighted log-excesses. J. Roy. Statist. Soc. B **70**(1), 31–52 (2008)
13. Haan, L. de, Peng, L.: Comparison of tail index estimators. Stat. Neerl. **52**, 60–70 (1998)
14. Hill, B.M: A simple general approach to inference about the tail of a distribution. Ann. Statist. **3**, 1163–1174 (1975)

Extremal Behavior of the Generalized Integer-Valued Random Coefficient Autoregressive Process

Luísa Canto e Castro, Dulce Gomes, and Maria da Graça Temido

Abstract
A stationary generalized random coefficient integer auto-regressive model of order 1 (Generalized RCINAR(1)), based on a thinning random operation, is presented. It is proved that the process satisfies a long-range condition as well as a local dependence condition, which are appropriate extensions of the well-known $D(u_n)$ and $D'(u_n)$ conditions of Leadbetter. Assuming that the marginal discrete distribution function belongs to Anderson's class, and then it does not belong to the domain of attraction of any max-stable distribution, the limit in distribution of the maximum of k_n random variables, being $\{k_n\}$ a geometric growing sequence, is obtained. This limit is a discrete max-semistable distribution function usually called discretized Gumbel.

1 Introduction

The analysis of integer-valued time series has become an important area of research in the last decades partially because its wide applicability to real data analysis. Within the reasonably large spectrum of integer-valued models proposed in the

L. Canto e Castro
CEAUL and DEIO, Fac. Ciências, University of Lisbon, Lisbon, Portugal
e-mail: luisa.loura@fc.ul.pt

D. Gomes (✉)
DMAT and CIMA-UE, University of Évora, Évora, Portugal
e-mail: dmog@uevora.pt

M. da Graça Temido
CMUC and FCTUC, University of Coimbra, Coimbra, Portugal
e-mail: mgtm@mat.uc.pt

J. Lita da Silva et al. (eds.), *Advances in Regression, Survival Analysis, Extreme Values, Markov Processes and Other Statistical Applications*, Studies in Theoretical and Applied Statistics, DOI 10.1007/978-3-642-34904-1_9,

literature, rather little is known about its extremal properties. Most of the work done deals with stationary sequences.

In this work we consider a strictly stationary process, named Generalized RCINAR(1) (Generalized Random Coefficient Integer AutoRegressive). The Generalized RCINAR(1) model has a form similar to that of the INAR(1) (INteger AutoRegressive) process proposed by [4], which, in turn, has a similar form to the well-known AR(1) process. As mentioned in [2] "the class of models presented here extends a Generalized AR(1) structure from which many new ideas can be established. The binomial "thinning" operation is replaced by a Generalized thinning operation and the vector of fixed covariates by a sequence of independent and identically distributed (i.i.d.) random variables (r.v's). In fact, the mentioned operation is defined in such a way that a large value can follow a previous one ("expanding" and not "thinning" it). However, in order to obtain stationarity of the time series the probability of this kind of occurrences must be very small (the mean and variance of thinning coefficient must be less than one)."

Note now that the fact of not having to require each coefficient to have support in [0,1], allows the use of this type of models in situations where time series present sporadic large peaks being therefore important to the analysis of their extremal behavior. As in [3], a basic assumption is that the innovations $\{Z_t\}_{t\in\mathbb{Z}}$ have common distribution function (d.f.) belonging to the class of Anderson. In this context we prove that the maximum, under linear normalization, has a non degenerate limiting distribution which is a discrete max-semistable d.f..

2 Generalized Random Coefficient Integer AutoRegressive Process

We start by presenting a Generalized Random Coefficient Integer AutoRegressive model of order 1 (Generalized RCINAR(1)). This model has a similar form to the one of the INAR(1) (INteger AutoRegressive) process proposed by [4]. Although with this similarity, they have two main differences: the binomial thinning operation is replaced by a generalized thinning operation and the coefficients are random.

The generalized thinning operation, represented by $\circ^G$, between two r.v's α and Y was defined by [2] as follows: given two r.v's, α and Y, and a family of distribution functions (d.f's) $G(\mu,\sigma)$, parameterized by the mean μ and standard deviation σ, the operation $\circ^G$ was defined imposing the following condition on the random variable (r.v.) $\alpha \circ^G Y$:

$$\alpha \circ^G Y|\alpha, Y \frown G(\alpha Y, \sigma)\ , \tag{1}$$

that is, the r.v. $\alpha \circ^G Y$, conditional to the values of α and Y, has a d.f. G with mean $\mu = \alpha Y$ and standard deviation σ. Note that the standard deviation σ can also depend on α and Y. We would like to emphasize that the binomial thinning operation used in the INAR(1) model is a particular case of the previous one, where the parameter α is real in [0, 1] and $G(\mu,\sigma)$ is a family of d.f's of a binomial r.v's.

The construction of models for count data, based on operation $\circ^G$, can then be made in a similar way to what has been done using the binomial thinning operation. More precisely, a stochastic process $\{Y_t\}_{t\in\mathbb{Z}}$ of discrete parameter and with support on nonnegative integers is said to be a Generalized RCINAR(1) if there exist a sequence of i.i.d. r.v's $\{\alpha_t\}_{t\in\mathbb{Z}}$, with support on $\mathbb{R}_+$ and finite second moment, and an uncorrelated sequence of nonnegative integer-valued r.v's, $\{Z_t\}_{t\in\mathbb{Z}}$, with mean μ_Z and finite variance σ_Z^2, such that $\{Y_t\}_{t\in\mathbb{Z}}$ satisfies the stochastic difference equation

$$Y_t = \alpha_t \circ^G Y_{t-1} + Z_t, \quad t \in \mathbb{Z}, \tag{2}$$

where:

1. For each t, the r.v. $\alpha_t \circ^G Y_{t-1}|\alpha_t, Y_{t-1}$ is independent of Y_{t-1-k}, α_{t-k} and Z_{t-k} for all $k \geq 1$.
2. The variable $Y_t|\alpha_t$ is independent of α_{t+k} and Z_{t+k} for all $k \geq 1$.

Gomes and Canto e Castro [2] gave necessary conditions on $\{\alpha_t\}$ such that Eq. (2) is a second-order process. Specifically, it is proved that a stationary solution exists for Eq. (2) since the second moment of α_t is less than one.

In what follows we assume that $\{Z_t\}$ is an i.i.d. sequence and denote by α and Z the r.v's with the d.f. of $\{\alpha_t\}$ and $\{Z_t\}$, respectively.

Proposition 1. *Suppose that the process $\{Y_t\}$ satisfies, for all $t \in \mathbb{Z}$, $\alpha \circ^G (\alpha \circ^G Y_t + Z_t) \stackrel{d}{=} \alpha \circ^G (\alpha \circ^G Y_t) + \alpha \circ^G Z_t$ and $\alpha \circ^G Y_t \leq Y_t$. Then the Generalized RCINAR(1) process $\{Y_t\}_{t\in\mathbb{Z}}$ is strictly stationary. A strictly stationary solution of Eq. (2) is given by*

$$Y_t = \sum_{i=1}^{+\infty} \alpha_t \circ^G \cdots \circ^G (\alpha_{t-i+1} \circ^G Z_{t-i}) + Z_t. \tag{3}$$

Proof. Since $\{\alpha_t\}$ and $\{Z_t\}$ are i.i.d. sequences it is easily proved, from the expression of the $P(Y_t = i|Y_{t-1} = i-1)$ [[2], p. 4093], that $\{Y_t\}$ is an irreducible and aperiodic Markov chain. Moreover following similar arguments to the ones of the proof of Proposition 2.2 of [8], we get that $\{Y_t\}$ is a positive recurrent Markov chain. Consequently it is strictly stationary.

Moreover, from Eq. (2), we deduce $Y_t \stackrel{d}{=} \alpha_t \circ^G \cdots \circ^G (\alpha_1 \circ^G Y_0) + \sum_{i=1}^{+\infty} V_i(t) + Z_t$ with $\{V_i(t) := \alpha_t \circ^G \cdots \circ^G (\alpha_{t-i+1} \circ^G Z_{t-i})\}_{i\geq 1}$. Due to the fact that $\alpha \circ^G Y_t \leq Y_t$, $\forall t$, we prove that $\alpha_t \circ^G \cdots \circ^G (\alpha_1 \circ^G Y_0)$ converges almost surely to zero. Since $\sum_{i=1}^{+\infty} V_i(t)$ is almost surely convergent it represents a mensurable function of the strictly stationary process $\{V_i(t)\}$ and so $\sum_{i=1}^{+\infty} V_i(t)$ is also strictly stationary. The same holds with $\sum_{i=1}^{+\infty} V_i(t) + Z_t$.

Thus, $\sum_{i=1}^{+\infty} V_i(t) + Z_t$ exists, is a solution of Eq. (2) and defines a strictly stationary process. □

3 The Anderson's Class and the Max-Semistable Class

As we said before, little is known about the extremal properties of many integer-valued models. In part, this is due to the fact that many integer-valued distributions do not belong to the domain of attraction on any max-stable (MS) distribution. Anderson [1] gave an important contribution to solve this problem by obtaining asymptotic upper and lower bounds for the distribution of the maximum of some particular i.i.d. sequences. Indeed, Anderson proved that an integer-valued d.f. F, with infinite right endpoint, satisfies

$$\lim_{n\to+\infty} \frac{1-F(n-1)}{1-F(n)} = r,\ r \in]1,+\infty[, \tag{4}$$

if and only if $\limsup_{n\to\infty} F^n(x+b_n) \le \exp(-r^{-x})$ and $\liminf_{n\to\infty} F^n(x+b_n) \ge \exp(-r^{-(x-1)})$, for any real x and b_n appropriately chosen. We shall say that a d.f. belongs to Anderson's class if and only if it satisfies Eq. (4). An example is the negative binomial d.f.. In order to overcome these limiting bounds and establish of a well-defined limiting distribution, Temido [6] proved that Eq. (4) is necessary and sufficient for the existence of a nondecreasing positive integer sequence $\{k_n\}$ satisfying

$$\lim_{n\to+\infty} \frac{k_{n+1}}{k_n} = r, \quad r \in]1,+\infty[, \tag{5}$$

and of a real sequence $\{u_n\}$ such that $k_n(1-F(u_n)) \to \tau > 0$, $n \to +\infty$, for some $\tau > 0$. So, considering the maximum of k_n r.v's, where $\{k_n\}$ satisfies Eq. (5), we can obtain a nondegenerate limiting distribution for the maximum M_{k_n}. The limit is not a MS distribution but belongs to a larger class of distributions, introduced by [5], known as max-semistable (MSS) distributions. This class includes the class MS and discrete or multimodal continuous d.f's. Following [5] we say that a real d.f. G is MSS if there are reals $r > 1$, $a = a(r) > 0$, and $b = b(r)$ such that $G(x) = G^r(\text{ax}+\text{b})$, $x \in \mathbb{R}$, or equivalently, if there exist a sequence of i.i.d. r.v's with d.f. F and two real sequences $\{a_n > 0\}$ and $\{b_n\}$ for which $\lim_{n\to+\infty} F^{k_n}(a_n x + b_n) = G(x)$, for each continuity point of G, where $\{k_n\}$ is a nondecreasing positive integer sequence satisfying the limit in Eq. (5) with $r \ge 1$. A d.f. in the class MSS can be written as follows:

$$G_{\gamma,\nu}(x) = \begin{cases} \exp\left\{-(1+\gamma x)^{-1/\gamma}\nu(\log(1+\gamma x))\right\} & x \in \mathbb{R},\ 1+\gamma x > 0 \text{ and } \gamma \neq 0 \\ \exp\{-e^{-x}\nu(x)\} & x \in \mathbb{R} \text{ and } \gamma = 0 \end{cases},$$

where ν is a positive, bounded, and periodic function. If ν is a suitable constant, we get the class MS.

Temido [6] proved that if F is an integer-valued d.f. with infinite upper endpoint and there are $\{k_n\}$ satisfying Eq. (5), $\{a_n > 0\}$, and $\{b_n\}$ such that $F^{k_n}(a_n x + b_n) \to G(x)$, $n \to +\infty$, then $G(x) = \exp(-\eta r^{-[x]})$, $x \in \mathbb{R}$, for some $\eta > 0$, if and only if

Eq. (4) holds. As we can transform $\{k_n\}$ and $\{b_n\}$ in order to obtain $a_n = 1$, $\eta = 1$, we have

$$\lim_{n\to+\infty} F^{k_n}(x+b_n) = \exp(-r^{-[x]}), \quad x \in \mathbb{R}\backslash\mathbb{Z}. \tag{6}$$

The d.f. $G(x) = \exp(-\eta r^{-[x]})$, $\eta > 0$, usually called discretized Gumbel, is MSS. Using these results, the limiting distribution of the maximum M_{k_n} of a large class of stationary integer-valued models can be obtained. Indeed, considering that these models satisfy a suitable extension of the long-range dependence condition of Leadbetter, $D(u_n)$, the limiting distribution of M_{k_n} can be inferred from the limiting behavior of the maximum of k_n i.i.d. r.v's with the some d.f. (see [3, 7]). This extended long-range dependence condition was introduced in [7] and was denoted by $D_{k_n}(u_n)$. According to these authors, the sequence $\{X_t\}$ satisfies $D_{k_n}(u_n)$ if, for any nondecreasing sequence of positive integers $\{k_n\}$ and for any integers $1 \le i_1 < \ldots < i_p < j_1 < \ldots < j_q \le k_n$ with $j_1 - i_p > \ell_n$, we have

$$\left|P\left(\bigcap_{s=1}^{p}\{X_{i_s} \le u_n\}, \bigcap_{m=1}^{q}\{X_{j_m} \le u_n\}\right) - P\left(\bigcap_{s=1}^{p}\{X_{i_s} \le u_n\}\right) P\left(\bigcap_{m=1}^{q}\{X_{j_m} \le u_n\}\right)\right| \le \alpha_{n,\ell_n},$$

where $\lim_{n\to+\infty} \alpha_{n,\ell_n} = 0$ for some sequence $\ell_n = o_n(k_n)$.

Considering stationary sequences $\{X_t\}$ satisfying $D_{k_n}(u_n)$ with $\{k_n\}$ satisfying Eq. (5), Temido and Canto e Castro [7] prove that the limit in distribution of M_{k_n}, under linear normalization, is MSS, whenever it exists. As well as long-range dependence conditions, the study of extremal properties of stationary sequences of r.v's is frequently based on the establishment of appropriate local dependence conditions. The generalized RCINAR(1) process satisfies the condition $D'_{k_n}(u_n)$ (introduced in [6]), which holds for stationary sequences $\{X_t\}$ if there is a sequence of positive integers $\{s_n\}$ such that $k_n/s_n \to +\infty$, $s_n \alpha_{n,l_n} \to 0$ and

$$\lim_{n\to+\infty} k_n \sum_{t=2}^{[k_n/s_n]} P(X_t > u_n, X_1 > u_n) = 0.$$

As it is expected, under $D_{k_n}(u_n)$ and $D'_{k_n}(u_n)$, the limiting distribution of M_{k_n}, it is the one of the associated i.i.d. sequence.

4 The Main Result

In this section we consider that the d.f. F of $\{Z_t\}$ belongs to the Anderson's class and that $\{Y_t\}$ is strictly stationary. We write $W_t := \alpha_t \circ^G Y_{t-1}$ and $W_t^{a,y} := W_t | \{\alpha_t = a, Y_t = y\}$ and denote by W a r.v. with the d.f. of $\{W_t\}$.

Lemma 1. *If $E(r^W)$ exists and $\{u_n\}$ are normalized levels for $\{Z_t\}$, then $\{u_n\}$ are also normalized levels for $\{Y_t\}$, with $\tau_Y = E(r^W)\tau_Z$.*

Proof. Since $\tau_Z := \tau_Z(x) = r^{-[x]}$, for n large enough we have

$$
\begin{aligned}
&k_n P\left(\alpha_t \circ^G Y_{t-1} + Z_t \geq u_n\right) \\
&= k_n \int_{S_{\alpha_t}} \sum_{y \in S_{Y_{t-1}}} P\left(W_t + Z_t \geq u_n \mid \alpha_t = a, Y_{t-1} = y\right) P\left(Y_{t-1} = y\right) f_{\alpha_t}(a) da \\
&= \int_{S_{\alpha_t}} \sum_{y \in S_{Y_{t-1}}} \sum_{w \in S_{W_t^{a,y}}} k_n P\left(Z_t \geq u_n - w\right) P\left(W_t^{a,y} = w\right) P\left(Y_{t-1} = y\right) f_{\alpha_t}(a) da \\
&= \int_{S_{\alpha_t}} \sum_{y \in S_{Y_{t-1}}} \sum_{w \in S_{W_t^{a,y}}} \left(r^{-[x]+w} + o_n(1)\right) P\left(W_t^{a,y} = w\right) P\left(Y_{t-1} = y\right) f_{\alpha_t}(a) da \\
&= r^{-[x]} \int_{S_{\alpha_t}} \sum_{y \in S_{Y_{t-1}}} \sum_{w \in S_{W_t^{a,y}}} r^w P\left(W_t^{a,y} = w\right) P\left(Y_{t-1} = y\right) f_{\alpha_t}(a) da + o_n(1) \\
&= r^{-[x]} E\left(r^W\right) + o_n(1) \to \tau_Y, \quad n \to +\infty.
\end{aligned}
$$

□

Observe now that from the last lemma we conclude that if there are $\{b_n\}$ and $\{k_n\}$ such that Eq. (6) holds, then for the d.f. H of $\{Y_t\}$ we get

$$\lim_{n \to +\infty} H^{k_n}(x + b_n) = \exp\left(-E(r^W) r^{-[x]}\right), \; x \in \mathbb{R} \backslash \mathbb{Z}. \tag{7}$$

Lemma 2. *If the d.f. of Z belongs to Anderson's class then the same holds with H.*

Proof. Due to

$$\frac{k_n(1 - H(u_n))}{k_n(1 - F(u_n))} = \frac{1 - H(u_n)}{1 - H(u_{n+1})} \times \frac{k_n(1 - H(u_{n+1}))}{k_n(1 - F(u_{n+1}))} \times \frac{1 - F(u_{n+1})}{1 - F(u_n)}$$

and taking into account that the first, the third, and the fourth quotients converge, we conclude that the second quotient also converges and the limit is r. □

We need the following lemma to prove that condition $D'_{k_n}(u_n)$ holds:

Lemma 3. *Under conditions above, for all $j \in \mathbb{N}_0$, we have*

1. $$\sum_{j_2, \ldots, j_{t-1} \in \mathbb{N}_0} P\left(Y_t > u_n, Y_{t-1} = j_{t-1}, \ldots, Y_2 = j_2, Y_1 = j\right) \leq \sum_{j_{t-1} \in \mathbb{N}_0} P\left(Y_t > u_n \mid Y_{t-1} = j_{t-1}\right) P\left(Y_1 = j\right),$$

2. $$k_n P\left(Y_t > u_n \mid Y_{t-1} = j_{t-1}\right) = r^{-[x]} \int_{S_\alpha} E\left(r^{W^{a, j_{t-1}}}\right) f_\alpha(a) + o_n(1),$$
$n \to +\infty$.

Proof. Taking into account that $\sum_{j_m \in \mathbb{N}_0} P\left(Y_m = j_m \mid Y_{m-1} = j_{m-1}\right) = 1$, we have

$$\sum_{j_2,\ldots,j_{t-1}\in\mathbb{N}_0} P\left(Y_t > u_n, Y_{t-1} = j_{t-1},\ldots,Y_2 = j_2, Y_1 = j\right)$$

$$= \sum_{j_2,\ldots,j_{t-1}\in\mathbb{N}_0} P\left(Y_t > u_n \mid Y_{t-1} = j_{t-1}\right)\left\{\prod_{l=3}^{t-1} P\left(Y_l = j_l \mid Y_{l-1} = j_{l-1}\right)\right\}\times$$

$$\times P\left(Y_2 = j_2, Y_1 = j\right)$$

$$= \sum_{j_{t-1}\in\mathbb{N}_0} P\left(Y_t > u_n \mid Y_{t-1} = j_{t-1}\right)\sum_{j_{t-2}\in\mathbb{N}_0} P\left(Y_{t-1} = j_{t-1} \mid Y_{t-2} = j_{t-2}\right)\times$$

$$\times \sum_{j_{t-3}\in\mathbb{N}_0} P\left(Y_{t-2} = j_{t-2} \mid Y_{t-3} = j_{t-3}\right)\times\cdots\times\sum_{j_3\in\mathbb{N}_0} P\left(Y_4 = j_4 \mid Y_3 = j_3\right)\times$$

$$\times \sum_{j_2\in\mathbb{N}_0} P\left(Y_3 = j_3 \mid Y_2 = j_2\right) P\left(Y_2 = j_2, Y_1 = j\right)$$

$$\leq \sum_{j_{t-1}\in\mathbb{N}_0} P\left(Y_t > u_n \mid Y_{t-1} = j_{t-1}\right)\sum_{j_{t-3}\in\mathbb{N}_0}\sum_{j_{t-2}\in\mathbb{N}_0} P\left(Y_{t-2} = j_{t-2} \mid Y_{t-3} = j_{t-3}\right)\times$$

$$\times\cdots\times\sum_{j_3\in\mathbb{N}_0} P\left(Y_4 = j_4 \mid Y_3 = j_3\right)\sum_{j_2\in\mathbb{N}_0} P\left(Y_3 = j_3 \mid Y_2 = j_2\right) P\left(Y_2 = j_2, Y_1 = j\right)$$

$$= \sum_{j_{t-1}\in\mathbb{N}_0} P\left(Y_t > u_n \mid Y_{t-1} = j_{t-1}\right)\times$$

$$\times \sum_{j_{t-4}\in\mathbb{N}_0}\sum_{j_{t-3}\in\mathbb{N}_0} P\left(Y_{t-3} = j_{t-3} \mid Y_{t-4} = j_{t-4}\right) P\left(Y_{t-4} = j_{t-4} \mid Y_{t-5} = j_{t-5}\right)\times$$

$$\times\cdots\times\sum_{j_3\in\mathbb{N}_0} P\left(Y_4 = j_4 \mid Y_3 = j_3\right)\sum_{j_2\in\mathbb{N}_0} P\left(Y_3 = j_3 \mid Y_2 = j_2\right) P\left(Y_2 = j_2, Y_1 = j\right)$$

$$= \sum_{j_{t-1}\in\mathbb{N}_0} P\left(Y_t > u_n \mid Y_{t-1} = j_{t-1}\right)\sum_{j_2\in\mathbb{N}_0}\sum_{j_3\in\mathbb{N}_0} P\left(Y_3 = j_3 \mid Y_2 = j_2\right)\times$$

$$\times P\left(Y_2 = j_2 \mid Y_1 = j\right) P\left(Y_1 = j\right)$$

$$= \sum_{j_{t-1}\in\mathbb{N}_0} P\left(Y_t > u_n \mid Y_{t-1} = j_{t-1}\right) P\left(Y_1 = j\right).$$

To prove 2, observe that

$$k_n P\left(Y_t > u_n \mid Y_{t-1} = j_{t-1}\right)$$

$$= \int_{S_{\alpha_t}} k_n P\left(\alpha_t \circ^G Y_{t-1} + Z_t \geq u_n \mid Y_{t-1} = j_{t-1}, \alpha_t = a\right) f_{\alpha_t}(a)da$$

$$= \int_{S_{\alpha_t}} P\left(W^{a,j_{t-1}} + Z_t \geq u_n \mid Y_{t-1} = j_{t-1}, \alpha_t = a\right) f_{\alpha_t}(a)da$$

$$= \int_{S_{\alpha_t}} \sum_{w\in W^{a,j_{t-1}}} k_n P\left(Z_t > u_n - w\right) P\left(W^{a,j_{t-1}} = w \mid Y_{t-1} = j_{t-1}, \alpha_t = a\right) f_{\alpha_t}(a)da$$

$$= r^{-[x]} \int_{S_{\alpha_t}} \sum_{w \in W^{a,j_{t-1}}} r^w P\left(W^{a,j_{t-1}} = w \mid Y_{t-1} = j_{t-1}, \alpha_t = a\right) f_{\alpha_t}(a) da$$

$$= r^{-[x]} \int_{S_{\alpha_t}} E\left(r^{W^{a,j_{t-1}}}\right) f_{\alpha_t}(a) da + o_n(1), \quad n \to +\infty.$$

□

Now we can present our main result.

Theorem 1. *For the stationary Generalized RCINAR(1) process assume that F belongs to Anderson's class and $E(r^W) < +\infty$. Let $\{k_n\}$ be a nondecreasing integer sequence satisfying Eq. (4) and $\{x + b_n\}$ be a sequence of normalized levels for F:*

1. The process $\{Y_t\}$ satisfies condition $D_{k_n}(x + b_n)$.

2. If $\sum_{k \in \mathbb{N}_0} \int_{S_\alpha} E\left(r^{W^{a,k}}\right) f_\alpha(a) da < +\infty$, then $\{Y_t\}$ satisfies condition $D'_{k_n}(x+b_n)$.

3. Consequently $\lim_{n \to +\infty} P(M_{k_n} \leq x + b_n) = \exp\left(-E(r^W) r^{-[x]}\right)$, $x \in \mathbb{R} \backslash \mathbb{Z}$.

Proof. Due to the fact that the process $\{Y_t\}$ is regenerative and aperiodic it is strong mixing. So $D_{k_n}(v_n)$ holds for any real sequence $\{v_n\}$. Write $u_n := x + b_n$. In order to prove that $D'_{k_n}(u_n)$ occurs, note that, by Lemma 3, we get

$$P(Y_t > u_n, Y_1 > u_n) = \sum_{j=[u_n]+1}^{+\infty} P(Y_t > u_n, Y_1 = j) =$$

$$= \sum_{j=[u_n]+1}^{+\infty} \sum_{j_2, \ldots, j_{t-1} \in \mathbb{N}_0} P(Y_t > u_n, Y_{t-1} = j_{t-1}, \ldots, Y_2 = j_2, Y_1 = j)$$

$$\leq \sum_{j=[u_n]+1}^{+\infty} \sum_{j_{t-1} \in \mathbb{N}_0} P(Y_t > u_n \mid Y_{t-1} = j_{t-1}) P(Y_1 = j).$$

Hence, using the second part of Lemma 3, we deduce

$$k_n \sum_{t=2}^{[k_n/s_n]} P(Y_t > u_n, Y_1 > u_n) \leq k_n \sum_{t=2}^{[k_n/s_n]} \sum_{j=[u_n]+1}^{+\infty} \sum_{k \in \mathbb{N}_0} P(Y_t > u_n \mid Y_{t-1} = k) P(Y_1 = j)$$

$$\leq r^{-[x]} \left\{ \sum_{k \in \mathbb{N}_0} \int_{S_{\alpha_t}} E\left(r^{W^{a,k}}\right) f_\alpha(a) da \right\} \frac{k_n}{s_n} P(Y_1 > u_n) + o_n(1) = o_n(1), n \to +\infty.$$

Since $\{u_n := x + b_n\}$ are also normalized levels for H (by Lemma 1) we obtain Eq. (7). Finally, with $D_{k_n}(u_n)$ and $D'_{k_n}(u_n)$, the maximum M_{k_n} has the limit distribution of the associated i.i.d. sequence. □

Example 1. Considering $\{\alpha_t\}$ with distribution $\mathscr{U}(0,1)$ and G binomial, we deduce that $\sum_{k\in\mathbb{N}_0}\int_{S_\alpha} E\left(r^{W_{a,k}}\right) f_{\alpha_t}(a)da < +\infty$ which implies $E(r^W) < +\infty$.

Acknowledgments This research was partially supported by Center of Mathematics and Applications of the University of Évora, by Center of Statistics and Applications of the University of Lisbon and by Center for Mathematics of the University of Coimbra, through the Program FCT/POCTI and POCI/FEDER.

References

1. Anderson, C.W.: Extreme value theory for a class of discrete distribution with applications to some stochastic processes. J. Appl. Prob. **7**, 99–113 (1970)
2. Gomes, D., Canto e Castro, L: Generalized integer-valued random coefficient for a first order structure autoregressive (RCINAR) process. J. Stat. Plann. Infer. **139**, 4088–4097 (2009)
3. Hall, A., Temido, M.G.: On the maximum term of MA and Max-AR models with margins in Anderson's class. Theor. Probab. Appl. **51**, 291–304 (2007)
4. McKenzie, E.: Some simple models for discrete variate time series. Water Resour. Bull. **21**, 645–650 (1985)
5. Pancheva, E.: Multivariate max-semistable distributions. Theor. Probab. Appl. **18**, 679–705 (1992)
6. Temido, M.G.: Classes de leis limite em teoria de valores extremos-estabilidade e semiestabilidade. Ph.D. Thesis, University of Coimbra (2000)
7. Temido, M.G., Canto e Castro, L.: Max-semistable laws in extremes of stationary random sequences. Theor. Probab. Appl. **47**, 365–374 (2003)
8. Zheng, H., Basawa, D, Datta, S.: First-order random coefficient integer-valued autoregressive processes. J. Stat. Plann. Infer. **173**, 212–229 (2007)

Models of Individual Growth in a Random Environment: Study and Application of First Passage Times

Clara Carlos, Carlos A. Braumann, and Patrícia A. Filipe

Abstract
We study the first-passage times for models of individual growth of animals in randomly fluctuating environments. In particular, we present results on the mean and variance of the first-passage time by a high threshold value (higher than the initial size). The models considered are stochastic differential equations of the form $dY(t) = \beta(\alpha - Y(t))\,dt + \sigma\,dW(t)$, $Y(t_0) = y_0$, where $Y(t) = g(X(t))$ is a transformed size, g being a strictly increasing C^1 function of the actual animal size $X(t)$ at time t, σ measures the effect of random environmental fluctuations on growth, $W(t)$ is the standard Wiener process, and y_0 is the transformed size (assumed known) at the initial instant t_0. Results are illustrated using cattle weight data, to which we have applied the Bertalanffy-Richards ($g(x) = x^c$) and the Gompertz ($g(x) = \ln x$) stochastic models.

1 Introduction

In previous work [1, 3, 4], in order to study the extinction of populations growing in random environments, we have obtained results about the first-passage time through a low threshold (below the initial population). We will now study the first-passage

C. Carlos (✉)
Escola Superior de Tecnologia do Barreiro, Instituto Politécnico de Setúbal, Rua Américo da Silva Marinho, 2890-001 Lavradio, Portugal
e-mail: clara.carlos@estbarreiro.ips.pt

C.A. Braumann · P.A. Filipe
Universidade de Évora, Centro de Investigação em Matemática e Aplicações, Rua Romão Ramalho, 59, 7000-671 Évora, Portugal
e-mail: braumann@uevora.pt; pasf@uevora.pt

J. Lita da Silva et al. (eds.), *Advances in Regression, Survival Analysis, Extreme Values, Markov Processes and Other Statistical Applications*, Studies in Theoretical and Applied Statistics, DOI 10.1007/978-3-642-34904-1_10,

time through a high threshold value (higher than the initial size) for somewhat similar models, namely models of individual growth in random environments.

Let $X(t)$ be the size of an individual (animal or plant) at age t, for example, weight, length, height, or volume. Many of the deterministic models proposed in the literature for the growth of an individual animal from birth to maturity can be written in the form

$$dY(t) = \beta(\alpha - Y(t))dt, \tag{1}$$

where $Y(t) = g(X(t))$ is a rescaled or modified size, with g a strictly increasing continuously differentiable function. We consider $\alpha = g(A)$, where A is the asymptotic (maturity) size and β describes how fast is the approach to this asymptotic value. We will assume the initial size $X(t_0) = x_0$ to be known and denote $y_0 = g(x_0)$ the initial modified size. Two of the most commonly used models are the Gompertz model and the Bertalanffy-Richards model, corresponding to $g(x) = \ln x$ and $g(x) = x^c$, $c > 0$, respectively.

In a randomly fluctuating environment, we propose (see [5–8]) as a general model of individual growth stochastic differential equations (SDE) of the form

$$dY(t) = \beta(\alpha - Y(t))dt + \sigma dW(t), \tag{2}$$

with $Y(t_0) = y_0$, where σ is an environmental noise intensity parameter and $W(t)$ is a standard Wiener process.

In Sect. 2 we present the basic properties of these models.

In Sect. 3, we present explicit results on the mean and variance of first-passage times for ergodic solutions of sufficiently regular autonomous SDE with a stationary density. We have obtained these results by solving known differential equations (see [10]) that are satisfied by the first-passage time non-centered moments, considering two thresholds (high and low). To obtain the solution, we have applied some changes of variable and other algebraic manipulations and then obtained the limit when a threshold is approaching the lower (or higher) boundary of the state space. Our expression for the variance is considerably simpler them the one presented by [12]. We then considered the particular case of our individual growth models.

In Sect. 4, we present an application of these results. We use data on the weight of Mertolengo cattle of the rosilho strain (provided by Carlos J. Roquete) and consider two models (the Bertalanffy-Richards and the Gompertz models), taking as parameters the maximum likelihood estimates (see [5, 6, 8]) based on the available data. Models of this type have also been used to describe tree growth (see [9]) and fish growth (see [11]), for example. We characterize the time for an animal to reach a given weight, for example, the weight at which the animal is sold to the meat market, determining the mean and standard deviation of the time required for the animal to reach such weight.

Section 5 presents the conclusions.

2 Stochastic Differential Equations Growth Models

The solution of model (2) is a homogeneous diffusion process with drift coefficient

$$a(y) = \beta(\alpha - y) \tag{3}$$

and diffusion coefficient

$$b^2(y) = \sigma^2. \tag{4}$$

The drift coefficient represents the mean speed of growth described by $Y(t)$ and the diffusion coefficient gives a measure of the local magnitude of the fluctuations.

Putting $Z(t) = Y(t)e^{\beta t}$, we can write $dZ(t) = (dY(t) + Y(t)\beta dt)e^{\beta t} = \alpha\beta e^{\beta t}dt + \sigma e^{\beta t}dW(t)$. Integrating between t_0 and t, we get $Z(t) = Z(t_0) + \alpha(e^{\beta t} - e^{\beta t_0}) + \sigma\int_{t_0}^{t} e^{\beta u}dW(u)$, and, inverting the change of variable Z, we obtain the explicit solution of Eq. (2):

$$Y(t) = \alpha - (\alpha - y_0)e^{-\beta(t-t_0)} + \sigma e^{-\beta t}\int_{t_0}^{t} e^{\beta u}dW(u). \tag{5}$$

From here, we can see that $Y(t)$ is Gaussian with mean $\alpha - (\alpha - y_0)e^{-\beta(t-t_0)}$ and variance $\frac{\sigma^2}{2\beta}(1 - e^{-2\beta(t-t_0)})$. Letting $t \to +\infty$, we see that the modified weight $Y(t)$ is asymptotically Gaussian with mean α and variance $\frac{\sigma^2}{2\beta}$.

We shall need the scale and speed measures of $Y(t)$, which can be characterized by their densities, defined in the interior of the state space. For any homogeneous diffusion process with sufficiently regular drift $a(y)$ and diffusion $b^2(y)$ coefficients, the scale density is defined up to a multiplicative constant by

$$s(y) := \exp\left(-\int_{y^*}^{y}\frac{2a(\theta)}{b^2(\theta)}d\theta\right) = c\exp\left(-\frac{2\beta\alpha}{\sigma^2}y + \frac{\beta}{\sigma^2}y^2\right), \tag{6}$$

where c is constant and y^* is an arbitrary, but fixed point in the interior of the state space and the speed density is defined by

$$m(y) := \frac{1}{s(y)b^2(y)} = \frac{1}{\sigma^2 s(y)}. \tag{7}$$

The distribution functions of these measures are the scale and speed functions defined by $S(y) = \int_{y^{**}}^{y} s(v)dv$ and $M(y) = \int_{y^{**}}^{y} m(v)dv$, where y^{**} is an arbitrary point in the interior of the state space.

3 First-Passage Time

Let us consider thresholds q^* and Q^*, one low and one high, for the animal size $X(t)$. We are interested in the time required for the animal to reach size q^* and Q^* for the first time. Since $X(t)$ and $Y(t)$ are related by the strictly increasing function g, this is the first-passage time of the modified size $Y(t)$ by $q = g(q^*)$ and $Q = g(Q^*)$. Let us denote it by T_q and T_Q, where

$$T_q = \inf\{t > 0 : Y(t) = q\} \quad \text{and} \quad T_Q = \inf\{t > 0 : Y(t) = Q\}, \tag{8}$$

and assume that $-\infty < q < y_0 < Q < +\infty$, with q and Q both in the interior of the state space of Y. Let $T_{qQ} = min\{T_q, T_Q\}$ be the first-passage time of $Y(t)$ through either of the thresholds q and Q. One can see in [10] that the probability of $Y(t)$ to reach Q before reaching q is

$$u(y_0) = P[T_Q < T_q | Y(0) = y_0] = \frac{\int_q^{y_0} s(z)dz}{\int_q^Q s(z)dz}. \tag{9}$$

The k-th order moment of the first-passage time of $Y(t)$ by q or Q is

$$M_{qQ}^{(k)}(y_0) = E[(T_{qQ})^k | Y(0) = y_0] \tag{10}$$

and satisfies the differential equation (see [10])

$$\frac{1}{2}b^2(y_0)\frac{d^2 M_{qQ}^{(k)}(y_0)}{dy_0^2} + a(y_0)\frac{dM_{qQ}^{(k)}(y_0)}{dy_0} + k\, M_{qQ}^{(k-1)}(y_0) = 0, \tag{11}$$

for $q < y_0 < Q$, with $M_{qQ}^{(k)}(q) = M_{qQ}^{(k)}(Q) = 0$ $(k = 1, 2, \ldots)$. The solution is

$$\begin{aligned} E[(T_{qQ})^k | Y(0) = y_0] = 2\Big\{u(y_0) \int_{y_0}^{Q} \int_{\zeta}^{Q} s(\xi)d\xi\, k\, M_{qQ}^{(k-1)}(\zeta)m(\zeta)d\zeta \\ +(1 - u(y_0)) \int_{q}^{y_0} \int_{q}^{\zeta} s(\xi)d\xi\, k\, M_{qQ}^{(k-1)}(\zeta)m(\zeta)d\zeta\Big\}. \end{aligned} \tag{12}$$

From $M_{qQ}^{(0)}(y_0) = 1$, one can iteratively obtain the moments of any order of T_{qQ}.

Since the process $Y(t)$ is ergodic, we can obtain $E[(T_q)^k | Y(0) = y_0]$ as the limiting case of $E[(T_{qQ})^k | Y(0) = y_0]$ when $Q \to +\infty$. One gets

$$M_q^{(k)}(y_0) := E[(T_q)^k | Y(0) = y_0] = 2\int_q^{y_0} s(\zeta) \int_{\zeta}^{+\infty} k\, M_q^{(k-1)}(\theta)m(\theta)d\theta d\zeta. \tag{13}$$

Using $M_q^{(0)}(y_0) = 1$ and expression (13) with $k = 1$, one obtains the mean time

$$M_q^{(1)}(y_0) = E[T_q|Y(0) = y_0] = 2\int_q^{y_0} s(\zeta)\int_\zeta^{+\infty} m(\theta)d\theta d\zeta$$

$$= \int_q^{\zeta} 2s(\xi)(M(+\infty) - M(\xi))d\xi. \tag{14}$$

Using Eq. (13) with $k = 2$, one gets

$$M_q^{(2)}(y_0) = 2\int_q^{y_0} s(\zeta)\int_\zeta^{+\infty} 2M_q^{(1)}(\theta)m(\theta)d\theta d\zeta = H_1 + H_2, \tag{15}$$

with

$$H_1 = 2\int_q^{y_0} s(\zeta)\int_\zeta^{+\infty} 2\left(M_q^{(1)}(\theta) - M_q^{(1)}(\zeta)\right)m(\theta)d\theta d\zeta, \tag{16}$$

$$H_2 = 2\int_q^{y_0} 2s(\zeta)M_q^{(1)}(\zeta)\int_\zeta^{+\infty} m(\theta)d\theta d\zeta = 2\int_q^{y_0} 2s(\zeta)M_q^{(1)}(\zeta)\,(M(+\infty)-M(\zeta))\,d\zeta. \tag{17}$$

Using Eq. (14) and $M_q^{(1)}(q) = 0$, one obtains

$$H_2 = \int_q^{y_0} 2M_q^{(1)}(\zeta)\frac{dM_q^{(1)}(\zeta)}{d\zeta}d\zeta = \left(M_q^{(1)}(y_0)\right)^2. \tag{18}$$

So, from Eqs. (15) and (18), one gets $Var[T_q|Y(0) = y_0] = M_q^{(2)}(y_0) - \left(M_q^{(1)}(y_0)\right)^2 = H_1$. Using Eqs. (14) and (16), it results

$$Var[T_q|Y(0) = y_0] = 2\int_q^{y_0} 2s(\zeta)\int_\zeta^{+\infty} 2\int_\zeta^{\theta} s(\xi)(M(+\infty)-M(\xi))d\xi m(\theta)d\theta d\zeta. \tag{19}$$

Exchanging the order of integration between ξ and θ, the variance becomes $2\int_q^{y_0} 2s(\zeta)\int_\zeta^{+\infty} 2s(\xi)(M(+\infty) - M(\xi))\int_\xi^{+\infty} m(\theta)d\theta d\xi d\zeta$, which simplifies to

$$Var[T_q|Y(0) = y_0] = 8\int_q^{y_0} s(\zeta)\int_\zeta^{+\infty} s(\xi)\left(\int_\xi^{+\infty} m(\theta)d\theta\right)^2 d\xi d\zeta. \tag{20}$$

Similarly, making $q \to -\infty$, we obtain the k-th order moments of the first-passage time of $Y(t)$ by Q:

$$M_Q^{(k)}(y_0) := E[(T_Q)^k|Y(0) = y_0] = 2\int_{y_0}^{Q} s(\zeta)\int_{-\infty}^{\zeta} k\,M_Q^{(k-1)}(\theta)m(\theta)d\theta d\zeta. \tag{21}$$

Adapting the reasoning above for q, one obtains for the mean and variance of T_Q:

Table 1 Maximum likelihood estimates and asymptotic half-width of 95 % confidence intervals for the parameters A, β, and σ using data from 97 Mertolengo cows

	A	β	σ
Gompertz	411.2 ± 8.1	1.676 ± 0.056	0.302 ± 0.009
Bertalanffy-Richards (c=1/3)	425.7 ± 9.5	1.181 ± 0.056	0.597 ± 0.019

$$E[T_Q|Y(0) = y_0] = 2\int_{y_0}^{Q} s(\zeta)\int_{-\infty}^{\zeta} m(\theta)d\theta d\zeta, \tag{22}$$

$$Var[T_Q|Y(0) = y_0] = 8\int_{y_0}^{Q} s(\zeta)\int_{-\infty}^{\zeta} s(\xi)\Big(\int_{-\infty}^{\xi} m(\theta)d\theta\Big)^2 d\xi d\zeta. \tag{23}$$

The expressions obtained above are valid for sufficiently regular homogeneous ergodic diffusion processes with drift coefficient $a(y)$ and diffusion coefficient $b^2(y)$. For the particular case of our models (2), using Eqs. (3) and (4), we obtain, making the change of variables $y = \sqrt{2\beta}(\theta - \alpha)/\sigma$ and $z = \sqrt{2\beta}(\zeta - \alpha)/\sigma$ and denoting by Φ and ϕ the distribution function and the probability density function of a standard normal random variable,

$$E[T_Q|Y(0) = y_0] = \frac{1}{\beta}\int_{\frac{\sqrt{2\beta}}{\sigma}(y_0-\alpha)}^{\frac{\sqrt{2\beta}}{\sigma}(Q-\alpha)} \frac{\Phi(z)}{\phi(z)}dz, \tag{24}$$

$$Var[T_Q|Y(0) = y_0] = \frac{2}{\beta^2}\int_{\frac{\sqrt{2\beta}}{\sigma}(y_0-\alpha)}^{\frac{\sqrt{2\beta}}{\sigma}(Q-\alpha)} \frac{1}{\phi(z)}\int_{-\infty}^{z} \frac{\Phi^2(y)}{\phi(y)}dydz. \tag{25}$$

Notice that T_Q is both the time required for the modified weight $Y(t)$ to reach Q and the time required for the actual weight $X(t)$ to reach the $Q^* = g^{-1}(Q)$.

4 Application

The data we have used for illustration was taken from cattle, namely Mertolengo cattle breed, and was collected by C. J Roquete. This cattle breed is considered by many as the Portuguese breed with higher progression in terms of population increment and market potential. These data come from animals raised in "Herdade da Abóboda" in Serpa region at the left margin of the Guadiana river. The animals were raised in pasture, together with their mothers during nursing, and later supplemented witch silage when pasture was in shortage (from August till January). In [2], we have applied the maximum likelihood estimation method for the parameters of model (2) to the weights (in kg) of 97 Mertolengo cows at several ages, for which we have a total of 2,129 observations. The results are shown in Table 1.

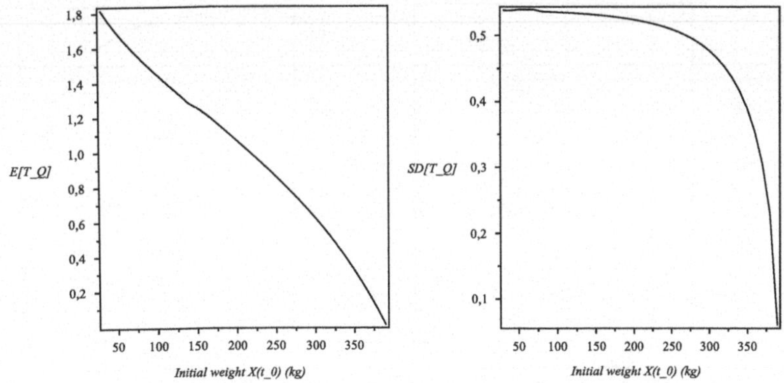

Fig. 1 Gompertz model: expected value and standard deviation of the time, in years, required for a Mertolengo cow to reach 390.6 Kg (95 % of the average asymptotic size) for the first time as a function of the initial weight $X(t_0)$. We have used as parameters the estimates presented in Table 1

For example, using the Kg as weight unit, let us consider the case of a Mertolengo cow with weight at birth $X(0) = 40$ and assume we want to determine the time required for the animal to reach 95 % of the asymptotic weight. Assuming the cow grows according to a Gompertz model with $A = 411.2$, $\beta = 1.676$ per year, and $\sigma^2 = (0.302)^2$ per year, we have $Q^* = 390.6$, $Q = \ln 390.6$, $y_0 = \ln 40$, and $\alpha = \ln 411.2$. We have obtained 1.75 years for the mean time to reach the desired threshold and 0.54 years for the standard deviation of that time. If we assume that the cow grows according to a Bertalanffy-Richards (with $c = 1/3$) model with $A = 425.7$, $\beta = 1.181$ per year, and $\sigma^2 = (0.597)^2$ per year, we have $Q^* = 404.4$, $Q = \sqrt[3]{404.4}$, $y_0 = \sqrt[3]{40}$, and $\alpha = \sqrt[3]{425.7}$, and the mean time to reach the desired threshold is 2.22 years, the standard deviation of that time being 0.53 years. If we have instead a non-newborn cow having now a weight $X(0) = 200$ and want to determine the time required for the animal to reach the asymptotic size under the same Bertalanffy-Richards model, we have $Q^* = 425.7$, $Q = \sqrt[3]{425.7}$, $y_0 = \sqrt[3]{200}$, and $\alpha = \sqrt[3]{425.7}$, obtaining for the mean and standard deviation of the time to reach the desired threshold 1.80 years and 0.65 years, respectively.

Using the maximum likelihood estimates of the parameters given in Table 1, we show in Figs. 1 and 2, for both Gompertz and Bertalanffy-Richards models, the mean and standard deviation of the time required for a Mertolengo cow to reach for the first time 95 % of the asymptotic weight as a function of the initial weight $X(t_0)$.

5 Conclusions

We have determined explicit expressions for the mean and variance of the first-passage times in models describing individual growth in a randomly varying environment. These results can be very important from the economic point of view, since they can be used to study the time it takes for an animal to achieve a certain

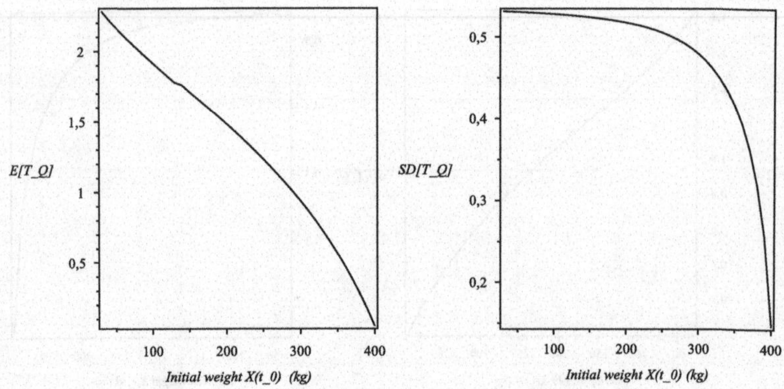

Fig. 2 Bertalanffy-Richards model: Expected value and standard deviation of the time, in years, required for a Mertolengo cow to reach 404.4 Kg (95 % of the average asymptotic size) for the first time as a function of the initial weight $X(t_0)$. We have used as parameters the estimates presented in Table 1

weight of interest, for instance, a weight appropriate to sell the animal for the meat market.

Acknowledgements The authors are members of the Centro de Investigação em Matemática e Aplicações, the research center of the Universidade de Évora financed by FCT (Fundação para a Ciência e Tecnologia). This work was partially financed by FCT within the research project PTDC/MAT/64297/2006.

References

1. Braumann, C.A.: Growth and extinction of populations in randomly varying environments. Comput. Math. Appl. **56**, 631–644 (2008)
2. Braumann, C.A., Filipe, P.A., Carlos C., Roquete, C.J.: Growth of individuals in randomly fluctuating environments. In: Vigo-Aguiar, J., Alonso, P., Oharu, S., Venturino, E., Wade, B. (eds.) Proceedings of the International Conference in Computational and Mathematical Methods in Science e Engineering, pp. 201–212. Gijon (2009)
3. Carlos, C., Braumann, C.A.: Tempos de extinção para populações em ambiente aleatório. In: Braumann, C.A., Infante, P., Oliveira, M.M., Alpízar-Jara, R., Rosado, F. (eds.) Estatística Jubilar, pp. 133–142. Edições SPE (2005)
4. Carlos, C., Braumann, C.A.: Tempos de extinção para populações em ambiente aleatório e cálculos de Itô e Stratonovich. In: Canto e Castro, L., Martins, E.G., Rocha, C., Oliveira, M.F., Leal, M.M., Rosado, F. (eds.) Ciência Estatística, pp. 229–238. Edições SPE (2006)
5. Filipe, P.A., Braumann, C.A.: Animal growth in random environments: estimation with several paths. Bull. Int. Stat. Institute LXII, 5806–5809 (2007)
6. Filipe, P.A., Braumann, C.A.: Modelling individual animal growth in random environments. In: Eilers, P.H.C. (ed.) Proceedings of the 23rd International Workshop on Statistical Modelling, pp. 232–237. Utrecht (2008)
7. Filipe, P.A., Braumann, C.A., Roquete, C.J.: Modelos de crescimento de animais em ambiente aleatório. In: Ferrão, M.E., Nunes, C., Braumann, C.A. (eds.) Estatística Ciência Interdisciplinar, pp. 401–410. Edições SPE (2007)

8. Filipe, P.A., Braumann C.A., Roquete, C.J.: Crescimento individual em ambiente aleatório: várias trajectórias. In: Hill, M.M., Ferreira, M.A., Dias, J.G., Salgueiro, M.F., Carvalho, H., Vicente, P., Braumann, C.A. (eds.) Estatística da Teoria à Prática, pp. 259–268. Edições SPE (2008)
9. Garcia, O.: A stochastic differential equation model for the height of forest stands. Biometrics **39**, 1059–1072 (1983)
10. Karlin, S., Taylor, H.M.: A Second Course in Stochastic Processes. Academic, New York (1981)
11. Qiming, Lv., Pitchford, J.: Stochastic Von Bertalanffy models, with applications to fish recruitment. J. Theor. Biol. **244**, 640–655 (2007)
12. Thomas, M.U.: Some mean first-passage time approximations for the Ornstein-Uhlenbeck process. J. Appl. Prob. **12**, 600–604 (1975)

Generalized Linear Mixed Effects Model in the Analysis of Longitudinal Discrete Data

Eunice Carrasquinha, M. Helena Gonçalves, and M. Salomé Cabral

Abstract

In many cancer studies and clinical research, repeated observations of response variables are taken over time for each subject in one or more treatment groups. Such research is commonly referred to longitudinal studies and the repeated observations of each vector response are likely to be correlated. The autocorrelation structure for the repeated data plays a significant role in the analysis of such data. The generalized linear mixed effects model (GLMM) is one of the approaches used to analyze discrete longitudinal data, where the use of random effects in the linear predictor accounts for the within-subject association. The goal of this chapter is to introduce this model in the analysis of longitudinal discrete data, taking into account the theoretical and computational difficulties as well as the problems related to parameters interpretation. The methodology is illustrated by analyzing data sets containing longitudinal measures of number of tumors in an experiment of carcinogenesis to study the influence of lipids in the development of breast cancer. The library `lme4` [Bates, D., Maechler, M., Bolker, B.: lme4: Linear mixed-effects models using S4 classes. R package version 0.999375-39. http://CRAN.R-project.org/package=lme4 (2011)] in `R` software is used.

E. Carrasquinha (✉)
Departamento de Estatística e Investigação Operacional da FCUL, Portugal
e-mail: nicecarrasquinha@hotmail.com

M.H. Gonçalves
CEAUL and Departamento de Matemática, FCT, UAlg, Portugal
e-mail: mhgoncal@ualg.pt

M.S. Cabral
CEAUL, Departamento de Estatística e Investigação Operacional da FCUL, Portugal
e-mail: mscabral@fc.ul.pt

J. Lita da Silva et al. (eds.), *Advances in Regression, Survival Analysis, Extreme Values, Markov Processes and Other Statistical Applications*, Studies in Theoretical and Applied Statistics, DOI 10.1007/978-3-642-34904-1_11,

1 Introduction

The generalized linear mixed effects model [2], usually denoted by GLMM, is an extension of the generalized linear model (GLM) that allows additional components of variability due to unobserved effects. The basic premise underlying GLMM for longitudinal data is the assumption of heterogeneity across individuals in the study population in a subset of the regression coefficients from the GLM. That is, a subset of the regression coefficients is assumed to vary across individuals according to some distribution [7]. The GLMMs are GLMs that permit both fixed and random effects in the linear predictor rather than only fixed effects. The use of random effects in the linear predictor accounts for the within-subject association. The correlation among the repeated observations on an individual can be thought of as arising from sharing a set of underlying random effects $\mathbf{b}_i$. The basic premise of the GLMMs is to describe the change in the mean response of each individual. As a result, the goal of these models is to make inferences about individuals rather than the study population [7]. The nonlinear link function and the presence of random effects lead to problems related to the interpretation of the regression coefficients as well as computational ones. In this chapter the methodology of the GLMMs is introduced, focused either on theoretical and computational aspects, and it is illustrated by analyzing discrete longitudinal data for the count of the number of tumors, from an experiment of carcinogenesis to study the influence of lipids in the development of breast cancer. The library `lme4` in `R` software is used [1].

2 Generalized Linear Mixed Effects Model

2.1 Model Structure

Let y_{it} be the response value at time t $(t = 1, \ldots, T_i)$ for subject i $(i = 1, \ldots, n)$ and Y_{it} its generating random variable. Associated to each observation time and each subject, a set of p covariates is available, denoted by $\mathbf{x}_{it}$ and $\boldsymbol{\beta}$ as the $p \times 1$ vector of unknown parameters. Let $\mathbf{z}_{it}$ be a $(q \times 1)$ vector of covariates (in general a subset of $\mathbf{x}_{it}$) associate to a $q \times 1$ vector of random effects $\mathbf{b}_i$. The equation of the GLMM assumes the form

$$g\{E(Y_{it}|\mathbf{b}_i)\} = \mathbf{x}_{it}^T\boldsymbol{\beta} + \mathbf{z}_{it}^T\mathbf{b}_i, \tag{1}$$

where conditional on $\mathbf{b}_i$ $Y_{i1}, \ldots, Y_{iT_i}$ are mutually independent and their distribution comes from an exponential family with density function $f(y_{it}|\mathbf{b}_i) = \exp[\frac{\omega_{it}}{\phi}(y_{it}\theta_{it} - c(\theta_{it})) + d(y_{it}, \phi)]$. The conditional mean value and variance are given, respectively, by $E(Y_{it}|\mathbf{b}_i) = \mu_{it}^b = c'(\theta_{it})$ and $Var(Y_{it}|\mathbf{b}_i) = \upsilon_{it}^b = c''(\theta_{it})\frac{\phi}{\omega_{it}}$, and satisfy $g(\mu_{it}^b) = \mathbf{x}_{it}^T\boldsymbol{\beta} + \mathbf{z}_{it}^T\mathbf{b}_i$ and $\upsilon_{it}^b = V(\mu_{it}^b)\frac{\phi}{\omega_{it}}$, where g and V are known as link and variance functions, respectively, ϕ is a dispersion or scale parameter, and ω_{it} are known constants. The random effects are mutually independent with $\mathbf{b}_i \sim N(\mathbf{0}, \mathbf{D})$.

2.2 Consequences of Having Random Effects

Since the model specification in Eq. (1) is conditional on the value of $\mathbf{b}_i$ we now derive aspects of the marginal distribution of Y_{it} related with the mean value, $E(Y_{it})$, and correlation between Y_{it} and $Y_{it'}$ (t and t' two distinct time points to subject i) [9]. As for the mean value, $E(Y_{it})$ is given by

$$E(Y_{it}) = E[E(Y_{it}|\mathbf{b}_i)] = E[g^{-1}(\mathbf{x}_{it}^T\boldsymbol{\beta} + \mathbf{z}_{it}^T\mathbf{b}_i)];$$

this, in general, cannot be simplified due to the nonlinear function $g^{-1}(\cdot)$. A direct relation between the marginal and GLMM parameters only exists in some special cases [10]. Assuming conditional independence of the elements of $\mathbf{Y}_i = (Y_{i1}, \ldots, Y_{iT_i})^T$, we have ([9], Sect. 8.3)

$$\begin{aligned} Cov(Y_{it}, Y_{it'}) &= Cov(E[Y_{it}|\mathbf{b}_i], E[Y_{it'}|\mathbf{b}_i]) + E[Cov(Y_{it}, Y_{it'}|\mathbf{b}_i)] \\ &= Cov(\mu_{it}^{b_i}, \mu_{it'}^{b_i}) + E[\mathbf{0}] \\ &= Cov[g^{-1}(\mathbf{x}_{it}^T\boldsymbol{\beta} + \mathbf{z}_{it}^T\mathbf{b}_i), g^{-1}(\mathbf{x}_{it'}^T\boldsymbol{\beta} + \mathbf{z}_{it'}^T\mathbf{b}_i)]. \end{aligned}$$

Another aspect to take into account is the interpretation of the regression parameters, $\boldsymbol{\beta}$, in the GLMM. In this case we must have in mind that the insertion of the random terms alters the meaning of the $\boldsymbol{\beta}$'s with respect to their meaning in a model with fixed effects only, as the marginal model. The covariates and random effects determine the person-specific or conditional mean and the regression parameters $\boldsymbol{\beta}$ can therefore be interpreted as subject-specific or conditional effects of covariates $\mathbf{x}_{it}$ given the random effect $\mathbf{b}_i$. Conditional effects express comparisons holding the subject-specific random effects (and covariates) constant. For this reason, the components of the fixed effects, $\boldsymbol{\beta}$, are often referred to as *subject-specific* regression coefficients. A way of interpreting estimated standard deviations of the random effects is to produce percentiles of the effects based on the normality assumption and plotting the μ_{it}^b for given covariates values [8].

2.3 Likelihood Inference and Approaches to Estimation

For the n individuals the likelihood is given by

$$L^R(\boldsymbol{\beta}, \phi, \mathbf{D}) = \prod_{i=1}^{n} \int \prod_{t=1}^{T_i} f_{y_{it}|\mathbf{b}_i}(y_{it}|\mathbf{b}_i, \boldsymbol{\beta}, \phi) f_{\mathbf{b}_i}(\mathbf{b}_i|\mathbf{D}) d\mathbf{b}_i, \tag{2}$$

where $L^R(\boldsymbol{\beta}, \phi, \mathbf{D})$ indicates the likelihood for the random-effects model. As $\mathbf{b}_i \sim N(\mathbf{0}, \mathbf{D})$, the log-likelihood for the whole sample is given by

$$l^R(\boldsymbol{\beta}, \phi, \mathbf{D}) = \frac{1}{(2\pi)^{q/2}} \sum_{i=1}^{n} log \int L_i^F(\boldsymbol{\beta}, \phi|\mathbf{D})|\mathbf{D}|^{-1/2} \times \exp(-\frac{1}{2}\mathbf{b}_i^T\mathbf{D}^{-1}\mathbf{b}_i) d\mathbf{b}_i, \tag{3}$$

where $L_i^F(\boldsymbol{\beta}, \phi, \mathbf{D})$ indicates the likelihood for the fixed effects model. The key problem in the maximization of Eq. (3) is the presence of n integrals over the q-dimensional random effects $\mathbf{b}_i$, which, in general, does not have an analytic solution and so likelihood inference requires numerical evaluation.

The computational approach has also several limitations due to the complexity of the integrals. The numerical approximations can be subdivided in three categories. We remark those that are based on the approximation of data and those that are based on the approximation of the integral itself [10].

Those that are based on the approximation of data share the same idea, the decomposition of the data into the mean, and an appropriate error term, with a Taylor series expansion of the mean that is a nonlinear function of the linear predictor. Two of these methods are the method of penalized quasi-likelihood (PQL) and the method of marginal quasi-likelihood (MQL). The essential difference between PQL and MQL is that the later does not incorporate the random effects $\mathbf{b}_i$ in the linear predictor. The methods based on the approximation of the integral itself, i.e., numerical integration, proof to be the more appropriate. In the context of random-effects models, so-called adaptive quadrature rules can be used. In these methods the numerical integration is centered around empirical Bayes (EB) estimates of random effects, and the number of quadrature points (Q) is then selected in terms of desired accuracy. In fitting GLMMs the adaptive Gaussian quadrature is very powerful. In general, the higher the order Q, the better the approximation will be of the n integrals in the likelihood. Convergence problems are present when q is high dimensional. We refer to [10] for more details. As the maximum likelihood (ML) method was used to fit GLMM, inferences for the parameters are obtained from classical maximum likelihood theory. The Wald test is used to test the value of a fixed parameter or of a linear combination of fixed parameters. To compare nested models with the same random effects the likelihood ratio test is used. When the interest is testing for the presence of random effects themselves the null hypothesis is on the boundary of the parameter space and standard asymptotic results on the null distribution on the likelihood ratio test do not hold. Taking into account [11], the null distribution is a mixture of chi-squared distributions.

3 An Illustrative Example

3.1 Data and Models

To illustrate the method, we use data from a study of the influence of lipids on the development of cancer in which fifty-seven 22-day-old virgin female Sprague-Dawley rats were housed four per cage and maintained in an environmentally controlled room at 24 ± 1 C, 50 % humidity in a 12 h light/12 h dark cycle. Upon arrival, the rats were fed *ad libitum* one of two different semisynthetic diets, either low-fat ($N3$), $n = 38$, or high-fat ($HL20$), $n = 19$. At 53 days of age, animals were

each given a single dose of 5 mg of carcinogen (7,12-dimethylbenz (α) anthracene; DMB, Sigma) per rat, administered in corn oil by means of a gastric gavage. One day post-administration of the carcinogen, 19 animals from the low-fat-diet group were fed this same diet for the whole study (Diet 1/group 1), 19 animals from the high-fat-diet group were fed this diet (Diet 2/group 2), and the remaining nineteen animals, initially fed on a low-fat-diet, were permanently transferred to the high-fat-diet (Diet 3/group 3) [4, 5]. The rats were examined and palpated for mammary tumors once per week during 25 weeks. When a tumor was first detected, the date and tumors location were recorded. At the end of the study, 201 days after carcinogen administration, the rats were decapitated. At necropsy, tumors were rapidly removed, measured, rinsed in normal saline, and divided for histopathology. Only confirmed mammary adenocarcinomas were reported in the results. The data analyzed in this chapter are the number of tumors. Each group is designed by the respective diet: Diet 1 (low-fat/low-fat-diet); Diet 2 (high-fat/high-fat-diet); Diet 3 (low-fat/high-fat-diet). Following the methodology presented in the chapter, four models were considered to analyze the data:
(i) *Model I*

$$\begin{aligned} log[E(Y_{it}|\mathbf{b}_i)] &= (\beta_0 + b_{0i}) + \beta_1(Time - 1) + \beta_2(Time - 1)^2 \\ &+ \beta_3 Diet1 + \beta_4 Diet3, \end{aligned}$$

(ii) *Model II*

$$\begin{aligned} log[E(Y_{it}|\mathbf{b}_i)] &= (\beta_0 + b_{0i}) + \beta_1(Time - 1) + \beta_2(Time - 1)^2 \\ &+ \beta_3 Diet1 + \beta_4 Diet3 + \beta_5((Time - 1) \times Diet1) \\ &+ \beta_6((Time - 1) \times Diet3) + \beta_7((Time - 1)^2 \times Diet1) \\ &+ \beta_8((Time - 1)^2 \times Diet3), \end{aligned}$$

(iii) *Model III*

$$log[E(Y_{it}|\mathbf{b}_i)] = (\beta_0 + b_{0i}) + \beta_1(Time - 1) + \beta_2 Diet1 + \beta_3 Diet3,$$

(iv) *Model IV*

$$\begin{aligned} log[E(Y_{it}|\mathbf{b}_i)] &= (\beta_0 + b_{0i}) + \beta_1(Time - 1) + \beta_2 Diet1 + \beta_3 Diet3 \\ &+ \beta_4((Time - 1) \times Diet1) + \beta_5((Time - 1) \times Diet3), \end{aligned}$$

where Y_{it} is a random variable, which has a Poisson distribution, that gives the number of tumors of subject i at time t; *Diet*1 is a binary variable taking the value 1 if the ith rat receives Diet 1; *Diet*3 is a binary variable for Diet 3; and *Diet*2 is the reference diet in the models. The variable time was centered at 1 to allow that the comparison among diets was made in the intercept, 1 week after administration

Table 1 Log-likelihood, reduction in deviance and p-value between the models considered

Model	LogL	Δ D	p-value
I	−348.20		
II	−341.78	12.84	0.01206
III	−389.67		
IV	−388.8	1.7297	0.4211
IV	−388.8		
II	−341.78	94.048	$\simeq 0$
III	−389.67		
II	−341.78	95.778	$\simeq 0$

Table 2 Parameter estimates, standard errors, Wald test, and p-value for Model II

Parameter	Estimate	SE	W-test	p-value
β_0	−3.501465	0.558922	−6.255	3.74e−10
β_1	0.290707	0.040942	7.101	1.24e−12
β_2	−0.005288	0.001300	−4.068	4.75e−05
β_3	−4.630175	1.176438	−3.936	8.29e−05
β_4	−2.106021	0.843074	−2.498	0.01249
β_5	0.341514	0.116943	2.920	0.00350
β_6	0.140857	0.065469	2.152	0.03144
β_7	−0.009707	0.003464	−2.802	0.00507
β_8	−0.004093	0.002030	−2.016	0.04381
d_{11}	3.9844			

of the carcinogen. In all models the vector $\mathbf{b}_i$ consists only of a random intercept, $b_{0i} \sim N(0, d_{11})$.

3.2 Results and Discussion

The analysis was performed using the `glmer` function of the `lme4` library [1]. A brief study was carried out to know the order Q to evaluate the adaptive Gauss-Hermite approximation to the log-likelihood and $Q = 25$ was chosen [3]. Based on the results of Table 1, Model II was selected and the results of this fit are given in Table 2.

The residual analysis of Model II (Fig. 1) consisted of the identification of outliers and the half-normal plot [6] was used.

The numbers 576 and 1,310, that appear in Fig. 1, are observations corresponding to rat 53 (Diet 3). Model II was fitted again to the data without rat 53 and no alteration was reported in the parameters estimates, standard errors, Wald test's values, or p-values.

The ML estimates of the regression parameters given in Table 2 provide evidence, at the level of significance 5 %, that the subject-specific number of tumors increase

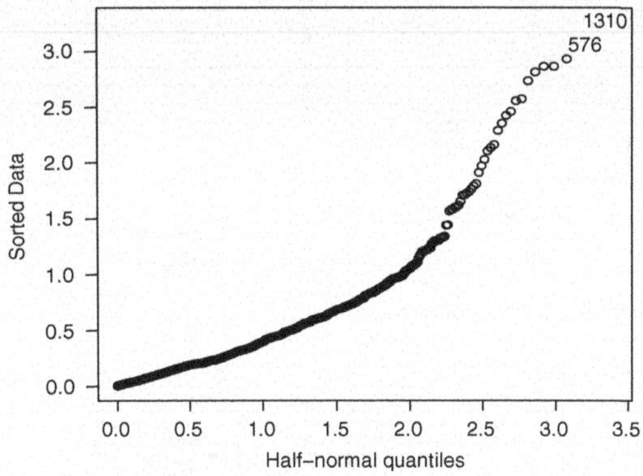

Fig. 1 Half-normal plot of Pearson residuals to Model II

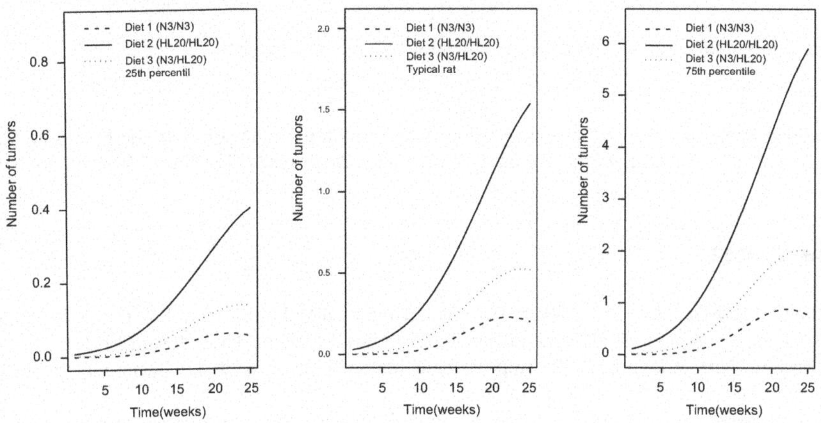

Fig. 2 Estimated conditional expected trajectories for a rat at 25th percentile, a "typical" rat, and at 75th percentile

over the 25 weeks of the experiment and depend on the diet. Figure 2 displays the plots of the estimate conditional expected trajectories for three hypothetical rats, one from each diet, at different percentiles. The 25th and 75th percentiles of the random effect ($b_{0i} \sim N(0, d_{11})$) correspond to a predicted random effect of $b_{0i(0.25)} = -0.67449\sqrt{3.9844} = -1.34635$ and $b_{0i(0.75)} = 0.67449\sqrt{3.9844} = 1.34635$, respectively. A "typical" rat is a rat with unobserved random effect $b_{0i} = 0$ (the mean and median of the distribution of b_{0i}). In all cases the rats in Diet 1 reveal lower time evolution of the expected number of tumors. Note that, in all the three diets, the rats at the 25th percentile do not developed tumors until the end of the experiment. To illustrate how the model fitted the data, Fig. 3 gives the individual mean profiles for two rats in each one of the diets as well as the observed data. The rats were chosen to provide different profiles in the same diet.

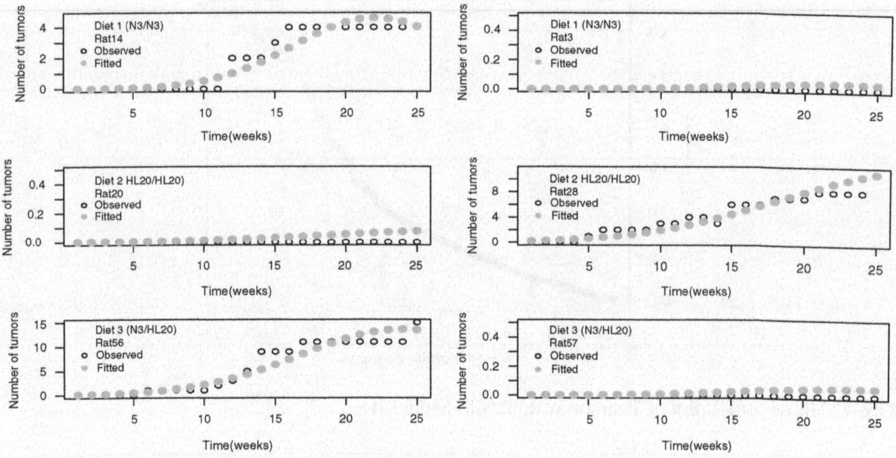

Fig. 3 Individual mean and observed profiles for two rats in each diet

Acknowledgements Research partially sponsored by national funds through the Fundação Nacional para a Ciência e Tecnologia, Portugal—FCT under the project (PEst-OE/MAT/UI0006/2011).

References

1. Bates, D., Maechler, M., Bolker, B.: lme4: Linear mixed-effects models using S4 classes. R package version 0.999375-39. http://CRAN.R-project.org/package=lme4 (2011)
2. Breslow, N.E., Clayton, D.G.: Approximate inference in generalized linear mixed Models. J. Am. Stat. Assoc. **88**, 9–25 (1993)
3. Carrasquinha, E.I.: Análise de dados longitudinais discretos: uma aplicação ao estudo da influência de lípidos no adenocarcinoma mamário. Mestrado em Bioestatística. FCUL, Lisboa (2009)
4. Escrich, E., Solanas, M., Segura, R.: Experimental diets for the study of lipid influence on the induced mammary carcinoma in rats: I-diet definition. Int. J. in vivo Res. **8**, 1099–1106 (1994)
5. Escrich, E., Solanas, M., Ruiz de Villa, M.C., Ribalta, T., Muntané, J., Segura, R.: Experimental diets for the study of lipid influence on the induced mammary carcinoma in rats: suitability of the diets definition. Int. J. in vivo Res. **8**, 1107–1112 (1994)
6. Faraway, J.J.: Extending the linear Model with R. Generalized Linear, Mixed Effects and Nonparametric Regression Models. Texts in Statistical Science. Chapman and Hall/CRC, Boca Raton (2006)
7. Fitzmaurice, G.M., Laird, N.M., Ware, J.H.: Applied Longitudinal Analysis. Wiley, New York (2004)
8. Fitzmaurice, G.M., Davidian, M., Verbeke, G., Molenberghs, G.: Longitudinal Data Analysis. Chapman & Hall, Boca Raton (2008)
9. Mcculloch, C.E., Searle, S.R.: Generalized Linear and Mixed Models. Wiley, New York (2001)
10. Molenberghs, G., Verbeke, G.: Models for Discrete Longitudinal Data. Springer, New York (2005)
11. Self, S.G., Liang, K.Y.: Asymptotic properties of maximum likelihood estimators and likelihood ratio tests under nonstandard conditions. J. Am. Stat. Assoc. **82**, 605–610 (1987)

Risk Assessment on *Campylobacter* in Broiler Meat at Slaughter Level in Portugal

Marta Castel-Branco, Marília Antunes, Patrícia Inácio, and Miguel Cardo

Abstract

A logistic regression model is used to assess the association between *Campylobacter* contamination in broiler meat at the slaughter level and general operating and hygienic aspects of the slaughterhouses and characteristics of the batches. *Campylobacter* was found in 62.6 % of the carcasses. The presence of *Campylobacter* in the caeca (RR= 1.47), the presence of hepatitis or perihepatitis (RR= 1.71), and conspurcation of carcasses with faeces during slaughter (RR= 1.59) increase the risk of *Campylobacter* contamination. The relative risk of contamination associated to the slaughterhouse location and capacity vary from 0.95 to 3.14, depending on the combination of the two characteristics.

1 Introduction

Campylobacteriosis is a zoonosis caused by thermophilic *Campylobacter* spp. and is referred as an important public health problem in most areas of the world [5]. According to the European Community summary report of zoonoses and zoonotic agents published in 2010, *Campylobacter* is one of the most commonly reported gastrointestinal bacterial pathogen in humans in the EU, with 190,566 reported

M. Castel-Branco (✉) · M. Antunes
Faculty of Sciences, University of Lisbon, Lisbon, Portugal

CEAUL, Lisbon, Portugal
e-mail: mcastelbranco@gmail.com; marilia.antunes@fc.ul.pt

P. Inácio · M. Cardo
DGAV - Direção Geral de Alimentação e Veterinária
(General Directorate for food and veterinary Affairs)
e-mail: miguelcardo@dgav.pt; pinacio@dgav.pt

J. Lita da Silva et al. (eds.), *Advances in Regression, Survival Analysis, Extreme Values, Markov Processes and Other Statistical Applications*, Studies in Theoretical and Applied Statistics, DOI 10.1007/978-3-642-34904-1_12,

confirmed human campylobacteriosis cases in 2008 (40.7 cases per 100,000 pop.) [7]. Humans can be infected by direct contact with contaminated animals or animal carcasses or indirectly through the ingestion of contaminated food or drinking water [5]. Most case-control studies have identified poultry meat as an important risk factor for human campylobacteriosis [1, 4, 7].

Broilers are commonly colonized by *Campylobacter* spp., being symptomless intestinal carriers of the organism. Studies in Europe indicate flock prevalence ranging from 18 % to over 90 %, with northern countries showing a lower proportion of positive flocks [1, 6]. Poultry meat becomes contaminated with *Campylobacter* during slaughter by faecal material from *Campylobacter* colonized broilers. These birds usually have high numbers of *Campylobacter* in the intestine, and also harbour *Campylobacter* spp. on the outer surface due to spread of faecal material during rearing and transport. The hygienic standards during processing influence the numbers of *Campylobacter* found on the final chicken meat product and thereby the human exposure to *Campylobacter* spp. through poultry meat [5].

2 Materials and Methods

2.1 Sampling Design

For the purpose of estimating the prevalence of *Campylobacter* in poultry carcasses and to investigate the risk factors associated with the contamination at slaughter level, it was decided to analyse carcasses for *Campylobacter* presence, collected at the slaughterhouses. The number of carcasses to be sampled was calculated considering a prevalence of *Campylobacter* equal to 50 %, a confidence level of 95 % and a precision of 0.05. Assuming that the population is infinite, the number of samples to be collected was 384, increased by 10 % to prevent problems arising from nonresponse. This resulted in a sample size of 422.

Analysed data come from a monitoring programme on *Salmonella* and *Campylobacter* in broiler meat carried out, in 2008, by the European Commission, advised by the European Food Safety Authority (EFSA). The sampling plan of the programme is based on a multistage design. The first stage represents the slaughterhouses and, for practical reasons, the largest slaughterhouses were selected for sampling until their combined slaughter capacity covers at least 80 % of the annual kill in the country. EFSA justifies the "80 % approach" with the argument that it "provides a reasonable indication of the exposure to these hazards on a population risk basis" [6]. Accordingly, the 35 existing Portuguese slaughterhouses were ranked by decreasing slaughter capacity and the 15 top ones, responsible for 81 % of the previous year's slaughter, were selected.

The second stage corresponds to the number of carcasses to be sampled from each slaughterhouse. The allocation of the number of samples between the selected slaughterhouses was made proportionally to the number of broiler chickens processed annually, with the programme requiring sampling to be distributed evenly over a 12 month period to allow for observation of any variation in risk during the year. In Portugal, due to implementation difficulties, the Portuguese Veterinary Authority decided that sampling would take place only from June to December (excluding August), with one-fifth of the total sample being collected per month [3]. For each selected slaughterhouse, the day (or days) of the month, the batch, carcass and caeca to be sampled were selected randomly [6].

2.2 The Data

The 422 collected carcasses and caeca (both from the same batch but different birds) correspond to 422 different batches. Data refer to characteristics of the slaughterhouses, hygienic and health conditions of the batches and *Campylobacter* presence in caeca (reflecting the prevalence of *Campylobacter* in the poultry farms) and in carcasses (the outcome variable). Test positiveness of the carcass reflects the contamination at the slaughter level, since the test positiveness of the caeca does not necessarily imply that the meat is contaminated. A batch is considered positive if the sampled carcass tested positive. The 12 risk factors evaluated in this study are described in Table 1. They concern the characteristics of the slaughterhouse and of the batch and reasons for *post-mortem* condemnation in the batch.

2.3 Statistical Analysis

We applied a logistic regression model for the probability of *Campylobacter* positiveness of a carcass (and hence of the corresponding batch). A three-stage procedure was used to assess the relationship between the explanatory variables and the *Campylobacter* status of the carcass, according to the method described by Hosmer and Lemeshow [8]. In the first stage, a univariable analysis was performed to relate *Campylobacter* contamination of the carcass to each explanatory variable. Only factors for which Wald test p-value was smaller than 0.25 were considered as candidates to a multivariable model. The second stage involved a logistic multiple regression model with the contribution of each factor being tested using a likelihood-ratio χ^2 test through a stepwise procedure. In the third stage, the model containing the selected variables from the stepwise procedure was compared with the models containing the same principal effects plus each of the possible interactions, again using likelihood-ratio test.

Pearson and deviance residuals were used to detect possible outliers. Goodness of fit of the final model was assessed using Pearson χ^2 and deviance statistics. The influence of the batches with a particular covariate pattern in Pearson's χ^2 and in deviance statistics was evaluated using ΔX_j^2 and ΔD_j, respectively [8]. Covariate patterns showing ΔX_j^2 or ΔD_j above $\chi^2_{0.95}(1) = 3.84$ are considered to be influential, since the difference in the goodness-of-fit statistics caused by their exclusion from the model building is statistically significant. Cook's distance was used to evaluate the influence of each case in the model. Observations presenting values above 1.0 were considered to have a significant effect on the estimated

Table 1 Summary of potential risk factors available for analysis

Categorical variable	Description	Value	Number of batches	% of batches	% *Camp.* +[a]
lvt	Slaughterhouse in LVT[b] region	Yes	267	63.3	94.2
		No	155	36.7	44.2
caphcat	Slaughter capacity (/hour)	≥ 7,000	174	41.2	71.3
		<7,000	248	58.8	56.5
eviscmec	Mechanical evisceration	Yes	350	82.9	63.4
		No	72	17.1	58.3
chillier	Type of carcass chillier	Air	290	68.7	60.7
		Air+spray	132	31.3	66.7
hom	Homogeneity of the batch	Yes	337	79.9	61.1
		No	85	20.1	68.2
crop	Full crop	Yes	17	4.0	76.5
		No	397	84.1	61.4
		Missing	8	1.9	–
dirt	Dirty feathers	Yes	55	13.0	89.1
		No	360	85.3	57.8
		Missing	7	1.7	–
consp	Conspurcation with faeces during slaughter	Yes	25	5.9	96.0
		No	397	94.1	60.5
campycaeca	Presence of *Campylobacter* in the caeca	Yes	351	83.2	67.8
		No	71	16.8	36.6
h.ph	Presence of hepatitis or perihepatitis	Yes	65	15.4	98.5
		No	357	84.6	56.0
fpl	Presence of fibrinopurulent lesions	Yes	33	7.8	90.9
		No	389	92.2	60.2
peric	Presence of pericarditis	Yes	41	9.7	95.1
		No	381	90.3	59.0

[a] *Campylobacter*-positive (contaminated) batches for each level of the categorical variables
[b] LVT: Lisbon and Tagus Valley (Greater Lisbon, Setúbal Peninsula, Middle Tagus, and Lezíria West Coast). All other slaughterhouses in this study are located in the north or centre regions

coefficients [8]. Sensitivity, specificity and discriminating ability of the model were evaluated through a ROC curve. For classification purposes, a cutpoint was chosen such that both sensitivity and specificity were maximized. The predictive ability of the model, considering the referred cutpoint, was assessed using leave-one-out cross-validation [9].

Odds ratio (OR$= (p_1/(1-p_1))/(p_0/(1-p_0))$, p_1 and p_0 being the probability of *Campylobacter* positiveness in the exposed and the unexposed group, respectively), is useful to evaluate the strength of the association between each risk factor and the outcome. OR describes the ratio of the odds of *Campylobacter* positiveness in the exposed group ($p_1/(1-p_1)$) and the odds in the unexposed group ($p_0/(1-p_0)$), being interpreted as a multiplicative factor of the risk of positiveness when exposed. However, it is well known that ORs always overestimate the strengths of relative risk (RR= p_1/p_0), especially when the outcome is not considered rare [2]. According to Beaudeau and Fourichon [2], the estimates of OR and RR ($\hat{OR}$ and $\hat{RR}$) relate through

$$n_{1.}\hat{RR}^2 + \left[n_{0.} - n_{.1}(1 - \hat{OR}) - n_{1.}\hat{OR}\right]\hat{RR} - n_{0.}\hat{OR} = 0,$$

which is a second-degree equation on $\hat{RR}$, where $n_{1.}$ and $n_{0.}$ are the number of exposed and not exposed to the risk factor, respectively, and $n_{.0}$ is the number of *Campylobacter* positive. The above equation allows to estimate the RR from the OR estimates issued from logistic regression.

3 Results

Data were available for 422 batches. The total number of *Campylobacter* spp. positive batches (carcasses) was 264 (62.6 %; $CI_{95\,\%}$: 57.9 %, 67.2 %). Crude descriptive statistics for the explanatory variables can be found in Table 1. The final multivariable model, presented in Table 2, included five significant (p-value<0.05) main-effects (lvt, campycaeca, caphcat, consp and h.ph) and one interaction (lvt×caphcat). Because all the explanatory variables are categorical, data resumed to 17 covariate patterns (Table 3).

Deviance residuals for the individual observations are shown in the first graph in Fig. 1, with the covariate pattern label indicated for the cases presenting values outside the interval $(-2, 2)$. The number of individual observations laying outside $(-2, 2)$ is equal to 4, 3, 1 and 1 for covariate patterns 3, 9, 15 and 16, respectively. Except for the case from covariate pattern 16, all the cases correspond to *Campylobacter* spp. negative batches that belong to covariate patterns with large number of positive cases and high estimated probability of positiveness. Values from Pearson χ^2 ($\chi^2 = 15.27$; df = 10; p-value = 0.122) and deviance ($D = 12.01$; df = 10, p-value= 0.284) statistics calculated for the covariate patterns indicate that the fitted model is adequate.

The observation from covariate pattern 16 (case 127) is a particular case which deserves some attention. Although it is considered influential and it has a large residual, we decided not to exclude this case from the analysis. It corresponds to the only batch in the study (batch 127) for which hepatitis or perihepatitis was found (h.ph=Yes) and *Campylobacter* tested negative (campycar=No) and hence, to exclude this observation raises a quasi-separability problem with implications in the convergence of the estimation procedure, making the inclusion of h.ph in the model not viable. Moreover, because h.ph is considered biologically relevant and because it is a plausible occurrence, we decided to keep this observation in the

Table 2 Risk factors for *Campylobacter*-positive batches from a multivariable generalized linear model (logistic regression)[a]

Variable	Description	$\hat{\beta}$	$\hat{SE}(\hat{\beta})$	*p*-value	95 % CI
lvt	Slaughterhouse in LVT region	3.65	0.4966	0.000	2.68,4.62
consp	Conspurcation with faeces	2.72	1.0712	0.011	0.62,4.82
h.ph	Hepatitis or perihepatitis	3.05	1.0418	0.003	1.01,5.09
campycaeca	*Campylobacter* in the caeca	0.87	0.3354	0.010	0.21,1.53
caphcat	Slaughter capacity $\geq 7,000$/hour	0.76	0.2918	0.009	0.19,1.33
lvt×caphcat	LVT× slaughter capacity	−1.64	0.7699	0.033	−3.15,−0.13

[a] Intercept = −1.60, model deviance = 355.73, null deviance = 558.10; d.f. = 7 (p-value< 0.001), AIC = 369.73

Table 3 Covariate patterns[a]

id	campy-caeca	consp	h.ph	lvt	caphcat ≥ 7000	caphacat ×lvt	Nr. *Campy-lobacter* +	Number of batches	Estimated probability[b]
1	0	0	0	0	1	0	7	23	0.3031
2	0	0	0	0	0	0	7	32	0.1686
3	1	0	0	1	0	0	82	86	0.9488
4	1	0	0	0	0	0	36	113	0.3261
5	1	0	0	0	1	0	29	58	0.5093
6	1	0	1	1	1	1	24	24	0.9939
7	1	1	0	0	1	0	12	12	0.9404
8	1	1	1	0	1	0	11	11	0.9970
9	1	0	0	1	1	1	18	21	0.8853
10	1	0	1	0	1	0	15	15	0.9562
11	1	0	1	1	0	0	11	11	0.9974
12	0	1	0	0	0	0	0	1	0.7551
13	0	0	1	1	1	1	2	2	0.9855
14	0	0	0	1	1	1	5	6	0.7639
15	0	0	0	1	0	0	4	5	0.8860
16	0	0	1	0	1	0	0	1	0.9014
17	0	1	1	0	1	0	1	1	0.9929

[a] 0:No; 1:Yes

[b] Estimated probability of *Campylobacter* spp. positiveness for each covariate pattern

study. The other influential observation (batch 33, covariate pattern 12) did not show a particularly large residual.

The ROC curve, showing the model sensitivity (70.1 %) and specificity (93.7 %) for a cutpoint of 0.7635, as well as the overall ability of the model to discriminate between negative and positive batches (AUC = 0.859), is displayed in Fig. 2. Considering the referred cutpoint, the accuracy of the classification rule (positive if fitted value>0.7635) is 78.93 %. This value was confirmed by the leave-one-out cross-validation procedure, which led to an estimate of 78.91 % for the accuracy.

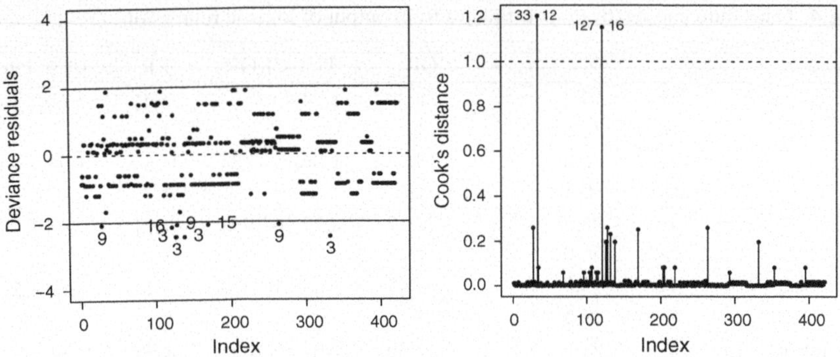

Fig. 1 Deviance residuals for the individual observations, with the covariate pattern label for the cases with values outside (−2, 2) interval; Cook's distance, with observation label on the *left side* and covariate pattern label on the *right side* of the dots above 1.0 threshold

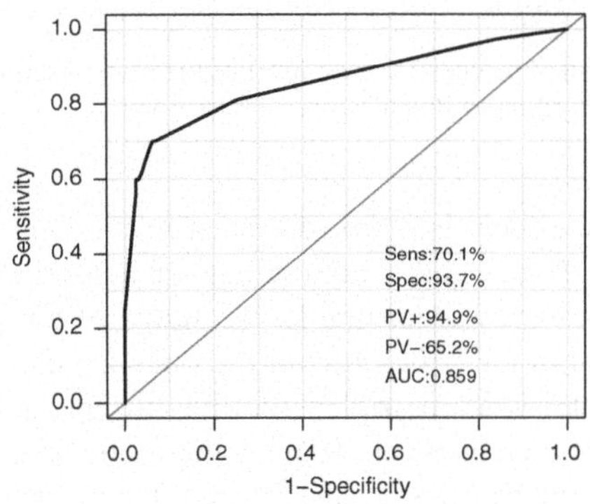

Fig. 2 ROC curve

4 Discussion

In this study 62.6 % of the analysed broiler carcasses were contaminated with *Campylobacter* spp. Because carrier flocks introduce large numbers of *Campylobacter* into the processing plant, equipment and surfaces, process water and the hands of operatives readily become contaminated. During defeathering and evisceration, an increase in contamination usually occurs as a consequence of expulsion of faecal matter or viscera rupture. Cross-contamination between birds within a flock and between *Campylobacter*-positive and negative flocks is inevitable [4, 5]. These aspects are reflected in the values of the relative risk associated to the variables campyceco (RR= 1.47) and consp (RR= 1.59). The presence of *Campylobacter* in the caeca and conspurcation of the carcasses with faeces increase the probability of positiveness of the batch in 47 % and 59 %, respectively. The existence of hepatitis or perihepatitis increase the probability of positiveness in

Table 4 Odds ratio and relative risk estimates from output of logistic regression

Variable		% *Camp.* +[a]	$\hat{OR}$	95 % CI(OR)	$\hat{RR}$[b]	95 % CI(RR)
campycaeca	Yes	67.8	2.39	1.24,4.61	1.47	1.09,2.13
	No	36.6	1	–	1	–
consp	Yes	96.0	15.18	1.86,124.13	1.59	1.22,1.65
	No	60.5	1	–	1	–
h.ph	Yes	98.5	21.12	2.73,161.99	1.71	1.35,1.78
	No	56.0	1	–	1	–
lvt|(caphcat = 0)	Yes	73.5	38.47	14.48,101.39	3.14	2.58,3.51
	No	29.5	1	–	1	–
lvt|(caphcat = 1)	Yes	92.5	7.46	0.62,89.33	1.49	0.87,1.68
	No	80.2	1	–	1	–
caphcat|(lvt = 0)	Yes	80.2	2.14	1.21,3.80	1.52	1.11,2.07
	No	29.5	1	–	1	–
caphcat|(lvt = 1)	Yes	92.5	0.41	0.05,3.32	0.95	0.86,1.06
	No	73.5	1	–	1	–

[a] *Campylobacter* contaminated batches for each level of the categorical variables
[b] Relative risk obtained according to Beaudeau and Fourichon [2]

71 % (RR= 1.71), since focal hepatitis are seen in infected broilers [4]. Because variables concerning location and slaughter capacity interact, the relative risk for each of these variables varies according to the status of the other. Comparing the slaughterhouses located in LVT region with the ones from centre or north regions, the probability of contamination triples for the lower capacity slaughterhouses (RR= 3.14), whereas it increases for 50 % for the high capacity ones (RR= 1.49). When comparing slaughterhouses by their slaughter capacity, if located in LVT region, there is no significant difference on the probability of contamination (RR= 0.95), whereas outside LVT region the farthest ones have the probability of positiveness increased by 50 % (RR= 1.52). See Table 4. The risk management options available for the processing level, like good hygienic practices and techniques to reduce faecal spread (feed withdrawal, cloacal plugging, higher water pressure and longer washing after evisceration), can reduce the concentration of organisms and thereby reduce the exposure to consumers [5].

Acknowledgements The authors acknowledge the DGV for granting permission to use the data. Research supported by CEAUL and FCT projects FCT/PTDC/MAT/ 64353/2006 and FCT/OE.

References

1. Barrios, P.R. et al.: Risk factors for *Campylobacter* spp. colonization in broiler flocks in Iceland. Prev. Vet. Med. **74**, 264–278 (2006)
2. Beaudeau, F., Fourichon, C.: Estimating relative risk of disease from outputs of logistic regression when disease is not rare. Prev. Vet. Med. **74**, 4, 243–256 (1998)

3. DGV, Portugal Final Report of Baseline Study for Campylobacter in Broilers and Campylobacter/Salmonella in broilers carcasses (Commission Decision 2007/516/EC) (2009)
4. FAO/WHO: Risk assessment of *Campylobacter* spp. in broiler chickens. Technical Report. Microbiological Risk Assessment Series. Geneva. **12**, p. 132 (2009)
5. EFSA: Opinion of the Scientific Panel on Biological Hazards on "*Campylobacter* in animals and foodstuffs". EFSA J. **173**, 1–10 (2005)
6. EFSA: Report of Task Force on Zoonoses Data Collection on proposed technical specifications for a coordinated monitoring programme for *Salmonella* and *Campylobacter* in broiler meet in the EU. EFSA J. **92**, 1–33 (2006)
7. EFSA: The community Summary Report on Trends and Sources of Zoonosis, Zoonotic Agents and Food-borne outbreaks in the EU in 2008. EFSA J. **1496**, (2010)
8. Hosmer, D.W., Lemeshow, S.: Applied Logistic regression. Wiley, New York (2000)
9. Picard, R.R., Cook, R.D.: Cross-Validation of Regression Models. J. Am. Stat. Assoc. **79**(387), 575–583 (1984)

Predicting and Treating Missing Data with Boot.EXPOS

Clara Cordeiro and M. Manuela Neves

Abstract

The Boot.EXPOS procedure is an algorithm that combines the use of exponential smoothing methods with the bootstrap methodology for obtaining forecasts. It starts with the selection of an exponential smoothing method and evolves to a bootstrapping design based on the residuals. The time series is reconstructed and forecasts are obtained. That procedure, now extended to "predict" missing values, is named NABoot.EXPOS.

1 Introduction

The most interesting and ambitious task in time series analysis is to forecast future values. Models are commonly fitted in order to predict future values of a time series. Exponential smoothing methods (EXPOS) are the most widely used forecasting methods because of its versatility and with very few prerequisites. Through EXPOS application, time series pattern (trend and/or seasonality) and error term are obtained. Thus, the preliminary phase of this study is the selection of the "best" EXPOS model, Table 1, using the AIC criterion.

The procedure proposed by the authors, Boot.EXPOS, is inspired on the sieve bootstrap; see [1]: an AR(p) is used to filter the random series and the centered

C. Cordeiro (✉)
University of Algarve, Campus de Gambelas, FCT, 8005-139 Faro, Portugal

CEAUL, Lisboa, Portugal
e-mail: ccordei@ualg.pt

M.M. Neves
ISA, Technical University of Lisbon, Tapada da Ajuda, 1349-017 Lisboa, Portugal

CEAUL, Lisboa, Portugal
e-mail: manela@isa.utl.pt

J. Lita da Silva et al. (eds.), *Advances in Regression, Survival Analysis, Extreme Values, Markov Processes and Other Statistical Applications*, Studies in Theoretical and Applied Statistics, DOI 10.1007/978-3-642-34904-1_13,

Table 1 Classification of the exponential smoothing methods

	Seasonal component		
Trend component	N (none)	A (additive)	M (multiplicative)
N (none)	N,N	N,A	N,M
A (additive)	A,N	A,A	A,M
Ad (additive damped)	Ad,N	Ad,A	Ad,M
M (multiplicative)	M,N	M,A	M,M
Md (multiplicative damped)	Md,N	Md,A	Md,M

residuals are resampled. Then the procedure works backwards: using the previous bootstrap residuals series, an AR(p) is obtained recursively. Adding the patterns found in the preliminary phase, a time series sample path is obtained. This series is forecasted using the EXPOS model and the smoothing parameters estimates obtained in the preliminary phase. The process is repeated B times and h step-ahead forecasts are obtained. At the end with a matrix $B \times h$, the mean is taken for each column. To perform an empirical evaluation of the Boot.EXPOS procedure, a case study with six time series, with different patterns, is used.

Given the good performance of the Boot.EXPOS procedure, the authors have extended its use to the missing data case. A procedure that detects, estimates, and replaces missing data is planned. NABoot.EXPOS is the designation of this procedure. It is applied to the complete time series previously used in the application of the Boot.EXPOS algorithm. For each time series some blocks and also some isolated observations are randomly removed. Then NABoot.EXPOS and two well-known functions in R environment, [15], for imputing missing data in time series, na.interp() and amelia(), are used and compared.

This chapter is organized as follows: Sect. 2 describes the bootstrap and EXPOS methodologies. Boot.EXPOS and NABoot.EXPOS are described in Sect. 3. Examples on using both procedures are presented in Sect. 4 and some closing comments in Sect. 5.

2 Methodologies

2.1 EXPOS Methods

EXPOS refers to a set of methods that can be used to model and to obtain forecasts. This is a versatile approach that continually updates a forecast emphasizing the most recent experience, that is, recent observations are given more weight than the older observations; see [6]. Many researchers have investigated and developed the EXPOS methods in a total of fifteen methods; see Table 1 [6, 11]. For each method in the framework, additive error and multiplicative error versions are considered.

Then a total of thirty EXPOS models are available and the selection is made by minimizing the AIC criterion. The estimates of the smoothing parameters are obtained by minimizing the mean squared error of the one-step-ahead forecasts errors, over the fitted period. In a previous study (see [5]), only four EXPOS methods were considered in Boot.EXPOS algorithm: single EXPOS, Holt's linear trend, and Holt–Winters seasonal smoothing with either additive or multiplicative seasonality. These methods were applied to the M3 competition data set, with 3,003 time series of different patterns and time intervals. Our procedure stayed within the six best among twenty-four procedures; see [13]. By that time, a large set of EXPOS methods was included in the R environment through the ets() function, [10]. A selection set of thirty EXPOS models were then incorporated in our algorithm.

2.2 Bootstrap

In previous works (see [2–4]) the authors have studied and analyzed the possibility of joining EXPOS methods and the bootstrap methodology. From these studies the idea behind the sieve bootstrap (see [1]) permitted to connect those two procedures. Sieve bootstrap considers first an autoregressive process that is fitted to a stationary time series. A bootstrap model-based approach, which resamples from approximately i.i.d. residuals (see [1, 12]) can be applied.

3 Computational Procedures

Boot.EXPOS is an automatic procedure developed by the authors for modeling and forecasting. For a time series with missing observations the procedure can not de applied. For replacing the missing values in the series the authors propose the extension of Boot.EXPOS. This extension, denoted NABoot.EXPOS, allows to detect and to impute missing observations. A sequential inspection of the time series is performed. Whenever an observation or a sequence of observations is missing the NABoot.EXPOS calls the Boot.EXPOS for "predicting." Below is the description of both procedures.

3.1 Boot.EXPOS: To Forecast

The **initial step** before applying the Boot.EXPOS procedure, is to select the "best" EXPOS method. Time series patterns and the optimized smoothing parameters are obtained and kept for later use, while the residual series $r_1, \cdots, r_n$ is used in the Boot.EXPOS procedure (it is now the input time series).

Due to the sieve bootstrap inspiration, this algorithm starts also by fitting an autoregressive model to a stationary time series. In autoregressive time series models the presence of a unit root means that the time series is nonstationary.

Much of the relevant literature concentrates on the unit roots in the AR polynomial; see [14]. The most common parametric unit root test is the augmented Dickey–Fuller (ADF) test. This test considers as the null hypothesis the nonstationary (random walk) *v.s.* the stationarity (an AR(p)).

In our automatic procedure, similar to what is done in auto.arima() function (that permits the possibility of choosing among ADF, KPSS, and PP) (see [9]), the ADF test is applied to $r_1, \cdots, r_n$ and in case of rejection of nonstationarity the Boot.EXPOS will be applied, as described next:

1. Use AIC criterion to select an AR(p) to the EXPOS residuals $(r_1, \cdots, r_n)$.
2. Obtain the AR residuals and center them.
3. Draw a random sample from the centered residuals.
4. Use AR model recursively for obtaining a bootstrap residuals series.
5. Add the bootstrap residuals series and the EXPOS patterns. The time series is now reconstructed.
6. Obtain h step-ahead forecasts for the time series using the EXPOS model selected in the **initial step** and the smoothing parameter estimates.
7. Repeat step 3 to step 6, B times.
8. Calculate the mean for each column of the $B \times h$ matrix.

3.2 NABoot.EXPOS: To Detect and Replace Missing Data

The basic idea of NABoot.EXPOS is to detect and to impute missing observations, through the application of Boot.EXPOS on the past observations with non-missing values. Let $\{y_1, y_2, \cdots, y_n\}$ be a time series with missing observations. The procedure starts by detecting the first missing observation, for example, y_i. Let $k \geq 0$ the number of consecutive missing observations, $\{y_i, y_{i+1}, \cdots, y_{i+k}\}$. Use Boot.EXPOS to obtain the estimated missing values $\{\hat{y}_i, \hat{y}_{i+1} \cdots, \hat{y}_{i+k}\}$ and replace them in the series. Proceed to the next missing observation and the cycle goes on until there is no missing observation in the time series.

4 Case Studies

The objective of this section is to use six time series to empirically evaluate the performance of Boot.EXPOS and NABoot.EXPOS. Some accuracy measures are used. Let y_t denote the observation at time t and $\hat{y}_t$ the forecast of y_t, $t = 1, \cdots, n$. The forecast error is defined by $e_t = y_t - \hat{y}_t$. The forecasts are computed for a hold-out period. Thus the out-of-sample forecasts $\{\hat{y}_n(1), \cdots, \hat{y}_n(h)\}$, where h is the forecast horizon, are computed based on the data. Note that all time series are split into a sample set $\{y_1, \cdots, y_{n-h}\}$ and a validation set $\{y_{n-h+1}, \cdots, y_n\}$ for the procedures' evaluation. The accuracy measures here considered are defined in Table 2.

4.1 In Forecasting

Figure 1 shows the different behaviour of the time series, available in [8] and R basis, described in Table 3.

Table 2 Accuracy measures

Acronyms	Definition	Formula
RMSE	Root mean squared error	$\sqrt{mean\,(y_t - \hat{y}_t)^2}$
MAE	Mean absolute error	$mean(\lvert y_t - \hat{y}_t \rvert)$
MAPE	Mean absolute percentage error	$mean\left(100\lvert \frac{y_t - \hat{y}_t}{y_t} \rvert\right)$

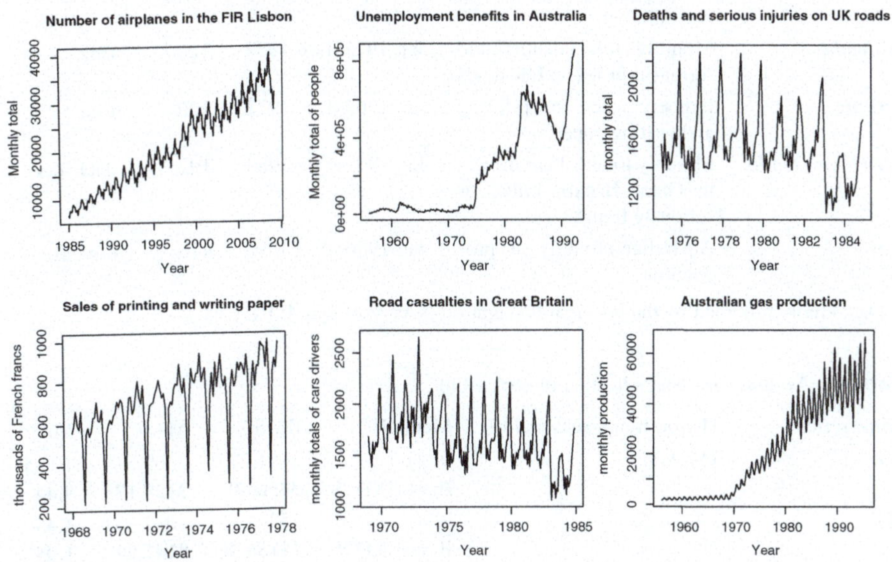

Fig. 1 Time series path

Each time series is forecasted with a time horizon $h = 12$, using Boot.EXPOS and the R function ets() for EXPOS; see [9]. Table 4 gives the model selected using Hyndmann et al. [11] terminology (second column) and the accuracy measures obtained for the forecasts (lower values in bold).

4.2 In Missing Data

For the complete series shown in Fig. 1, some blocks and also some isolated observations are randomly removed. For the shortest time series, *ukdeaths* and *writing*, one individual observation and two blocks of length 6 and 12 were removed. For time series *UKDriverDeaths* and *nav* two blocks of size 6 and 12 and two individual values were removed. For the longest time series, *dole* and *gas*, three blocks of length 6 and 12 and three individual observations were chosen as missing. The true values are kept for using as a validation set in the determination of the out-of-sample measures. In R software two functions can be used for imputing missing data in time series: na.interp() that uses linear interpolation and amelia() [7] that

Table 3 The data set description

Time series	Description	Time period	Length	R package
nav	Monthly total of airplanes in flight information region of Lisbon	Jan 1985–Mar 2009	291	[a]
dole	Monthly total of people on unemployment benefits in Australia	Jan 1965–Jul 1992	439	fma
ukdeaths	Monthly total deaths and serious injuries on UK roads	Jan 1975–Dec 1984	120	fma
writing	Industry sales for printing and writing paper	Jan 1963–Dec 1972	120	fma
UKDriverDeaths	Monthly totals of car drivers in Great Britain killed or seriously injured	Jan 1969–Dec 1984	192	data sets
gas	Australian monthly gas production	Jan 1956–Aug 1995	476	forecast

[a] Data kindly provided by the Portugal Navigation-NAV Portugal, E.P.E

Table 4 Accuracy measures results: in forecasting

Time series	(Error, trend, seasonality)	Method	RMSE	MAE	MAPE
nav	(M,A,M)	ets()	3661.23	3369.51	10.15
		Boot.EXPOS	**3456.60**	**3128.17**	**9.44**
dole	(A,Ad,A)	ets()	15271.15	10927.08	1.45
		Boot.EXPOS	**11156.44**	**8147.94**	**1.05**
ukdeaths	(M,N,M)	ets()	156.84	143.16	10.13
		Boot.EXPOS	**89.84**	**71.28**	**4.89**
writing	(A,A,A)	ets()	58.61	44.96	5.97
		Boot.EXPOS	**57.21**	**43.95**	**5.92**
UKDriverDeaths	(M,N,A)	ets()	205.63	198.49	14.68
		Boot.EXPOS	**87.78**	**70.60**	**5.09**
gas	(M,Md,M)	ets()	2773.72	2097.73	4.22
		Boot.EXPOS	**2348.16**	**1908.15**	**3.84**

uses the bootstrap with the EM algorithm. NABoot.EXPOS procedure is applied and the three procedures' performance is evaluated.

Figure 2 shows the six series with missing values and Table 5 presents the accuracy measures after the application of the above procedures to impute unobserved data (lower values in bold). We see that in series *dole* the linear interpolation presented better results. Perhaps it can be sensible to make a previous analysis of the series behavior in the neighbourhood of the missing observations to choose the method to be applied. This is a point for future research.

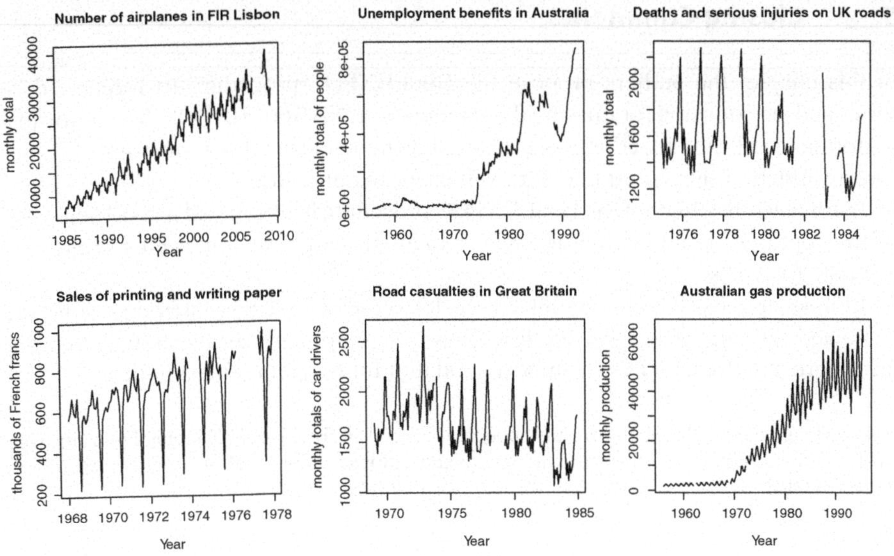

Fig. 2 Block of missing observations

Table 5 Accuracy measures results: in missing data

Time series	Method	RMSE	MAE	MAPE
nav	NABoot.EXPOS	**242.77**	**46.72**	**0.16**
	na.interp	415.10	91.23	0.35
	amelia	897.86	191.39	0.69
dole	NABoot.EXPOS	17433.24	3260.22	0.78
	na.interp	**5642.78**	**907.20**	**0.31**
	amelia	36411.61	7063.74	11.21
ukdeaths	NABoot.EXPOS	**83.49**	**29.78**	**2.10**
	na.interp	131.63	55.50	3.74
	amelia	173.82	64.30	4.11
writing	NABoot.EXPOS	**21.33**	**6.48**	**0.86**
	na.interp	79.64	21.72	4.13
	amelia	81.83	25.70	3.96
UKDriverDeaths	NABoot.EXPOS	**46.36**	**11.36**	**0.62**
	na.interp	89.88	22.04	1.19
	amelia	119.02	33.28	1.84
gas	NABoot.EXPOS	**344.92**	**49.83**	**0.27**
	na.interp	1415.73	198.61	0.93
	amelia	1946.46	357.95	8.69

5 Closing Comments

In this chapter the authors propose the Boot.EXPOS procedure to forecast time series and an extension for missing data imputation, NABoot.EXPOS. The empirical performance of the procedures is evaluated using some accuracy measures. Therefore a validation set is used to obtain values for that measures.

In forecasting situation the Boot.EXPOS procedure has revealed a good behavior, so the "optimal" combination of EXPOS methods and bootstrap seems to provide accurate forecasts.

In missing data, the authors have calculated the accuracy of methods using the estimated missing values and the true ones. The empirical results suggest that our procedure, NABoot.EXPOS, can be a good tool for replacing missing data.

Acknowledgements Research partially supported by DM/FCT/Ualg and National Funds through **FCT**—Fundação para a Ciência e a Tecnologia, project PEst-OE/MAT/UI0006/2011, and PTDC/FEDER.

References

1. Bühlmann, P.: Sieve bootstrap for time series. Bernoulli **3**, 123–148 (1997)
2. Cordeiro, C., Neves, M.: The Bootstrap methodology in time series forecasting. In: Rizzi, A., Vichi, M. (eds.) Proceedings of CompStat2006, pp. 1067–1073. Springer, Heidelberg (2006)
3. Cordeiro, C., Neves, M.: Resampling techniques in time series prediction: a look at accuracy measures. In: Gomes, M.I., Pestana, D., Silva, P. (eds.) ISI-2007 - Book of Abstracts, vol. 353. CEAUL, INE and ISI Editions, Lisbon (2007)
4. Cordeiro, C., Neves, M.: Bootstrap and exponential smoothing working together in forecasting time series. In: Brito, P. (ed) Proceedings in Computational Statistics (COMPSTAT 2008), pp. 891–899. Physica, Heidelberg (2008)
5. Cordeiro, C., Neves, M.: Forecasting time series with Boot.EXPOS procedure. REVSTAT **7**(2), 135–149 (2009)
6. Gardner Jr, E.S.: Exponential smoothing: the state of the art-part II. Int. J. Forecasting **22**, 637–666 (2006)
7. Honaker, J., King, G., Blackwell, M.: **AMELIA II**: A program for missing data. R package version 1.5-3. http://cran.r-project.org/package=Amelia
8. Hyndman, R.: **fma**: Data sets from forecasting: methods and applications. In: Makridakis, Wheelwright, Hydman R package version 1.21. http://cran.r-project.org/package=forecasting
9. Hyndman, R.: **forecast**: Forecasting functions for time series. In: R package version 1.21. http://cran.r-project.org/package=forecasting
10. Hyndman, R., Khandakar, Y.: Automatic time series forecasting: the forecast package. R. J. Stat. Software **27**(3) (2008)
11. Hyndman, R., Koehler, A., Ord, J., Snyder, R.: Forecasting with Exponential Smoothing: The State Space Approach. Springer, New York (2008)
12. Lahiri, S.N.: Resampling Methods for Dependent Data. Springer, New York (2003)
13. Makridakis, S., Hibon, M.: The M3 competition: results, conclusions and implications. Int. J. Forecasting **16**, 451–476 (2000)
14. Patterson, K.: Unit Root Tests in Time Series. Palgrave macmillan, New York (2011)
15. R Development core team: A language and environment for statistical computing. Software available at http://www.r-project.org

Bayesian Genetic Mapping of Binary Trait Loci

César Correia, Nuno Sepúlveda, and Carlos Daniel Paulino

Abstract
Genetic mapping aims to find genomic regions affecting a given phenotype. This task is typically made by means of likelihood-ratio tests carried out on a large data set of genetic markers. As an alternative we present some Bayesian methods to map binary trait loci (BTL). All methods are based on (1) a mixture probability structure relating a single or two adjacent markers to the putative BTL and (2) Bayes factors to detect the set of markers most associated with the phenotype. As an example of application, we perform a genetic mapping analysis on experimental cerebral malaria susceptibility.

1 Introduction

The goal of experimental genetic mapping is to identify the genomic regions (loci) controlling a certain phenotype of interest. Two animal strains with distinct phenotypes are commonly crossed up to the second generation. Using data on a

C. Correia
Instituto Superior Técnico, Lisbon, Portugal
e-mail: cafcorreia@gmail.com

N. Sepúlveda (✉)
London School of Hygiene and Tropical Medicine, London, UK

Center of Statistics and Applications of University of Lisbon, Lisbon, Portugal
e-mail: nuno.sepulveda@lshtm.ac.uk

C.D. Paulino
Department of Mathematics, Instituto Superior Técnico, Portugal

Center of Statistics and Applications of University of Lisbon, Lisbon, Portugal
e-mail: dpaulino@math.ist.utl.pt

J. Lita da Silva et al. (eds.), *Advances in Regression, Survival Analysis, Extreme Values, Markov Processes and Other Statistical Applications*, Studies in Theoretical and Applied Statistics, DOI 10.1007/978-3-642-34904-1_14,

large set of genetic markers, one typically uses likelihood-ratio tests to find the loci most associated with the phenotype [6, 7]. Statistical significance of a given genomic region (or marker) is established by stringent thresholds for the p-value aiming to control the global significance level due to multiple testing. Thus, the success of genetic mapping under a frequentist framework is intimately dependent on the underlying sample size and the number of markers considered.

Recent years have revealed Bayesian analysis as a good alternative to tackle genetic mapping problems [12, 13]. Current proposals use powerful simulation techniques allied to genetic models with increased complexity. This chapter aims then to present some Bayesian genetic mapping methods to map binary trait loci (BTL). All methods are based on a mixture probability structure describing the recombination rates between markers and BTL and the underlying penetrance (the probability of phenotypic expression given the genotype of the true BTL). To contemplate different genetic actions, penetrance of the putative BTL is described through single-locus allelic penetrance models [9, 10]. The strength of association between markers and the phenotype is assessed by Bayes factors estimated through different methods, as discussed in Sepúlveda et al. [10]. As an example of application, a genetic mapping concerning experimental cerebral malaria susceptibility is performed. A detailed description of this work can be found elsewhere [3].

2 Genetic Mapping in Experimental Populations

Genetic mapping in experimental populations usually contemplates the crossing of two animal strains exhibiting distinct phenotypes. Such experimental design leads to a diallelic system in the sense that each genetic marker considered has only two possible alleles—one from each strain—segregating in the cross. To detect a putative BTL, let us first consider the analysis of two adjacent markers M (with alleles M_1 and M_2) and N (with alleles N_1 and N_2). Let us also assume the existence of a single putative BTL Q (with alleles Q_1 and Q_2) between M and N. The core of genetic mapping modelling is based on the fact that recombination events might occur between M, Q and N during gamete formation and, thus, the genotype of the markers and the putative BTL may differ. The probability at which those events happen is called recombination rate and is included in the statistical modelling to describe the probabilistic relationship between the genotypes of the markers and the putative BTL.

The detection of a BTL is based on the following decomposition of the probability for the expression of the phenotype given a marker genotype

$$P(Y = 1|G_m) = \sum_g P(Y = 1|G_t = g)P(G_t = g|G_m), \quad (1)$$

where Y is the random variable regarding the expression of the phenotype, G_t and G_m are the genotypes of BTL and markers, respectively, and $P(Y = 1|G_t = g)$

is the so-called penetrance of the BTL with genotype g. It is worth noting that the above equation assumes that the phenotypic expression does not depend on the markers under analysis.

The probabilities $P(G_t = g|G_m)$ are defined according to the recombination events that might occur during gamete formation. Under the assumption of a single recombination event (crossover) occurring between M, N and Q, one can easily derive these probabilities for intercrosses or backcrosses; a detailed discussion on how to obtain these probabilities can be found elsewhere [3, 7].

Given the above equation, different genetic mapping methods can be obtained by modelling the penetrance $P(Y = 1|G_t = g)$ accordingly. The most general model for penetrance is to consider the following parametric structure:

$$P(Y = 1|G_t = g) = \begin{cases} p_1, \text{ if } g = Q_1Q_1, \\ p_2, \text{ if } g = Q_1Q_2, \\ p_3, \text{ if } g = Q_2Q_2. \end{cases} \tag{2}$$

If no BTL is present at a given position between M and N, one expects that all genotypic penetrances would be the same. Therefore, detecting the presence of a BTL is done by testing $H_0 : p_1 = p_2 = p_3$ against $H_1 : \exists_{i,j}\ p_i \neq p_j$. It is worth noting that the above model provides little information on the genetic nature of the putative BTL. Such limitation can be easily surpassed by describing penetrance properly through models based on known genetic concepts.

The allelic penetrance approach has been recently proposed to model different gene actions acting upon a complex binary trait [9, 10]. This approach embodies the idea that penetrance may have two components: internal and external. The internal component models the stochastic expression of the alleles composing the BTL genotype towards the phenotype (the allelic penetrance). The external penetrance describes the probability of phenotypic expression by the action of factors other than the BTL under analysis. Assuming independence between these two components, the penetrance of a putative BTL can be decomposed into

$$P(Y = 1|G_t) = \pi_{G_t}^{int} + \pi^{ext} - \pi_{G_t}^{int}\pi^{ext}, \tag{3}$$

where $\pi_{G_t}^{int}$ and π^{ext} are the internal and external components, respectively. Different genetic actions for a single or two loci can be obtained by modelling $\pi_{G_t}^{int}$ appropriately [9]. For the matter of genetic mapping, the allelic penetrance models for a single-locus action are applied as it was assumed above the existence of a single BTL between any two adjacent markers.

Let us assume that Q_1 is a dominant allele towards the phenotype. Under the allelic penetrance approach, the dominance model is defined by the condition that the phenotype is observed by the expression of at least one dominant allele in the genotype. Under the assumption of independent allelic expressions, the internal component of penetrance of a putative BTL is given by

$$\pi_{G_t}^{int} = \begin{cases} \theta_1^2 + 2\theta_1(1-\theta_1), & \text{if } G_t = Q_1Q_1, \\ \theta_1, & \text{if } G_t = Q_1Q_2, \\ 0, & \text{if } G_t = Q_2Q_2, \end{cases} \tag{4}$$

where θ_1 is the penetrance of allele Q_1 towards the phenotype.

Two definitions are possible for the action of a recessive allele, both compatible with classical Mendelian recessiveness inheritance [10]: (1) the expression of both recessive alleles are required to observe the phenotype (type I recessiveness), or (2) the expression of a single recessive allele is enough to observe phenotype as long as the dominant allele is not active (type II recessiveness). In this work, type II recessive allele model is applied to genetic mapping because of its generality [10]. In the same line of the allelic dominance model, the internal component of penetrance for the recessive allele model is described by

$$\pi_{G_t}^{int} = \begin{cases} \theta_1^2 + 2\theta_1(1-\theta_1), & \text{if } G_t = Q_1Q_1, \\ \theta_1(1-\theta_2), & \text{if } G_t = Q_1Q_2, \\ 0, & \text{if } G_t = Q_2Q_2, \end{cases} \tag{5}$$

where θ_2 stands for the penetrance of allele Q_2.

In both dominant and recessive allele models, the detection of a BTL requires testing the hypothesis of no expression of the phenotype-conferring allele Q_1 against its opposite hypothesis (that is, $H_0 : \theta_1 = 0$ against $H_1 : \theta_1 \neq 0$).

3 Bayesian Analysis

Two-marker data is typically represented by a $G \times 2$ frequency table, where G stands for the number of possible genotypic combinations between any two markers ($G = 4$ and $G = 9$ for backcross and intercross experiments, respectively) and the two columns refer to the binary trait under study. The respective sampling model is given by the product of G independent binomial distributions $\{Bin(m_g, P(Y = 1|G_m = g))\}$ where m_g is the frequency of individuals with joint genotype g of the two markers and $P(Y = 1|G_m = g)$ is modelled by Eq. (1).

Detection of a putative BTL between any two markers is carried out as follows: (1) estimate the recombination rate between the markers and consider it throughout as a fixed constant, (2) assume the left marker as the putative BTL and draw inferences over model parameters, (3) increment the position of the BTL by a small value and make inferences again and (4) repeat previous step until the location of the putative BTL coincides with the right marker. For each location considered for the putative BTL, Bayes factors favouring H_1 are calculated using different estimation methods. A large enough Bayes factor shows evidence for a putative BTL; Congdon [2] provides some guidelines for Bayes factor analysis. A non-informative Bayesian analysis is followed in all genetic mapping models. Thus, for the general and allelic penetrance models, uniform prior distributions can be specified for the respective parameters modelling penetrance of the putative BTL.

For Bayes factor calculations, three methods are applied to estimate the predictive prior probability (PPP) of each competing hypothesis: ordinary Monte Carlo (MC), Markov Chain Monte Carlo (MCMC) and numerical integration, which might be considered as the most "exact" method [10]. Ordinary MC-based estimates are given by the average of the likelihood function evaluated at the parameter values simulated from their prior distributions. In this regard, it was shown that PPP estimates based on this method are not accurate when analysing allelic penetrance models [10]. According to Raftery et al. [8], a robust and stable PPP estimate can be obtained as below

$$\log P\hat{P}P_{BIC-MC}(\{n_g\}|H_i) = \bar{l} - s_{\bar{l}}^2(\log m - 1), \qquad (6)$$

where $m = \sum_g m_g$ is the sample size, $\bar{l}$ and $s_{\bar{l}}^2$ are the posterior mean and variance of the log-likelihood function, respectively.

4 Application

Bayesian genetic mapping is now illustrated with a data set regarding the genetic control of experimental cerebral malaria in two mouse strains, one susceptible and another resistant to the disease [1]. Phenotypic data refers to 190 F_2 backcrossed animals generated from a first progeny bred with the susceptible parental strain. Marker data encompasses about 130 genetic markers scattered around the mouse genome. Using Pearson's independence test for two-way tables, two putative BTL were previously identified at chromosomes 1 and 11 [1]. In the same line of analysis, Correia [3] applied a simple Bayesian homogeneity test using uniform prior distributions for the "penetrances" of the markers. This analysis detected the same BTL at chromosomes 1 and 11 and another at chromosome 14. It is worth noting that both analyses provide a crude estimate for the true location of three BTL. Interval genetic mapping can be then applied to refine previous results.

Since data at hand refers to a backcross experiment, there are four combined genotypes for any two adjacent markers. The probabilities $P(G_t|G_m)$ in Eq. (1) can be easily derived (Table 1) as one allele in each marker genotype is fixed by experimental design due to backcrossing. Penetrances of the putative BTL are given by Eqs. (2) and (3) with π_g^{int} given by Eqs. (4) and (5), but considering only the ones for genotypes Q_1Q_1 and Q_1Q_2. Bayesian interval mapping was then performed on data regarding chromosomes 1, 11, and 14. All calculations were done in the R software using several packages, namely, the `cubature` and `R2WinBUGS` for numerical integration and posterior simulation, respectively.

Figure 1 shows the Bayes factors for the genetic mapping of chromosome 1 using different models. In this application, the MC-based Bayes factors are very close to the "exact" ones, in contrast to those obtained through BIC-MC approximation. In fact, the Bayes factors based on this approximation seem to underestimate the "exact" ones, as opposed to previous findings when dissecting genetic interaction of two loci [10].

Table 1 Marginal, joint and conditional probabilities of genotypic inheritance between two markers M and N and a putative BTL Q in a backcross between strains 1 and 2, where the first progeny was crossed again with parental strain 1. Recombination rates between the respective markers M and N, and the putative BTL are given by λ_{MQ} and λ_{NQ}. The subscripts in the alleles indicate the strains where the alleles are derived from

Genotype G_m	$P(G_m)$	Genotype G_t	$P(G_t, G_m)$	$P(G_t \mid G_m)$
M_1M_1/N_1N_1	$\frac{1}{2}(1-\lambda_{MN})$	Q_1Q_1	$\frac{1}{2}(1-\lambda_{MQ})(1-\lambda_{NQ})$	$\frac{(1-\lambda_{MQ})(1-\lambda_{NQ})}{1-\lambda_{MN}}$
		Q_1Q_2	$\frac{1}{2}\lambda_{MQ}\lambda_{NQ}$	$\frac{\lambda_{MQ}\lambda_{NQ}}{1-\lambda_{MN}}$
M_1M_1/N_1N_2	$\frac{1}{2}\lambda_{MN}$	Q_1Q_1	$\frac{1}{2}(1-\lambda_{MQ})\lambda_{NQ}$	$\frac{(1-\lambda_{MQ})\lambda_{NQ}}{\lambda_{MN}}$
		Q_1Q_2	$\frac{1}{2}\lambda_{MQ}(1-\lambda_{NQ})$	$\frac{\lambda_{MQ}(1-\lambda_{NQ})}{\lambda_{MN}}$
M_1M_2/N_1N_1	$\frac{1}{2}\lambda_{MN}$	Q_1Q_1	$\frac{1}{2}\lambda_{MQ}(1-\lambda_{NQ})$	$\frac{\lambda_{MQ}(1-\lambda_{NQ})}{\lambda_{MN}}$
		Q_1Q_2	$\frac{1}{2}(1-\lambda_{MQ})\lambda_{NQ}$	$\frac{(1-\lambda_{MQ})\lambda_{NQ}}{\lambda_{MN}}$
M_1M_2/N_2N_2	$\frac{1}{2}(1-\lambda_{MN})$	Q_1Q_1	$\frac{1}{2}\lambda_{MQ}\lambda_{NQ}$	$\frac{\lambda_{MQ}\lambda_{NQ}}{1-\lambda_{MN}}$
		Q_1Q_2	$\frac{1}{2}(1-\lambda_{MQ})(1-\lambda_{NQ})$	$\frac{(1-\lambda_{MQ})(1-\lambda_{NQ})}{1-\lambda_{MN}}$

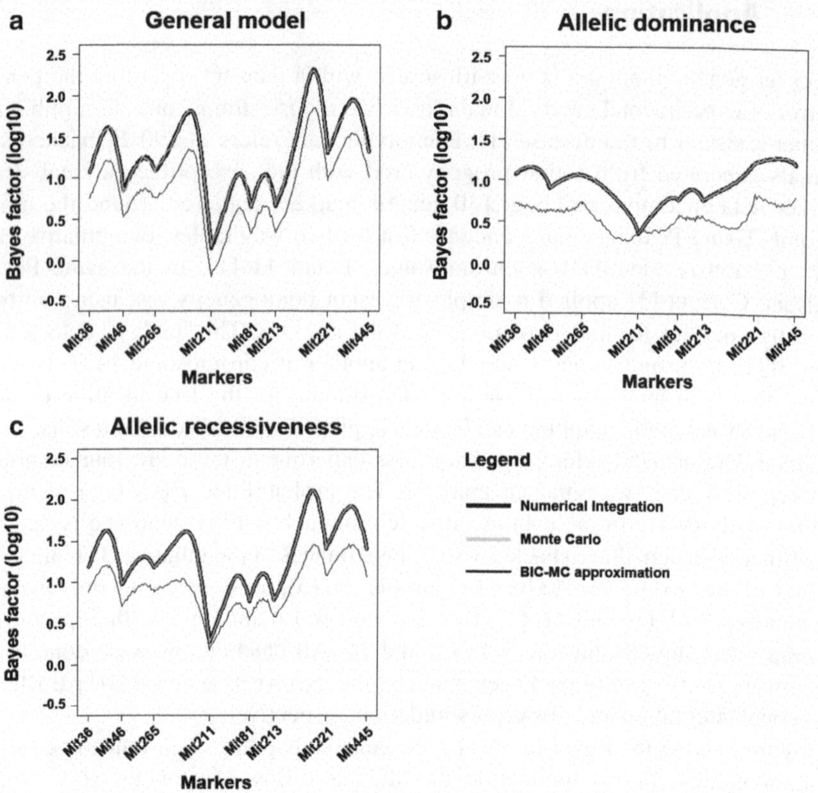

Fig. 1 Interval mapping of chromosome 1 using Bayes factors (towards H_1): (**a**) General model; (**b**) Allelic dominance model; (**c**) Allelic recessiveness model. Ordinary Monte Carlo estimates were calculated according to 100,000 generated values. Good convergence of MCMC-generated chains was obtained using 1,010,000 iterations with a burn-in period of 10,000 iterations and a lag of ten iterations

Table 2 Genomic positions of the maximum Bayes factors (BF) for chromosomes 1, 11, and 14, using numerical integration for the calculations. Distance from the nearest left marker is given in centimorgan

Chromosome	Left marker	Distance	Model	$\log_{10}(\mathrm{BF}_{H_1})$
1	Mit213	28	General	2.31
	Mit221	17	Allelic dominance	1.29
	Mit213	29	Allelic recessiveness	2.12
11	Mit199	12	General	1.13
	Mit100	0	Allelic dominance	1.12
	Mit199	12	Allelic recessiveness	1.21
14	Mit37	0	General	1.33
	Mit37	0	Allelic dominance	1.26
	Mit37	0	Allelic recessiveness	1.45

The general and the recessive allele models lead to similar Bayes factor profiles along chromosome 1. According to these models, a putative BTL might be present around Mit213 and Mit221. In contrast, Bayes factors obtained from the dominant allele model provide a weaker signal for BTL detection than those from the remaining models. This is in agreement with previous results that cerebral malaria susceptibility in these mice might be controlled by a cumulative action model requiring the expression of at least three alleles derived from the susceptible strain at different loci [9, 10].

The remaining results are summarized in Table 2. All models agree that BTL at chromosomes 11 and 14 show weaker signals of detection than the one(s) at chromosome 1. Moreover, maximum Bayes factors implied by these models are close to each other, which suggests a complex inheritance of the trait under analysis.

Concluding Remarks

Different penetrance models were applied to Bayesian interval genetic mapping. The results show that the location and strength of BTL detection are dependent on the model used. One way to overcome this problem is to undertake a Bayesian model averaging analysis, as previously suggested for weather forecasting [11]. Another solution is to take into account available information of genes known to affect the phenotype. On the one hand, it seems possible to learn the location of BTL for certain phenotypes, such as the HbS gene in human malaria [5] and the MHC locus in autoimmune diseases [4]. On the other hand, it seems difficult to elicit prior information on the genetic effects or allelic penetrances from the experts.

Finally, most interesting phenotypes are affected by a large number of genes and, thus, the estimation of the number of underlying BTL may be of interest. In a Bayesian framework one can consider the overall number of putative BTL as a model parameter. This modelling task has already been carried out for a liability-based model [13]. Similar exercise remains to be done for the proposed methodology.

Acknowledgments The work of NS and CDP was financed by National Funds through FCT - Fundação para a Ciência e a Tecnologia - in the scope of project PEst-OE/MAT/UI0006/2011

References

1. Bagot, S., Campino, S., Penha-Gonçalves, C., Pied, S., Cazenave, P.: Identification of two cerebral malaria resistance loci using an inbred wild-derived mouse strain. Proc. Natl. Acad. Sci. USA **99**, 9199–9923 (2002)
2. Congdon, P.: Bayesian statistical modelling. Wiley, Chichester (2001)
3. Correia, C.: Mapeamento Bayesiano para fenótipos binários complexos. Master's thesis, Instituto Superior Técnico, Lisbon (2009)
4. Hauptmann, G., Bahram, S.: Genetics of the central MHC. Curr. Opin. Immunol. **16**, 668–672 (2004)
5. Kwiatkowski, D.P.: How malaria has affected the human genome and what human genetics can teach us about malaria. Am. J. Hum. Genet. **77**, 171–92 (2005)
6. Lander, E.S., Botstein, D.: Mapping Mendelian factors underlying quantitative traits using RFLP linkage maps. Genetics **121**, 185–199 (1989)
7. McIntyre, L.M., Coffman, C.J., Doerge, R.W.: Detection and localization of a single binary trait locus in experimental populations. Genet. Res. **78**, 79–92 (2001)
8. Raftery, A.E., Newton, M.A., Satagopan, S.M., Krivitsky, P.N.: Estimating the integrated likelihood via posterior simulation using the harmonic mean identity (with discussion). In: Bernardo, J.M., Bayarri, M.J., Berger, J.O., Dawid, A.P., Heckerman, D., Smith, A.F.M., West, M. (eds.) Bayesian Statistics, vol. 8, pp. 371–416. Oxford University Press, Oxford (2007)
9. Sepúlveda, N., Paulino, C.D., Carneiro, J., Penha-Gonçalves, C.: Allelic penetrance approach as a tool to model two-locus interaction in complex binary traits. Heredity **99**, 173–184 (2007)
10. Sepúlveda, N., Paulino, C.D., Penha-Gonçalves, C.: Bayesian analysis of allelic penetrance models for complex binary traits. Comp. Stat. Data Anal. **53**, 1271–1283 (2009)
11. Sloughter, J.M., Raftery, A.E., Gneiting, T., Fraley, C.: Probabilistic quantitative precipitation forecasting using Bayesian model averaging. Mon. Wea. Rev. **135**, 3209–3220 (2007)
12. Yi, N., Shriner, D.: Advances in Bayesian multiple quantitative trait loci mapping in experimental crosses. Heredity **100**, 240–252 (2008)
13. Yi, N., Xu, S.: Bayesian mapping of quantitative trait loci for complex binary traits. Genetics **155**, 1391–1403 (2000)

Concomitant Latent Class Models Applied to Mathematics Education

Maria Eugénia Ferrão and José G. Dias

Abstract

The research project *School Effectiveness in Teaching-Learning of Mathematics* allowed a longitudinal study in the primary, elementary and lower secondary education which was conducted between 2004 and 2009 in Portugal. It stated as one of the specific objectives the development and promotion of quantitative methods in education, particularly in mathematics education. This chapter presents a latent class model with concomitant variables applied to the data of a paired sample (data collected at the beginning and at the end of the academic year) of 276 students enrolled in the 7th grade. The response variable represents whether learning has or has not occurred during the year and the concomitant variables are scores to assess the level of fluid intelligence components. Model parameter estimates suggest that there are two distinct latent classes explained by verbal and spatial reasoning.

1 Introduction

The field of research on the learning of mathematics deals with "what happens around, in and with students who engage in acquiring such knowledge, skills, among others, with particular regard to the processes and products of learning. A closely related area of investigation is the outcomes (results and consequences) of the teaching and the learning of mathematics, respectively" [14]. There is a general

M.E. Ferrão (✉)
University of Beira Interior and CEMAPRE, Portugal
e-mail: meferrao@ubi.pt

J.G. Dias
Instituto Universitário de Lisboa (ISCTE-IUL), BRU, Portugal
e-mail: jose.dias@iscte.pt

J. Lita da Silva et al. (eds.), *Advances in Regression, Survival Analysis, Extreme Values, Markov Processes and Other Statistical Applications*, Studies in Theoretical and Applied Statistics, DOI 10.1007/978-3-642-34904-1_15,

consensus that prior achievement is an important predictor of mathematics learning. Primi et al. [16] applied a multilevel growth curve model to a Portuguese data set, where learning is considered over two academic years for a sample of students ranging in age from 11 to 14. Results confirm that initial math ability predicts math ability 2 years later. The authors also showed that fluid intelligence [3, 4, 12] and numerical reasoning predict initial math ability; that there are significant between-student differences in the growth rate of mathematics learning across 2 years; and that the rate of growth was higher for those higher in fluid intelligence. Considering a sample of 11-year-old students, Irwin and Irwin [11] applied a latent class model to identify two groups of students: those who have not developed certain mathematical abilities (addition, multiplication and proportional reasoning) over one academic year (called *stayers*) and those who have (called *movers*). However, the authors do not present what students' attributes explain differences between *stayers* and *movers*.

This chapter serves two main purposes. The first is to illustrate how concomitant variables in a latent class model could help understand the population heterogeneity towards the fluid intelligence components such as numerical, abstract, verbal and spatial reasoning. The second is to determine the number of latent classes underlying that heterogeneity and to estimate the odds ratio for a correct answer to every item of the mathematics test.

Latent class models with concomitant variables are applied to data collected in a representative random sample of students enrolled at the 7th grade (lower secondary education) in the region of Cova da Beira, Portugal. Data collection took place in the context of a school effectiveness research project [7] with the support of the Portuguese Ministry of Science Technology and Higher Education and the Calouste Gulbenkian Foundation.

The structure of this chapter is the following: Sect. 2 presents the methodology; Sect. 3 summarizes the main results of the statistical analysis; and Sect. 4 contains a discussion of the results and concluding remarks.

2 Methodology

2.1 The Sample

The sample comprises 276 students (52.7 % boys) enrolled in the 7th grade in the region of Cova da Beira, Portugal. The survey design is longitudinal. Data were collected at the beginning and at the end of academic years 2005/2006, 2006/2007 and 2007/2008. Two cohorts of students were considered. In 2005/2006 the 1st, 3rd, 5th, 7th and 8th grade students were involved. They were monitored in the 2nd, 4th, 6th, 8th and 9th years, respectively, and a new cohort at the 1st, 3rd, 5th, and 7th years was surveyed. In 2007/8 these students were monitored again. The random sample is representative at county and region levels [17]. Data considered for the purpose of this chapter were collected at the beginning and at the end of academic year 2005/2006. Students' age varied from 11 to 17, and 62 % were 11- or 12-year-old students when they started the 7th grade.

2.2 Assessment Instruments

Math Tests. 3EMat is a battery of tests designed for the assessment of math skills, knowledge, and abilities throughout primary, elementary and lower secondary education [8]. Each test includes around 30 multiple-choice selected items covering the core curriculum for each grade and adjacent grades. Item calibration (the estimation of coefficients of discrimination, difficulty and its contribution for test information) was done during the pretest at the end of 2004/2005. The test booklets include common items (about 30 %) from adjacent grades in order to allow posterior vertical equating [5]. Standard norms on development, revision and administration of tests were followed, including those norms concerning test security [2]. In the 7th grade the distribution of items per subject is approximately as follows: geometry 24 %; numbers 36 %; equations 27 %; statistics 13 %. Response or manifest variable represents whether learning has or has not occurred during the academic year. The learning outcomes were measured considering common items administered at the beginning and at the end of academic year. Primary data on test items correction were recoded in order to get a categorical variable (1—an incorrect answer at the beginning followed by an incorrect answer at the end; 2—an incorrect followed by a correct answer; 3—a correct followed by an incorrect answer; 4—a correct followed by a correct answer). This is the response scale for each item included in the model presented below. Thus, the response variable represents whether learning has or has not occurred during the year.

Intelligence Tests. Cognitive abilities were assessed using the *Differential Reasoning Tests Battery* [1]. Although tests are based on analogy or series tasks combining different contents, the same cognitive operation—reasoning or fluid intelligence—is evaluated for each of the different components: numerical reasoning (NR) consisting of 30 numerical series items involving simple arithmetic operations; abstract reasoning (AR) consisting of 40 items involving abstract analogies of geometric figures; verbal reasoning (VR) consisting of 40 items involving verbal analogies; and spatial reasoning (SR) consisting of 30 spatial series related to the rotation of the six faces of a cube. There is a score related to each domain which is included as concomitant variables in the model presented in the following section.

2.3 Latent Class Model with Concomitant Variables

The popularity of latent class models has recently increased, mainly due to the availability of software implementations of this type of models [6]. The latent class model assumes a discrete latent variable with S classes, in which units' class membership is unknown. The latent class concomitant model [18] is a finite mixture model in which the prior probabilities or cluster sizes are regressed on some variables known as concomitant variables. Let us have a sample of size n. An observation is denoted by i $(i = 1, \ldots, n)$ and is measured by J items. Let $\mathbf{y}_i$ be the vector of the J items defined by $\mathbf{y}_i = (y_{i1}, \ldots, y_{iJ})$. Then, the latent class concomitant model with S latent classes for $\mathbf{y}_i$ is defined by the density

$$f(\mathbf{y}_i; \boldsymbol{\varphi}, \mathbf{w}_i) = \sum_{s=1}^{S} \pi_{is}(\mathbf{w}_i, \boldsymbol{\gamma}_s) f_s(\mathbf{y}_i; \boldsymbol{\theta}_s), \tag{1}$$

where the discrete latent variable, z_i, has a multinominal distribution, such that $z_i \sim Multi_{S-1}(\boldsymbol{\pi}_i)$, with $\boldsymbol{\pi}_i = (\pi_{i1}(\mathbf{w}_i, \boldsymbol{\gamma}_1), \ldots, \pi_{i,S-1}(\mathbf{w}_i, \boldsymbol{\gamma}_{S-1}))$, $\pi_{is}(\mathbf{w}_i, \boldsymbol{\gamma}_s) > 0$ and $\sum_{s=1}^{S} \pi_{is}(\mathbf{w}_i, \boldsymbol{\gamma}_s) = 1$. The vector of the K concomitant variables is $\mathbf{w}_i = (w_{i1}, \ldots, w_{iK})$. Each manifest variable has nominal scale with four categories. Thus, the distribution of the items within in latent class s, $f_s(\mathbf{y}_i; \boldsymbol{\theta}_s)$, is a product of J conditionally independent multinominal distributions. The parameters of the model are defined by $\boldsymbol{\varphi} = (\boldsymbol{\gamma}_1, \ldots, \boldsymbol{\gamma}_{S-1}, \boldsymbol{\theta}_1, \ldots, \boldsymbol{\theta}_S)$, where $\boldsymbol{\gamma}_s$ and $\boldsymbol{\theta}_s$ are the vector of parameters in each class s. McHugh [13] and Goodman [10] give sufficient conditions for the identifiability of the latent class model. All our models are identified. The maximum likelihood estimation of latent class models is not available in close-form, being the *expectation–maximization* (EM) algorithm, a popular iterative procedure in this context. This algorithm allows maximum likelihood estimation with incomplete data by reintroducing the additivity in the log-likelihood function, using data augmentation. This algorithm has two steps, first the E-step, that consists of associating each individual observation with its conditional expectation of class membership, given the observed values. The next M-step consists in maximizing the full data log-likelihood function using the complete data as the observed data. The optimal number of latent classes is traditionally identified as the model that minimizes the Bayesian information criterion (BIC)

$$BIC_s = -2\ell_s(\hat{\varphi}; \mathbf{y}, \mathbf{w}) + N_s \cdot log(m), \tag{2}$$

where N_s represents the number of parameters in the model and m is the sample size. We consider two possibilities for m: (1) n, the sample size of individuals; (2) $n \cdot J$, the total number of answers. Model estimation was implemented in MATLAB. As the log-likelihood surface is extremely complex and with many local maxima, we report the solution with the maximum log-likelihood value out of 300 runs for each model (random starting parameters). Convergence tolerance is 10^{-6}.

3 Results

Thirteen models were fitted with varying number of latent classes, number of items and number of concomitant variables (Table 1). The gradual reduction in the number of items is justified by the exclusion of those that belong to the 6th and 8th grades.

Table 1 Model selection and goodness of fit

Models	# Classes	# Items	# Concomitants	LL	# Param.	*BIC*(*n*)	*BIC*(*nJ*)
I. Original data							
Model I.1	1	28	6	−8459.350	84	17387.71	17668.56
Model I.2	2	28	6	−7982.043	183	16991.45	17597.71
II. Reduced number of concomitant variables							
Model II.1	1	28	3	−8556.669	84	17583.29	17863.20
Model II.2	2	28	3	−8132.977	181	17278.60	17881.72
III. Exclusion of non-significant concomitants							
Model III.1	1	28	2	−8556.669	84	17583.29	17863.20
Model III.2	2	28	2	−8141.506	171	17239.71	17809.51
Model III.3	3	28	2	−7998.570	258	17440.58	18300.28
IV. Exclusion of non-significant items							
Model IV.1	1	23	2	−7282.655	69	14951.34	15181.27
Model IV.2	2	23	2	−6878.707	141	14546.27	15016.11
Model IV.3	3	23	2	−6748.453	213	14688.58	15398.34
V. Exclusion of items of the previous year							
Model V.1	1	17	2	−5370.189	51	11025.71	11195.65
Model V.2	2	17	2	−5117.232	105	**10821.91**	**11171.79**
Model V.3	3	17	2	−5049.552	159	10988.66	11518.48

Table 2 Correlation structure between concomitant variables

	w1	w2	w3	w4	w5
Correlation matrix					
Score7 (w1)	1.00	0.59 ***	0.46 ***	0.45 ***	0.50 ***
Numerical reasoning (w2)		1.00	0.50 ***	0.53 ***	0.53 ***
Verbal reasoning (w3)			1.00	0.43 ***	0.45 ***
Spatial reasoning (w4)				1.00	0.53 ***
Abstract reasoning (w5)					1.00
Principal components					
1 component loadings (exp. var. = 59.93 %)	0.78	0.82	0.72	0.76	0.79
2 components (exp. var. = 72.01 %)					
Component 1 loading (varimax)	0.4	0.54	0.16	**0.86**	0.77
Component 2 loading (varimax)	0.7	0.63	**0.87**	0.21	0.34

***$p < 0.001$

The first group of models (I.1 and I.2) included two scores in math test and four components of fluid intelligence as concomitant variables. The second math score was dropped from the analysis as it was too correlated with the first one (score7). Correlations and principal component analysis (Table 2) show that the remaining concomitant variables are moderately correlated and that 72 % of total variance is explained by two linear-dependent components of verbal reasoning (factor loading is 0.87) and spatial reasoning (factor loading is 0.86). Thus, only these components of fluid intelligence were retained in the analysis.

Table 3 Latent class profiles

Concomitant variables	Cluster 1	Cluster 2	Aggregate
Verbal reasoning			
1–8	0.356	0.057	0.216
9–11	0.234	0.202	0.219
12–13	0.149	0.172	0.160
14–17	0.179	0.241	0.208
18–24	0.081	0.328	0.197
Mean	15.474	19.871	17.539
Spatial reasoning			
1–7	0.328	0.049	0.197
8–10	0.248	0.148	0.201
11–13	0.200	0.170	0.186
14–16	0.149	0.267	0.205
17–25	0.076	0.365	0.212
Mean	8.930	13.749	11.193

The Bayesian Information Criterion was used for model selection, implying the Model V.2 (see bold values in Table 1) as the best. The results suggest that there are two latent classes with dimensions 0.53 and 0.47. Class profiles (Table 3) show that pupils belonging to class 2 tend to have higher levels of verbal and spatial reasoning.

The odds ratio of being a *stayer* (incorrect answer at the beginning and incorrect answer at the end of the academic year) in class 2 compared with class1 was calculated for every item of the test. Results obtained are 0.422, 0.331, 0.149, 0.26, 0.391, 0.308, 0.002, 0.160, 0.233, 0.332, 0.021, 0.162, 0.082, 0.065, 0.463, 0.099 and 0.5 giving evidence that is more plausible to find *stayers* in class 1 than in class 2.

4 Discussion and Conclusion

The methodological approach presented reveals itself as a powerful method to a better understanding of how learning happens. We illustrated the use of a latent class model with concomitant variables that represent verbal and spatial reasoning. Model parameter estimates allowed us to calculate the probability of being a *stayer* in mathematics learning over an academic year. Results showed that the students who belong to the cluster with higher verbal and spatial reasoning are less likely to be classified as *stayers* (pupils who have not developed expected mathematical abilities over the academic year 2005/2006) since the odds ratio determined for test items is lower than one. The underlying relationship between mathematics learning and verbal or spatial reasoning corroborates some findings from quasi-experiments or qualitative methodological approaches on the topic, such as those described in [9, 15]. However, further work should be conducted considering other relevant variables potentially related to the heterogeneity of population, such as socioeconomic status [11]. It would also constitute an important contribution if we could associate patterns of curriculum content to the dimensions of fluid intelligence. In other words, depending on items used we might expect different abilities to be predictive—e.g. numerical ability for statistics, numbers or equations, and spatial ability for geometry. This requires extending the method across every grade and cohorts in the longitudinal study.

References

1. Almeida, L.S.: Baterias de Provas de Raciocínio Diferencial (BPRD): Manual. [Battery of Differential Reasoning Tests. Manual]. Universidade do Minho, Braga (1992)
2. American Educational Research Association, American Psychological Association, National Council on Measurement in Education: Standards for educational and psychological testing. American Educational Research Association, Washington (1999)
3. Cattell, R.B.: Theory of fluid and crystallized intelligence: a critical experiment. J. Educ. Psychol. **54**, 1–22 (1963)
4. Cattell, R.B.: Abilities, Their Structure, Growth, and Action. Houghton Mifflin, Boston (1971)
5. Costa, P., Oliveira, P.N., Ferrão, M.E.: Modelo de resposta ao item de dois parâmetros: construção de uma escala vertical de desempenho em Matemática. In: Hill, M.M., Ferreira, M.A., Dias, J.G., Salgueiro, M.F., Carvalho, H., Vicente, P., Braumann, C.A. (eds.) Estatística da Teoria à Prática. Actas do XV Congresso Anual da Sociedade Portuguesa de Estatística, pp. 155–166. Edições SPE, Lisboa (2008)
6. Dias, J.G.: Finite Mixture Models: Review, Applications, and Computer-intensive Methods (Ph.D. Thesis). Ridderprint, The Netherlands (2004)
7. Ferrão, M.E., Loureiro, M.J., Simões, M.F., Calmão, M.J., Guedes, P.: À Procura da Escola Eficaz - Referencial Teórico no Ensino da Matemática. UBI, Covilhã (2005)
8. Ferrão, M.E., Costa, P., Dias, V., Dias, M.: Medição da Competência dos Alunos do Ensino Básico em Matemática: 3EMat, uma proposta [Measuring Mathematics Skills of Students in Compulsory Education: 3EMat, a proposal]. In: Machado, C. Almeida, L., Guisande, M.A., Gonçalves, M., Ramalho, V. (eds.) Actas da XI Conferência Internacional de Avaliação Psicológica: Formas e Contextos [Proceedings of the XI International Conference on Psychological Evaluation]. Psiquilibrios Edições, Braga (2006)
9. Garfield, J.: The challenge of developing statistical reasoning. J. Stat. Educ. **10**(3), 1–11. Online:amstat.org/publications/jse/ (2002)
10. Goodman, L.A.: Exploratory latent structure analysis using both identifiable and unidentifiable models. Biometrika **61**(2), 215–231 (1974)
11. Irwin, K.C., Irwin, R.J.: Assessing development in numeracy of students from different socio-economic areas: A Rasch analysis of three fundamental tasks. Educ. Stud. Math. **58**, 283–298 (2005)
12. Kane, M.J., Gray, J.R.: Fluid intelligence. In: Salkind, N.J. (ed.) Encyclopedia of Human Development, vol. 3, pp. 528–529. Erlbaum, NJ (2005)
13. McHugh, R.B.: Efficient estimation and local identification in latent class analysis. Psychometrika **21**(1), 331–347 (1956)
14. Niss, M.: Aspects of the nature and state of research in mathematics education. Educ. Stud. Math. **40**, 1–24 (1999)
15. Phelps, E., William, D.: Problem solving with equals: peer collaboration as a context for learning mathematics and spatial concepts. J. Educ. Psychol. **81**(4), 639–646 (1989)
16. Primi, R., Ferrão, M.E., Almeida, L.: Fluid intelligence as a predictor of learning: a longitudinal multilevel approach applied to mathematics. Learn. Individ. Differ. **20**(5), 446–451 (2010)
17. Vicente, P.: Plano amostral do projecto 3EM - eficácia escolar no ensino da matemática. In: Ferrão, M.E., Nunes, C., Braumann, C.A. (eds.) Estatística Ciência Interdisciplinar. Actas do XIV Congresso Anual da Sociedade Portuguesa de Estatística, pp. 847–856. Edições SPE, Lisboa (2007)
18. Wedel, M.: Concomitant variables in finite mixture models. Stat. Neerl. **56**(3), 362–375 (2002)

Evaluating Discriminant Analysis Results

Ana Sousa Ferreira and Margarida Cardoso

Abstract

In discrete discriminant analysis (DDA) different models often exhibit different classification performances. Therefore, the idea of combining models has increasingly gained importance. In the present work we focus on the evaluation of alternative DDA models, including combined models. The proposed approach uses not only the classic indicators of classification precision but also indices of agreement that regard the relationship between the actual classes and the ones predicted by discriminant analysis. The performance of the DDA methods is analyzed based on simulated binary data, using small and moderate sample sizes. The results obtained illustrate the potential of combining DDA models, offering different evaluation perspectives.

1 Introduction

In discrete discriminant analysis (DDA) different models often exhibit different classification performances for different individuals or observations. This seems to be a particularly relevant issue in the small or moderate sample setting and when the classes are not well separated. Therefore, the idea of combining models currently

A.S. Ferreira (✉)
LEAD, FP, Universidade de Lisboa, CEAUL and UNIDE, Alameda da Universidade, 1649-013 Lisboa, Portugal
e-mail: asferreira@fp.ul.pt

M. Cardoso
Department of Quantitative Methods and UNIDE, ISCTE-Lisbon University Institute, Avenida das Forças Armadas, 1600-083 Lisboa, Portugal
e-mail: mgsc@iscte.pt

J. Lita da Silva et al. (eds.), *Advances in Regression, Survival Analysis, Extreme Values, Markov Processes and Other Statistical Applications*, Studies in Theoretical and Applied Statistics, DOI 10.1007/978-3-642-34904-1_16,

appears in an increasing number of DDA papers, in an attempt to obtain more robust and stable models.

In this chapter we compare the performance of the full multinomial model (FMM) [9] and the first-order independence model (FOIM) [9] with a model based on the two referred models that produce an intermediate model between them. In order to deal with the multi-class case we use the hierarchical coupling model (HIERM) (e.g., [3, 15]) that enables to reduce the problem into several bi-class problems embedded in a binary tree. The comparison is extended to the results of the classification and regression trees (CART) algorithm [2], a classical approach within the classification domain.

The performance of the alternative models considered is compared based on simulated data. To evaluate this performance we consider several measures of precision including traditional classification indices and indices of agreement between the actual classes and the ones predicted by the DDA methods. Results obtained refer to two-fold cross-validation.

2 Methodological Approach

In the present study, a new methodology is proposed for the evaluation of DDA results. It enables the comparison of DDA classical models with the DDA combining models approach. The proposed methodology relies on indices of agreement between the actual and predicted (by DDA) classes and is illustrated using simulated data according to the Bahadur model.

2.1 Indices for Evaluating Classification Results

When evaluating results from classification we focus on the $K \times K$ confusion matrix $\mathbf{M} = [n_{ij}]$ which is a contingency table of the actual classes (lines refer to partition $\prod_a^K$ with K classes) by the ones predicted by discriminant analysis (columns refer to partition $\prod_b^K$ with K classes). The row totals are $n_{i.}$, $(i = 1, \dots, K)$.

Some commonly used indices depend only on the diagonal of the referred matrix, which adds up to the number of correctly classified observations (see Table 1). The percent agreement varies between 0 (null classification precision) and 100 % (perfect classification precision). The Cohen's Kappa deducts agreement by chance and the Huberty index deducts the percentage of correctly classified by default (majority class rule).

In this work we suggest further exploring the confusion matrix to evaluate the agreement between $\prod_a^K$ and $\prod_b^K$. We thus consider the indices on Tables 1 and 2 (see [4], for further details). The Cramer's V statistic quantifies simple agreement and variation of information (VI) considers entropy and mutual information. They vary between 0 and 1 (0 indicating null agreement for Cramer's V and perfect agreement for the normalized VI). The Adjusted Rand quantifies paired agreement deducting agreement by chance. A priori, the advantage of using these indices is to complement the evaluation of agreement between partitions $\prod_a^K$ and $\prod_b^K$.

Table 1 Indices of agreement based on the diagonal of the confusion matrix

Indices	Definition
Percent agreement	Perc-agree$(\prod_a^K,\prod_b^K) = (\sum_{k=1}^K n_{kk})/n$
Cohen's Kappa [6]	Kappa$(\prod_a^K,\prod_b^K) = (\sum_{k=1}^K n_{kk} - \sum_{k=1}^K n_{k.}n_{.k}/n)/(n - \sum_{k=1}^K n_{k.}n_{.k}/n)$
Huberty [12]	Huberty$(\prod_a^K,\prod_b^K) = ((\sum_{k=1}^K n_{kk})/n - \max_i n_{i.}/n)/(1 - \max_i n_{i.}/n)$, where $n_{i.}$, $(i = 1, \ldots, K)$ are the row totals

Table 2 Indices of agreement based on the complete confusion matrix

Indices	Definition
Cramer's V [7]	$\mathrm{V}(\prod_a^K,\prod_b^K) = \sqrt{Chi - sq(\prod_a^K,\prod_b^K)/(nK - n)}$ where $Chi - sq(\prod_a^K,\prod_b^K) = \sum_{k=1}^K \sum_{q=1}^K (n_{kq} - \frac{n_{k.}n_{.q}}{n})^2 / \frac{n_{k.}n_{.q}}{n}$
Adjusted rand [11]	Adj-Rand$(\prod_a^K,\prod_b^K) = \frac{\sum_{k=1}^K \sum_{q=1}^K C_2^{n_{kq}} - \sum_{k=1}^K C_2^{n_{k.}} \sum_{q=1}^K C_2^{n_{.q}} / C_2^n}{\frac{1}{2}[\sum_{k=1}^K C_2^{n_{k.}} + \sum_{q=1}^K C_2^{n_{.q}}] - \sum_{k=1}^K C_2^{n_{k.}} \sum_{q=1}^K C_2^{n_{.q}} / C_2^n}$
Normalized variation of information [14]	$N\_VI(\prod_a^K,\prod_b^K) = [\mathrm{H}(\prod_a^K) + \mathrm{H}(\prod_b^K) - 2\mathrm{I}(\prod_a^K,\prod_b^K)]/\log n$ where H indicates the entropy $\mathrm{H}(\prod^K) = \sum_{k=1}^K \frac{n_{.k}}{n} \log \frac{n_{.k}}{n}$ and I indicates the mutual information $I(\prod_a^K,\prod_b^K) = \sum_{k=1}^K \sum_{q=1}^K \frac{n_{kq}}{n} \log \frac{n_{kq}}{n_{k.}n_{.q}/n}$

2.2 Simulated Data

The performance of the DDA methods is analyzed based on simulated binary data. We use the Bahadur model, as proposed in Godstein and Dillon [5, 9], to simulate the predictive binary variables' values. This model representation defines class conditional probabilities for class C_k, $(k = 1, \ldots, K)$ as

$$P(\mathbf{x}|C_k) = \prod_p \theta_{kp}^{x_p}(1 - \theta_{kp})^{(1-x_p)}[1 + \sum_{g \neq p} \rho_k(p, g) \mathrm{Z}_{kp}\mathrm{Z}_{kg}] \tag{1}$$

where X_{kp} is a Bernoulli variable with parameter $\theta_{kp} = E(X_{kp}), p = 1, \ldots, P$ such that

$$\mathrm{Z}_{kp} = \frac{X_{kp} - \theta_{kp}}{[\theta_{kp}(1 - \theta_{kp})]^2} \quad \text{and} \quad \rho_k(p, g) = E(\mathrm{Z}_{kp}\mathrm{Z}_{kg}). \tag{2}$$

We consider two types of population structures with P = 6 variables and for illustrative purposes, let us consider the case of K = 2 classes and one of the most usual multi-class case, K = 4 classes. Location parameters are described on Table 3.

For each structure, data sets generated have small sample sizes (60 observations for each class) and moderate sample sizes (200 observations for each class). The training and test samples represent 50 % of the total of observations.

Table 3 Parameters for simulated Bernoulli variables

K = 2	K = 4
$\theta_1 = (0.6, 0.4, 0.6, 0.5, 0.5, 0.6)$	$\theta_1 = (0.6, 0.4, 0.6, 0.5, 0.5, 0.6)$
$\theta_2 = (0.5, 0.3, 0.5, 0.4, 0.4, 0.5)$	$\theta_2 = (0.5, 0.3, 0.5, 0.4, 0.4, 0.5)$
	$\theta_3 = (0.6, 0.3, 0.6, 0.4, 0.5, 0.5)$
	$\theta_4 = (0.6, 0.4, 0.6, 0.5, 0.5, 0.6)$

The first structure, denoted IND (Independent), is generated according to FOIM ($\rho_k(p,p) = 1$ and $\rho_k(p,g) = 0$, if $p \neq g$, $k = 1,\ldots,K$; $p,g = 1,\ldots,6$), for all classes.

The second one, called DIF (Different), is implemented considering the existence of different relations among the variables, for different classes:

- In the bi-class case $\rho_1(p,p) = 1$ and $\rho_1(p,g) = 0.2$, if $p \neq g$, $p,g = 1,\ldots,6$; $\rho_2(p,p) = 1$ e $\rho_2(p,g) = 0.4$, if $p \neq g$, $p,g = 1,\ldots,6$.
- In the multi-class case $\rho_k(p,p) = 1$ and $\rho_k(p,g) = 0.1$, if $p \neq g$, $k = 1,2,3$; $p,g = 1,\ldots,6$; and $\rho_4(p,p) = 1$ and $\rho_4(p,g) = 0.3$, if $p \neq g$, $p,g = 1,\ldots,6$.

The prior probabilities are considered equal.

2.3 Discrete Discriminant Analysis

In discrete classification problems the most natural model is the FMM where the conditional probabilities are estimated by the observed frequencies [9]. This model involves $2^P - 1$ parameters to be estimated in each class. Hence, even for moderate P (e.g., ten binary variables lead to 1,023 parameters to be estimated), generally, not all of the parameters are identifiable.

One way to deal with this problem consists in reducing the number of parameters to be estimated. The FOIM assumes that the P binary variables are independent in each class C_k, $k = 1,\ldots,K$ [9]. Then, the number of parameters to be estimated for each class is reduced from $2^P - 1$ to P.

Since we are mainly concerned with small or moderate sample sizes, we may encounter a problem of sparseness in which some of the multinomial cells may have no data in the training sets. Therefore, we suggest to smooth the observed frequencies of model FMM as follows:

$$P(\mathbf{x}|\lambda) = \frac{1}{n}\sum_{i=1}^{n} \lambda^{P-\|\mathbf{x}-x_i\|}(1-\lambda)^{\|\mathbf{x}-x_i\|}, 0 < \lambda \leq 1 \tag{3}$$

where $\lambda = 1.00$, $\lambda = 0.99$, $\lambda = 0.95$ or $\lambda = 0.90$ according to the training sample size.

In this work, taking into account the size of our samples, we consider $\lambda = 1.00$ (no smoothing) or $\lambda = 0.95$ (moderate smoothing) for all samples.

Note that according to Hand [10], we opt for a computationally less demanding method since the choice of the smoothing method is not particularly important.

FMM and FOIM provide different classifications in many circumstances. Therefore, we expect a combining model (using a single coefficient β for the linear combination of FMM and FOIM) to yield better results.

There are several strategies to estimate the coefficient β (e.g., [3, 15]) that combines the two referred models. A natural way of deriving this coefficient is by minimizing the fitting error using a least squares criterion [15, 16]. For the two classes case, we use an approach to estimate the coefficient β using a least squares regression (LSR) criterion:

$$\widehat{\beta}_{LSR} = \frac{\sum_{i=1}^{n}(l_2(x_i) - l_1(x_i))l_2(x_i) - \sum_{i=1}^{n} y_i(l_2(x_i) - l_1(x_i))}{\sum_{i=1}^{n}(l_2(x_i) - l_1(x_i))^2} \tag{4}$$

where y_i denotes an indicator of class membership for observation i and l_1 and l_2 represent, respectively, the log ratio of the class conditional probabilities for model FMM and FOIM (denoted by LSR1) or the a posteriori probabilities of the first class for FOIM and FMM models (denoted by LSR2), estimated by cross-validation in a sample of size n.

In the multi-class case, we use the HIERM, inspired by Friedman's approach [8], for reducing the multi-class problem into several bi-class problems embedded in a binary tree. HIERM needs two decisions at each level:

1. Selecting the hierarchical coupling among the $2^{K-1} - 1$ possible classes couple
2. In each node of the tree, selecting the combining model that gives the best classification rule for the chosen couple

At the beginning we have K classes that we want to reorganize into two classes. So, we propose to select the two new classes that are the most separable. The basic affinity coefficient [1, 13] can be used to select the hierarchical coupling at each level of the tree.

Denoting $F_1 = q_j^1$ and $F_2 = q_j^2$, $j = 1, \ldots, P$ two discrete distributions defined in the same space, the affinity coefficient is defined by

$$\rho(F_1, F_2) = \sum_j \sqrt{q_j^1}\sqrt{q_j^2}, j = 1, \ldots P \tag{5}$$

and is easily computed in our classification problem. The individual vector $\mathbf{x}$ is assigned to the class associated with the last node of the tree on which $\mathbf{x}$ falls.
The main aim of this approach is to obtain a better prediction performance and improve results stability.

3 Experimental Results

After running discriminant analysis for the simulated data we obtain the results presented in Tables 4 and 5.

Table 4 Small samples results/cross-validation (two-fold results)

Data	Methods	Perc-agree (%)	Kappa (%)	Huberty (%)	Cramer's V	Adj-Rand	N-VI
IND	CART	52	5	−7	0.048	−0.019	0.355
K=2	FMM	31	21	−2	0.226	0.040	0.340
	FOIM	58	16	4	0.198	0.013	0.328
	LSR2	60	21	11	0.222	0.025	0.320
DIF	CART	77	54	48	0.559	0.291	0.242
K=2	FMM	65	50	30	0.520	0.286	0.245
	FOIM	58	17	0	0.165	0.004	0.335
	LSR2	76	52	46	0.400	0.097	0.278
IND	CART	28	5	−1	0.156	−0.005	0.536
K=4	FMM	0	*	0	*	*	*
	FOIM	30	6	3	0.173	0.005	0.534
	LSR2	50	34	30	0.505	0.208	0.368
DIF	CART	23	−1	−6	*	−0.010	*
K=4	FMM	10	−20	−23	0.347	0.083	0.472
	FOIM	32	12	6	0.241	0.036	0.510
	LSR1	48	31	29	0.426	0.135	0.474

*Not defined (null observed frequency in denominator)

Table 5 Moderate samples results/cross-validation (two-fold results)

Data	Methods	Perc-agree (%)	Kappa (%)	Huberty (%)	Cramer's V	Adj-Rand	N-VI
IND	CART	54	8	6	0.078	0.004	0.258
K=2	FMM	55	14	10	0.139	0.014	0.259
	FOIM	59	17	15	0.172	0.025	0.255
	LSR2	60	19	17	0.195	0.031	0.253
DIF	CART	69	37	36	0.398	0.138	0.219
K=2	FMM	61	32	23	0.346	0.120	0.208
	FOIM	50	−1	−3	0.039	−0.022	0.261
	LSR2	63	30	24	0.333	0.100	0.224
IND	CART	33	11	9	0.154	0.016	0.447
K=4	FMM	0	*	0	*	*	*
	FOIM	35	13	12	0.225	0.043	0.429
	LSR2	44	26	25	0.327	0.093	0.407
DIF	CART	29	6	4	0.105	0.002	0.425
K=4	FMM	11	−18	−20	0.221	0.039	0.431
	FOIM	35	13	12	0.220	0.038	0.433
	LSR1	46	28	27	0.393	0.130	0.362

*Not defined (null observed frequency in denominator)

When referring to the combining models we simply present the results yielded by the best strategy (LSR1 or LSR2). For the sake of simplicity, we only report the best FMM results (smoothed or not).

In these results, the DDA methods seem to perform similarly for the small- and moderate-sized samples. Except for the case of DIF and K=2 (where the best results are attained by CART) the combined models evidence the best performances.

Table 6 Pearson correlations (r)

Methods	Perc-agree	Kappa	Huberty	Cramer's V	Adj-Rand	N-VI
Perc-agree	1					
Kappa	0.807	1				
Huberty	0.790	0.952	1			
Cramer's V	0.339	0.709	0.699	1		
Adj-Rand	0.436	0.739	0.711	0.948	1	
N-VI	−0.807	−0.516	−0.464	−0.181	−0.307	1

4 Discussion and Perspectives

In general, the best DDA results are obtained using the combining models approach, with the LSR2 strategy where the a posteriori probabilities characterize the class conditional probabilities.

The various indicators used to evaluate DDA results offer different insights regarding the confusion matrix and the corresponding results do not necessarily agree (see correlations in Table 6). Note that we consider small and moderate size samples when computing correlations, since they exhibit similar (correlation) patterns.

The percent agreement index is strongly related with the normalized variation of information index which has the advantage of quantifying not only the correctly classified cases but also the relationship between the incorrectly classified ones. The Cramer's V statistic and the Adjusted Rand index are strongly related as well as the Kappa and the Huberty indices. These indicators offer a different perspective, quantifying simple agreement and paired agreement between the actual classes and the predicted ones.

In future research, the advantages of using indices of agreement for evaluating DDA results should be further explored. In addition, real data should be used to further illustrate the utility of the proposed approach.

References

1. Bacelar-Nicolau, H.: The affinity coefficient in cluster analysis. Meth. Oper. Res. **53**, 507–512 (1985)
2. Breiman, L., Friedman, J.H., Olshen, R.A., Stone, C.J.: Classification and Regression Trees. Wadsworth, California (1984)
3. Brito, I., Celeux, G., Sousa Ferreira, A.: Combining methods in supervised classification: A comparative study on discrete and continuous problems. REVSTAT - Statist. J. **4**(3), 201–225 (2006)
4. Cardoso, M.G.M.S.: Clustering and cross-validation. In: Paper Presented at the IASC 07 - Statistics for Data Mining, Learning and Knowledge Extraction (2007)
5. Celeux, G., Mkhadri, A.: Discrete regularized discriminant analysis. Statist. Comput. **2**(3), 143–151 (1992). doi:10.1007/BF01891,206

6. Cohen, J.: A coefficient of agreement for nominal scales. Educ. Psychol. Meas. **20**, 37–46 (1960)
7. Cramér, H.: Mathematical Methods of Statistics. Princeton University Press, Princeton (1946)
8. Friedman, J.: Another approach to polychotomous classification. In: Technical Report, Stanford University (1996)
9. Goldstein, M., Dillon, W.: Discrete Discriminant Analysis. Wiley, New York (1978)
10. Hand, D.: Kernel Discriminant Analysis. Research Studies Press, Wiley, Chichester (1982)
11. Hubert, L., Arabie, P.: Comparing partitions. J. Classif. **2**, 193–218 (1985)
12. Huberty, C.J., Olejnik, S.: Applied MANOVA and Discriminant Analysis. Wiley-Interscience, Wiley, New York (2006)
13. Matusita, K.: Decision rules based on distance for problems of fit, two samples and estimation. Ann. Inst. Stat. Math. **26**(4), 631–640 (1955)
14. Meila, M.: Comparing clusterings-an information based distance. J. Multivar. Anal. **98**(5), 873–895 (2007). doi:10.1007/BF01891,206
15. Sousa Ferreira, A.: Combinação de modelos em análise discriminante sobre variáveis qualitativas. Ph.D. thesis, University of Nova de Lisboa (2000)
16. Sousa Ferreira, A.: Classification as a tool for research, studies in classification, data analysis, and knowledge organization, chap. In: Locarek-Junge, H., Weihs, C. (eds.) A Comparative Study on Discrete Discriminant Analysis Through a Hierarchical Coupling Approach. Springer, Berlin (2009)

Distribution of the Number of Losses in Busy-Periods of $M^X/G/1/n$ Systems

Fátima Ferreira, António Pacheco, and Helena Ribeiro

Abstract
This chapter addresses $M^X/G/1/n$ queues, i.e., single server batch Markovian arrival queues with finite customer waiting space of size n. Taking profit of the Markov regenerative structure of these systems, we develop an efficient recursive procedure to compute the probability mass function of the number of losses in busy-periods starting with an arbitrary number of customers in the system. The derived computational procedure is easy to implement and leads to a fast numerical computation of the loss probabilities. To illustrate the effectiveness of the procedure, loss probabilities are computed for a wide variety of queues, with different capacities, batch size distributions, and arrival and service parameters.

1 Introduction

Queues, or waiting lines, in which customers arrive, wait for service, are served, and then leave the system are a familiar feature of daily life. In the pioneer queueing problems it was assumed that customers arrive single at a service facility and find an

F. Ferreira (✉)
Department of Mathematics, CM-UTAD and CEMAT, University of Trás-os-Montes e Alto Douro, Ed. Ciências Florestais, Apartado 1013, Quinta de Prados 5000-801, Vila Real, Portugal
e-mail: mmferrei@utad.pt

A. Pacheco
Department of Mathematics and CEMAT, Instituto Superior Técnico - TU Lisbon, Av. Rovisco Pais, 1049-001 Lisboa, Portugal
e-mail: apacheco@math.ist.utl.pt

H. Ribeiro
Department of Mathematics and CEMAT, ESTG - Polytechnic Institute of Leiria, Morro do Lena - Alto do Vieiro, 2411-901 Leiria, Portugal
e-mail: helena.ribeiro@ipleiria.pt

J. Lita da Silva et al. (eds.), *Advances in Regression, Survival Analysis, Extreme Values, Markov Processes and Other Statistical Applications*, Studies in Theoretical and Applied Statistics, DOI 10.1007/978-3-642-34904-1_17,

infinite waiting room. However, these assumptions are violated in many real world queueing situations. Letters arriving to a post office, customers arriving to a terminal gate of an airport, and data files arriving to a computer system are a few examples of queuing situations in which, in general, customers do not arrive in a single form but in batches (of fixed or random size). In turn, while infinite (waiting space) queuing systems are analytically easier to handle, finite queues are more realistic as infinite waiting spaces do not exist in the real-world. The relevance of finite capacity batch arrival queues for applications is well reflected by the abundance of studies of such systems in the literature (see, e.g., [1, 2, 6, 7, 12] and the references therein).

The study of queues is performed either from the user's perspective or from the operator's perspective. While measures such as the distribution of the customer waiting time in the system, the number of customers in queue, and the loss probability are oriented toward the user's perspective, the analysis of such quantities during busy-period cycles, i.e., during effective system utilization periods, is relevant from the operator's point of view. In fact, the analysis of busy-period characteristics such as the busy-period length, the number of customers served, or the number of customers lost (due to overflow) during the busy-period, can provide crucial useful information for the management of congested systems. An intensive care unit of a hospital, where losses in system may result in losses of lives, is an example of such situation.

In recent years, there has been in fact an increased interest in the study of the number of losses in busy-periods. In this context, Abramov [1], Wolff [16], and Peköz et al. [12] showed the interesting phenomenon that the mean number of losses during busy-periods in $M^X/G/1/n$ queues is invariant in the queue capacity (n) when the traffic intensity (ρ) is unitary, varying with n otherwise. In particular, case $\rho = 1$,

$$E[L_{i,n}] = i \quad \text{and} \quad E[L_n] = \beta, \text{ for all } n \geq 1,$$

where $L_{i,n}$ denotes the number of losses during a busy-period initiated with i customers and L_n the corresponding measure for a busy-period initiated with a random number of customers, X, with mean β. As shown in [12], this invariance property for mean losses when $\rho = 1$ does not extend to general arrival processes. For $M^X/G/1/n$, if $\rho < 1$ ($\rho > 1$), these quantities are decreasing (increasing) functions of n for all i and $n \geq 1$, and, in particular, for $n \geq 1$,

$$E[L_{i,n}] < i \quad \text{and} \quad E[L_n] < \beta, \text{ case } \rho < 1,$$

$$E[L_{i,n}] > i \quad \text{and} \quad E[L_n] > \beta, \text{ case } \rho > 1.$$

Moreover, it was shown in [12] that if a $GI/M/1/n$ queue is such that the mean number of losses during a busy-period is unitary and invariant on the system capacity, then it must be an $M/M/1/n$ queue with $\rho = 1$. Wolff [16] showed the validity of the above relations for $GI^X/G/1/n$ systems, under certain specific conditions on the interarrival times.

Moments of higher order for the number of losses during a busy-period for the $M/G/1/n$ and $GI/M/1/n$ queues were derived in Peköz [11]. At the same time, Righter [13] showed that the loss probabilities during a busy-period initiated with a single customer may be used recursively to compute the mean number of losses in busy-periods of $M/G/1/m$ systems, $m = 1, 2, \ldots$, all with the same parameters except the queue capacity. Additionally, Pacheco and Ribeiro [8–10] obtained recursive procedure on the system capacity to compute the probability of consecutive losses in busy-periods of $M/G/1/n$ and $GI/M(m)/n$ queueing systems and moments of the duration of busy-periods of $M^X/G/1/n$ systems.

To our knowledge, the literature on losses in busy-periods of $M^X/G/1/n$ systems is confined to studies of the mean number of losses during such periods. The main contribution of this work is to provide a recursive procedure to compute the distribution of the number of losses in busy-periods of these queueing systems.

The structure of the remaining sections of this chapter is the following. In Sect. 2 the $M^X/G/1/n$ queue model is presented, with the accompanying definitions and notation. Section 3 discusses the proposed approach to compute the distribution of the number of losses in a busy-period. Finally, to illustrate the computational procedure, we compute the distribution of the number of losses in busy-periods of several $M^X/G/1/n$ systems in Sect. 4.

2 The $M^X/G/1/n$ Queue

In this chapter we investigate customer loss characteristics of an $M^X/G/1/n$ queueing system. Customers arrive into such a system (in batches) according to a compound Poisson process with batch arrival rate λ, and the batch sizes are independent random variables identically distributed to a random variable X with probability mass function (p.m.f.) $P(X = i) = b_i, i = 1, 2, \ldots$, with finite mean $\bar{b}$.

The queue has finite capacity of size n, including the customer in service (if any), so that at any time a maximum of n customers can be present in the system. The batches which upon arrival are unable to find enough space in the buffer for all the customers of the batch are partially rejected—partial batch rejection policy. Specifically, if at arrival of a batch of l customers there are only m, $m < l$, free positions available in the system, then m customers of the batch enter the system and the remaining $l - m$ customers of the batch are blocked. Customers accepted in system are served by a single server. The service time is characterized by the (general) distribution function A with mean μ^{-1}. Customer service times are independent of the customer arrival process and of previous customer service times.

The process of interest is $Y = \{Y(t), t \geq 0\}$, where $Y(t)$ denotes the number of customers in the system at time t. This process is non-Markovian, but it is a well-known fact that it constitutes a Markov regenerative process associated with the renewal sequence $(T_n)_{n\in\mathbb{N}}$ of customer post-departure epochs (i.e., instants immediately after customer service completions); see, e.g., [3, 5] for more details on Markov regenerative processes. Therefore, the embedded process at customer post-departure epochs, $\{Y(T_n+), n \geq 1\}$ will be considered.

The main purpose of this chapter is to derive the p.m.f. of the number of losses in i-busy-periods, i.e., in busy-periods starting with i, $i \geq 1$, customers in the system. For that, we let C denote the total number of customer arrivals during the first service taking place in a 1-busy-period and $(r_j)_{j\in\mathbb{N}}$ denote its p.m.f. Conditioning on the number of batches that arrive during an interval of length t, the probability that a total of j customers arrive in such period is given by $r_j(t) = \sum_{l=0}^{j} e^{-\lambda t} \frac{(\lambda t)^l}{l!} b_j^{(l)}$ where $b_j^{(l)}$ denotes the probability that the total number of customers in l customer batches is equal to j, i.e., $b_j^{(0)} = \delta_{0j}$, and $b_j^{(l)} = \sum_{i=l-1}^{j-1} b_i^{(l-1)} b_{j-i}$, for $l \in \mathbb{N}$ and $j = l, l+1, \ldots$, where δ_{ij} is the Kronecker delta function, i.e., $\delta_{ij} = 1$ if $i = j$ and $\delta_{ij} = 0$ otherwise. Therefore, the conditional probability that exactly j customers arrive during the first service taking place in the 1-busy-period is $r_j = \sum_{l=0}^{j} \alpha_l b_j^{(l)}$ where $\alpha_l = \int_0^\infty e^{-\lambda t} \frac{(\lambda t)^l}{l!} A(\mathrm{d}t)$ denotes the l-th mixed Poisson probability with arrival rate λ and mixing distribution A, that, for many distributions, can be computed in a fast recursive way (cf., e.g., [15]).

3 Customer Loss Probability Distribution

In this section, it is assumed that the system is in steady state and, to address the computation of customer loss probabilities in busy-periods of an $M^X/G/1/n$ system, we let $L_{i,n}$ denote the number of customer losses in an i-busy-period of the system. The Markov regenerative property at post-departure epochs implies that on $\{Y(0) = i, Y(0^-) = 0\}$, with $i > 1$, the time the system takes to reach state 1—from state i—and the subsequent time it takes to reach state 0—from state 1—are independent. Therefore, by putting apart one of the customers initially present in the system and supposing that such a customer will start being served only when being alone in the system, we can straightforwardly argue that on $\{Y(0) = i, Y(0^-) = 0\}$, with $i > 1$, the time the system takes to reach state 1—from state i—has the same distribution as the duration of an $(i-1)$-busy-period of an $M^X/G/1/n-1$ system with the same parameters as the original $M^X/G/1/n$ system, except for the capacity of the system (that now has one position less). Thus, the following result holds.

Lemma 1. *For* $1 \le i \le n$,

$$L_{i,n} \overset{\mathrm{d}}{=} L_{i-1,n-1} \oplus L_{1,n} \tag{1}$$

where $\overset{\mathrm{d}}{=}$ *denotes equality in distribution,* $\oplus$ *denotes the sum of independent random variables, and* $L_{0,m} = 0$.

As a result, that follows immediately by induction, we have $L_{i,n} \overset{\mathrm{d}}{=} \bigoplus_{j=n+1-i}^{n} L_{1,j}$. This shows that the distribution of the number of losses in an i-busy-period of an $M^X/G/1/n$ system is a direct function of the distribution of the number of customer losses in 1-busy-periods of $M^X/G/1/m$ systems with smaller or equal system capacity, namely, $n+1-i \le m \le n$, but otherwise with the same parameters as the former system. Accordingly, we will next focus explicitly on the characterization of the p.m.f. of the number of customer losses in 1-busy-periods of $M^X/G/1/n$ systems.

For that, conditioning on the number of customers that arrive to the system during the service of the customer that initiates a 1-busy-period, we express the loss p.m.f. as a mixture of the conditional distributions of i-busy-periods given each possible number of customer that may arrive to the system during the service of the customer that initiates the 1-busy-period, as next explained. If no customers arrive to the system during the service time of the customer that starts the 1-busy-period, the busy-period ends at the departure of that customer. In this case no costumer is lost. Otherwise, the customers that arrive to the system during the service time of that customer and are not blocked initiate at his departure from the system a multiple busy-period that is part of the busy-period under consideration. As a consequence,

$$[L_{1,n}|C=l] \overset{d}{=} \begin{cases} 0 & l=0 \\ L_{l,n} & 1 \leq l \leq n-1\,, \\ l-(n-1)+L_{n-1,n} & l \geq n \end{cases} \quad \text{for } n \geq 1. \tag{2}$$

Taking into account (2), from the total probability law, it follows that

$$P[L_{1,n}=k] = r_0\delta_{0k} + \sum_{l=1}^{n-1} P[L_{l,n}=k]r_l + \sum_{l=n}^{n-1+k} P[l-(n-1)+L_{n-1,n}=k]r_l, \tag{3}$$

which allows one to establish the following result.

Theorem 1. *The customer loss probabilities in i-busy-periods of $M^X/G/1/n$ systems are such that*

$$P[L_{1,1}=k] = r_k, \quad k \in \mathbb{N}, \tag{4}$$

and, for $n \geq 2$:

$$P[L_{1,n}=k] = \frac{r_0\,\delta_{0k} + \sum_{l=1}^{n-1} \Phi_{l,k}^{(n)} r_l + \sum_{j=1}^{k} P[L_{n-1,n}=k-j]r_{j+(n-1)}}{1-\sum_{l=1}^{n-1} P[L_{l-1,n-1}=0]r_l} \tag{5}$$

with

$$P[L_{l,n}=j] = \sum_{i=0}^{j} P[L_{1,n}=j-i]P[L_{l-1,n-1}=i] \tag{6}$$

for $2 \leq l \leq n$, where $L_{0,n}=0$ and

$$\Phi_{l,i}^{(n)} = \sum_{j=1}^{i} P[L_{1,n}=i-j]P[L_{l-1,n-1}=j]. \tag{7}$$

Proof. Rewriting the loss probabilities in Eq. (3) in the form

$$P[L_{1,n} = k] = r_0\,\delta_{0k} + \sum_{l=1}^{n-1} P[L_{1,n} = k]r_l + \sum_{j=1}^{k} P[L_{n-1,n} = k - j]r_{j+(n-1)} \quad (8)$$

taking into account (1), it follows that for $1 \leq l \leq n-1$,

$$P[L_{l,n} = k] = P[L_{l-1,n-1} \oplus L_{1,n} = k] = \sum_{j=0}^{k} P[L_{1,n} = k - j]P[L_{l-1,n-1} = j]. \quad (9)$$

Separating in Eq. (9) the $j = 0$ term from the remaining terms, we conclude that

$$P[L_{l,n} = k] = P[L_{1,n} = k]P[L_{l-1,n-1} = 0] + \Phi_{l,k}^{(n)}$$

with $\Phi_{l,k}^{(n)}$ defined in Eq. (7), and the statement (5) follows directly from the previous equation and Eq. (8). Finally, the statement (6) follows from Eq. (1). □

Theorem 1 has immediate application in the computation of the loss probabilities of an i-busy-period in an $M^X/G/1/n$ system, $P[L_{i,n} = K]$, $1 \leq i \leq n$ and $K \in \mathbb{N}$. This can be done recursively from Eqs. (4)–(7), using the following algorithm:

Compute $P[L_{1,1} = k]$, $k \in \mathbb{N}$, from Eq. (4)
For $k = 0, 1, \ldots, K$
  For $m = 2, 3, \ldots, n$
    Compute $P[L_{1,m} = k]$ from Eq. (5)
    Compute $P[L_{l,m} = k]$, $2 \leq l \leq m - 1$, from Eq. (6).

The algorithm computes the loss probabilities in i-busy-periods of $M^X/G/1/n$ systems using $O(n^3)$ operations. The highest computational effort is needed, in general, to compute the convolution probabilities of the customer batch size distribution, which are required to obtain the probability function $(r_i)_{i \geq 0}$ of the number of customer arrivals during the service of a customer.

4 Numerical Illustration

To end this chapter, the procedure derived in the previous section is applied to compute and analyze the sensitivity of costumer losses in busy-periods of $M^X/G/1/n$ systems with respect to the batch size and service time distributions. The results presented were computed with MATLAB algorithms using the recursions proposed in [15] for computing the mixed Poisson probabilities (α_l).

To investigate the influence of the batch size distribution on the loss probabilities, the latter were computed for $M^X/G/1/n$ systems with five different batch size distributions with common mean $\bar{b}$: deterministic with the constant $\bar{b}$, $D(\bar{b})$, geometric with success probability $1/\bar{b}$, $Geo(1/\bar{b})$, uniform discrete on the set

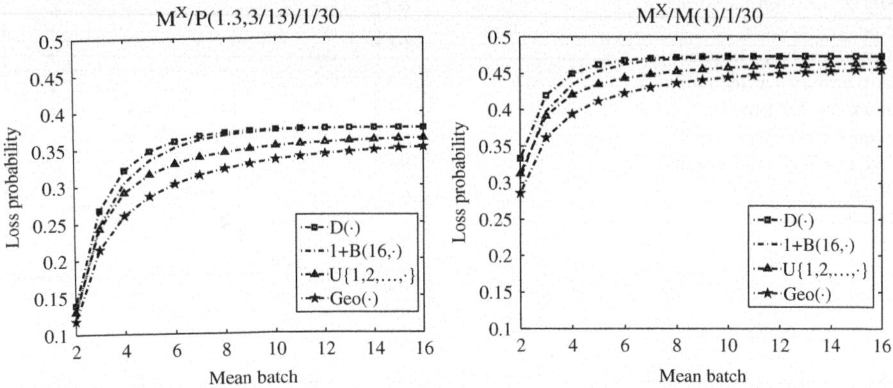

Fig. 1 Loss probability in 1-busy-periods of $M^X/G/1/30$ systems with traffic intensity $\rho = 0.9$ and unitary service rate, as a function of the mean batch size, for deterministic, geometric, discrete uniform, and shifted binomial batch size distributions

$\{1, 2, \cdots, 2\bar{b} - 1\}$, $U\{1, \cdots, 2\bar{b} - 1\}$, shifted binomial—a binomial with a trials—and success probability $(\bar{b} - 1)/a$ added of one unit, $1 + B(a, (\bar{b} - 1)/a)$. In turn, to study the sensitivity of loss probabilities on the service time distribution, four different service time distributions with common mean μ^{-1} were considered: exponential with rate μ, $M(\mu)$, deterministic, $D(\mu^{-1})$, uniform on the interval $(0, 2/\mu)$, $(U(0, 2/\mu))$, and Pareto with parameters β and k, $P(\beta, k)$, with $\beta > 1$ and $k = (\beta - 1)/\beta\mu$. The latter are specially useful, e.g., to model Internet traffic service [4, 14].

As expected, the results presented in Fig. 1 show that the loss probability (probability that at least one customer is lost during a busy-period) increases with the mean batch size. Nevertheless, we observe that, independently of the mean batch size, the batch size distribution may have great impact on the performance of the queue. The loss probability shows a tendency to decrease for batch size distributions with higher variability. In fact, among the service times and the batch size distributions considered, we observe that the systems with deterministic batch size distributions may experience a higher loss probability in contrast with the systems with geometric batch sizes, which present the smallest loss probabilities. Obviously, when the mean batch size becomes sufficiently high compared with the queue capacity, in which case all systems experience high loss probabilities, the effect of the batch size distribution becomes weak.

Figures 2 and 3 illustrate the sensitivity of the loss probability with respect to the traffic intensity and the service time distribution, respectively. Among the studied systems, one can observe that congested light tail service systems experience higher loss probabilities, in contrast with systems with heavy tail service time distribution—here represented by the Pareto service times with small value of β (cf. Fig. 2). In general, in all the studied systems we observe high probability of a small number of costumer losses and that the loss probabilities tend to

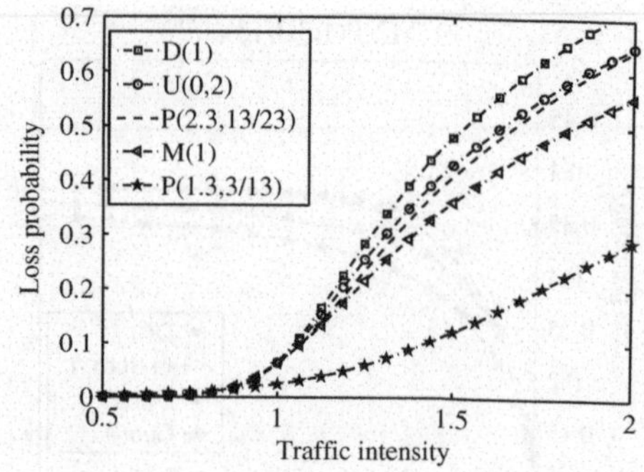

Fig. 2 Loss probabilities in 2-busy-periods of $M^{Geo(0.5)}/G/1/30$ systems with unitary service rate as a function of the traffic intensity for exponential, deterministic, uniform, and Pareto service time distributions

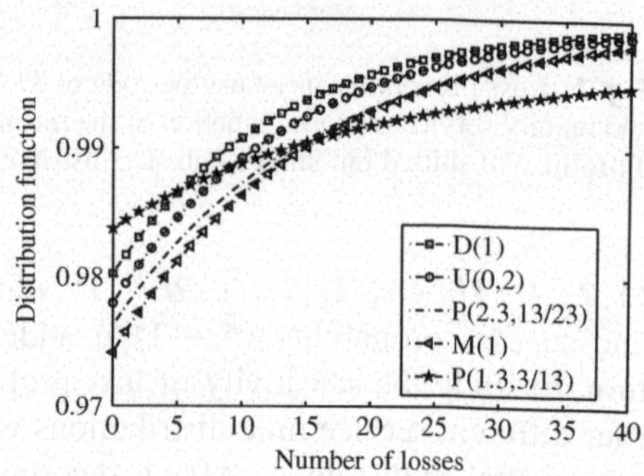

Fig. 3 Distribution function of the number of customer lost in 2-busy-periods of $M^{Geo(0.5)}/G/1/30$ systems with traffic intensity 0.9 and unitary service rate, for exponential, deterministic, uniform, and Pareto service time distributions

decrease as the number of customer lost increases (cf. Fig. 3). Among the light tail service time distributions considered, we observe that the lighter ones tend to assign more probability to small numbers of costumer lost and very small probabilities to experiencing a large number of costumer losses. In contrast, due to their heavy tail distribution, the $M^{Geo(0.5)}/P(\beta,.)/1/30$ queues with small values of β have the interesting behavior of having higher probability of experiencing a very small number of losses but also have small but persistent positive probabilities of experiencing a huge number of losses. This is explained by the fact that in these systems, in general, most of the customers require small service times causing a null or small number of losses but, with small probability, customers requiring huge service times may appear, causing in turn a large number of customer losses during their service times.

Acknowledgments The authors thank the support of the Portuguese Government through FCT (*Fundação para a Ciência e a Tecnologia*) under the projects Pest-OE/MAT/UI0822/2011, PEst-OE/MAT/UI4080/2011, and PTDC/EIA-EIA/115988/2009.

References

1. Abramov, V.M.: On a property of a refusals stream. J. Appl. Probab. **34**, 800–805 (1997)
2. Chaudhry, M.L., Templeton, J.G.C.: A First Course in Bulk Queues. Wiley, New York (1983)

3. Çinlar, E.: Introduction to Stochastic Processes. Prentice-Hall, Englewood Cliffs (1975)
4. Crovella, M.E., Taqqu, M.S., Bestavros, A.: Heavy-tailed probability distributions in the world wide web. In: A Practical Guide to Heavy Tails: Statistical Techniques and Applications, pp. 3–25. Birkhauser, Boston (1998)
5. Kulkarni, V.G.: Modeling and Analysis of Stochastic Systems. Chapman & Hall, London (1995)
6. Medhi, J.: Stochastic Models in Queueing Theory. Academic, Amsterdam (2003)
7. Miyazawa, M.: Complementary generating function for the $M^X/GI/1/k$ and $GI/M^Y/1/k$ queues and their application to the comparison for loss probabilities. J. Appl. Probab. **27**, 684–692 (1990)
8. Pacheco, A., Ribeiro, H.: Consecutive customer loss probabilities in $M/G/1/n$ and $GI/M(m)//n$ systems. In: Proceedings of Workshop on Tools for Solving Structured Markov Chains, Pisa, Italy, 10 Oct 2006
9. Pacheco, A., Ribeiro, H.: Consecutive customer losses in regular and oscillating $M^X/G/1/n$ systems. Queueing Syst. Theor. Appl. **58**(2), 121–136 (2008)
10. Pacheco, A., Ribeiro, H.: Moments of the duration of busy periods of $M^X/G/1/n$ systems. Probab. Eng. Inform. Sci. **22**, 347–354 (2008)
11. Peköz, E.A.: On the number of refusals in a busy period. Probab. Eng. Inform. Sci. **13**(1), 71–74 (1999)
12. Peköz, E.A., Righter, R., Xia, C.H.: Characterizing losses during busy periods in finite buffer systems. J. Appl. Probab. **40**(1), 242–249 (2003)
13. Righter, R.: A note on losses in $M/GI/1/n$ queues. J. Appl. Probab. **36**(4), 1240–1243 (1999)
14. Willinger, W., Paxson, V., Taqqu, M.S.: Self-similarity and heavy tails: structural modeling of network traffic. In: A Practical Guide to Heavy Tails, pp. 27–51. Birkhauser, Boston (1998)
15. Willmot, G.E.: On recursive evaluation of mixed-Poisson probabilities and related quantities. Scand. Actuar. J. **2**, 114–133 (1993)
16. Wolff, R.W.: Losses per cycle in a single-server queue. J. Appl. Probab. **39**(4), 905–909 (2002)

17. [illegible]

Misleading Signals in Simultaneous Residual Schemes for the Process Mean and Variance of AR(1) Processes: A Stochastic Ordering Approach

Patrícia Ferreira Ramos, Manuel Cabral Morais, and António Pacheco

Abstract
Assessing the performance of simultaneous schemes for the process mean and variance requires the use of the probability of misleading signals (PMS). This chapter discusses the impact of autocorrelation on the PMS of simultaneous Shewhart and EWMA residual schemes for the mean and the variance of a stationary autoregressive process of order 1, AR(1). The assessment of this impact is done numerically and by means of stochastic ordering.

1 Introduction

When we want to monitor both the mean and the variance of a process it is common to run two individual charts at the same time, one for the mean (μ) and another one for the variance (σ^2). The schemes that make use of two individual charts are the popular simultaneous (or joint) schemes. According to [11,17], when a simultaneous scheme is at use, the following misleading signals (MS) are likely to happen:

Since the assignable causes on charts for μ can differ from those on charts for σ^2, the diagnostic procedures that follow a signal can differ depending on whether the signal is given by the chart for μ or the one for σ^2 ([8], p. 189). Thus, misleading signals are valid signals that can lead the quality control operator or engineer to

P.F. Ramos (✉)
CEMAT, Instituto Superior Técnico, Technical University of Lisbon, Av. Rovisco Pais 1, 1049-001 Lisboa, Portugal
e-mail: patriciaferreira@ist.utl.pt

M.C. Morais · A. Pacheco
CEMAT and Mathematics Department, Instituto Superior Técnico, Technical University of Lisbon, Av. Rovisco Pais 1, 1049-001 Lisboa, Portugal
e-mail: maj@math.ist.utl.pt; apacheco@math.ist.utl.pt

J. Lita da Silva et al. (eds.), *Advances in Regression, Survival Analysis, Extreme Values, Markov Processes and Other Statistical Applications*, Studies in Theoretical and Applied Statistics, DOI 10.1007/978-3-642-34904-1_18,

Type of MS	μ	σ^2	First chart to signal
III	On-target	Off-target	Chart for μ
IV	Off-target	On-target	Chart for σ^2

misdiagnose assignable causes and deploy incorrect actions to bring the process back to target. Awareness of this phenomenon can be traced back to [17], and it has been addressed for i.i.d. and Gaussian output by some authors [9, 11–14]. More recently, [2, 7] presented a numerical discussion on simultaneous residual schemes for the process mean and variance of $AR(1)$ output.

This chapter discusses and assesses the monotone behaviour of the PMS in simultaneous residual schemes. Let us remind the reader that a residual scheme is a traditional scheme were residuals of a time-series model are plotted instead of the original data [1].

2 Simultaneous Residual Schemes and the Phenomenon of Misleading Signals

Let us denote by $\{Y_{i,j}\}$ the target process where i is the sample number and the index j is the number of the observation within the sample. The sample size is fixed and equal to n, and we will assume that different samples are independent. However, $(Y_{i,1}, \ldots, Y_{i,n})$ is a stationary $AR(1)$ process with known mean μ_0 and autocovariance function $\{\gamma_0, \gamma_1, \ldots, \gamma_{n-1}\}$:

$$Y_{i,j} = \mu_0 + \phi(Y_{i,j-1} - \mu_0) + \varepsilon_{i,j}, \quad (1)$$

where $-1 < \phi < 1$ and $\varepsilon_{i,j} \overset{i.i.d.}{\sim} \mathcal{N}(0, \sigma_\varepsilon^2)$.

The observed process is denoted by $X_{i,j}$ and is related to the target process as follows:

$$X_{i,j} = \mu_0 + \delta\sqrt{\gamma_0} + \theta(Y_{i,j} - \mu_0), \; i = 1, 2, \ldots, \quad (2)$$

where $\delta \geq 0$ (resp. $\theta \geq 1$) represents the magnitude of the shift in the process mean (resp. variance).

Following [5], the standardized residuals $\{\hat{\varepsilon}_{i,j}\}$ for a fixed i are obtained assuming that the process is in control ($\delta = 0, \theta = 1$). Since the best linear predictor for $X_{i,j}$ given $X_{i,j-1}, \ldots, X_{i,1}$ is $\hat{X}_{i,1} = \mu_0$ and $\hat{X}_{i,j} = \mu_0 + \phi(X_{i,j-1} - \mu_0)$ ($j = 2, \ldots, n$) we get

$$\hat{\varepsilon}_{i,j} = \begin{cases} \frac{\sqrt{1-\phi^2}(X_{i,1}-\mu_0)}{\sigma_\varepsilon} & \text{for } j = 1, \\ \frac{X_{i,j}-\mu_0-\phi(X_{i,j-1}-\mu_0)}{\sigma_\varepsilon} & \text{for } j = 2, \ldots, n. \end{cases} \quad (3)$$

In this chapter we consider Shewhart and EWMA residual charts with the control statistics and limits in Table 1. It is important to note that upper one-sided residual

Table 1 Control statistics and upper control limits for the upper one-sided Shewhart ($S-\mu$, $S-\sigma$) and EWMA ($E-\mu$, $E-\sigma$) individual residual charts (the lower control limits are all equal to zero), where $\gamma_{\{S-\mu,S-\sigma,E-\mu,E-\sigma\}}$ are the critical values chosen such that the ARL of the individual residual charts are equal and $\lambda_{\{\mu,\sigma\}}$ are the smoothing parameters of the EWMA charts)

Control statistic	Upper control limit
$\max\{0, \bar{\hat{\varepsilon}}_i\}$	$UCL_{S-\mu} = \frac{\gamma_{S-\mu}}{\sqrt{n}}$
$\hat{S}_i^2$	$UCL_{S-\sigma} = 1 + \gamma_{S-\sigma}\sqrt{\frac{2}{n-1}}$
$W_{\bar{\hat{\varepsilon}},i} = \begin{cases} 0, & i = 0 \\ \max\{0, (1-\lambda_\mu)W_{\bar{\hat{\varepsilon}},i-1} + \lambda_\mu \bar{\hat{\varepsilon}}_i\}, & i > 0 \end{cases}$	$UCL_{E-\mu} = \gamma_{E-\mu}\sqrt{\frac{\lambda_\mu}{n(2-\lambda_\mu)}}$
$W_{\hat{S}_i^2,i} = \begin{cases} 1, & i = 0 \\ (1-\lambda_\sigma)W_{\hat{S}_i^2,i-1} + \lambda_\sigma \hat{S}_i^2, & i > 0 \end{cases}$	$UCL_{E-\sigma} = 1 + \gamma_{E-\sigma}\sqrt{\frac{2\lambda_\sigma}{(n-1)(2-\lambda_\sigma)}}$

charts for μ are at use, so the corresponding limits and statistics result from an adaptation of the ones in [7]. According to [6], the sample mean ($\bar{\hat{\varepsilon}}_i$) and variance ($\hat{S}_i^2$) of the standardized residuals are independent and

$$\bar{\hat{\varepsilon}}_i \overset{i.i.d.}{\sim} \mathcal{N}\left(\tfrac{\delta}{n}\left(1+(n-1)\sqrt{\tfrac{1-\phi}{1+\phi}}\right), \tfrac{\theta^2}{n}\right) \quad \text{and} \quad \tfrac{(n-1)\hat{S}_i^2}{\theta^2} \overset{i.i.d.}{\sim} \chi^2_{n-1,\nu}, \tag{4}$$

where $\chi^2_{n-1,\nu}$ denotes the noncentral χ^2 distribution with $n-1$ degrees of freedom and noncentrality parameter equal to

$$\nu = \frac{n-1}{n}\left(\frac{\delta}{\theta}\right)^2\left(1-\sqrt{\frac{1-\phi}{1+\phi}}\right)^2. \tag{5}$$

Capitalizing on the distributional properties of $\bar{\hat{\varepsilon}}_i$ and $\hat{S}_i^2$, we conclude that the RL of the individual Shewhart residual charts for μ ($RL_{S-\mu}$) and for σ^2 ($RL_{S-\sigma}$) have geometric distributions with parameters

$$\xi_{S-\mu}(\delta,\theta,\phi) = 1 - \Phi\left[\tfrac{1}{\theta}\left(\gamma_{S-\mu} - \tfrac{\delta}{\sqrt{n}}\left(1+(n-1)\sqrt{\tfrac{1-\phi}{1+\phi}}\right)\right)\right], \tag{6}$$

$$\xi_{S-\sigma}(\delta,\theta,\phi) = 1 - F_{\chi^2_{n-1,\nu}}\left[\tfrac{n-1}{\theta^2}\left(1+\gamma_{S-\sigma}\sqrt{\tfrac{2}{n-1}}\right)\right]. \tag{7}$$

Since $\bar{\hat{\varepsilon}}_i$ and $\hat{S}_i^2$ are independent, the RL of the simultaneous residual scheme ($S-\mu,\sigma$) also has geometric distribution with parameter

$$\xi_{S-\mu,\sigma}(\delta,\theta,\phi) = \xi_{S-\mu}(\delta,\theta,\phi) + \xi_{S-\sigma}(\delta,\theta,\phi) - \xi_{S-\mu}(\delta,\theta,\phi) \times \xi_{S-\sigma}(\delta,\theta,\phi). \tag{8}$$

As for the EWMA individual and simultaneous scheme, the Markov chain approach [3] provides the following approximations to the survival functions of the run lengths $RL_{E-\mu}(\delta,\theta,\phi)$, $RL_{E-\sigma}(\delta,\theta,\phi)$ and $RL_{E-\mu,\sigma}(\delta,\theta,\phi)$:

$$\overline{F}_{RL_{E-\mu}(\delta,\theta,\phi)}(m) \simeq \mathbf{e}_{\mu}^{T}[\mathbf{Q}_{\mu}(\delta,\theta,\phi;x_{\mu})]^{m}\mathbf{1}_{\mu}, \tag{9}$$

$$\overline{F}_{RL_{E-\sigma}(\delta,\theta,\phi)}(m) \simeq \mathbf{e}_{\sigma}^{T}[\mathbf{Q}_{\sigma}(\delta,\theta,\phi;x_{\sigma})]^{m}\mathbf{1}_{\sigma}, \tag{10}$$

$$\overline{F}_{RL_{E-\mu,\sigma}(\delta,\theta,\phi)}(m) = \overline{F}_{RL_{E-\mu}(\delta,\theta,\phi)}(m) \times \overline{F}_{RL_{E-\sigma}(\delta,\theta,\phi)}(m), \tag{11}$$

for $m = 0, 1, 2, \ldots$, where

- $\mathbf{e}_{\mu}$ (resp. $\mathbf{e}_{\sigma}$) denotes the first [resp. $(x_{\sigma} + 1)/UCL_{E-\sigma}$)th] vector of the orthonormal basis for $\mathbb{R}^{x_{\mu}+1}$ (resp. $\mathbb{R}^{x_{\sigma}+1}$), associated with the state related to the initial value of the control statistic.
- $\mathbf{1}_{\mu}$ (resp. $\mathbf{1}_{\sigma}$) is a column vector of $(x_{\mu} + 1)$ [resp. $(x_{\sigma} + 1)$] ones.
- The entries of the sub-stochastic matrix $\mathbf{Q}_{\mu}(\delta,\theta,\phi;x_{\mu})$ [resp. $\mathbf{Q}_{\sigma}(\delta,\theta,\phi;x_{\sigma})$] follow from an adaptation of (resp. are equal to) the ones defined in [7].

Moreover, the left partial row sums of the entries of $\mathbf{Q}_{\mu}(\delta,\theta,\phi;x_{\mu})$ and $\mathbf{Q}_{\sigma}(\delta,\theta,\phi;x_{\sigma})$ are given, respectively, by

$$a_{\mu,ij}(\delta,\theta,\phi;x_{\mu}) = \Phi\left\{\frac{1}{\theta}\left[\frac{\gamma_{E-\mu}[(j+1)-(1-\lambda_{\mu})(i+1/2)]}{(x_{\mu}+1)\sqrt{\lambda_{\mu}(2-\lambda_{\mu})}} - \frac{\delta}{\sqrt{n}}\left(1+(n-1)\sqrt{\frac{1-\phi}{1+\phi}}\right)\right]\right\}, \tag{12}$$

for $i, j = 0, \ldots, x_{\mu}$,

$$a_{\sigma,ij}(\delta,\theta,\phi;x_{\sigma}) = F_{\chi^2_{n-1,\nu}}\left[\frac{(n-1)[(j+1)-(1-\lambda_{\sigma})(i+1/2)]}{\theta^2\lambda_{\sigma}(x_{\sigma}+1)}\left(1+\gamma_{E-\sigma}\sqrt{\frac{2\lambda_{\sigma}}{(n-1)(2-\lambda_{\sigma})}}\right)\right], \tag{13}$$

for $i, j = 0, \ldots, x_{\sigma}$.

2.1 Probability of a Misleading Signal

When dealing with simultaneous schemes for μ and σ^2, the main question is not whether MS will occur or not but rather how frequently they occur. This obviously suggests the use of an additional performance measure—the probability of misleading signal (PMS)—whose definition and (stochastic) monotonicity properties depend on those of the RL of the two individual residual charts. According to the definition of MS of types III and IV, the corresponding PMS can be written as

$$PMS_{III}(\theta,\phi) = P[RL_{\mu}(0,\theta,\phi) < RL_{\sigma}(0,\theta,\phi)] \tag{14}$$

$$= \sum_{i=1}^{+\infty}\left[\overline{F}_{RL_{\mu}(0,\theta,\phi)}(i-1) - \overline{F}_{RL_{\mu}(0,\theta,\phi)}(i)\right] \times \overline{F}_{RL_{\sigma}(0,\theta,\phi)}(i), \tag{15}$$

for $\theta > 1$, and

$$PMS_{IV}(\delta,\phi) = P[RL_{\sigma}(\delta,1,\phi) < RL_{\mu}(\delta,1,\phi)] \tag{16}$$

$$= \sum_{i=1}^{+\infty}\left[\overline{F}_{RL_{\sigma}(\delta,1,\phi)}(i-1) - \overline{F}_{RL_{\sigma}(\delta,1,\phi)}(i)\right] \times \overline{F}_{RL_{\mu}(\delta,1,\phi)}(i), \tag{17}$$

for $\delta > 0$.

A close inspection of the noncentrality parameter in Eq. (5) and of expressions (6) and (12) leads to the conclusion that, when $\delta = 0$, the RL of the individual residual charts for μ and σ^2 do not depend on the autocorrelation parameter ϕ. Therefore, we will denote the PMS of type III as $PMS_{III}(\theta)$.

It is worthy of note that we are able to obtain exact values for the PMS of the Shewhart simultaneous residual schemes and approximate values in the EWMA case. The exact expressions of the PMS of the simultaneous Shewhart residual schemes are obtained by plugging in the expressions of the survival functions of the RL into Eqs. (15) and (17):

$$PMS_{III-S}(\theta) = \frac{\xi_{S-\mu}(0,\theta,\phi) \times [1 - \xi_{S-\sigma}(0,\theta,\phi)]}{\xi_{S-\mu,\sigma}(0,\theta,\phi)}, \tag{18}$$

$$PMS_{IV-S}(\delta,\phi) = \frac{[1 - \xi_{S-\mu}(\delta,1,\phi)] \times \xi_{S-\sigma}(\delta,1,\phi)}{\xi_{S-\mu,\sigma}(\delta,1,\phi)}. \tag{19}$$

As for the EWMA schemes, we use the approximate expressions of the survival functions of the RL and truncate the series Eqs. (15) and (17).

In the next subsection we present some numerical results for the PMS of types III and IV for simultaneous Shewhart and EWMA residual schemes. These results suggest some monotonicity properties of this performance measure and provide insights into how the simultaneous residual schemes work in practice.

2.1.1 Numerical Results

The numerical values presented in this subsection were obtained by considering:

- Sample size equal to $n = 5$.
- Nominal values of the process mean and variance equal to $\mu_0 = 0$ and $\sigma_0^2 = 1$.
- $\lambda_\mu = \lambda_\sigma = \lambda \in \{1, 0.5, 0.05\}$, allowing the comparative assessment of a Shewhart and two EWMA schemes.
- $x_\mu + 1 = x_\sigma + 1 = 101$ transient states used in the Markov approach.
- Magnitude of the shift in the process standard deviation $\theta \in \{1.01, 1.5, 2, 3, 4\}$.
- Magnitude of the shift in the process mean $\delta \in \{0.05, 0.5, 1, 1.5\}$.
- Autoregressive parameter $\phi \in \{-0.9, -0.3, 0, 0.3, 0.9\}$.

The control limits of the individual residual charts depend on the so called critical values. These are obtained so that the in-control average run length (ARL) of the simultaneous residual schemes would be approximately 500 samples; and the individual residual charts would have the same in-control ARL. The critical values for the individual residual charts for σ^2 were obtained from [5]. The remaining critical values were calculated using the *regula falsi* method. The critical values for both individual residual charts are condensed in Table 2. Interestingly enough, these critical values can be used regardless of the value of the autocorrelation parameter.

The values of PMS of types III and IV are summarized in Tables 3 and 4 and surely deserve some comments. The PMS of type III can be larger than 0.48 for very small shifts in σ. Moreover, this PMS seems to be larger in the simultaneous Shewhart residual schemes than in the EWMA ones and it tends to decrease with

Table 2 Critical values for the individual residual charts

λ	γ_μ	γ_σ
1.00	3.0901	5.1144
0.50	3.1598	4.6439
0.05	2.8296	2.9103

Table 3 PMS of type III for simultaneous Shewhart ($\lambda = 1$) and EWMA residual schemes

	λ		
θ	1.00	0.50	0.05
1.01	0.48434	0.47557	0.41372
1.50	0.17627	0.11108	0.01046
2.00	0.11093	0.07190	**0.00471**
3.00	0.05402	0.04169	**0.00359**
4.00	0.02762	0.02395	**0.00442**

Table 4 PMS of type IV for simultaneous Shewhart ($\lambda = 1$) and EWMA residual schemes

	ϕ					
δ	−0.9	−0.3	0	0.3	0.9	λ
0.05	0.21989	0.38296	0.40806	0.42711	0.46452	1
	0.13380	0.32932	0.36466	0.39207	0.44700	0.5
	0.03591	0.17686	0.22642	0.27166	0.37966	0.05
0.5	0.00233	0.01979	0.03862	0.06979	0.24436	1
	0.00340	0.00680	0.01325	0.02654	0.15325	0.5
	0.00184	*0.00132*	*0.00269*	0.00570	**0.05015**	0.05
1	0.00000	0.00171	0.00407	0.01099	0.14811	1
	0.00000	0.00108	0.00182	0.00412	0.07071	0.5
	0.01131	*0.00013*	*0.00024*	0.00074	**0.03010**	0.05
1.5	0.00000	0.00019	0.00066	0.00265	0.11218	1
	0.00000	0.00028	0.00054	0.00148	0.05459	0.5
	0.00009	0.00004	0.00005	0.00020	**0.03357**	0.05

θ for the Shewhart schemes. The values in **bold** in Table 3 indicate the absence of monotonicity behaviour in terms of θ for the EWMA case.

As for the PMS of type IV, we ought to state that these probabilities can be larger than 0.46 for small shifts in μ. Like $PMS_{III}(\theta)$, these probabilities appear to be larger in the Shewhart case when the autoregressive parameter ϕ is positive. $PMS_{IV}(\delta, \theta)$ seems to increase with nonnegative values of ϕ and shows signs of decreasing with δ for the Shewhart schemes. The values in **bold** (resp. *italic*) in Table 4 show that there is no monotonic behaviour in terms of δ for the EWMA schemes (resp. in terms of ϕ for nonpositive values of this parameter).

Proving the monotonicity properties suggested by the analysis of these tables requires the use of stochastic ordering [16], as shown in the next section.

3 Stochastic Monotonicity Properties

We start the section with some auxiliary results which play a major role in the assessment of the stochastic monotonicity properties of the RL of the individual residual charts and therefore in the monotonicity properties of the PMS.

Lemma 1 ([9], p. 16). *Let $RL(\Delta) \sim geometric(\xi(\Delta))$ be the RL of a Shewhart-type control chart, where $\xi(\Delta)$ represents the probability of detecting a shift of magnitude Δ. Then, if $\xi(\Delta)$ increases ($\uparrow$) with Δ, $RL(\xi(\Delta))$ stochastically decreases ($\downarrow_{st}$) with Δ, i.e., $\overline{F}_{RL(\xi(\Delta))}(m)$ decreases ($\downarrow$) with Δ, for all m.*

Lemma 2 ([10]). *Let $\{S_N(\Delta),\ N \in \mathbb{N}_0\}$ be an absorbing Markov chain with single absorbing state $x+1$, state space $\{0, 1, \ldots, x, x+1\}$, governed by the transition matrix $P(\Delta) = [p_{ij}(\Delta)]$. If all left partial sums of $P(\Delta)$, $\sum_{j=1}^{k} p_{ij}(\Delta)$, decrease with i and decrease with Δ, then $RL(\Delta) \downarrow_{st}$ with Δ.*

Lemma 3. *Let $X_\nu \sim \chi^2_{n-1,\nu}$ be a continuous random variable with noncentral chi-squared distribution. Then, X_ν stochastically increases ($\uparrow_{st}$) with ν.*

Proof. According to ([4], p. 435), the distribution function of X can be written as an expected value $F_{\chi^2_{n-1,\nu}}(x) = E_Y[F_{\chi^2_{n-1+2Y_{\nu/2}}}(x)]$, where $Y_{\nu/2} \sim$ Poisson$(\nu/2)$. On one hand $\chi^2_\lambda \uparrow_{st}$ with λ, i.e., $F_{\chi^2_\lambda}(x) \downarrow$ with λ, for any fixed x. On the other hand $Y_{\nu/2}$ increases in likelihood ratio ($\uparrow_{lr}$) with ν ([15], p. 281). Thus, $E[\varphi(Y_\nu)] \uparrow$ with ν, for any nondecreasing function $\varphi(.)$, according to ([16], p. 4). Consequently, we conclude that $F_{\chi^2_{n-1,\nu}}(x) \downarrow$ with ν for any x, i.e., $X_\nu \uparrow_{st}$ with ν. □

3.1 RL of Individual Residual Charts

Making use of the previous lemmas, the monotonicity properties of the survival functions of the RL and the left partial sums of the matrices $\mathbf{Q}_\mu(\delta, \theta, \phi; x_\mu)$, and $\mathbf{Q}_\sigma(\delta, \theta, \phi; x_\sigma)$, we are able to state the following stochastic monotonicity properties of the RL.

Theorem 1. *The following results are, respectively, valid for the RL of the Shewhart and EWMA upper one-sided individual residual charts for the mean and variance of an $AR(1)$ process.*

Proof. Results (1.1), (1.2) and (1.3) follow in a straightforward manner from Lemma 1 in view of expressions (6) and (7).

As for the EWMA charts, result (1.8) [resp. (1.9)] follows from Lemma 2 due to the decreasing (resp. increasing) behaviour of the partial row sum Eq. (12) [resp. Eq. (13)].

Individual residual chart	μ	σ
Shewhart	(1.1) $RL_{S-\mu}(0,\theta,\phi) \downarrow_{st}$ with θ	(1.4) $RL_{S-\sigma}(\delta,\theta,\phi) \uparrow_{st}$ with θ
	(1.2) $RL_{S-\mu}(\delta,\theta,\phi) \downarrow_{st}$ with δ	(1.5) $RL_{S-\sigma}(\delta,\theta,\phi) \downarrow_{st}$ with δ
	(1.3) $RL_{S-\mu}(\delta,\theta,\phi) \uparrow_{st}$ with ϕ	(1.6) $RL_{S-\sigma}(\delta,\theta,\phi) \uparrow_{st}$ with $\phi \in (-1,0]$
		(1.7) $RL_{S-\sigma}(\delta,\theta,\phi) \downarrow_{st}$ with $\phi \in (0,1)$
EWMA	(1.8) $RL_{E-\mu}(\delta,\theta,\phi) \downarrow_{st}$ with δ	(1.10) $RL_{E-\sigma}(\delta,\theta,\phi) \downarrow_{st}$ with δ
	(1.9) $RL_{E-\mu}(\delta,\theta,\phi) \uparrow_{st}$ with ϕ	(1.11) $RL_{E-\sigma}(\delta,\theta,\phi) \uparrow_{st}$ with $\phi \in (-1,0]$
		(1.12) $RL_{E-\sigma}(\delta,\theta,\phi) \downarrow_{st}$ with $\phi \in (0,1)$

For both individual residual charts, results (1.4)–(1.7) and (1.10)–(1.12) follow from Lemmas 2 and 3, in view of Eq. (13). For example, to prove (1.6) and (1.11) it suffices to show that ν decreases with ϕ when $\phi \in (-1,0]$ and apply Lemma 3. In fact, $[1-\sqrt{(1-\phi)/(1+\phi)}]^2$ decreases with $\phi \in (-1,0]$, and so does ν, thus proving results (1.6) and (1.11). □

3.2 Probabilities of Misleading Signals

Using the definitions of PMS and on Theorem 1, we are able to derive the following monotonicity properties of PMS of types III and IV.

Theorem 2. *The following monotonicity properties are valid for the PMS of types III and IV of simultaneous Shewhart and EWMA schemes based on upper one-sided individual residual charts for the mean and the variance of an $AR(1)$ process.*

Simultaneous scheme	$PMS_{III}(\theta)$	$PMS_{IV}(\delta)$
Shewhart	(2.1) $PMS_{III-S}(\theta) \downarrow$ with θ	(2.2) $PMS_{IV-S}(\delta,\phi) \downarrow$ with δ
		(2.3) $PMS_{IV-S}(\delta,\phi) \uparrow$ with ϕ
EWMA	(C1) $PMS_{III-E}(\theta)$[a]	(C2) $PMS_{IV-E}(\delta,\phi)$[b]
		(C3) $PMS_{IV-E}(\delta,\phi)$[c]
		(2.4) $PMS_{IV-E}(\delta,\phi) \uparrow$ with $\phi \in (0,1)$

[a]Non-monotonic in θ
[b]Non-monotonic in δ
[c]Non-monotonic in $\phi \in (-1,0]$

Proof. To prove result (2.1) we need to rewrite $PMS_{III-S}(\theta)$ as follows:

$$\left(1-\frac{1-\left\{F_{\chi^2_{n-1}}\left[\frac{n-1}{\theta^2}\left(1+\gamma_{S-\sigma}\sqrt{\frac{2}{n-1}}\right)\right]\right\}^{-1}}{1-\Phi\left(\frac{\gamma_{S-\mu}}{\theta}\right)}\right)^{-1} \tag{20}$$

and remind the reader that $F_{\chi^2_{n-1}}\left[\frac{n-1}{\theta^2}\left(1+\gamma_{S-\sigma}\sqrt{\frac{2}{n-1}}\right)\right]$ and $\Phi\left(\frac{\gamma_{S-\mu}}{\theta}\right)$ decrease with θ.

Similarly, $PMS_{IV-S}(\delta,\phi)$ can be rewritten as

$$\left(1-\frac{1-\left\{\Phi\left[\gamma_{S-\mu}-\frac{\delta}{\sqrt{n}}\left(1+(n-1)\sqrt{\frac{1-\phi}{1+\phi}}\right)\right]\right\}^{-1}}{1-F_{\chi^2_{n-1,\nu}}\left[(n-1)\left(1+\gamma_{S-\sigma}\sqrt{\frac{2}{n-1}}\right)\right]}\right)^{-1}. \tag{21}$$

Since $F_{\chi^2_{n-1,\nu}}\left[(n-1)\left(1+\gamma_{S-\sigma}\sqrt{\frac{2}{n-1}}\right)\right]$ and $\Phi\left[\gamma_{S-\mu}-\frac{\delta}{\sqrt{n}}\left(1+(n-1)\sqrt{\frac{1-\phi}{1+\phi}}\right)\right]$ decrease with δ, we immediately conclude that $PMS_{IV-S}(\delta,\phi)\downarrow$ with δ.

To prove result (2.3) we need to consider two different cases:

- Case $\phi \in (0,1)$
  According to results (1.3) and (1.7) of Theorem 1, $RL_\mu(\delta,\theta,\phi)\uparrow_{st}$ with $\phi \in (0,1)$ and $RL_\sigma(\delta,\theta,\phi)\downarrow_{st}$ with $\phi \in (0,1)$. Having in mind these two results and Eq. (17), we conclude $PMS_{IV}(\delta,\phi)\uparrow$ with $\phi \in (0,1)$.
- Case $\phi \in (-1,0]$
  On one hand, when $\theta = 1$, the noncentrality parameter

$$\nu = \frac{n-1}{n}\delta^2\left(1-\sqrt{(1-\phi)/(1+\phi)}\right)^2 \downarrow \phi \in (-1,0]; \tag{22}$$

thus, from Lemma 3, $F_{\chi^2_{n-1,\nu}}(x)\uparrow$ with $\phi \in (-1,0]$. As on the other hand,

$$\Phi\left[\gamma_{S-\mu}-\frac{\delta}{\sqrt{n}}\left(1+(n-1)\sqrt{(1-\phi)(1+\phi)}\right)\right]\uparrow \text{ with } \phi \in (-1,0], \tag{23}$$

$PMS_{IV}(\delta,\phi)\uparrow$ with $\phi \in (-1,0]$ in view of Eq. (17).

Since $PMS_{IV}(\delta,\phi) = P[RL_\sigma(\delta,1,\phi) < RL_\mu(\delta,1,\phi)]$, and $RL_{\star-\mu}(\delta,\theta,\phi)\uparrow_{st}$ with ϕ and $RL_{\star-\sigma}(\delta,\theta,\phi)\downarrow_{st}$ with $\phi \in (0,1)$ by results (1.3), (1.7), (1.9) and (1.12) from Theorem 1, where $\star = S, E$, we prove results (2.3) and (2.4). □

We ought to end this subsection by noting that conjectures (C1), (C2) and (C3) in Theorem 2 refer to the nonexistence of monotonic behaviour in terms of θ, δ and $\phi \in (-1,0]$, respectively, and are supported by the numerical results in Tables 3 and 4.

4 Concluding Remarks

The results presented in this chapter show that MS of types III and IV are very likely to happen in simultaneous residual schemes, in particular for very small shifts in the process mean and variance. In this respect, we note that MS of type III are not affected by the autocorrelation of the process, as previously noted by [2, 7].

The use of stochastic ordering allowed us to make a qualitative assessment of the impact of the presence of autocorrelation on the performance of simultaneous residual schemes. For instance, we have proved that the PMS of type III is a decreasing function of the shift in σ which means that an underestimation of the magnitude of this shift results in an overestimation of the PMS. We have also shown that larger nonnegative values of the autocorrelation parameter are associated to more frequent MS of type IV.

Acknowledgements The authors thank the support of *Centro de Matemática e Aplicações* (CEMAT) and *Fundação para a Ciência e a Tecnologia* (FCT). The first author was supported also by grant SFRH/BD/35739/2007 of FCT and would like to thank all the people in the Department of Statistics of the European University Viadrina (Frankfurt Oder, Germany) for their hospitality.

References

1. Alwan, L.C., Roberts, H.V.: Time-series modeling for statistical process control. J. Bus. Econ. Stat. **6**, 87–95 (1988)
2. Antunes, C.: Avaliação do impacto da correlação em sinais erróneos de esquemas conjuntos para o valor esperado e variância (assessment of the impact of the correlation on misleading signals in joint schemes for the mean and variance). Master's thesis, Instituto Superior Técnico, Universidade Técnica de Lisboa (2009)
3. Brook, D., Evans, D.A.: An approach to the probability distribution of CUSUM run length. Biometrika **59**, 539–549 (1972)
4. Johnson, N.L., Kotz, S., Balakrishnan, N.: Continuous Univariate Distributions, vol. 2. Wiley, New York (1995)
5. Knoth, S., Schmid, W.: Monitoring the mean and the variance of a stationary process. Stat. Neerl. **56**, 77–100 (2002)
6. Knoth, S., Schmid, W., Schöne, A.: Simultaneous Shewhart-type charts for the mean and the variance of a time series. In: Lenz, H.J., Wilrich, P.-Th. (eds.) Frontiers of Statistical Quality Control, vol. 8, pp. 61–79. Physica, Heidelberg (2001)
7. Knoth, S., Morais, M.C., Pacheco, A., Schmid, W.: Misleading signals in simultaneous residual schemes for the mean and variance of a stationary process. Commun. Stat. Theor. Meth. **38**, 2923–2943 (2009)
8. Montgomery, D.C.: Introduction to Statistical Quality Control. Wiley, Hoboken (1985)
9. Morais, M.J.C.: Stochastic Ordering in the Performance Analysis of Quality Control Schemes. Ph.D. thesis, Instituto Superior Técnico, Universidade Técnica de Lisboa (2002)
10. Morais, M.C., Pacheco, A.: Two stochastic properties of one-sided exponentially weighted moving average control charts. Commun. Stat. Simul. Comput. **27**, 937–952 (1998)
11. Morais, M.C., Pacheco, A.: On the performance of combined EWMA schemes for μ and σ: a Markovian approach. Commun. Stat. Simul. Comput. **29**, 153–174 (2000)
12. Morais, M.C., Pacheco, A.: Misleading signals in joint schemes for μ and σ. In: Lenz, H.J., Wilrich, P.-Th. (eds.) Frontiers of Statistical Quality Control, vol. 16, pp. 100–122. Physica, Heidelberg (2006)
13. Reynolds Jr, M.R., Stoumbos, Z.G.: Monitoring the process mean and variance using individual observations and variable sampling intervals. J. Qual. Technol. **33**, 181–205 (2001)
14. Reynolds Jr, M.R., Stoumbos, Z.G.: Control charts and the efficient allocation of sampling resources. Technometrics **46**, 200–214 (2004)
15. Ross, S.: Stochastic Processes. Wiley, New York (1996)
16. Shaked, M., Shanthikumar, J.G.: Stochastic Orders and Their Applications. Academic, Boston (1994)
17. St.John, R.C., Bragg, D.J.: Joint X-bar and R-charts under shift in mu or sigma. ASQC Quality Congress Transactions — Milwaukee, 547–550 (1991)

Conditional EVT for VAR Estimation: Comparison with a New Independence Test

M.I. Fraga Alves and P. Araújo Santos

Abstract
We compare the out-of-sample performance of methods for value-at-risk (VaR) estimation, using a new exact independence test. This test is appropriate for detecting risk models with a tendency to generate clusters of violations and evaluating the performance under heteroscedastic time series. We focus the comparison on a two-stage hybrid method which combines a GARCH filter with an extreme value theory (EVT) approach, known as conditional EVT. Previous comparative studies show that this method performs better for VaR estimation. Our contributions are comparing the performance with the new exact independence test and considering recent developments in EVT involving bias reduction.

1 Introduction

The desire for a less fragile financial system, increase the demand for quantitative risk management tools. The value-at-risk (VaR) aggregates several components of risk into a single number and has emerged as the standard measure that financial

M.I.F. Alves (✉)
Departamento de Estatística e Investigação Operacional, Universidade de Lisboa, Lisboa, Portugal

CEAUL, Lisboa, Portugal
e-mail: isabel.alves@fc.ul.pt

P.A. Santos
Departamento de Informática e Métodos Quantitativos, Escola Superior de Gestão e Tecnologia, Instituto Politécnico de Santarém, Lisboa, Portugal

CEAUL, Lisboa, Portugal
e-mail: paulo.santos@esg.ipsantarem.pt

J. Lita da Silva et al. (eds.), *Advances in Regression, Survival Analysis, Extreme Values, Markov Processes and Other Statistical Applications*, Studies in Theoretical and Applied Statistics, DOI 10.1007/978-3-642-34904-1_19,

analysts use to quantify risk. In terms of regulation, the Basel Committee on banking and supervision [4] imposes capital requirements to banks and investment firms, based on VaR estimation; see, for example, Kuester et al. [24] and the references therein for a survey of competing methods. For a detailed discussion of VaR, see Jorion [23].

Let $R_{t+1} = \log(V_{t+1}/V_t)$ be the log returns at time $t+1$, where V_t is the value of the portfolio at time t. The one-day-ahead VaR forecasts made at time t for time $t+1$, $VaR_{t+1|t}(p)$, is defined by

$$P[R_{t+1} \leq VaR_{t+1|t}(p)|\Omega_t] = p,$$

where Ω_t is the information setup to time t and p is the *coverage rate*. In Sect. 2, we summarize some of the major approaches to VaR estimation. In Sect. 3, we present the results of our comparative out-of-sample study. The final section, provides final remarks.

2 VaR Methods

We consider, for the out-of-sample study, the following methods.

2.1 Historical Simulation

The simplest way to estimate $VaR_{t+1|t}(p)$ is to use the unconditional quantile of the past n_w returns:

$$VaR^{HS}_{t+1|t}(p) := \text{quantile}(\{R_s\}^{n_w}_{s=1}, 100p).$$

2.2 Filtered Historical Simulation (FHS)

Barone-Adesi et al. [3] proposed the combination of a volatility model and the historical simulation (HS) method:

$$VaR^{FHS}_{t+1|t}(p) := \hat{\mu}_{t+1|t} + \hat{\sigma}_{t+1|t}\text{quantile}(\{Z_s\}^{n_w}_{s=1}, 100p),$$

where Z_s are the standardized residuals using a AR(1) GARCH(1,1) process; $\hat{\mu}_{t+1|t}$ and $\hat{\sigma}_{t+1|t}$ are the conditional mean estimators and conditional volatility at time $t+1$. We denote this method with normal innovations Z_t (presented in Sect. 2.4) by **N-HS** and with *Skewed-t* innovations by **ST-HS**.

2.3 Unconditional Skewed t

We choose one popular parametric unconditional model, the *Skewed-t* (ST). With this model, we assume for the returns $R_t = \mu + \sigma Z_t$, where μ and σ are the unconditional mean and standard deviation and Z_t are iid to a random variable (r.v.) Z with *Skewed-t* (Fernandez and Steel [12]):

$$VaR^{ST}_{t+1|t}(p) := \hat{\mu}_{t+1} + \hat{\sigma}_{t+1} F_Z^{-1}(p).$$

Here $F_Z^{-1}(p) := \inf\{x : F_z(x) \geq p\}$ denotes the generalized inverse function of the distribution function $F_z(x)$ of the r.v. Z.

2.4 Heteroscedastic Parametric Models

We consider a model, denoted by AR(1) GARCH(1,1), such that $R_t = \mu_t + \epsilon_t = \mu_t + \sigma_t Z_t$, with $\mu_t = \phi R_{t-1}$, $\phi \in \Re$ and $\sigma_t^2 = \alpha_0 + \alpha_1 \epsilon_{t-1}^2 + \beta_1 \sigma_{t-1}^2$, $\alpha_0 > 0$, $\alpha_1 > 0$ and $\beta_1 > 0$. We denote this method with *Normal* innovations Z_t by **N-GARCH** and with *Skewed-t* innovations Z_t by **ST-GARCH**. Several studies showed excellent forecast results with GARCH type with *Skewed-t*. See, for example, Mittnik and Paolella [27] and Giot and Laurent [15].

2.5 Unconditional Peaks over Threshold from EVT

The peaks over threshold (POT) method is based on Balkema–de Haan–Pickands theorem on the distribution of excesses over a high threshold. See Balkema and de Haan [2] and Pickands [29] for details. We apply the method with the MLE implemented in the POT package [31] of the R software [30]:

$$VaR^{POT}_{t+1|t}(p) := u + \frac{\hat{\delta}}{\hat{\gamma}} \left\{ \left(\frac{k}{np}\right)^{\hat{\gamma}} - 1 \right\}$$

where n is the sample size, k is the number of excesses over u $\hat{\gamma}$ and $\hat{\delta}$ are estimates of the parameters γ and δ of the generalized Pareto distribution (GPD):

$$H_{\gamma,\delta}(x) = \begin{cases} 1 - (1 + \gamma x/\delta)^{-1/\gamma}, & 1 + \gamma x/\delta > 0, \ \gamma \neq 0 \\ 1 - \exp(-x/\delta), & \gamma = 0. \end{cases}$$

2.6 Unconditional Minimum Variance Reduced Bias from EVT

The classical Weissman estimator [32] is defined by

$$VaR^{W}_{t+1|t}(p) := X_{n-k:n} \left(\frac{k}{np}\right)^{\hat{\gamma}},$$

with $X_{n-k:n}$ the $(k+1)$ top order statistic of a random sample $\{X_i, 1 \leq i \leq n\}$ and $\hat{\gamma}$ some consistent estimator of the tail index $\gamma > 0$. The Hill estimator for the positive tail index [22],

$$H(k) := \frac{1}{k}\sum_{j=1}^{k} \log \frac{X_{n-j+1:n}}{X_{n-k:n}},$$

may exhibit a strong bias for moderate k, if the underlying model is not a strict Pareto model. Recent developments in EVT involve the reduction of bias. For example, Peng [28], Beirlant et al. [5], Gomes et al. [16, 17], Gomes and Martins [18], Caeiro and Gomes [8], among others. They achieved γ estimators with asymptotic variance equal or higher than $(\gamma(1-\rho)/\rho)^2 > \gamma^2$, with ρ the second-order parameter.

Recently, Caeiro et al. [9], Gomes and Pestana [19], Gomes et al. [20], and Gomes et al. [21] proposed minimum variance and reduced bias (MVRB) estimators for γ. They reduce bias without increasing the asymptotic variance, which is kept at the value γ^2. Here we consider the MVRB γ estimator introduced by Caeiro et al. [9]

$$\bar{H}(k) := H(k)\left\{1 - \frac{\hat{\beta}}{1-\hat{\rho}}\left(\frac{n}{k}\right)^{\hat{\rho}}\right\}$$

where $\hat{\rho}$ and $\hat{\beta}$ are estimates of the second-order parameters ρ and β. See Fraga Alves et al. [13] for ρ estimation and Gomes et al. [21] for β estimation. We obtain the estimates of ρ and β using the algorithm suggested in Gomes and Pestana [19].

2.7 Conditional EVT

Diebold et al. [11] proposed in a *first step* the standardization of the returns through the conditional standard deviations estimated with a volatility model. In a *second step*, estimation of a p quantile using the EVT and the standardized returns. McNeil and Frey [26] combine a AR(1) GARCH(1,1) assuming normal innovations with the POT method from EVT. This is the conditional EVT method. Formally

$$VaR^{cEVT}_{t+1|t}(p) := \hat{\mu}_{t+1|t} + \hat{\sigma}_{t+1|t}\hat{z}_p$$

where $\hat{\mu}_{t+1|t}$ and $\hat{\sigma}_{t+1|t}$ are the conditional mean estimates and conditional volatility at time $t+1$, obtained with a AR(1) GARCH(1,1) process. $\hat{z}_p$ is a quantile p estimate, obtained with an EVT method and the standardized residuals. Several studies conclude that conditional EVT is the method with better out-of-sample performance, to estimate $VaR_{t+1|t}(p)$, for example, McNeil and Frey [26], Bystrom [7], Bekiros and Georgoutsos [6], Kuester et al. [24], and Ghorbel and Trabelsi [14].

For the comparative study in Sect. 3, we combine the POT and MVRB methods with two filters: one involving *normal* innovations and the other, *Skewed-t* innovations, reaching four conditional EVT methods: **N-POT**, **ST-POT**, **N-MVRB**, and **ST-MVRB**. In our study we consider recent developments in EVT involving bias reduction, and not only the classical POT or Block Maxima methods. Additionally, we use in the backtesting a new independence test with several advantages.

3 Out-of-Sample Study

Considering a *violation* the event that a return on the portfolio is lower than the reported VaR, the "hit" function is defined by

$$I_{t+1}(p) = \begin{cases} 1 \text{ se } R_{t+1} < VaR_{t+1|t}(p) \\ 0 \text{ se } R_{t+1} \geq VaR_{t+1|t}(p). \end{cases}$$

Christoffersen [10] showed that a forecast model is accurate when the hit sequence, $\{I_t\}_{t=1}^{T}$, satisfies the unconditional coverage (UC) and independence properties (IND). UC hypothesis means $P[I_{t+1}(p) = 1] = p, \forall t_t$. IND hypothesis means that past information does not hold information about future violations. We test the UC hypothesis with the Kupiec [25] test. This test measures whether the number of violations is consistent with the coverage rate and is a likelihood ratio test and the test statistic is asymptotically chi-squared distributed with one degree of freedom.

Let us define the duration between two violations as $D_i := t_i - t_{i-1}$, where t_i denotes the day of violation number i and $t_0 = 0$. The IND hypothesis can be expressed as

$$D_i \overset{iid}{\sim} D \sim \text{Geometric}(\pi), \text{ with } 0 < \pi < 1.$$

Considering the order statistics $D_{1:N}, \ldots, D_{N:N}$ of durations $D_1, \ldots, D_N$, for testing the IND hypothesis versus tendency to clustering of violations, we apply the test statistic proposed in Araújo Santos and Fraga Alves [1]

$$R_{N,[N/2]} = \frac{D_{N:N} - 1}{D_{[N/2]:N}}. \tag{1}$$

The asymptotic distribution for the test is Gumbel and the exact distribution is given in Proposition 3.1 of Araújo Santos and Fraga Alves [1]. The methods were backtested with two historical series: 20326 log returns from Dow Jones Industrial Average index (October 2, 1928, to September 11, 2009) and 15019 log returns from Standard and Poor's 500 index (January 4, 1950, to September 11, 2009). The data come from the Web site http://finance.yahoo.com/. We calculate $VaR_{t+1|t}(p)$ using moving windows of size $n_w = 1,000$ days and $p = 0.05, 0.01, 0.0025, 0.001, 0.005$. As in previous studies, for the EVT methods, we choose the number of $(k+1)$ top order statistics with $k = 100$. The programs were written in the R language, with the fGarch (Wuertz et al. [33]) and POT (Ribatet [31]) packages. Tables 1 and 2 present the percentage of violations and the p values for the Kupiec test. Tables 3 and 4 present the observed values of the independence test statistic Eq. (1) and the p values computed with Monte Carlo simulations.

Table 1 Dow Jones industrial average index

100p	Model	% Viol.	Kupiec p value	Model	% Viol.	Kupiec p value	Model	% Viol.	Kupiec p value
5	HS	5.231	0.143	N-HS	5.195	0.216	ST-HS	5.159	0.313
1		1.325	0.000		1.133	0.068		1.185	0.012
0.1		0.279	0.000		0.207	0.000		0.217	0.000
5	ST	5.449	0.005	N-GARCH	5.211	0.182	ST-GARCH	5.329	0.037
1		1.242	0.001		1.718	0.000		1.164	0.025
0.1		0.160	0.015		0.595	0.000		0.124	0.306
5	POT	5.211	0.182	N-POT	5.143	0.326	ST-POT	5.117	0.455
1		1.252	0.001		1.004	0.957		1.035	0.628
0.1		0.181	0.001		0.160	0.015		0.160	0.015
5	MVRB	6.711	0.000	N-MVRB	6.540	0.000	ST-MVRB	6.556	0.000
1		1.842	0.000		1.557	0.000		1.511	0.000
0.1		0.098	0.941		0.083	0.435		0.083	0.435

UC Unconditional coverage

Table 2 Standard and Poor's 500 index

100p	Model	% Viol.	Kupiec p value	Model	% Viol.	Kupiec p value	Model	% Viol.	Kupiec p value
5	HS	5.849	0.000	N-HS	5.207	0.263	ST-HS	5.243	0.190
1		1.569	0.000		1.191	0.027		1.170	0.049
0.1		0.314	0.000		0.243	0.000		0.214	0.000
5	ST	6.077	0.000	GARCH-N	5.393	0.035	GARCH-ST	5.443	0.018
1		1.405	0.000		1.698	0.000		1.006	0.945
0.1		0.143	0.133		0.585	0.000		0.143	0.133
5	POT	5.877	0.000	N-POT	5.172	0.354	ST-POT	5.200	0.280
1		1.384	0.000		1.006	0.945		1.006	0.945
0.1		0.221	0.001		0.164	0.028		0.150	0.083
5	MVRB	7.276	0.000	N-MVRB	6.648	0.000	ST-MVRB	6.648	0.000
1		2.061	0.000		1.591	0.000		1.591	0.000
0.1		0.157	0.049		0.107	0.796		0.100	0.996

UC Unconditional coverage

4 Final Remarks

- We confirm the poor performance of the unconditional methods. In almost all cases, the new independence test rejects the IND hypothesis with high ratios (1).
- With $p = 0.01$, the conditional POT methods perform better than all other methods, in most cases.
- With the very small coverage rate $p = 0.001$, the MVRB and ST-GARCH methods are the best performers.

Table 3 Dow Jones industrial average index

100p	Model	r obs.	p value	Model	r obs.	p value	Model	r obs.	p value
5	HS	94.6	0.000	N-HS	10.2	0.589	ST-HS	16.6	0.011
1		76.0	0.000		7.6	0.692		7.5	0.736
0.1		24.9	0.000		4.5	0.831		4.5	0.822
5	ST	74.1	0.000	GARCH-N	10.2	0.594	GARCH-ST	10.2	0.596
1		64.5	0.000		10.0	0.293		8.6	0.467
0.1		10.1	0.075		15.3	0.007		9.5	0.096
5	POT	94.6	0.000	N-POT	10.1	0.603	ST-POT	16.6	0.011
1		70.6	0.000		7.2	0.729		7.7	0.634
0.1		9.2	0.148		3.7	0.920		3.5	0.945
5	MVRB	80.0	0.000	N-MVRB	10.3	0.634	ST-MVRB	11.4	0.378
1		89.8	0.000		9.3	0.410		9.1	0.439
0.1		8.3	0.199		3.9	0.711		4.0	0.676

IND Independence

Table 4 Standard and Poor's 500 index

100p	Model	r obs.	p value	Model	r obs.	p value	Model	r obs.	p value
5	HS	112.2	0.000	N-HS	19.4	0.001	ST-HS	19.4	0.001
1		89.8	0.000		9.7	0.227		7.4	0.630
0.1		36.8	0.000		5.4	0.587		5.5	0.552
5	ST	112.2	0.000	GARCH-N	19.4	0.001	GARCH-ST	21.2	0.000
1		70.6	0.000		13.8	0.024		14.1	0.016
0.1		15.6	0.018		15.1	0.007		9.0	0.113
5	POT	112.2	0.000	N-POT	19.4	0.001	ST-POT	19.4	0.001
1		76.0	0.000		13.9	0.017		12.9	0.032
0.1		19.2	0.000		3.7	0.881		3.5	0.892
5	MVRB	134.6	0.000	N-MVRB	12.0	0.213	ST-MVRB	12.0	0.217
1		89.8	0.000		11.6	0.095		9.7	0.268
0.1		10.3	0.095		13.8	0.039		10.0	0.082

IND Independence

Acknowledgements Research partially supported by National Funds through **FCT**—Fundação para a Ciência e a Tecnologia, FCT/PROTEC, project PEst-OE/MAT/UI0006/2011, and FCT /PTDC/MAT/101736/2008, EXTREMA project. The authors would like to thank the two referees for their comments and suggestions which lead to improvements of an earlier version of this chapter.

References

1. Araújo Santos, P., Fraga Alves, M.I.: A new class of independence tests for interval forecasts evaluation. Computational Statistics and Data Analysis, **56**(11), 3366–3380 (2012)
2. Balkema, A.A. e de Haan, L.: Residual life time at great age. Ann. Probab. **2**, 792–804 (1974)
3. Barone-Adesi, G., Bourgoin, F., Giannopoulos, K.: Dont look back. Risk **11**(8), 100–103 (1998)

4. Basel Committee on Banking Supervision: Supervisory Framework for the Use of Backtesting in Conjunction with the Internal Model-Based Approach to Market Risk Capital Requirements. BIS, Basel, Switzerland (1996)
5. Beirlant, J., Dierckx, G., Goegebeur, Y., Matthys, G.: Tail index estimation and exponential regression model. Extremes **2**, 177–200 (1999)
6. Bekiros, S.D., Georgoutsos, D.A.: Estimation of value-at-risk by extreme value and conventional methods: a comparative evaluation of their predictive performance. J. Int. Financ. Market. Inst. Money **15**(3), 209–228 (2005)
7. Bystrom, H.: Managing extreme risks in tranquil and volatile markets using conditional extreme value theory. Int. Rev. Financ. Anal. **13**, 133–152 (2004)
8. Caeiro, F., Gomes, M.I.: A class of asymptotically unbiased semi-parametric estimators of the tail index. Test **11**(2), 345–364 (2002)
9. Caeiro, F., Gomes, M.I., Pestana, D.: Direct reduction of bias of the classical Hill estimator. RevStat **3**(2), 111–136 (2005)
10. Christoffersen, P.: Evaluating intervals forecasts. Int. Econ. Rev. **39**, 841–862 (1998)
11. Diebold, F.X., Schuermann, T., Stroughair, J.D.: Pitfalls and opportunities in the use of extreme value theory in risk management. In: Working Paper, pp. 98–10, Wharton School, University of Pennsylvania (1998)
12. Fernandez, C., Steel, M.F.J.: On Bayesian modelling of fat tails and skewness. J. Am. Stat. Assoc. **93**, 359–371 (1998)
13. Fraga Alves, M.I., Gomes, M.I., de Haan, L.: A new class of semi-parametric estimators of the second order parameter. Portugaliae Mathematica **60**(1), 193–213 (2003)
14. Ghorbel, A., Trabelsi, A.: Predictive performance of conditional extreme value theory in value-at-risk estimation. Int. J. Monetary Econ. Finance. **1**(2), 121–147 (2008)
15. Giot, P., Laurent, S.: Modelling daily value-at-risk using realized volatility and ARCH type models. J. Empir. Finance **11**, 379–398 (2004)
16. Gomes, M.I., Martins, M.J., Neves M.M.: Alternatives to a semi-parametric estimator of parameters of rare events - the Jackknife methodology. Extremes **3**(3), 207–229 (2000)
17. Gomes, M.I., Martins, M.J., Neves M.M.: Generalized Jackknife semi-parametric estimators of the tail index. Portugaliae Mathematics **59**(4), 393–408 (2002)
18. Gomes, M.I., Martins, M.J.: Asymptotically unbiased estimators of the tail index based on the external estimation of the second order parameter. Extremes **5**(1), 5–31 (2002)
19. Gomes, M.I., Pestana D.: A sturdy reduced bias extreme quantile (VaR) estimator. J. Am. Statist. Assoc. **102**(477), 280–292 (2007)
20. Gomes, M.I., Martins, M.J., Neves, M.M.: Improving second order reduced-bias tail index estimator. Revstat **5**(2), 177–207 (2007)
21. Gomes, M.I., de Haan, L., Henriques Rodrigues, L.: Tail Index estimation for heavy-tailed models: accommodation of bias in weighted log-excesses. J. Roy. Stat. Soc. **B70**(1), 31–52 (2008)
22. Hill, B.M.: A simple general approach to inference about the tail of a distribution. Ann. Statist. **3**(5), 1163–1174 (1975)
23. Jorian, P.: Value at Risk: The New Benchmark for Managing Financial Risk. McGraw- Hill, New York (2000)
24. Kuester, K., Mittik, S., Paolella, M.S.: Value-at-risk prediction: a comparison of alternative strategies. J. Financ. Econometrics **4**(1), 53–89 (2006)
25. Kupiec, P.: Techniques for verifying the accuracy of risk measurement models. J. Deriv. **3**, 73–84 (1995)
26. McNeil, A.J., Frey, R.: Estimation of tail-related risk measures for heteroscedastic financial time series: an extreme value approach. J. Empir. Finance **7**, 271–300 (2000)
27. Mittnik, S., Paolella, M.S.: Conditional density and value-at-risk prediction of asian currency exchange rates. J. Forecast. **19**, 313–333 (2000)
28. Peng, L.: Asymptotically unbiased estimator for the extreme-value-index. Stat. Probab. Lett. **38**(2), 107–115 (1998)

29. Pickands III, J.: Statistical inference using extreme value order statistics. Ann. Statist. **3**, 119–131 (1975)
30. R Development Core Team: R: A language and environment for statistical computing. R Foundation for Statistical Computing, Vienna, Austria. ISBN 3-900051-07-0. http://www.R-project.org(2008)
31. Ribatet, M.: POT: Generalized Pareto Distribution and Peaks Over Threshold. R package version 1.0–9. http://people.epfl.ch/mathieu.ribatet, http://r-forge.r-project.org/projects/pot/ (2009)
32. Weissman, I.: Estimation of parameters and large quantiles based on the *k* largest observations. J. Am. Stat. Assoc. **73**, 812–815 (1978)
33. Wuertz, D., Chalabi, Y., Miklovic M.: fGarch: Rmetrics - Autoregressive Conditional Heteroskedastic Modelling, R package version 290.76. http://www.rmetrics.org (2008)

Asymptotic Distribution of the Maximum for a Chaotic Economic Model

Ana Cristina Moreira Freitas

Abstract
In this work we study the asymptotic distribution of the partial maximum of observable random variables evaluated along the orbits of the tent map, which is frequently used as a model in the economic literature.

1 Introduction

Chaos has become a subject of great interest in several domains like, for example, economics, biology, physics and meteorology. Its relevance in economics has been explored in models of consumer behaviour, growth cycles, overlapping generations and stock market behaviour, among others (see, e.g. [1,8,9] and references therein).

In this work, we study the tent map defined by $f(x) = 1 - 2|x|$, for $x \in [-1, 1]$, which has been used extensively in applied economics. It is well known that there is a unique, ergodic, f-invariant probability measure, absolutely continuous with respect to Lebesgue measure, that is given by $\mu(A) = \int_A 1/2dx$.

The tent map is chaotic and highly sensitive on initial conditions. Actually, after some iterates, the behaviour of most orbits becomes erratic and uniformly distributed on the set $[-1, 1]$. Hence, from the applications point of view, it is meaningful to study the statistical properties of this system. In fact, the statistical properties of one-dimensional chaotic systems, like the tent map, have been a subject of much interest and investigation. We refer, among many others, the works [3, 6, 9, 10], that give, in particular, some applications of the tent map in economic

A.C.M. Freitas (✉)
Centro de Matemática & Faculdade de Economia, Universidade do Porto, Rua Dr. Roberto Frias, 4200-464 Porto, Portugal
e-mail: amoreira@fep.up.pt

J. Lita da Silva et al. (eds.), *Advances in Regression, Survival Analysis, Extreme Values, Markov Processes and Other Statistical Applications*, Studies in Theoretical and Applied Statistics, DOI 10.1007/978-3-642-34904-1_20,

models. In this work, we are particularly concerned with the extreme type behaviour of the orbits of this system.

Given an observable $\varphi : [-1,1] \to \mathbb{R} \cup \{\pm\infty\}$ achieving a global maximum at 0, consider the stationary stochastic process $X_0, X_1, \ldots$ defined by

$$X_n = \varphi \circ f^n, \quad \text{for each } n \in \mathbb{N}. \tag{1}$$

Here, our main goal is to study of the statistical properties of the partial maximum $M_n := \max\{X_0, \ldots, X_{n-1}\}$, when properly normalised.

2 Extreme Value Laws

We are interested in knowing if there are normalising sequences $\{a_n\}_{n\in\mathbb{N}} \subset \mathbb{R}^+$ and $\{b_n\}_{n\in\mathbb{N}} \subset \mathbb{R}$ such that

$$\mu\left(\{x : a_n(M_n - b_n) \leq y\}\right) = \mu\left(\{x : M_n \leq u_n\}\right) \to H(y), \tag{2}$$

for some non-degenerate distribution function (d.f.) H, as $n \to \infty$. Here $u_n := u_n(y) = y/a_n + b_n$ is such that

$$n\mu(X_0 > u_n) \to \tau, \quad \text{as } n \to \infty, \tag{3}$$

for some $\tau = \tau(y) \geq 0$ and in fact $H(y) = H(\tau(y))$. When this happens we say that we have an *extreme value law* (EVL) for M_n.

Classical extreme value theory asserts that there are only three types of nondegenerate asymptotic distributions for the maximum of an independent and identically distributed (i.i.d.) sample under linear normalisation. They will be referred to as *classical* EVL and we denote them by:

Type 1: $EV_1(y) = \mathrm{e}^{-\mathrm{e}^{-y}}$, $y \in \mathbb{R}$; this is also known as the Gumbel extreme value distribution (e.v.d.).

Type 2: $EV_2(y) = \begin{cases} \mathrm{e}^{-y^{-\alpha}} & , y > 0 \\ 0 & , y \leq 0 \end{cases}$ $(\alpha > 0)$; this family of distribution functions is known as the *Fréchet* e.v.d.

Type 3: $EV_3(y) = \begin{cases} \mathrm{e}^{-(-y)^{\alpha}} & , y \leq 0 \\ 1 & , y > 0 \end{cases}$ $(\alpha > 0)$; this family of distribution functions is known as the *Weibull* e.v.d.

The study of the limiting behaviour for maxima of a stationary process can be reduced, under adequate conditions on the dependence structure, to the classical extreme value theory for sequences of i.i.d. random variables (r.v.). Hence, to the stationary process $X_0, X_1, \ldots$ we associate an independent sequence of r.v. denoted by $Z_0, Z_1, \ldots$, whose d.f. is the same of X_0. For each $n \in \mathbb{N}$, consider

$$\widehat{M}_n = \max\{Z_0, \ldots, Z_{n-1}\}. \tag{4}$$

Leadbetter et al. established, in Theorem 3.5.2 of [7], conditions on the dependence structure that allow us to relate the asymptotic distribution of M_n with that of $\widehat{M}_n$. They denoted those conditions by $D(u_n)$ and $D'(u_n)$.

Freitas and Freitas stated, in [5], that the usual condition $D(u_n)$ can be weakened in such a way that Theorem 3.5.2 of [7] still holds, that is, the asymptotic distributions of M_n and $\widehat{M}_n$ are the same. The condition proposed in [5] is the following mixing-type condition.

Condition ($D(u_n)$). We say that $D(u_n)$ holds for the sequence $X_0, X_1, \ldots$ if for any integers ℓ, t and n

$$|\mu\{(X_0 > u_n) \cap (X_t \leq u_n \cap \ldots \cap X_{t+\ell-1} \leq u_n)\} \\ -\mu\{X_0 > u_n\}\mu\{X_0 \leq u_n \cap \ldots \cap X_{\ell-1} \leq u_n\}| \leq \gamma(n,t),$$

where $\gamma(n,t)$ is nonincreasing in t for each n and $n\gamma(n,t_n) \to 0$ as $n \to \infty$ for some sequence $t_n = o(n)$, that is, for some sequence t_n verifying $t_n/n \to 0$ as $n \to \infty$.

While condition $D(u_n)$ is a long range type of dependence requirement, the next one imposes some restrictions on the dependence structure but on a short range scope.

Condition ($D'(u_n)$). We say that $D'(u_n)$ holds for the sequence $X_0, X_1, \ldots$ if

$$\lim_{k\to\infty} \limsup_{n\to\infty} n \sum_{j=1}^{[n/k]} \mu\{X_0 > u_n \cap X_j > u_n\} = 0.$$

This condition $D'(u_n)$ restricts the occurrence of a large number of "exceedances" of the level u_n close together in time.

The result of [5] that we present below establishes that M_n and $\widehat{M}_n$ share the same asymptotic distribution under conditions $D(u_n)$ and $D'(u_n)$, for any sequence $(u_n)_{n\in\mathbb{N}}$ such that $n\mu\{X_0 > u_n\} \to \tau$, as $n \to \infty$, for some $\tau \geq 0$.

Proposition 1 ([5], Theorem 1). *Let $(u_n)_{n\in\mathbb{N}}$ be a sequence such that $n\mu\{X_0 > u_n\} \to \tau$ as $n \to \infty$, for some $\tau \geq 0$. Assume that conditions $D(u_n)$ and $D'(u_n)$ hold for the stationary stochastic process $X_0, X_1, \ldots$. Then, $\lim_{n\to\infty} \mu\{M_n \leq u_n\} = \lim_{n\to\infty} \mu\{\widehat{M}_n \leq u_n\}$.*

Based on this proposition, in [4], we obtained a result for the particular case of Eq. (1), where $\varphi \equiv Id$. We stated that, in this case, under appropriate normalisation, the limiting law of M_n is the same as if $X_0, X_1, \ldots$ were independent with the same marginal d.f., that is, it is of type III (Weibull) with parameter equal to 1.

Theorem 1 ([4], Theorem 1). *For the stochastic process $X_0, X_1, \ldots$ given by Eq. (1) where $\varphi \equiv Id$, we have*

$$P\left\{\frac{n}{2}(M_n-1)\le x\right\}\to H(x)=\begin{cases}e^x & ,x\le 0\\ 1 & ,x>0\end{cases},\quad as\ n\to\infty.$$

In the next section we will generalise this result for general observables. The advantage of using these general forms for the observables is that in this way we allow any kind of behaviour near the point where the maximum is achieved and for which a nondegenerate limiting law is attainable.

3 Characterisation of the Observables and Main Result

Consider again the general case $X_n = \varphi \circ f^n$, for each $n \in \mathbb{N}$, for an observable $\varphi : [-1, 1] \to \mathbb{R} \cup \{\pm\infty\}$ achieving a global maximum at 0.

We assume that the observable φ is of the form

$$\varphi(x) = g(|x|), \tag{5}$$

where the function $g : [0, +\infty) \to \mathbb{R} \cup \{+\infty\}$ has a global maximum at 0, is a strictly decreasing bijection $g : V \to W$ in a neighbourhood V of 0 and has one of the following three types of behaviour:

Type 1: There exists some strictly positive function $p : W \to \mathbb{R}$ such that for all $y \in \mathbb{R}$

$$\lim_{s\to g_1(0)} \frac{g_1^{-1}(s + yp(s))}{g_1^{-1}(s)} = e^{-y}.$$

Type 2: $g_2(0) = +\infty$ and there exists $\beta > 0$ such that for all $y > 0$

$$\lim_{s\to +\infty} \frac{g_2^{-1}(sy)}{g_2^{-1}(s)} = y^{-\beta}.$$

Type 3: $g_3(0) = D < +\infty$ and there exists $\gamma > 0$ such that for all $y > 0$

$$\lim_{s\to 0} \frac{g_3^{-1}(D - sy)}{g_3^{-1}(D - s)} = y^{\gamma}.$$

Examples of each one of the three types are as follows: $g_1(x) = -\log x$, $g_2(x) = x^{-1/\alpha}$ for some $\alpha > 0$ and $g_3(x) = D - x^{1/\alpha}$ for some $D \in \mathbb{R}$ and $\alpha > 0$.

In the following result we characterise the limiting law of M_n, which depends on the form of the observable φ defined in Eq. (5).

Theorem 2. *Suppose the stochastic process $X_0, X_1, \ldots$ is given by Eq. (1), where φ is of the form (5) and g has one of the three types of behaviour defined above. Then we have an EVL for M_n which coincides with that of $\widehat{M}_n$. Moreover, if g is of type g_i, for some $i \in \{1, 2, 3\}$, then we have an EVL for M_n of type i.*

4 Proof of Theorem 2

In this section, we start by noting that, by Proposition 1, for proving Theorem 2 it is enough to prove the following three lemmas.

Lemma 1. *Consider the stochastic process $X_0, X_1, \ldots$ defined as in Theorem 2 and to this process associate a sequence of i.i.d. r.v. $Z_0, Z_1, \ldots$ whose d.f. is the same of X_0. Then, if g is of type g_i, for some $i \in \{1,2,3\}$, then we have an EVL for $\widehat{M}_n$ of type i.*

Lemma 2. *Let $(u_n)_{n\in\mathbb{N}}$ be a sequence such that $n\mu\{X_0 > u_n\} \to \tau$, as $n \to \infty$, for some $\tau \geq 0$. Then, condition $D(u_n)$ holds for the stochastic process $X_0, X_1, \ldots$ defined as in Theorem 2.*

Lemma 3. *Let $(u_n)_{n\in\mathbb{N}}$ be a sequence such that $n\mu\{X_0 > u_n\} \to \tau$, as $n \to \infty$, for some $\tau \geq 0$. Then, condition $D'(u_n)$ holds for the stochastic process $X_0, X_1, \ldots$ defined as in Theorem 2.*

Proof of Lemma 1. We start by noting that the f-invariant probability measure μ is such that, for $-1 \leq x \leq 1$, $\mu\{[-1,x]\} = \frac{x+1}{2}$.

Let F be the d.f. of X_0, that is, $F(x) = \mu\{X_0 \leq x\}$ and let φ be defined as in Eq. (5). We may write

$$1 - F(x) = \mu\{w : X_0(w) \geq x\} = \mu\{w : \varphi(w) \geq x\} = \mu\{w : g(|w|) \geq x\}$$
$$= \mu\{w : |w| \leq g^{-1}(x)\} = g^{-1}(x).$$

Moreover, $x_F := \sup_{x\in\mathbb{R}}\{x : F(x) < 1\} = g(0)$.

First suppose that $g = g_1$ is of type 1. Then, there exists some strictly positive function $p : W \to \mathbb{R}$ such that for all $y \in \mathbb{R}$

$$\lim_{s\to g_1(0)} \frac{g_1^{-1}(s + yp(s))}{g_1^{-1}(s)} = \mathrm{e}^{-y}.$$

So,

$$\lim_{s\to x_F} \frac{1 - F(s + yp(s))}{1 - F(s)} = \mathrm{e}^{-y}.$$

By Theorem 1.6.2 of [7], we conclude that we have an EVL for $\widehat{M}_n$ of type 1.

Suppose now that $g = g_2$ is of type 2. Then $g_2(0) = +\infty$ and there exists $\beta > 0$ such that for all $y > 0$

$$\lim_{s\to+\infty} \frac{g_2^{-1}(sy)}{g_2^{-1}(s)} = y^{-\beta}.$$

Thus, $x_F = +\infty$ and

$$\lim_{s\to+\infty} \frac{1-F(sy)}{1-F(s)} = y^{-\beta}.$$

By Theorem 1.6.2 of [7], we conclude that we have an EVL for $\widehat{M}_n$ of type 2.

Finally, suppose that $g = g_3$ is of type 3. Then $g_3(0) = D < +\infty$ and there exists $\gamma > 0$ such that for all $y > 0$

$$\lim_{s\to 0} \frac{g_3^{-1}(D-sy)}{g_3^{-1}(D-s)} = y^{\gamma}.$$

Consequently, $x_F = D < +\infty$ and

$$\lim_{s\to 0} \frac{1-F(D-sy)}{1-F(D-s)} = y^{\gamma}.$$

By Theorem 1.6.2 of [7], we conclude that we have an EVL for $\widehat{M}_n$ of type 3. □

The proofs of Lemmas 2 and 3 are similar to the proofs of Lemmas 2 and 3 of [4].

Proof of Lemma 2. For proving Lemma 2 we will use the fact that the tent map has exponential decay of correlations. In fact, by Theorem 8.3.2 of [2], for any functions $\varphi, \psi : [0,1] \to \mathbb{R}$ with bounded variation, there are positive constants C and $0 < r < 1$, independent of φ, ψ and t, such that

$$\left| \int \varphi \cdot (\psi \circ f^t) d\mu - \int \varphi d\mu \int \psi d\mu \right| \le C\,\mathrm{Var}(\varphi)\|\psi\|_\infty r^t, \quad \forall t \ge 0, \qquad (6)$$

where $\mathrm{Var}(\varphi)$ denotes the total variation of φ.

Taking $\varphi = \mathbf{1}_{\{X_0 > u_n\}}$ and $\psi = \mathbf{1}_{\{X_0 \le u_n \cap \ldots \cap X_{\ell-1} \le u_n\}}$ in Eq. (6), we obtain

$$\begin{aligned} &|\mu\{(X_0 > u_n) \cap (X_t \le u_n \cap \ldots \cap X_{t+\ell-1} \le u_n)\} \\ &\qquad -\mu\{X_0 > u_n\}\mu\{X_0 \le u_n \cap \ldots \cap X_{\ell-1} \le u_n\}| \\ &\le C\,\mathrm{Var}(\mathbf{1}_{\{X_0 > u_n\}})\|\mathbf{1}_{\{X_0 \le u_n \cap \ldots \cap X_{\ell-1} \le u_n\}}\|_\infty r^t \le 2Cr^t. \end{aligned}$$

Consider now the function $\gamma(t) = 2Cr^t$ and observe that it is decreasing in t and taking, for example, $t_n = \sqrt{n}$ we clearly have $n\gamma(t_n) \to 0$ as $n \to \infty$. Many other choices for t_n suit our purposes. In this way, we have just proved that condition $D(u_n)$ is valid for the stochastic process $X_0, X_1, \ldots$ □

Proof of Lemma 3. Let $(u_n)_{n\in\mathbb{N}}$ be a sequence such that

$$n\mu\{X_0 > u_n\} \to \tau, \text{ as } n \to \infty, \qquad (7)$$

for some $\tau \geq 0$. Consider the interval $U_\delta := (-\delta, \delta)$, for $\delta = \delta_n$ such that if $x \in U_\delta$ then $\varphi(x) > u_n$. By Eq. (7), we have, in particular, that $u_n \to 1$ as $n \to \infty$, and so $\delta = \delta_n \to 0$ as $n \to \infty$.

Fix $j \in \{1, \ldots, [n/k]\}$, let $w_i := \left[\frac{i-1}{2^j}, \frac{i}{2^j}\right]$ and $w_{-i} := \left[-\frac{i}{2^j}, -\frac{i-1}{2^j}\right]$, for $i = 1, \ldots, 2^j$, and consider the partition of $[-1, 1]$ defined by $\mathscr{P}_j := \{w_i : i = -2^j, \ldots, -1, 1, \ldots, 2^j\}$. Note that, for each $w_i \in \mathscr{P}_j$, we have that $f, f^2, \ldots f^j$ are one-to-one on w_i. Moreover, since $|f'(x)| = 2$ for all $x \neq 0$, we have

$$|f^j(w_i)| = 1. \tag{8}$$

Define $\ell := \max\{i \in \mathbb{N}_0 : \frac{i}{2^j} \leq \delta\}$.

If $\ell \geq 1$, we have

$$|U_\delta| = 2\left(\sum_{i=1}^{\ell} |w_i| + |w^*|\right),$$

where $w^* = [\frac{\ell}{2^j}, \delta]$. So,

$$|U_\delta| = 2\left(\sum_{i=1}^{\ell} \frac{1}{2^j} + \alpha\frac{1}{2^j}\right) = \frac{2(\ell + \alpha)}{2^j}, \tag{9}$$

where $0 \leq \alpha < 1$. We also have

$$|U_\delta \cap f^{-j}(U_\delta)| = 2\left(\sum_{i=1}^{\ell} |w_i \cap f^{-j}(U_\delta)| + |w^* \cap f^{-j}(U_\delta)|\right). \tag{10}$$

Since $|f'(x)| = 2$ for all $x \neq 0$ and f^j is one-to-one on w_i, we have

$$\frac{|w_i \cap f^{-j}(U_\delta)|}{|w_i|} = \frac{|f^j(w_i \cap f^{-j}(U_\delta))|}{|f^j(w_i)|} = \frac{|f^j(w_i) \cap U_\delta|}{|f^j(w_i)|} \leq \frac{|U_\delta|}{|f^j(w_i)|}.$$

Using now Eq. (8), we obtain $|w_i \cap f^{-j}(U_\delta)| \leq |U_\delta||w_i|$.

So, by Eq. (10), and noting that $w^* \subset w_{\ell+1}$, we have

$$|U_\delta \cap f^{-j}(U_\delta)| \leq 2\left(\sum_{i=1}^{\ell} |U_\delta||w_i| + |U_\delta||w_{\ell+1}|\right) = 2\left(\sum_{i=1}^{\ell} \frac{|U_\delta|}{2^j} + \frac{|U_\delta|}{2^j}\right) = 2(\ell+1)\frac{|U_\delta|}{2^j}.$$

Dividing by $|U_\delta|$, and using Eq. (9), we obtain

$$\frac{|U_\delta \cap f^{-j}(U_\delta)|}{|U_\delta|} \leq \frac{2(\ell + 1)\frac{|U_\delta|}{2^j}}{\frac{2(\ell+\alpha)}{2^j}} = \frac{\ell + 1}{\ell + \alpha}|U_\delta|.$$

So, for $\ell \geq 1$,

$$|U_\delta \cap f^{-j}(U_\delta)| \leq 2|U_\delta|^2. \tag{11}$$

Consider now the case where $\ell = 0$. We start by observing that $f(U_\delta) = (f(\delta), 1] = (u_n, 1]$ and, for $j \geq 2$, $f^j(U_\delta) = [-1, f^j(\delta))$. Choose now n sufficiently large such that $\delta < 1/2$. In this way, if $f^j(U_\delta) \cap U_\delta \neq \emptyset$, then $|f^j(U_\delta)| > 1/2$. Thus,

$$\frac{|U_\delta \cap f^{-j}(U_\delta)|}{|U_\delta|} = \frac{|f^j(U_\delta \cap f^{-j}(U_\delta))|}{|f^j(U_\delta)|} \leq \frac{|U_\delta|}{1/2},$$

and so, also for $\ell = 0$, we obtain the same estimate of Eq. (11).

Since, for $x \in [-1, 1]$, $\mu\{[x, 1]\} = \frac{x+1}{2}$, we have that

$$|U_\delta| = 2\mu\{U_\delta\} = 2\mu\{X_0 > u_n\}$$

and

$$|U_\delta \cap f^{-j}(U_\delta)| = 2\mu\{X_0 > u_n \cap X_j > u_n\}.$$

Now, rewriting Eq. (11), we obtain

$$2\mu\{X_0 > u_n \cap X_j > u_n\} \leq 8(\mu\{X_0 > u_n\})^2.$$

Consequently,

$$n \sum_{j=1}^{[n/k]} \mu\{X_0 > u_n \cap X_j > u_n\} \leq 4n[n/k](\mu\{X_0 > u_n\})^2 \leq \frac{4}{k}(n\mu\{X_0 > u_n\})^2.$$

Now, by Eq. (7), $\frac{4}{k}(n\mu\{X_0 > u_n\})^2 \to \frac{4}{k}\tau^2$ as $n \to \infty$.

So,

$$\lim_{k\to\infty} \limsup_{n\to\infty} n \sum_{j=1}^{[n/k]} \mu\{X_0 > u_n \cap X_j > u_n\} = \lim_{k\to\infty} \frac{4}{k}\tau^2 = 0.$$

□

Acknowledgements The author was supported by FCT grant SFRH/BPD/66174/2009 and also by FCT through CMUP and PTDC/MAT/099493/2008.

References

1. Baumol, W.J., Benhabib, J.: Chaos: significance, mechanism, and economic applications. J. Econ. Perspect. **3**, 77–107 (1989)
2. Boyarsky, A., Góra, P.: Laws of Chaos, Probability and Its Applications. Birkhäuser, Boston (1997)
3. Day, R.H., Pianigiani, G.: Statistics dynamics and economics. J. Econ. Behav. Organ. **65**, 37–83 (1991)

4. Freitas, A.C.M.: Statistics of the maximum for the tent map. Chaos Soliton. Fract. **42**, 604–608 (2009)
5. Freitas, A.C.M., Freitas, J.M.: On the link between dependence and independence in extreme value theory for dynamical systems. Stat. Probab. Lett. **78**, 1088–1093 (2008)
6. Hsieh, D.A.: Chaos and nonlinear dynamics: application to financial markets. J. Financ. **XLVI**, 1839–1877 (1991)
7. Leadbetter, M.R., Lindgren, G., Rootzén, H.: Extremes and related properties of stationary sequences and processes. In: Springer Series in Statistics. Springer, New York (1983)
8. Lorenz, H.W.: Nonlinear Dynamical Economics and Chaotic Motion. Springer, Berlin (1989)
9. Medio, A.: Nonlinear dynamics and chaos part II: ergodic approach. Macroecon. Dyn. **3**, 84–114 (1999)
10. Savit, R.: Nonlinearity and chaotic effects in option prices. J. Future Markets **9**, 507–518 (1989)

Adaptive Choice of Thresholds and the Bootstrap Methodology: An Empirical Study

M. Ivette Gomes, Fernanda Figueiredo, and M. Manuela Neves

Abstract
In this chapter, we discuss an algorithm for the adaptive estimation of a positive *extreme value index*, γ, the primary parameter in *Statistics of Extremes*. Apart from classical extreme value index estimators, we suggest the consideration of associated second-order corrected-bias estimators, and propose the use of bootstrap computer-intensive methods for the adaptive choice of *thresholds*.

1 Introduction and Outline of the Chapter

Heavy-tailed models appear often in practice in fields like telecommunications, insurance, finance, bibliometrics and biostatistics. We shall deal with the estimation of a positive *extreme value index* (EVI), γ, the primary parameter in *Statistics of Extremes*. Apart from the *classical Hill, moment* and *generalized-Hill* semi-parametric estimators of γ, detailed in Sect. 2, we shall consider the associated

M.I. Gomes (✉)
FCUL, Campo Grande, 1749-016 Lisboa, Portugal

CEAUL, Lisboa, Portugal
e-mail: ivette.gomes@fc.ul.pt

F. Figueiredo
FEP, Rua Dr. Roberto Frias, 4200-464 Porto, Portugal

CEAUL, Lisboa, Portugal
e-mail: otilia@fep.up.pt

M.M. Neves
ISA, Tapada da Ajuda, 1349-017 Lisboa, Portugal

CEAUL, Lisboa, Portugal
e-mail: manela@isa.utl.pt

J. Lita da Silva et al. (eds.), *Advances in Regression, Survival Analysis, Extreme Values, Markov Processes and Other Statistical Applications*, Studies in Theoretical and Applied Statistics, DOI 10.1007/978-3-642-34904-1_21,

classes of *second-order reduced-bias* estimators, based on an adequate estimation of generalized scale and shape second-order parameters, valid for a large class of heavy-tailed underlying parents, and appealing in the sense that we are able to reduce the asymptotic bias of a classical estimator without increasing its asymptotic variance. We shall call these estimators "*classical-variance reduced-bias*" (CVRB) estimators.

After the introduction, in Sect. 2, of a few technical details in the area of *extreme value theory* (EVT), related with the EVI-estimators under consideration in this chapter, we shall briefly discuss, in Sect. 3, the kind of second-order parameters' estimation which enables the building of *reduced-bias* estimators with the same asymptotic variance, σ_C^2, of the associated *classical* estimator. After the discussion, in Sect. 4, of the asymptotic behaviour of the estimators under consideration, we propose and discuss in Sect. 5, in the lines of [13] and, more recently, [14], an algorithm for the adaptive estimation of a positive EVI, through the use of bootstrap computer-intensive methods. The algorithm is described for a classical EVI-estimator and associated CVRB estimator, but it works similarly for the estimation of any other parameter of extreme events, like a high quantile, the probability of exceedance or the return period of a high level. Section 6 is entirely dedicated to the application of the algorithm to the analysis of environmental data, the number of hectares burned during all wildfires recorded in Portugal in the period 1999–2003.

2 The EVI-Estimators Under Consideration

In the area of EVT, and for large values, a model F is said to be *heavy-tailed* whenever the *right-tail function*, $\overline{F} := 1 - F$, is a regularly varying function with a negative index of regular variation, denoted $-1/\gamma$, i.e., if for all $x > 0$, there exists $\gamma > 0$, such that $\overline{F}(tx)/\overline{F}(t) \to x^{-1/\gamma}$, as $t \to \infty$. If this holds, we use the notation $\overline{F} \in RV_{-1/\gamma}$, and we are working in the whole *domain of attraction* (for maxima) of heavy-tailed models, denoted $\mathscr{D_M}\left(EV_\gamma\right)_{\gamma>0}$. Equivalently, with $U(t) := F^{\leftarrow}\left(1 - 1/t\right) = \inf\{x : F(x) \geq 1 - 1/t\}$, $F \in \mathscr{D_M}\left(EV_\gamma\right)_{\gamma>0} \iff \overline{F} \in RV_{-1/\gamma} \iff U \in RV_\gamma$, the so-called *first-order* conditions. For these heavy-tailed parents, given a sample $\underline{\mathbf{X}}_n := (X_1, \ldots, X_n)$ and the associated sample of ascending order statistics (o.s.'s), $(X_{1:n} \leq \cdots \leq X_{n:n})$, the classical EVI-estimator is the Hill estimator [16],

$$H_k \equiv H_{k,n} := \frac{1}{k} \sum_{i=1}^{k} \{\ln X_{n-i+1:n} - \ln X_{n-k:n}\}, \tag{1}$$

the average of the k log-excesses over a high random threshold $X_{n-k:n}$, an *intermediate* o.s., i.e., with k such that

$$k = k_n \to \infty \quad \text{and} \quad k/n \to 0, \text{ as } n \to \infty. \tag{2}$$

But the Hill-estimator H_k, in Eq. (1), reveals usually a high non-null asymptotic bias at optimal levels, i.e., levels k where the mean squared error (MSE) is minimum. This non-null asymptotic bias, together with a rate of convergence of the order of $1/\sqrt{k}$, leads to sample paths with a high variance for small k, a high bias for large k, and a very sharp MSE pattern, as function of k. Recently, several authors have been dealing with bias reduction in the field of *extremes* (for an overview, see [18], Chap. 6, as well as the more recent paper, [4]). For technical details, we then need to work in a region slightly more restrict than $\mathscr{D}_{\mathscr{M}}\left(EV_\gamma\right)_{\gamma>0}$. In this chapter, we shall consider parents such that, with $\gamma > 0$, $\rho < 0$ and $\beta \neq 0$,

$$U(t) = Ct^\gamma\big(1 + A(t)/\rho + O(A^2(t))\big), \text{ as } t \to \infty, \quad A(t) =: \gamma\beta t^\rho. \tag{3}$$

The most simple class of second-order minimum-variance reduced-bias (MVRB) EVI-estimators is the one in [3], used for a semi-parametric estimation of high quantiles in [10]. This class, here denoted $\overline{H} \equiv \overline{H}_k$, is the CVRB-estimator associated with the Hill estimator $H = H_k$, in Eq. (1), and depends upon the estimation of the second-order parameters (β, ρ), in Eq. (3). Its functional form is

$$\overline{H}_k \equiv \overline{H}_{k,n,\hat\beta,\hat\rho} := H_k\big(1 - \hat\beta(n/k)^{\hat\rho}/(1-\hat\rho)\big), \tag{4}$$

where $(\hat\beta, \hat\rho)$ is an adequate consistent estimator of (β, ρ). Algorithms for the estimation of (β, ρ) are provided, for instance, in [10], and will be reformulated in Sect. 3 of this chapter.

Apart from the *Hill* estimator, in Eq. (1), we suggest the consideration of two other classical estimators, valid for all $\gamma \in \mathbb{R}$, but taken here exclusively for heavy tails, the *moment* [6] and the *generalized-Hill* [1, 2] estimators. The *moment* estimator (M) has the functional expression

$$M_k \equiv M_{k,n} := M_{k,n}^{(1)} + \tfrac{1}{2}\left\{1 - \left(M_{k,n}^{(2)}/(M_{k,n}^{(1)})^2 - 1\right)^{-1}\right\}, \tag{5}$$

with $M_{k,n}^{(j)} := \frac{1}{k}\sum_{i=1}^k \left(\ln X_{n-i+1:n} - \ln X_{n-k:n}\right)^j$, $j \geq 1$ $\left(M_{k,n}^{(1)} \equiv H_k, \text{ in Eq. (1)}\right)$. The *generalized Hill* estimator (GH), based on $H_{k,n}$, in Eq. (1), is given by

$$GH_k \equiv GH_{k,n} := H_{k,n} + \frac{1}{k}\sum_{i=1}^k \left\{\ln H_{i,n} - \ln H_{k,n}\right\}. \tag{6}$$

The associated CVRB estimators have similar expressions, due to the same dominant component of asymptotic bias of the estimators in Eqs. (5) and (6), for a positive EVI. Denoting $\overline{W}$ either $\overline{M}$ or $\overline{GH}$, and with the notation W for either M or GH, we get

$$\overline{W}_k \equiv \overline{W}_{k,n,\hat\beta,\hat\rho} := W_k\big(1 - \hat\beta\,(n/k)^{\hat\rho}/(1-\hat\rho)\big) - \hat\beta\,\hat\rho\,(n/k)^{\hat\rho}/(1-\hat\rho)^2. \tag{7}$$

In the sequel, we generally denote C any of the classical EVI-estimators, in Eqs. (1), (5) and (6), and $\overline{C}$ the associated CVRB-estimator.

3 Estimation of Second-Order Parameters

The estimation of γ, β and ρ at the same value k leads to a high increase in the asymptotic variance of CVRB estimators $\overline{C}_{k,n,\hat{\beta},\hat{\rho}}$, which becomes $\sigma_C^2\,((1-\rho)/\rho)^4$ (see [17], among others). The external estimation of ρ at k_1, but the estimation of γ and β at $k = o(k_1)$, slightly decreases the asymptotic variance to $\sigma_C^2\,((1-\rho)/\rho)^2$, still greater than σ_C^2 (see [8], among others). The external estimation of both β and ρ at a level k_1 and the estimation of γ at a level $k = o(k_1)$, or even $k = O(k_1)$, can lead to a CVRB estimator with an asymptotic variance σ_C^2, provided we choose adequately k_1 (see [3,11,12]). Such a choice is theoretically possible (see [4], among others), but under conditions difficult to guarantee in practice. As a compromise between theoretical and practical results, we have so far advised any choice $k_1 = [n^{1-\epsilon}]$, with ϵ small and $[x]$ denoting the integer part of x. We shall consider here $\epsilon = 0.001$.

Algorithm. (second-order parameters' estimation) :
1. Given an observed sample $(x_1,\ldots,x_n)$, plot the observed values of $\hat{\rho}_\tau(k)$, the most simple estimator in [7], for the tuning parameters $\tau = 0$ and $\tau = 1$.
2. Consider $\{\hat{\rho}_\tau(k)\}_{k\in\mathscr{K}}$, with $\mathscr{K} = ([n^{0.995}],\ [n^{0.999}])$, compute their median, denoted η_τ, and compute $I_\tau := \sum_{k\in\mathscr{K}} (\hat{\rho}_\tau(k) - \eta_\tau)^2$, $\tau = 0, 1$. Next choose the *tuning parameter* $\tau^* = 0$ if $I_0 \le I_1$; otherwise, choose $\tau^* = 1$.
3. Work with $\hat{\rho} \equiv \hat{\rho}_{\tau^*} := \hat{\rho}_{\tau^*}(k_1)$ and $\hat{\beta} \equiv \hat{\beta}_{\tau^*} := \hat{\beta}_{\hat{\rho}_{\tau^*}}(k_1)$, with $k_1 = [n^{0.999}]$, being $\hat{\beta}_{\hat{\rho}}(k)$, the estimator in [8].

Remark 1. This algorithm leads usually to the *tuning parameter* $\tau = 0$ whenever $|\rho| \le 1$ and $\tau = 1$, otherwise. For details on this and similar algorithms, see [10].

4 Asymptotic Distributional Behaviour of the Estimators

In order to obtain a nondegenerate behaviour for any semi-parametric EVI-estimator, it is convenient to assume a *second-order* condition, measuring the rate of convergence in the first-order condition. Such a condition, valid for all $x > 0$, involves a parameter $\rho \le 0$, a rate function A, with $|A| \in RV_\rho$ and is given by

$$\lim_{t\to\infty} (U(tx)/U(t) - x^\gamma)/A(t) = x^\gamma\,(x^\rho - 1)/\rho. \tag{8}$$

In this chapter, and mainly because of the reduced-bias estimators in Eqs. (4) and (7), generally denoted $\overline{C}_k \equiv \overline{C}_{k,n,\hat{\beta},\hat{\rho}}$, we shall assume that Eq. (3) holds. Then, Eq. (8) holds, with $A(t) = \gamma\,\beta t^\rho$. Let C_k be the associated classical estimator of γ. Trivial adaptations of the results in the above-mentioned papers (see also [15]) enable us to state, without proof, the following theorem, again for models with $\gamma > 0$.

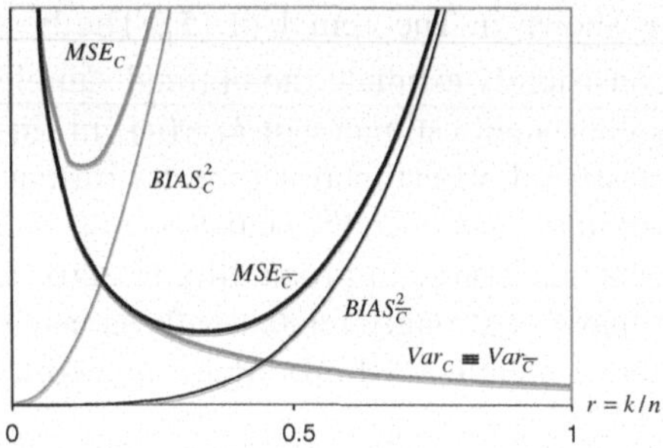

Fig. 1 General patterns of asymptotic variances (Var), squared bias ($BIAS^2$) and mean squared errors (MSE) of a classical EVI-estimator and associated CVRB estimator

Theorem 1. *Assume that condition* (8) *holds and let* $k = k_n$ *be an intermediate sequence, i.e., Eq.* (2) *holds. Then, there exist a sequence* Z_k^C *of asymptotically standard normal random variables,* $\sigma_C > 0$ *and real numbers* $b_{C,1}$ *such that* $C_k =^d \gamma + \sigma_C Z_k^C/\sqrt{k} + b_{C,1}\ A(n/k)\ (1 + o_p(1))$*. If we further assume that Eq.* (3) *holds and estimate* β *and* ρ *consistently through* $\hat\beta$ *and* $\hat\rho$*, in such a way that* $\hat\rho - \rho = o_p(1/\ln n)$*, we can guarantee that there exists a pair of real numbers* $(b_{\overline{C},1}, b_{\overline{C},2})$*, with* $b_{\overline{C},1} = 0$*, such that* $\overline{C}_{k,n,\hat\beta,\hat\rho} =^d \gamma + \sigma_C Z_k^C/\sqrt{k} + b_{\overline{C},1}\ A(n/k) + b_{\overline{C},2}\ A^2(n/k)\ (1 + o_p(1))$.

As $n \to \infty$, let $k = k_n$ be intermediate such that $\sqrt{k}\ A(n/k) \to \lambda$, finite, the levels k where the MSE of C_k is minimum. Let $\hat\gamma_k$ denote either C_k or $\overline{C}_k$. Then, we have $\sqrt{k}(\hat\gamma_k - \gamma) \overset{d}{\to} \text{Normal}(\lambda\ b_{\hat\gamma,1},\ \sigma_C^2)$, even if we work with CVRB EVI-estimators. If $\sqrt{k}\ A(n/k) \to \infty$, with $\sqrt{k}A^2(n/k) \to \lambda_A$, finite, the levels k where the MSE of $\overline{C}_k$ is minimum, $\sqrt{k}\ \left(\overline{C}_k - \gamma\right) \overset{d}{\to} \text{Normal}\left(\lambda_A b_{\overline{C},2},\ \sigma_C^2\right)$.We have $\sigma_H^2 = \gamma^2$, $\sigma_M^2 = \sigma_{GH}^2 = 1 + \gamma^2$,$b_{H,1} = 1/(1-\rho)$, $b_{M,1} = b_{GH,1} = (\gamma - \gamma\rho + \rho)/(\gamma(1-\rho)^2)$, and $b_{\overline{H},1} = b_{\overline{M},1} = b_{\overline{GH},1} = 0$. Consequently, since $b_{C,1} \neq 0$ whereas $b_{\overline{C},1} = 0$, the $\overline{C}$-estimators outperform the C-estimators for all k, as can be seen in Fig. 1.

5 The Bootstrap Methodology and Adaptive EVI-Estimation

With AMSE standing for "*asymptotic MSE*", and $k_0^{\hat\gamma}(n) := \arg\min_k MSE(\hat\gamma_k)$,

$$k_{0|\hat\gamma}(n) := \arg\min_k AMSE(\widehat{\gamma}_k) \tag{9}$$

$$= \arg\min_k \begin{cases} \left(\sigma_{\hat\gamma}^2/k + b_{\hat\gamma,1}^2\ A^2(n/k)\right) & (\text{if } \widehat\gamma = C) \\ \left(\sigma_{\hat\gamma}^2/k + b_{\hat\gamma,2}^2\ A^4(n/k)\right) & (\text{if } \widehat\gamma = \overline{C}) \end{cases} = k_0^{\hat\gamma}(n)(1 + o(1)),$$

as shown in Theorem 1 of [5]. The bootstrap methodology can thus enable us to consistently estimate the optimal sample fraction (OSF), $k_0^{\hat{\gamma}}(n)/n$, on the basis of a consistent estimator of $k_{0|\hat{\gamma}}(n)$, in Eq. (9), in a way similar to the one used for classical EVI-estimation (see, for instance, [9]). We shall here use the most obvious auxiliary statistics, the statistics $T_{k|\hat{\gamma}} \equiv T_{k,n|\hat{\gamma}} := \widehat{\gamma}_{[k/2]} - \widehat{\gamma}_k, \quad k = 2, \ldots, n-1$, which converge in probability to zero, for intermediate k, and have an asymptotic behaviour strongly related with the asymptotic behaviour of $\widehat{\gamma}_k$. Indeed, under the above-mentioned third-order framework in Eq. (3), we easily get

$$T_{k|\hat{\gamma}} \stackrel{d}{=} \frac{\sigma_{\hat{\gamma}}\, P_k^{\hat{\gamma}}}{\sqrt{k}} + \begin{cases} b_{\hat{\gamma},1}(2^{\rho}-1)\, A(n/k)(1+o_p(1)) & (\text{if } \widehat{\gamma} = C) \\ b_{\hat{\gamma},2}(2^{2\rho}-1)\, A^2(n/k)(1+o_p(1)) & (\text{if } \widehat{\gamma} = \overline{C}) \end{cases}$$

with $P_k^{\hat{\gamma}}$ asymptotically standard normal. Consequently, denoting $k_{0|T}(n) := \arg\min_k AMSE(T_{k|\hat{\gamma}})$, we have

$$k_{0|\hat{\gamma}}(n) = k_{0|T}(n) \begin{cases} (1-2^{\rho})^{\frac{2}{1-2\rho}}\,(1+o(1)) & (\text{if } \widehat{\gamma} = C) \\ (1-2^{2\rho})^{\frac{2}{1-4\rho}}\,(1+o(1)) & (\text{if } \widehat{\gamma} = \overline{C}). \end{cases}$$

How Does The Bootstrap Methodology Then Work?
Given the sample $\underline{\mathbf{X}}_n = (X_1, \ldots, X_n)$ from an unknown model F, and the functional $T_{k,n} \equiv T_{k,n|\hat{\gamma}} =: \phi_k(\underline{\mathbf{X}}_n)$, $1 < k < n$, consider for any $n_1 = O(n^{1-\epsilon})$, $0 < \epsilon < 1$, the bootstrap sample $\underline{\mathbf{X}}^*_{n_1} = (X_1^*, \ldots, X_{n_1}^*)$, from $F_n^*(x) = \frac{1}{n}\sum_{i=1}^{n} I_{[X_i \le x]}$, the empirical d.f. associated to the available sample, $\underline{\mathbf{X}}_n$. Next, associate to the bootstrap sample the corresponding bootstrap auxiliary statistic, $T^*_{k_1,n_1} := \phi_{k_1}(\underline{\mathbf{X}}^*_{n_1})$, $1 < k_1 < n_1$. Then, with $k^*_{0|T}(n_1) = \arg\min_{k_1} AMSE\big(T^*_{k_1,n_1}\big)$,

$$\frac{k^*_{0|T}(n_1)}{k_{0|T}(n)} = \left(\frac{n_1}{n}\right)^{-\frac{c\,\rho}{1-c\,\rho}}(1+o(1)), \qquad c = \begin{cases} 2 & (\text{if } \widehat{\gamma} = C) \\ 4 & (\text{if } \widehat{\gamma} = \overline{C}). \end{cases} \tag{10}$$

Consequently, for another sample size n_2 and for every $\alpha > 1$,

$$\frac{\left(k^*_{0|T}(n_1)\right)^{\alpha}}{k^*_{0|T}(n_2)} \left(\frac{n_1^{\alpha}\; n}{n^{\alpha}\; n_2}\right)^{-\frac{c\,\rho}{1-c\,\rho}} = \left\{k_{0|T}(n)\right\}^{\alpha-1}(1+o(1)).$$

It is then enough to choose $n_2 = [n\,(n_1/n)^{\alpha}]$, to have independence of ρ. If we put $n_2 = [n_1^2/n]$, i.e., $\alpha = 2$, we have $\left(k^*_{0|T}(n_1)\right)^2 / k^*_{0|T}(n_2) = k_{0|T}(n)(1+o(1))$, as $n \to \infty$. We are now able to estimate $k_0^{\hat{\gamma}}(n)$, on the basis of any estimate $\hat{\rho}$ of ρ. With $\hat{k}^*_{0|T}$ denoting the sample counterpart of $k^*_{0|T}$, and $\hat{\rho}$ the ρ-estimate in Step 3 of the algorithm, initiated in Sect. 3, we have the k_0-estimate

$$\hat{k}_0^{\hat{\gamma}}(n;n_1) := \min\left(n-1,\ [c_{\hat{\rho}}\,(\hat{k}^*_{0|T}(n_1))^2/\hat{k}^*_{0|T}([n_1^2/n]+1)]+1\right), \qquad (11)$$

with $c_{\hat{\rho}} = \left(1-2^{c\hat{\rho}/2}\right)^{\frac{2}{1-c\hat{\rho}}}$, c given in Eq. (10).

Again, with $\widehat{\gamma}$ denoting either C or $\overline{C}$, we proceed with the algorithm.

Algorithm. (cont.) (bootstrap adaptive estimation of γ):

4. Compute $\widehat{\gamma}_k$, $k = 1, 2, \ldots, n-1$
5. Next, consider the sub-sample size $n_1 = [n^{0.955}]$ and $n_2 = [n_1^2/n] + 1$
6. For l from 1 till $B = 250$, generate independently, from the empirical d.f. $F_n^*(x) = \frac{1}{n}\sum_{i=1}^{n} I_{[X_i \le x]}$, associated with the observed sample, B bootstrap samples $(x_1^*, \ldots, x_{n_2}^*)$ and $(x_1^*, \ldots, x_{n_2}^*, x_{n_2+1}^*, \ldots, x_{n_1}^*)$, of sizes n_2 and n_1, respectively
7. Denoting $T^*_{k,n}$ the bootstrap counterpart of $T_{k,n}$, obtain $(t^*_{k,n_1,l}, t^*_{k,n_2,l})$, $1 \le l \le B$, the observed values of the statistic T^*_{k,n_i}, $i = 1, 2$, compute $MSE^*(n_i, k) = \frac{1}{B}\sum_{l=1}^{B}\left(t^*_{k,n_i,l}\right)^2$, $k = 2, \ldots, n_i - 1$ and obtain $\hat{k}^*_{0|T}(n_i) := \arg\min_{1\le k\le n_i-1} MSE^*(n_i, k)$, $i = 1, 2$
8. Compute $\hat{k}_0^{\hat{\gamma}}(n;n_1)$ in Eq. (11)
9. Compute $\widehat{\gamma}^*_{n,n_1|T} := \widehat{\gamma}_{\hat{k}_0^{\hat{\gamma}}(n;n_1)}$

6 An Application to Burned Areas Data

Most of the wildfires are extinguished within a short period of time, with almost negligible effects. However, some wildfires go out of control, burning hectares of land and causing significant and negative environmental and economical impacts. The data we analyse here consists of the number of hectares, exceeding 100 ha, burnt during wildfires recorded in Portugal during 14 years (1990–2003). The data (a sample of size $n = 2,627$) do not seem to have a significant temporal structure, and we have used it as a whole, although we think also sensible, to try avoiding spatial heterogeneity, considering at least three different regions: the north, the centre and the south of Portugal (a study out of the scope of this note).

The box plot and a histogram of the available data provide evidence on the heaviness of the right tail. We shall next consider, for this type of data, the performance of the adaptive CVRB-EVI estimates $\overline{H}$, in Eq. (4), which are MVRB. These MVRB estimators exhibit stabler sample paths than H, as functions of k, and often enable us to take a decision upon the estimates to be used, even with the help of any heuristic stability criterion. The algorithm in this chapter, valid asymptotically, enables us to better and adaptively estimate the OSF associated with the MVRB or CVRB estimates. For a sub-sample size $n_1 = [n^{0.955}] = 1843$, and B = 250 bootstrap generations, we have got $\hat{k}_0^{\overline{H}^*} = 1319$ and the bootstrap MVRB-EVI-estimate $\overline{H}^* = 0.658$, the value pictured in Fig. 2, jointly with the above-mentioned

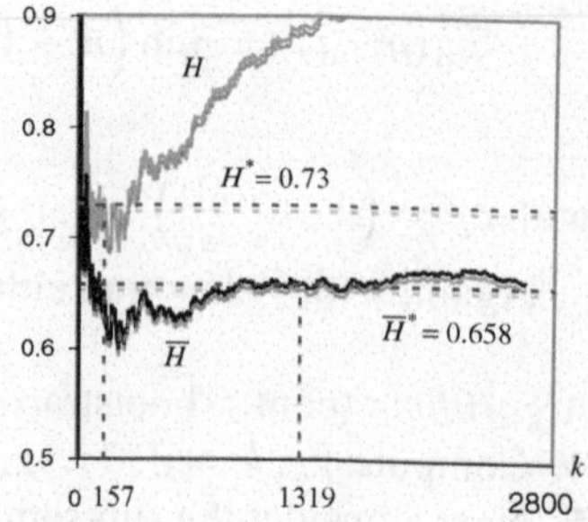

Fig. 2 Estimates of the EVI, γ, through the Hill estimator, H, in Eq. (1), and the MVRB estimator, $\overline{H}$, in Eq. (4), for the burned areas under analysis, together with the bootstrap adaptive estimates H^* and $\overline{H}^*$

adaptive bootstrap Hill estimate, $H^* = 0.73$. Note the fact that the MVRB EVI-estimators look practically "unbiased" for the data under analysis, but different patterns can occur for other data sets.

A few practical questions may be raised under the set-up developed: is the method strongly dependent on the choice of n_1? What is its sensitivity with respect to the choice of ρ-estimate? Although aware of the need of $n_1 = o(n)$, what happens if we choose $n_1 = n$? Answers to these questions are expected not to be a long way from the ones given for classical estimation (see [9]), have lightly been addressed in [13, 14], for reduced-bias estimation, but are out of the scope of this chapter.

Acknowledgements Research partially supported by National Funds through **FCT**—Fundação para a Ciência e a Tecnologia—projects PEst-OE/MAT/UI0006/2011 (CEAUL) and PTDC/FEDER, EXTREMA.

References

1. Beirlant, J., Vynckier, P., Teugels, J.: Excess functions and estimation of the extreme-value index. Bernoulli **2**(4), 293–318 (1996)
2. Beirlant, J., Dierckx, G., Guillou, A.: Estimation of the extreme-value index and generalized quantile plots. Bernoulli **11**(6), 949–970 (2005)
3. Caeiro, F., Gomes, M.I., Pestana, D.D.: Direct reduction of bias of the classical Hill estimator. Revstat **3**(2), 111–136 (2005)
4. Caeiro, F., Gomes, M.I., Henriques-Rodrigues, L.: Reduced-bias tail index estimators under a third order framework. Commun. Stat. Theor. Meth. **38**(7), 1019–1040 (2009)
5. Danielsson, J., Haan, L.de, Peng, L., de Vries, C.G.: Using a bootstrap method to choose the sample fraction in the tail index estimation. J. Multivar. Anal. **76**, 226–248 (2001)
6. Dekkers, A.L.M., Einmahl, J.H.J., Haan, L.de: A moment estimator for the index of an extreme-value distribution. Ann. Statist. **17**, 1833–1855 (1989)
7. Fraga Alves, M.I., Gomes M.I., Haan, L.de: A new class of semi-parametric estimators of the second order parameter. Portugaliae Mathematica **60**(2), 194–213 (2003)
8. Gomes, M.I., Martins M.J.: Asymptotically unbiased estimators of the tail index based on external estimation of the second order parameter. Extremes **5**(1), 5–31 (2002)
9. Gomes, M.I., Oliveira, O.: The bootstrap methodology in statistical extremes—choice of the optimal sample fraction. Extremes **4**(4), 331–358 (2001)
10. Gomes, M.I., Pestana, D.D.: A sturdy reduced-bias extreme quantile (VaR) estimator. J. Am. Stat. Assoc. **102**(477), 280–292 (2007)

11. Gomes, M.I., Martins, M.J., Neves, M.M.: Improving second order reduced-bias tail index estimation. Revstat **5**(2), 177–207 (2007)
12. Gomes, M.I., Haan, L.de, Henriques-Rodrigues, L.: Tail Index estimation for heavy-tailed models: accommodation of bias in weighted log-excesses. J. R. Stat. Soc. **B70**(1), 31–52 (2008)
13. Gomes, M.I., Mendonça, S., Pestana, D.D.: Adaptive reduced-bias tail index and value-at-risk estimation. In: Sakalauskas, L., Skiadas, C., Zavadskas, E.K. (eds.) Applied Stochastic Models and Data Analysis – ASMDA 2009, IMI & VGTU Editions, pp. 41–44 (2009)
14. Gomes, M.I., Figueiredo, F., Neves, M.M.: Adaptive Estimation of Heavy Right Tails: Resampling-Based Methods in Action. Extremes **15**, 463–489 (2012)
15. Haan, L.de, Ferreira, A.: Extreme Value Theory: An Introduction. Springer, New York (2006)
16. Hill, B.M.: A simple general approach to inference about the tail of a distribution. Ann. Statist. **3**, 1163–1174 (1975)
17. Peng, L., Qi, Y.: Estimating the first and second order parameters of a heavy tailed distribution. Aust. NZ J. Stat. **46**, 305–312 (2004)
18. Reiss, R.-D., Thomas, M.: Statistical Analysis of Extreme Values, 3rd edn. Birkhäuser, Boston (2007)

Distributional Properties of Generalized Threshold ARCH Models

E. Gonçalves and N. Mendes-Lopes

Abstract
The aim of this chapter is to give a contribution for the estimation of the law of stationary generalized threshold ARCH (GTARCH) processes. Firstly we present bounds for the marginal distribution of a threshold ARCH process, ε, with an independent generator process Z, as well as for the laws of finite dimension of the absolute value of this process. The results are illustrated by a simulation study considering several distributions for Z, in particular with different behavior in what concerns the tails' height, and estimating the distribution function of the model by the empirical one. Secondly, with the same goal we establish a dependence property for strictly stationary GTARCH processes from which, as an application of Berkes and Horváth [Ann. Appl. Probab. **11**(2), 789–809 (2001)] results, the convergence in law and the almost sure uniform convergence of the empirical process are obtained.

1 Introduction

In this chapter, we present distributional properties of generalized threshold autoregressive conditionally heteroscedastic (GTARCH) processes (Zakoian [8]). We observe that this class of nonlinear models has advantage over the GARCH one (Engle [3], Bollerslev [2]). In fact, contrary to what happens with GARCH models, its conditional variance depends, not necessarily symmetrically, on the past observations, being therefore more adequate to take into account the asymmetries in the volatility so often found in financial time series.

E. Gonçalves (✉) · N. Mendes-Lopes
CMUC, Department of Mathematics, University of Coimbra, FCTUC, Apartado 3008, EC Universidade, 3001 454 Coimbra, Portugal
e-mail: esmerald@mat.uc.pt; nazare@mat.uc.pt

J. Lita da Silva et al. (eds.), *Advances in Regression, Survival Analysis, Extreme Values, Markov Processes and Other Statistical Applications*, Studies in Theoretical and Applied Statistics, DOI 10.1007/978-3-642-34904-1_22,

The studies here presented concern stationary processes. Namely, the theoretical results are presented in a weakly stationary frame while the simulation analysis use the more general frame, in this case, of strictly stationary processes.

A real stochastic process, $\varepsilon = (\varepsilon_t, t \in \mathbb{Z})$, follows a GTARCH model with orders p and q, GTARCH (p, q), if

$$\begin{cases} \varepsilon_t = \sigma_t Z_t \\ \sigma_t = \alpha_0 + \sum_{i=1}^{q} \alpha_i \varepsilon_{t-i}^{+} - \sum_{i=1}^{q} \beta_i \varepsilon_{t-i}^{-} + \sum_{j=1}^{p} \gamma_j \sigma_{t-j} \end{cases} \tag{1}$$

where $\alpha_0 > 0$, $\alpha_i \geq 0$, $\beta_i \geq 0$, $\gamma_j \geq 0$, $Z = (Z_t, t \in \mathbb{Z})$ is a sequence of independent and identically distributed real random variables, with zero mean and unit variance and such that Z_t is independent of $\underline{\varepsilon}_{t-1} = \sigma(\varepsilon_{t-1}, \varepsilon_{t-2}, \ldots)$ and where $\varepsilon_t^{+} = \varepsilon_t \mathbf{I}_{\{\varepsilon_t \geq 0\}}$, $\varepsilon_t^{-} = \varepsilon_t \mathbf{I}_{\{\varepsilon_t < 0\}}$. We note that Z is called the generating process of ε. We say that ε follows a TARCH (q) model if $\gamma_j = 0$, $j = 1, \ldots, p$.

The stationarity (weak and strict) of the general model (1) is equivalent to the stationarity of the vectorial process of $\mathbb{R}^{p+2q-2}$, $X = (X_t, t \in \mathbb{Z})$, defined by

$$X_t = \left(\sigma_t, \sigma_{t-1}, \ldots, \sigma_{t-p+1}, \varepsilon_{t-1}^{+}, \ldots, \varepsilon_{t-q+1}^{+}, -\varepsilon_{t-1}^{-}, \ldots, -\varepsilon_{t-q+1}^{-}\right)^T,$$

satisfying the autoregressive equation $X_{t+1} = A_t X_t + B$, with $B = (\alpha_0, 0, \ldots, 0)^T$ and $(A_t, t \in \mathbb{Z})$ a specific sequence of independent and identically distributed random square matrices of $p + 2q - 2$ order. Sufficient conditions for the existence of the strictly stationary and ergodic solution of the general model (1) are stated in Gonçalves and Mendes-Lopes [4, 6].

In some particular cases the stationarity study of ε may be undertaken without the vectorial frame. For example, let us consider the TARCH (q) model defined by

$$\begin{cases} \varepsilon_t = \sigma_t Z_t \\ \sigma_t = \alpha_0 + \alpha_q \varepsilon_{t-q}^{+} - \beta_q \varepsilon_{t-q}^{-}. \end{cases} \tag{2}$$

Taking $X_t = \sigma_t$ and the random variables $A_t = \alpha_q Z_t^{+} - \beta_q Z_t^{-}$, we clearly have $X_{t+q} = A_t X_t + \alpha_0$. A necessary and sufficient condition of strict stationarity of ε is $E[\log(A_t)]$ exists and $E[\log(A_t)] < 0$ (Gonçalves and Mendes-Lopes [4]). Moreover, a sufficient condition of weak (and strict) stationarity is $E\left(A_t^2\right) < 1$. When the Z law is symmetrical, this condition is equivalent to $(\alpha_q)^2 + (\beta_q)^2 < 2$. In particular, if the process Z follows the standard Gaussian law, we have

$$E[\log(A_t)] < 0 \Leftrightarrow \alpha_q \beta_q < \frac{1}{2} \exp\left(-\Psi\left(\frac{1}{2}\right)\right) \simeq 3.569$$

where Ψ is the Euler function. Moreover, if ε is weakly stationary, its variance is given by

$$\sigma_\varepsilon^2 = \frac{2\alpha_0^2}{2-\left(\alpha_q^2+\beta_q^2\right)} \frac{\sqrt{2\pi}+\left(\alpha_q+\beta_q\right)}{\sqrt{2\pi}-\left(\alpha_q+\beta_q\right)}.$$

Starting from the theoretical study held in Gonçalves and Mendes-Lopes [5], we develop in Sects. 2 and 3 a simulation analysis to illustrate both its applicability in the estimation of the process distribution and its behavior in a more general framework. So, we present bounds for the corresponding marginal distribution function; bounds for the finite dimensional laws distribution functions of the process $|\varepsilon|$ are also evaluated. Some examples and simulation results are considered in both Sects.; in particular, in Sect. 2 we discuss the hypothesis of ε weak stationarity, imposed by the theoretical results. For clarity and simplicity, this simulation study is undertaken only for TARCH models.

The simulation analysis makes use of the empirical distribution function of TARCH processes. The consistence of that estimation is analyzed by studying the asymptotic behavior of the empirical process associated. In this sense in Sect. 4, we state, for the general class of GTARCH processes, a m-dependence property, which allows to reduce that behavior study to that of independent clusters of random variables. As in Berkes and Horváth [1], the strong approximation for the empirical process of n observed elements of GTARCH processes is then obtained. As consequences, the weak convergence of the empirical process and the law of the iterated logarithm are deduced.

2 TARCH Processes: Bounds for the Distribution Function

In the following theorem, bounds for the marginal distribution function of ε are established showing that the law of ε is, in certain regions, strongly controlled by the law of the white noise associated. This fact is very relevant as we know that these laws have in general quite different characteristics (e.g., the marginal law of ε is leptokurtic even if it doesn't happen with the Z marginal law).

Theorem 1. *Let $\varepsilon = (\varepsilon_t,\, t \in \mathbb{Z})$ be a weakly stationary TARCH(q) process with $V(\varepsilon_t) = \sigma_\varepsilon^2$. If Z_t is absolutely continuous with a differentiable density of probability f_{Z_t}, we have for every $t \in \mathbb{Z}$*

(a) $F_{Z_t}\left(\frac{x}{\alpha_0}\right) \le F_{\varepsilon_t}(x) \le F_{Z_t}\left(\frac{x}{\alpha_0+\sigma_\varepsilon\sqrt{ck}}\right)$, *if* $x<0$ *and* $h(x) \ge 0$

(b) $F_{Z_t}\left(\frac{x}{\alpha_0+\sigma_\varepsilon\sqrt{ck}}\right) \le F_{\varepsilon_t}(x) \le F_{Z_t}\left(\frac{x}{\alpha_0}\right)$, *if* $x>0$ *and* $h(x) \ge 0$

where

$$h(x) = 2 f_{Z_t}\left(\frac{x}{m}\right) + \frac{x}{m} f'_{Z_t}\left(\frac{x}{m}\right),$$

$c = \sum_{i=1}^{q} \left(\alpha_i^2 + \beta_i^2\right)$, $k = \begin{cases} 1, & q = 1 \\ q-1, & q \geq 2 \end{cases}$, $m = \alpha_0 + y, y \geq 0$, *and* f'_{Z_t} *the derivative of* f_{Z_t}.

The proof of this result in a more general setting may be found in Gonçalves and Mendes-Lopes [5]. We point out that the inequalities $F_{Z_t}\left(\frac{x}{\alpha_0}\right) \leq F_{\varepsilon_t}(x)$ for $x < 0$ and $F_{\varepsilon_t}(x) \leq F_{Z_t}\left(\frac{x}{\alpha_0}\right)$ for $x > 0$ are verified without demanding the weak stationarity of the process. Nevertheless, in the other bounds this hypothesis plays an important role, namely on the application of Jensen's inequality.

The result presented is valid for a large class of probability laws of the process Z. In order to evaluate the sets of $\mathbb{R}$ where the bounds obtained for the point value of the marginal distribution function of ε are valid, we consider in the following two distributions for Z with different characteristics, namely in what concerns the behavior of the corresponding tails.

Examples. For Z_t distributed according to the standard Gaussian law, we have

$$h(x) = \frac{1}{\sqrt{2\pi}} f_{Z_t}\left(\frac{x}{m}\right)\left[2 - \left(\frac{x}{m}\right)^2\right],$$

and so $h(x) \geq 0$ if $x \in \left[-\alpha_0\sqrt{2}, \alpha_0\sqrt{2}\right]$. If Z_t follows a Cauchy law, $\mathscr{C}(0,1)$, we get $h(x) = \frac{2}{\pi}\frac{1}{\left[1+\left(\frac{x}{m}\right)^2\right]^2}$, which is strictly positive for every $x \in \mathbb{R}$.

We illustrate now the theoretical result established in Theorem 1, using 10,000 simulated realizations of a particular TARCH (2) model, defined by Eq. (2).

Firstly, we suppose that Z follows the standard Gaussian law and choose $\alpha_0 = 10, \alpha_2 = \beta_2 = 0.8$. The process ε is strict and weakly stationary and $\sigma_\varepsilon \simeq 35.6$.

The distribution function of ε_t is estimated, using the generated trajectory, by the empirical one (represented by DISTE in Fig. 1a). The distribution function of Z_t on $\frac{x}{\alpha_0}$ (DISTZMIN) and on $\frac{x}{\alpha_0 + \sigma_\varepsilon\sqrt{ck}}$ (DISTZMIN1) is also plotted in Fig. 1a. The bounds for the distribution function of ε_t stated in Theorem 1 are clearly respected, in particular in the interval $\left[-10\sqrt{2}, 10\sqrt{2}\right]$ obtained in the Examples.

Let us remark that the bounds obtained for the marginal distribution of ε involve the variance of ε_t. In order to explore the need of this moment, we consider now two situations where the strict stationarity is preserved but ε_t has no variance.

Under the previous setting, we take now $\alpha_0 = 10.0$, $\alpha_2 = \beta_2 = 1.2$. The process ε is still strictly stationary, but the parameter σ_ε is no longer the standard deviation of ε. Nevertheless, it may be interpreted as a scale parameter. It is this interpretation that we use to "pseudo" estimate σ_ε taking here $\widehat{\sigma} = \frac{1}{2}\left(\widehat{Q}_{0.841} - \widehat{Q}_{0.159}\right)$ with $\widehat{Q}_a$ denoting the $100a\,\%$ quantile of the empirical distribution of ε. In Fig. 1b we plot the estimated distribution function of ε_t on x (DISTE), the distribution function of Z_t on $\frac{x}{\alpha_0}$ (DISTZMIN) and on $\frac{x}{\alpha_0 + \widehat{\sigma}\sqrt{ck}}$ (DISTZMIN1). We see that the order

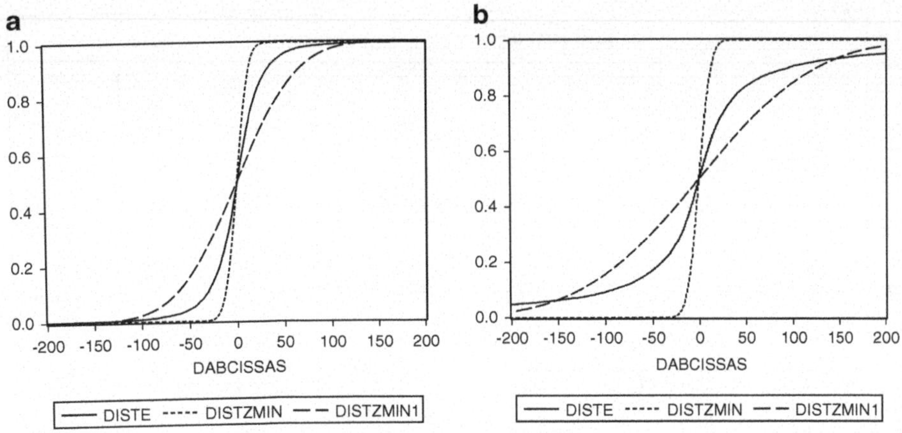

Fig. 1 (**a**) Plots of $F_{Z_t}\left(\frac{x}{\alpha_0}\right)$, (DISTZMIN), the estimation of $F_{\varepsilon_t}(x)$, (DISTE), and $F_{Z_t}\left(\frac{x}{\alpha_0+\sigma_\varepsilon\sqrt{ck}}\right)$, (DISTZMIN1). (**b**) Plots of $F_{Z_t}\left(\frac{x}{\alpha_0}\right)$, (DISTZMIN), the estimation of $F_{\varepsilon_t}(x)$, (DISTE), and $F_{Z_t}\left(\frac{x}{\alpha_0+\widehat{\sigma}\sqrt{ck}}\right)$, (DISTZMIN1)

relation between $F_{Z_t}\left(\frac{x}{\alpha_0}\right)$ and the estimation of $F_{\varepsilon_t}(x)$ is coherent with the result of Theorem 1 for all x. The relation between $F_{Z_t}\left(\frac{x}{\alpha_0+\widehat{\sigma}\sqrt{ck}}\right)$ and the estimation of $F_{\varepsilon_t}(x)$ is still coherent with that result but only for small or moderate values of $|x|$, in particular for $|x| \leq 10\sqrt{2}$.

Finally we consider that the generating process Z_t follows a Cauchy law, $\mathscr{C}(0,1)$. The strict stationarity of ε is, in this case, equivalent to $\alpha_2\beta_2 < 1$. So, we consider $\alpha_0 = 10.0$, $\alpha_2 = \beta_2 = 0.2$, and we use the analogous interquantile interval to estimate the parameter σ_ε. In the simulations we expect to find the same coherence with the result of Theorem 1, for all x, in the order relation between $F_{Z_t}\left(\frac{x}{\alpha_0}\right)$ and the estimation of $F_{\varepsilon_t}(x)$, as these relations are satisfied for all GTARCH processes even those without moments. Nevertheless, in Fig. 2 we observe the coherence with that result in the set $]-\infty, -6.31[\, \cup\,]0, +\infty[$. In what concerns $F_{Z_t}\left(\frac{x}{\alpha_0+\widehat{\sigma}\sqrt{ck}}\right)$ and the estimation of $F_{\varepsilon_t}(x)$, for which the second order moment is needed, the simulations are also coherent in the same domain. As the result fails for the two bounds in the same region, it is likely to believe that the noncoherence in $[-6.31, 0.0]$ is not due to the non existence of the second-order moments but, eventually, to the estimate of σ_ε. The use of inequalities not requiring the existence of moments in the statement of bounds for F_{ε_t} when ε is not a weakly stationary process certainly deserves further analysis.

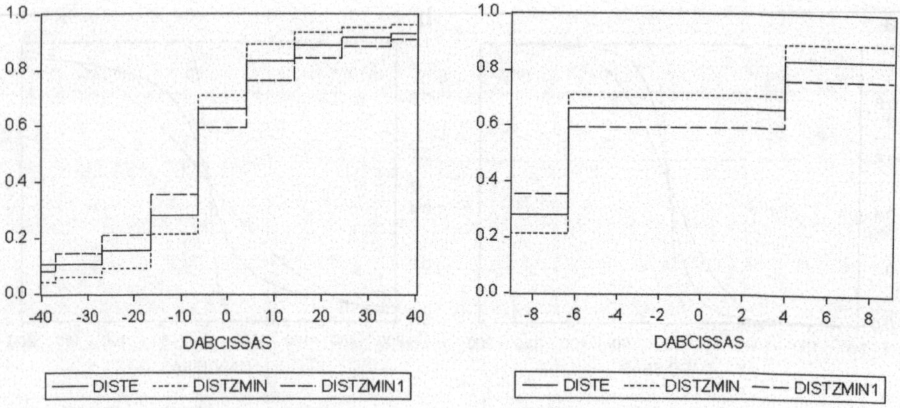

Fig. 2 Plots of $F_{Z_t}\left(\frac{x}{\alpha_0}\right)$, (DISTZMIN), the estimation of $F_{\varepsilon_t}(x)$, (DISTE), and $F_{Z_t}\left(\frac{x}{\alpha_0+\sigma\sqrt{ck}}\right)$, (DISTZMIN1)

3 TARCH Processes: Bounds for a Related Distribution Function

In accordance to Pawlak and Schmid [7], in certain problems related to assessing the performance of control charts, it is important to evaluate the probabilities $P\left(|\varepsilon_t| \leq x_t, t = 1, \ldots, n\right)$. In this sense, we deduce an upper and a lower bound for these probabilities when ε is a TARCH(q) process. The proof of the following result in a more general setting is in Gonçalves and Mendes-Lopes [5].

Theorem 2. *Let $\varepsilon = (\varepsilon_t, t \in \mathbb{Z})$ be a weakly stationary* TARCH(q)*process with $V(\varepsilon_t) = \sigma_\varepsilon^2$. We suppose Z_t absolutely continuous with a differentiable density of probability f_{Z_t} and denote by $f'_{|Z_t|}$ the derivative of $f_{|Z_t|}$. If $q \leq n$, then*[1]

$$F_{|Z_1|}\left(\frac{x_1}{\alpha_0 + \sigma_\varepsilon \sum_{i=1}^{q} \alpha_i}\right) \prod_{t=2}^{q} F_{|Z_t|}\left(\frac{x_t}{\alpha_0 + \sum_{i=1}^{t-1} \alpha_i x_{t-i} + \sigma_\varepsilon \sum_{i=t}^{q} \alpha_i}\right)$$

$$\times \prod_{t=q+1}^{n} F_{|Z_t|}\left(\frac{x_t}{\alpha_0 + \sum_{i=1}^{q} \alpha_i x_{t-i}}\right)$$

$$\leq F_{(|\varepsilon_1|,\ldots,|\varepsilon_n|)}(x_1, \ldots, x_n) \leq \prod_{t=1}^{n}\left[F_{|Z_t|}\left(\frac{x_t}{\alpha_0}\right)\right]$$

[1] If $q = 1$ we take $\prod_{t=2}^{q} g(t) = 1$.

for every $(x_1, \ldots, x_n) \in]0, +\infty[^n$ *such that* $h_t(x_t) \geq 0,\ t = 1, \ldots, q$, *where*

$$h_t(x) = 2 f_{|Z_t|}\left(\frac{x}{m_t}\right) + \frac{x}{m_t} f'_{|Z_t|}\left(\frac{x}{m_t}\right),\ t \in \{1, \ldots, q\}$$

$$m_t = \begin{cases} u, & t = 1,\ u \geq 0 \\ \sum_{i=1}^{t-1} \alpha_i x_{t-i} + u, & t = 2, \ldots, q,\ u \geq 0. \end{cases}$$

Following Pawlak and Schmid [7], these bounds for $F_{(|\varepsilon_1|,\ldots,|\varepsilon_n|)}(x_1, \ldots, x_n)$ may be related with the run length of control charts. Let us suppose, for example, that ε follows a TARCH(q) process and that $q \leq n$. Let us consider $x_1 = \ldots = x_n = x$. Then, for x under the conditions stated in Theorem 2,we obtain

$$P\left(\max_{1 \leq t \leq n} |\varepsilon_t| \leq x\right)$$

$$\geq F_{|Z_1|}\left(\frac{x}{\alpha_0 + \sigma_\varepsilon \sum_{i=1}^{q} \alpha_i}\right) \prod_{t=2}^{q} F_{|Z_t|}\left(\frac{x}{\alpha_0 + \sigma_\varepsilon \sum_{i=1}^{q} \alpha_i + \sum_{i=1}^{t-1} \alpha_i (x - \sigma_\varepsilon)}\right)$$

$$\prod_{t=q+1}^{n} F_{|Z_t|}\left(\frac{x}{\alpha_0 + \sum_{i=1}^{q} \alpha_i x}\right).$$

In particular, if $x < \sigma_\varepsilon$, $P(\max_{1 \leq t \leq n} |\varepsilon_t| \leq x) \geq \left[F_{|Z_1|}\left(\frac{x}{\alpha_0 + \sigma_\varepsilon \sum_{i=1}^{q} \alpha_i}\right)\right]^n$.

Then $\left[F_{|Z_1|}\left(\frac{\sigma_\varepsilon x}{\alpha_0 + \sigma_\varepsilon \sum_{i=1}^{q} \alpha_i}\right)\right]^n \leq P\left(\max_{1 \leq t \leq n} \frac{|\varepsilon_t|}{\sigma_\varepsilon} \leq x\right) \leq \left[F_{|Z_1|}\left(\frac{\sigma_\varepsilon}{\alpha_0} x\right)\right]^n$.

The last inequality may have an interesting interpretation in statistical quality control. In fact, as $P\left(\max_{1 \leq t \leq n} \frac{|\varepsilon_t|}{\sigma_\varepsilon} \leq x\right)$ is the probability of no alarm until time n in the in-control state, $\left[F_{|Z_1|}\left(\frac{\sigma_\varepsilon}{\alpha_0} x\right)\right]^n$ is the analogical probability if the process becomes independent ($\alpha_i = 0,\ \beta_i = 0,\ i = 1, \ldots, q$). Thus, for a TARCH process, the probability of no alarm shall be evaluated, for certain values of x, using the independent generator process Z.

We illustrate now this result in Fig. 3 considering $n = 3$ and using 5,000 realizations of the $\max_{1 \leq t \leq 3} |\varepsilon_t|$ where ε is the TARCH (2) process with $\sigma_t = 10.0 + 0.8\varepsilon_{t-2}^+ - 0.8\varepsilon_{t-2}^-$ and Z following the standard Gaussian law. The sample was generated using the previous 10,000 observations of the process and choosing three-dimensional subsamples of consecutive indices, the first one of them randomly selected. We estimate the distribution function of $\max_{1 \leq t \leq 3} \frac{|\varepsilon_t|}{\sigma_\varepsilon}$ on x (represented by DISTMAXEREDUZ). We also plot the functions $\left[F_{|Z_1|}\left(\frac{\sigma_\varepsilon x}{\alpha_0 + \sigma_\varepsilon \sum_{i=1}^{q} \alpha_i}\right)\right]^3$ (DISTINF) and $\left[F_{|Z_1|}\left(\frac{\sigma_\varepsilon}{\alpha_0} x\right)\right]^3$ (DISTSUP). The announced bounds for the distribution function of $\max_{1 \leq t \leq 3} \frac{|\varepsilon_t|}{\sigma_\varepsilon}$ are clearly respected.

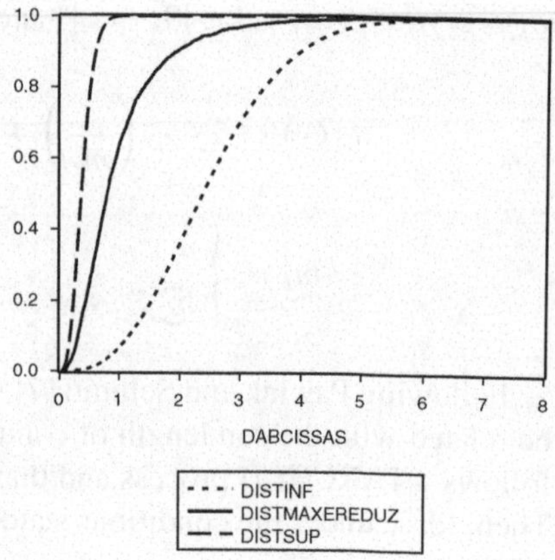

Fig. 3 Plots of $\left[F_{|Z_1|}\left(\frac{\sigma_\varepsilon x}{\alpha_0+\sigma_\varepsilon\sum_{i=1}^{q}\alpha_i}\right)\right]^3$, (DISTINF), the estimation of $P\left(\max_{1\leq t\leq 3}\frac{|\varepsilon_t|}{\sigma_\varepsilon}\leq x\right)$, (DISTMAXEREDUZ), and $\left[F_{|Z_1|}\left(\frac{\sigma_\varepsilon}{\alpha_0}x\right)\right]^3$, (DISTSUP)

4 GTARCH Processes: The Empirical Process

The study of the empirical process associated to a GTARCH process is a usual way to consistently estimate the process distribution. This study is here particularly relevant as in Sects. 2 and 3 empirical distribution functions were used systematically to estimate the distributions related with the process ε. Thus, in this section the results of Berkes and Horvath [1] for GARCH models are applied to GTARCH ones to analyze the asymptotical behavior of the empirical process. We begin by stating the existence of a m-dependent sequence $(\varepsilon_n^{'})$ that is close to (ε_n), where $m = [n^\rho]$, $0<\rho<1$. So, variables $\varepsilon_n^{'}$ with indices differing more than $[n^\rho]$ are independent.

Let us consider the strictly stationary GTARCH (p,q) process ε defined by (1), the corresponding matrices $(A_t, t\in\mathbb{Z})$, and the random vectorial process $X = (X_t, t\in\mathbb{Z})$, defined in Sect. 1. Considering any norm in $\mathbb{R}^{p+2q-2}$ and the corresponding induced norm on the set of the square matrices of order $p+2q-2$, we also assume the existence of $E\left(\log^+\|A_0\|\right)^\mu$ for some $\mu>2$. Let's define $X' = \left(X_n^{'}, n\in\mathbb{N}\right)$ by $X_n^{'} = B+\sum_{k=1}^{[n^\rho]}A_{n-1}A_{n-2}\ldots A_{n-k}B$, with some $0<\rho<1$.

Property 3. *Under the previous hypotheses, there exist measurable functions* $f_n : \mathbb{R}^{[n^\rho]}\to\mathbb{R}$ $(n=1,2,\ldots)$ *such that, setting* $\varepsilon_n^{'} = f_n\left(Z_n, Z_{n-1},\ldots,Z_{n-[n^\rho]}\right)$, *we have, for some* $c>0$,

$$P\left\{\left|\varepsilon_n-\varepsilon_n^{'}\right| > cn^{-\rho(\upsilon-2)/2}\right\}\leq cn^{-\rho(\upsilon-2)/2}.$$

Proof. Let $\varepsilon_n^{'}$ be such that $\left(\varepsilon_n^{'}\right)^+$ is the $p+1$ component of $X_{n+1}^{'}$ and $-\left(\varepsilon_n^{'}\right)^-$ is the $p+q$ component of $X_{n+1}^{'}$. So, $\varepsilon_n^{'}$ is a measurable function of $Z_n, Z_{n-1},\ldots,Z_{n-[n^\rho]}$. Moreover, taking into account Lemma 2.3 of Berkes and Horvath [1], we have

$$P\left(\left\|X_{n+1}-X_{n+1}^{'}\right\|>c\,(n+1)^{-\rho(\mu-2)/2}\right)\leq c\,(n+1)^{-\rho(\mu-2)/2}.$$

Denoting by $X_{i,n+1}$ the i-component of X_{n+1}, this is equivalent to

$$P\left(\max_{1\leq i\leq p+2q-2}\left|X_{i,n+1}-X_{i,n+1}^{'}\right|>a_{n,\rho}\right)\leq a_{n,\rho},$$

where $a_{n,\rho}=c\,(n+1)^{-\rho(\mu-2)/2}$. So, for every $i\in\{1,\ldots,p+2q-2\}$,

$$P\left(\left|X_{i,n+1}-X_{i,n+1}^{'}\right|>a_{n,\rho}\right)\leq a_{n,\rho},$$

in particular for $P\left(\left|\varepsilon_n^{+}-\left(\varepsilon_n^{'}\right)^{+}\right|>a_{n,\rho}\right)$ and for $P\left(\left|-\varepsilon_n^{-}-\left[-\left(\varepsilon_n^{'}\right)^{-}\right]\right|>a_{n,\rho}\right)$. Finally,

$$\begin{aligned}P&\left(\left|\varepsilon_n-\varepsilon_n^{'}\right|>c\,(n+1)^{-\rho(\mu-2)/2}\right)\\&\leq P\left(\left\{\left|\varepsilon_n^{+}-\left(\varepsilon_n^{'}\right)^{+}\right|>\frac{a_{n,\rho}}{2}\right\}\cup\left\{\left|\varepsilon_n^{-}-\left(\varepsilon_n^{'}\right)^{-}\right|>\frac{a_{n,\rho}}{2}\right\}\right)\leq a_{n,\rho}.\end{aligned}$$

□

The strong approximation for the empirical process of n observed elements of GTARCH processes is deduced from this property via standard blocking techniques as in Berkes and Horváth [1] for GARCH models, taking into account that most of their lemmas apply in our case. Let us summarize the final form of this result.

The empirical process of n observations, $\varepsilon_1,\varepsilon_2,\ldots,\varepsilon_n$, of the strictly stationary solution, $\varepsilon=(\varepsilon_t,\ t\in\mathbb{Z})$, of the model GTARCH defined in Eq. (1), is

$$R\,(s,t)=\sum_{i=1}^{t}\left[1_{\{\varepsilon_i\leq s\}}-F_{\varepsilon_1}(s)\right],\ s\in\mathbb{R},\ t\in\{1,\ldots,n\}$$

with F_{ε_1} the distribution function of ε_1. Let $\varepsilon_k\,(s)=1_{\{\varepsilon_k\leq s\}}-F_{\varepsilon_1}(s)$, $k\in\{1,\ldots,n\}$ and $\Gamma\,(s,s')=E\left[\varepsilon_0\,(s)\,\varepsilon_0\left(s^{'}\right)\right]+\sum_{n=1}^{+\infty}E\left[\varepsilon_0\,(s)\,\varepsilon_n\left(s^{'}\right)\right]+\sum_{n=1}^{+\infty}E\left[\varepsilon_0\left(s^{'}\right)\varepsilon_n\,(s)\right]$.

The strong approximation for the empirical process is stated now.

Theorem 4. *Let us assume the strict stationarity of* ε *defined on* (1). *If*

- $|F_{Z_1}(t)-F_{Z_1}(s)|\leq C\,|t-s|^{\theta}$, *with some* $0<C<+\infty$, $0<\theta\leq 1$ *where* F_{Z_1} *is the distribution function of* Z_1.
- $E\left(\log^{+}\|A_0\|\right)^{\mu}$ *exists with some* $\mu>2+\frac{16}{\theta}$.

then the series $\Gamma(s,s')$ *is absolutely convergent for any* $-\infty < s, s' < +\infty$ *and there is a Gaussian process* $K(s,t)$ *with* $EK(s,t) = 0$, $EK(s,t)K(s',t') = \min(t,t)\,\Gamma(s,s')$ *such that, with some* $\lambda > 0$,

$$\sup_{0\le t\le T}\ \sup_{-\infty<s<+\infty} |R(s,t) - K(s,t)| = 0\left(T^{\frac{1}{2}}(\log T)^{-\lambda}\right),\ a.s.$$

The weak convergence of the empirical process and the law of the iterated logarithm follow from this result. In fact, if $\widehat{K}(s)$ is a Gaussian process with $E\widehat{K}(s) = 0$ and $E\left[\widehat{K}(s)\widehat{K}\left(s'\right)\right] = \Gamma(s,s')$, then $n^{\frac{1}{2}}\left[\frac{1}{n}\sum_{i=1}^{n}\left[1_{\{\varepsilon_i\le s\}} - F_{\varepsilon_1}(s)\right]\right] \overset{D[-\infty,+\infty]}{\rightarrow} \widehat{K}(s)$, as $n \to +\infty$, where $\overset{D[-\infty,+\infty]}{\rightarrow}$ denotes the weak convergence of a random variable in the Skorokhod space $[-\infty,+\infty]$. Moreover,

$$\lim_{n\to+\infty}\sup\left(\frac{n}{2\log\log n}\right)^{\frac{1}{2}}\sup_{-\infty<s<+\infty}\left|\frac{1}{n}\sum_{i=1}^{n}\left[1_{\{\varepsilon_i\le s\}} - F_{\varepsilon_1}(s)\right]\right| = c,\ a.s.,$$

with some $0 < c < +\infty$.

Acknowledgements We thank the referees for the constructive suggestions.

References

1. Berkes, I., Horváth, L.: Strong approximation of the empirical process of GARCH sequences. Ann. Appl. Probab. **11**(2), 789–809 (2001)
2. Bollerslev, T.: Generalized autoregressive conditional heteroskedasticity. J. Econom. **31**, 307–327 (1986)
3. Engle, R.F.: Autoregressive conditional heteroskedasticity with estimates of the variance of the United Kingdom inflation. Econometrica **50**, 987–1008 (1982)
4. Gonçalves, E., Mendes-Lopes, N.: Stationarity of GTARCH processes. Statistics **28**, 171–178 (1996)
5. Gonçalves, E., Mendes-Lopes, N.: On the distribution of generalized threshold ARCH stochastic processes. Int. J. Pure Appl. Math. **35**(3), 397–419 (2007)
6. Gonçalves, E., Mendes-Lopes, N.: On the structure of generalized threshold ARCH processes. Stat. Probab. Lett. **80**, 573–580 (2010)
7. Pawlak, M., Schmid, W.: On the distributional properties of GARCH processes. J. Time Ser. Anal. **22**(3), 339–352 (2001)
8. Zakoian, J.M.: Threshold heteroskedasticity models. J. Econ. Dyn. Control. **18**, 931–955 (1994)

Preliminary Results on Confidence Intervals for Open *Bonus Malus*

Gracinda R. Guerreiro, João T. Mexia, and Maria F. Miguens

Abstract
Considering open portfolios, we analyze *bonus–malus* systems (BMS) under a realistic approach, as we already did in Guerreiro and Mexia (Discuss. Math. Probab. Stat. 24(2):197–213, 2004). Using stochastic vortices model we are now able to predict long-run distribution through confidence intervals.

1 Introduction

A *bonus–malus* system (BMS), in automobile insurance, is a rating system from which Insurers, through premiums, are able to, simultaneously, penalize drivers who are responsible for accidents and reward claim-free policyholders. It is, in fact, an a posteriori classification from which the a priori premium is adjusted, according to past experience information. This a posteriori premium aims to better measure the risk that the policyholder represents to the insurer: in the long run, he will pay the premium corresponding to his claim frequency.

The design and evaluation of BMS is based on Markov chains (for detailed presentations of BMS techniques, see [9]). Many authors proposed models for the study of BMS. However, most models are based on the assumption of closed portfolios with a pre-defined entry class for all new policyholders. In Portugal, as stated in [2, 3, 6], there are many movements among different insurers and frequently, due to commercial goals, a priori discounts are given to new policyholders. This facts clearly reveal unrealistic restrictions in classic models. We consider that analyzing

G.R. Guerreiro (✉) · J.T. Mexia · M.F. Miguens
Departamento de Matemática, Universidade Nova de Lisboa-Faculdade de Ciências e Tecnologia, Campus de Caparica, 2829-516 Caparica, Portugal
e-mail: grg@fct.unl.pt; jtm@fct.unl.pt; mfvm@fct.unl.pt

J. Lita da Silva et al. (eds.), *Advances in Regression, Survival Analysis, Extreme Values, Markov Processes and Other Statistical Applications*, Studies in Theoretical and Applied Statistics, DOI 10.1007/978-3-642-34904-1_23,

BMS under open portfolio approaches renders a more realistic perspective. For open portfolio models, see, for instance, [2, 6].

In this chapter, availing ourselves of the stochastic vortices (SV) model (see [6–8]), we estimate long-run distribution through confidence intervals. In this way, we are able to obtain intervals for bonus scales which can be useful to define optimal and competitive premiums. SV model has already been developed for populations with complex characteristics. In this paper, we focus on BMS application, so the presented model is congruent to it's structure. For general results, see [8].

2 Stochastic Vortices Model for Bonus Malus Systems

2.1 Transition Matrix

Let us consider:

- A BMS with s bonus classes in one Markov chain communication class
- One recurrent state, representing the withdrawals of policyholders

The one-step transition matrix of the Markov chain will be

$$\boldsymbol{P} = \begin{bmatrix} \boldsymbol{K} & \boldsymbol{q}_1 \\ 0 & 1 \end{bmatrix} \tag{1}$$

with

$\boldsymbol{K}$—$s \times s$ transition matrix between bonus classes

$\boldsymbol{q}_1$—s components vector of annulment probabilities

With $\boldsymbol{q}_n = \sum_{j=0}^{n-1} \boldsymbol{K}^j \, \boldsymbol{q}_1$, $n \in \mathbb{N}$, the n step transition matrix will be

$$\boldsymbol{P}^n = \begin{bmatrix} \boldsymbol{K}^n & \boldsymbol{q}_n \\ 0 & 1 \end{bmatrix}, \quad n \in \mathbb{N}. \tag{2}$$

2.2 Policyholders Entries

We assume that entries into the portfolio occur at the beginning of time periods, which we will consider as years. Moreover, we assume that:

- Numbers of new policyholders in year i, E_i , $i \in \mathbb{N}$, are independent and Poisson-distributed random variables with means λ'_i , $i \in \mathbb{N}$.
- Mean values λ'_i are given by

$$\lambda'_i = a + b\,\theta^i \ , \ a, b \in \mathbb{R} \, , \ 0 < \theta < 1 \, , \ i \in \mathbb{N}. \tag{3}$$

Note that we are focusing on $0 < \theta < 1$, but the model was developed for $\theta > 0$; see [7]. Equation (3) represents a quite general assumption and applies to a variety of population entries. We point out the next example: when $a = 0$,

(3) represents a population with a geometric growth on entrances; if $b = -a$ and $\theta = e^{-\delta}$, $\delta > 0$, (3) will represent a population with an asymptotic growth on entrances; for the situations $\theta = 1$ or $b = 0$, (3) reflects a constant rate on entrances.

- New policyholders are subject to an initial classification. Elements entered in the ith year will be allocated to any of the bonus classes, according to the components of probabilities vector $\boldsymbol{c}_i, i \in \mathbb{N}$. We assume that new elements do not leave the portfolio immediately after initial classification; thus $\boldsymbol{c}_i^T = \left[\boldsymbol{t}_i^T \mid 0\right]$, with $\boldsymbol{t}_i$ corresponding to the probabilities of a new element entering into transient states, in year i. For details about initial classification criteria, see [7].

2.3 Expected Subpopulations Dimension

Let $\boldsymbol{N}_i$ be the number of policyholders initially placed in each bonus class in time period i, $i \in \mathbb{N}$.

The next proposition (see [4]) has a fundamental role in our developments:

Proposition 1. *If* $E \sim$ *Poisson*(μ) *and* $(\boldsymbol{X} \mid E = e) \sim$ *Multinomial*$(e, \boldsymbol{c})$ *with* $\boldsymbol{c}^T = (c_1, \ldots, c_k)$*, then* $\boldsymbol{X}$ *is a random vector whose margins,* $X_1, \ldots, X_k$*, are independent and Poisson-distributed random variables with mean values* $(\mu_1, \ldots, \mu_k)^T = (\mu c_1, \ldots, \mu c_k)^T$*, respectively.*

We will say that $\boldsymbol{X}$ has a multivariate Poisson distribution with mean vector $\boldsymbol{\mu} = (\mu_1, \ldots, \mu_k)^T$ and will be represented by $\boldsymbol{X} \sim \text{Poisson}(\boldsymbol{\mu})$.

Theorem 1. *Consider a population with* k *subpopulations and that the numbers* $E_i, i \in \mathbb{N}$*, of new elements arriving to the population in year* i *are Poisson distributed with mean value* $\lambda'_i, i \in \mathbb{N}$*. New elements are allocated in subpopulations, in year* $i, i \in \mathbb{N}$*, according to probabilities vector* $\boldsymbol{c}_i, i \in \mathbb{N}$*. After entry, future periodic re-classifications follow stable probability transition matrices. In a time period* m*, the number* $\boldsymbol{N}_{i,m}$ *of elements in each sub-population, entered in the* i*th year, will have been subject to* $m - i$ *reclassifications and are Poisson distributed with parameter*

$$\boldsymbol{\lambda}_{i,m}^T = \lambda'_i \, \boldsymbol{c}_i^T \, \boldsymbol{P}^{m-i}. \tag{4}$$

Proof. According to Proposition 1, we may acknowledge that $\boldsymbol{N}_i$, number of elements initially placed in each subpopulation in year i, is Poisson distributed with mean vector $\lambda'_i \, \boldsymbol{c}_i$, $\boldsymbol{N}_i \sim \text{Poisson}(\lambda'_i \, \boldsymbol{c}_i)$.

In each time period $m, m \geq i$, the $N_{i,j}, i \in \mathbb{N}, j = 1, \ldots, k$, elements entered in year i, and initially placed in subpopulation j, have been subject to $m - i$ reclassifications and distributed over the subpopulations according to transition matrix $\boldsymbol{P}$. The vector of the number of elements $\boldsymbol{N}_{i,m,j}^T = (N_{i,m,j,1}, \ldots, N_{i,m,j,k})$, will, according once again to Proposition 1, also be Poisson distributed,

$$\boldsymbol{N}_{i,m,j} \sim P\left(\lambda_i' c_{i,j} \boldsymbol{\delta}_j^T \boldsymbol{P}^{m-i}\right), \; i, m \in \mathbb{N}, j = 1, \ldots, k,$$

with $\boldsymbol{\delta}_j$ a vector whose components are null, except the jth one, which is 1.

Since the components of $\boldsymbol{N}_i$, $i \in \mathbb{N}$, are independent random variables, vectors $\boldsymbol{N}_{i,m,j}$, $i, m \in \mathbb{N}$, $j = 1, \ldots, k$, will also be independent. To complete the proof we only need to point out that the vector of total sizes, in time period m, will be given by $\boldsymbol{N}_{i,m} = \sum_{j=1}^k \boldsymbol{N}_{i,m,j}$. Thus, due to Poisson distribution reproducibility, we obtain $\boldsymbol{N}_{i,m} \sim P(\boldsymbol{\lambda}_{i,m})$ with $\boldsymbol{\lambda}_{i,m}^T = \sum_{j=1}^k \lambda_i' \, c_{i,j} \, \boldsymbol{\delta}_j^T \, \boldsymbol{P}^{m-i} = \lambda_i' \, \boldsymbol{c}_i^T \, \boldsymbol{P}^{m-i}$. □

Using (4) (see [7]), we are able to estimate bonus classes dimension, according to stochastic vortices model. The estimator for bonus classes dimension now reflects entrances intensities, initial classification, transition, and annulment probabilities.

For total number of policyholders in each bonus class in time period m, we have

$$\boldsymbol{N}_m^{++} = \sum_{i=1}^{m} \boldsymbol{N}_{i,m} \sim P(\boldsymbol{\lambda}_m^{++}) \tag{5}$$

with

$$\boldsymbol{\lambda}_m^{++T} = \left[\sum_{i=1}^m \lambda_i' \, \boldsymbol{t}_i^T \, \boldsymbol{K}^{m-i} \;\middle|\; \sum_{i=1}^m \lambda_i' \, \boldsymbol{t}_i^T \, \boldsymbol{q}_{m-i}\right]. \tag{6}$$

2.4 Asymptotic Results for Transient States

The existence of *stochastic vortices* in transient states implies stable limit relative dimension for the bonus classes.

Let us assume that sub-matrix $\boldsymbol{K}$ is a $s \times s$ diagonalizable matrix. Under very general conditions (see [11]), we will have

$$\boldsymbol{K} = \sum_{j=1}^{s} \eta_j \, \boldsymbol{\alpha}_j \, \boldsymbol{\beta}_j^T \qquad \boldsymbol{K}^m = \sum_{j=1}^{s} \eta_j^m \, \boldsymbol{\alpha}_j \, \boldsymbol{\beta}_j^T \tag{7}$$

with $\eta_j \left[\boldsymbol{\alpha}_j, \boldsymbol{\beta}_j^T\right]$, $j = 1, \ldots, s$ matrix $\boldsymbol{K}$ eigenvalues [left and right eigenvectors].

Considering the first block of (6), as well as assumption (3), let

$$\boldsymbol{\lambda}_m^{+T} = \sum_{i=1}^{m} \lambda_i' \, \boldsymbol{t}_i^T \, \boldsymbol{K}^{m-i} = \sum_{i=1}^{m} (a + b\,\theta^i) \, \boldsymbol{t}_i^T \, \boldsymbol{K}^{m-i} \tag{8}$$

be the mean vector for transient states (bonus classes), in time period m.

Using (7), we identified conditions for convergence of (8). Proposition 2 is established in [7]. For computational simplicity, we obtained last expression in [8].

Proposition 2. *If entries intensities are given by* $\lambda'_i = a + b\,\theta^i$, $i \in \mathbb{N}, (a,b) \in \mathbb{R}^2$, $0 < \theta < 1$ *and* $\lim_{i\to+\infty} t_{i,j} = t_j$, $j = 1, \ldots, s$, *we then have*

$$\boldsymbol{\lambda}_{\infty}^{+T} = \lim_{m\to+\infty} \boldsymbol{\lambda}_m^{+T} = \sum_{j=1}^{s} \frac{\boldsymbol{t}^T\, \boldsymbol{\alpha}_j\, a}{1-\eta_j}\, \boldsymbol{\beta}_j^T = a\, \boldsymbol{t}^T\, (\boldsymbol{I}_s - \boldsymbol{K})^{-1}\,. \tag{9}$$

This proposition guarantees, under general conditions, the existence of finite limits for the parameters vector of sub-populations in transient states, if $0 < \theta < 1$.

Long-run distribution of BMS corresponds to limit relative dimensions for transient states, which will be stable as $m \to +\infty$ (see [7]) and given by

$$\pi_{\infty,j} = \lim_{m\to+\infty} \pi_{m,j} = \lim_{m\to+\infty} \frac{\lambda^+_{m,j}}{\sum_{j=1}^{s} \lambda^+_{m,j}} = \frac{\lambda^+_{\infty,j}}{\sum_{j=1}^{s} \lambda^+_{\infty,j}}\ , \quad j = 1, \ldots, s \tag{10}$$

so a *stochastic vortex* is established in transient states and long-run distribution for BMS can be easily obtained. Note that initial classification will not interfere in long-run distribution. However, regarding weighted distributions (see [1]), initial classification renders more realistic models.

2.4.1 Confidence Intervals for Bonus Classes

Due to (5), for large portfolios we obtain level q confidence intervals for $\lambda^+_{m,j}$, $j = 1, \ldots, s$, $m \in \mathbb{N}$:

$$\mathbb{P}\left[N^+_{m,j} - z_{q/2}\sqrt{N^+_{m,j}} \le \lambda^+_{m,j} \le N^+_{m,j} - z_{q/2}\sqrt{N^+_{m,j}}\right] = 1 - \frac{q}{2}$$

where $z_{q/2}$ is the upper $1 - \frac{q}{2}$ critical value for standard normal distribution.

Using delta method (see [12]), we obtain level q confidence intervals for $\pi_{m,j}$, $j = 1, \ldots, s$, $m \in \mathbb{N}$:

$$\mathbb{P}\left[\frac{N^+_{m,j}}{\sum_{j=1}^{k} N^+_{m,j}} - z_{q/2}\sqrt{V_j\, N^+_{m,j}} \le \pi_{m,j} \le \frac{N^+_{m,j}}{\sum_{j=1}^{k} N^+_{m,j}} + z_{q/2}\sqrt{V_j\, N^+_{m,j}}\right] = 1 - \frac{q}{2} \tag{11}$$

with

$$V_j = \left(\lambda^+_{\infty}\right)^{-2}\left[(1-\pi_{m,j})^2 + C^{-2}\,\pi^2_{m,j} - 2C^{-1}\,\pi_{m,j}(1-\pi_{m,j})\right] \tag{12}$$

or

$$V_j = \left(\lambda^+_{\infty}\right)^{-2}\,\pi^2_{m,j}\,. \tag{13}$$

considering $\lambda^+_{\infty} = \sum_{j=1}^{s} \lambda^+_{\infty,j}$ and $S_j = \sum_{i\neq j}^{s} \hat{\lambda}^+_{m,j}$.

Equation (12) holds if $\lim_{m\to+\infty} \sigma_{\lambda^+_{\infty,j}}/\sigma_{S_j} = C > 0$ and (13) holds if, for class j, we have $\lim_{m\to+\infty} \sigma_{\lambda_j^{++}}/\sigma_{S_j} = 0$.

Table 1 Number of new policyholders per year

1997	1998	1999	2000	2001	2002	2003	2004	2005	2006
4,107	9,607	15,829	22,443	29,216	34,770	39,686	32,588	46,692	49,283

3 An Example

3.1 Transition Rules and Claim Frequency

Consider data from a Portuguese insurer portfolio. BMS has $s = 20$ classes with premium increasing with class index. For each claim-free year, index decreases by one. The first [each of the next] claim increases the class index by three [five]. The zero bonus/malus class is the tenth.

To $C(t)$, the number of claims in $[0, t)$, we adjusted a mixed Poisson distribution with Gamma structural distribution: $C(t) \sim P(\Lambda)$, $\Lambda \sim \text{Gamma}(\alpha, \beta)$. From the data we obtained the ML estimates $\hat{\alpha} = 0.70523$ and $\hat{\beta} = 10.10695$.

3.2 New Policyholders Estimation

Table 1 resumes insurer information about new policies for automobile insurance.

Let $(E_1, \ldots, E_m)$ be the random sample of the number of entries in m consecutive years. Let us assume that $E_i \sim P(\lambda_i')$ with $\lambda_i' = \tau(1 - e^{-\delta i})$, $(\tau, \delta) \in \mathbb{R}^2$. Note that this is a particular case of (3) with $\theta = e^{-\delta}$ and $\tau = -b = a$.

ML estimators for τ and δ are the solutions of

$$\hat{\tau} = \frac{\sum_{i=1}^{m} e_i}{m - \sum_{i=1}^{m} e^{-\hat{\delta} i}} \tag{14}$$

$$\hat{\tau} \sum_{i=1}^{m} i\, e^{-\hat{\delta} i} = \sum_{i=1}^{m} \frac{i\, e^{-\hat{\delta} i}}{1 - e^{-\hat{\delta} i}}\, e_i . \tag{15}$$

From (14) and (15), ML estimates were obtained: $\hat{\tau} = 212109$ and $\hat{\delta} = 0.026692$. This implies, for general model (3), that $\hat{\theta} = e^{-\hat{\delta}} = 0.973661$.

We note that the ML estimate for τ is unrealistic for this insurer. Due to the Portuguese market and the insurer's quota share, it is not likely that they attain such number of new annual entries. An alternative estimate relies on fix τ as the insurer's long run perspective on growth and estimate δ, based on that assumption.

3.3 Initial Classification and Annulment Probabilities

Initial classification and annulment probabilities were considered not depending on year and estimated from data. Results are presented in Table 2.

Table 2 Initial classification and annulment probabilities per bonus class

j	1	2	3	4	5	6	7	8	9	10
$c(j)$	0.2394	0.0537	0.1914	0.0696	0.1886	0.0061	0.0342	0.0104	0.0625	0.1424
$q(j)$	0.1043	0.1275	0.1542	0.1833	0.2248	0.2179	0.2473	0.2350	0.2375	0.4533
j	11	12	13	14	15	16	17	18	19	20
$c(j)$	0.0006	0.0004	0.0003	0.0002	0.0002	$2 \cdot 10^{-5}$	$3 \cdot 10^{-5}$	$3 \cdot 10^{-5}$	$4 \cdot 10^{-6}$	$2 \cdot 10^{-5}$
$q(j)$	0.3909	0.4718	0.5621	0.5964	0.5703	0.7353	0.9487	0.4815	0.7364	0.8276

Table 3 Long run distributions and optimal bonus scales—S. vortices and C. model approaches

j	$\pi_C(j)$	$b_C^G(j)$	$\underline{\pi}_S(j)$	$\pi_S(j)$	$\bar{\pi}_S(j)$	$b_S^G(j)$
1	0.7943	0.0531	0.6879157898	0.6879163605	0.6879169313	0.06086
2	0.0412	0.0764	0.0735734532	0.0736314416	0.0736894300	0.07082
3	0.0462	0.0996	0.0854663157	0.0854859575	0.0855055993	0.08078
4	0.0134	0.1228	0.0435894748	0.0435998164	0.0436101579	0.09074
5	0.0113	0.1461	0.0428224948	0.0428325975	0.0428427002	0.10070
6	0.0081	0.1693	0.0152794230	0.0152815824	0.0152837419	0.11067
7	0.0077	0.1926	0.0158469193	0.0158492070	0.0158514948	0.12063
8	0.0074	0.2158	0.0111921268	0.0111934917	0.0111948565	0.13059
9	0.0055	0.2390	0.0101670336	0.0101682137	0.0101693938	0.14055
10	0.0050	0.2623	0.0103512971	0.0103525116	0.0103537261	0.15051
11	0.0045	0.2855	0.0013487107	0.0013487707	0.0013488308	0.16047
12	0.0044	0.3088	0.0009259372	0.0009259720	0.0009260068	0.17043
13	0.0043	0.3320	0.0004858670	0.0004858807	0.0004858945	0.18039
14	0.0044	0.3552	0.0003050116	0.0003050186	0.0003050255	0.19035
15	0.0047	0.3785	0.0002083131	0.0002083171	0.0002083212	0.20031
16	0.0051	0.4017	0.0001247669	0.0001247688	0.0001247707	0.21027
17	0.0058	0.4250	0.0001169801	0.0001169818	0.0001169835	0.22023
18	0.0069	0.4482	0.0000559784	0.0000559790	0.0000559796	0.23019
19	0.0086	0.4714	0.0000377499	0.0000377502	0.0000377506	0.24016
20	0.0113	0.4947	0.0000793801	0.0000793812	0.0000793823	0.25012

Note that policyholders entered through all classes and a large number of insured nullified his policy when arrived to *maluses* classes. This highlights that assuming closed models and a "starting class" for all new policyholders in rather unrealistic.

3.4 Long Run Distribution and Optimal Bonus Scale—Stochastic Vortices and Closed Model Approach

Using (10) for SV model and classical results for BMS (see [9]), long-run distributions were obtained. For SV model we are able to predict long-run distribution through confidence intervals using (11). Following [5] after [10], an optimal bonus scale was obtained for each approach.

Table 3 presents results for both models. Indexes C and S refer to closed and SV model, respectively. b^G represents Gilde and Sundt's linear optimal bonus scale and $\underline{\pi}_S(j)$ and $\bar{\pi}_S(j)$ the 95 % confidence intervals. Note that long-run distributions differ significantly and closed model overestimates probabilities in maluses classes as well as in higher discount class. This, naturally, has impacts on optimal bonus scales.

With the ML estimates for τ and δ, the SV model converges slowly to stationarity. This implies that Borgan et al. [1], optimal bonus scale should be implemented instead of Norberg's [10]. In this chapter we illustrate Norberg's optimal bonus scale in order to evaluate portfolio performance in a long-run perspective.

Acknowledgments This work was partially supported by Financiamento Base 2009 ISFL-1-297 from FCT/MCTES/PT.

References

1. Borgan, Ø., Hoem, J., Norberg, R.: A nonasymptotic criterion for the evaluation of automobile bonus system. Scand. Actuar. J. **3**, 165–178 (1981)
2. Centeno, L., Andrade e Silva, J.: Bonus systems in open portfolio. Insurance Math. Econom. **28**, 341–350 (2001)
3. Denuit, M., Dhaene, J.: Bonus-malus scales using exponential loss functions. Blatter der DGVFM **25**(1), 13–27 (2001)
4. Feller, W.: An Introduction to Probability Theory and it's Applications, 2nd edn. Wiley, New York (1966)
5. Gilde, V., Sundt, B.: On bonus systems with credibility scales. Scand. Act. J. **2**, 13–22 (1989)
6. Guerreiro, G.R., Mexia, J.T.: An alternative approach to bonus malus. Discuss. Math. Probab. Stat. **24**(2), 197–213 (2004)
7. Guerreiro, G.R., Mexia, J.T.: Stochastic vortices in periodically reclassified populations. Discuss. Math. Probab. Stat. **28**(2), 209–227 (2008)
8. Guerreiro, G.R., Mexia, J.T, Miguens, M.F.: A model for open populations subject to periodic re-classifications. J. Stat. Theory Pract. **4**(2), 303–321 (2010)
9. Lemaire, J.: Bonus-Malus Systems in Automobile Insurance. Kluwer, Boston (1995)
10. Norberg, R.: A credibility theory for automobile bonus system. Scand. Act. J. **2**, 92–107 (1976)
11. Schott, J.R.: Matrix Analysis for Statistics, Wiley Series in Probability and Statistics. Wiley, New York (1997)
12. Tiago de Oliveira, J.: The delta-method for obtention of asymptotic distributions-Applications. Publ. de l'Inst. de Stat. de l'Univ. de Paris **27**, 49–70 (1982)

Study of the Electrocardiographic Fluctuations on Brugada Syndrome Screening

Carla Henriques, Ana Cristina Matos, and Luís Ferreira dos Santos

Abstract

Brugada syndrome (BS) is a cardiologic disorder which favours cardiac arrhythmias and is thought to be responsible for about 20–50 % of sudden cardiac death (SCD) in individuals with a structurally normal heart. There are three electrocardiogram (ECG) characteristic patterns associated with BS. From an index case, 130 family members were screened for BS and data collected in order to identify possible influential factors for the manifestation, and fluctuations, of Brugada patterns in ECG results. Moreover, data collected from family members were analysed in order to evaluate the necessity for more than one ECG in screening for BS. Also, the effect of displacing ECG electrodes in the sensitivity and specificity of the exam was analysed.

1 Introduction

Brugada syndrome (BS) is a recent clinical cardiologic entity, described for the first time in 1992 (Brugada and Brugada [1]), which is reputed to be responsible for

C. Henriques (✉)
CMUC and Escola Superior Tecnologia e Gestão, Instituto Politécnico de Viseu, Campus Politécnico de Repeses, 3504-510, Viseu, Portugal
e-mail: carlahenriq@estv.ipv.pt

A.C. Matos
Escola Superior Tecnologia e Gestão, Instituto Politécnico de Viseu, Campus Politécnico de Repeses, 3504-510, Viseu, Portugal
e-mail: amatos@estv.ipv.pt

L.F. dos Santos
Serviço de Cardiologia, Hospital São Teotónio, Viseu, Portugal
e-mail: luisferreirasantos@gmail.com

J. Lita da Silva et al. (eds.), *Advances in Regression, Survival Analysis, Extreme Values, Markov Processes and Other Statistical Applications*, Studies in Theoretical and Applied Statistics, DOI 10.1007/978-3-642-34904-1_24,

about 4–12 % of all cases of sudden cardiac death (SCD) and for 20–50 % of SCD in subjects with a structurally normal heart (Brugada et al. [3]). This syndrome is characterized by a dysfunction of a cardiac ionic channel, which favours cardiac arrhythmias. It is an inherited disorder, but there are also fortuitous cases (absent in other relatives). For the diagnosis of BS, a special characteristic pattern in an electrocardiogram (ECG) must be found. In fact, there are three special ECG patterns associated with BS—type 1, type 2 and type 3 ECG patterns—but only type 1 is considered to be diagnostic of the syndrome. Type 2 and type 3 are regarded as Brugada suggestive patterns. The diagnosis is made when a type 1 ECG pattern is found and the subject presents one or more of some clinical symptoms which are easily detected, such as nocturnal agonal respiration, syncope and family history of sudden death at < 45 years old. However, the diagnosis is not as simple as it might appear. In fact, Brugada ECG patterns are dynamic, meaning that a Brugada patient may exhibit intermittently normal ECGs and Brugada pattern ECGs. Genetic tests may be done, but these are too expensive and the mutations related to the syndrome are not easy to identify. In this study we aim to identify possible influential factors for the manifestation and fluctuations of Brugada ECG patterns. Also we analyse some measures that confirm the importance of more than one ECG on screening for BS. Another interesting question which is analysed in this study, is related to the known important fact that raising the ECG electrodes, with respect to the conventional positions, affects the ECG patterns. It is consensual that this procedure increases the type 1 ECG patterns found, thus leading to a greater sensitivity. However, some authors question the benefit of this procedure because there is no guarantee that this does not increase the number of false positives, thus decreasing the specificity (cf. p. 287 of [2]). We analyse this issue in the final section of this chapter by comparing the sensitivity and the specificity of ECGs obtained for three different positions of the electrodes.

Our data were collected from an index case to whom BS was diagnosed. Given the hereditary character of the syndrome, family members were screened providing useful information which allowed the analysis made in this study. We note that, given the family history of sudden death of the individuals under study, a type 1 ECG pattern is enough to diagnose BS. For the statistical analysis, the SPSS version 15 software was used.

2 Determinant Factors for the Manifestation of Brugada ECG Patterns

As said before, ECG results in patients with BS often fluctuate between diagnostic (type 1) and non-diagnostic patterns, making the diagnosis more difficult. In this section, age, body mass index (BMI) and gender are considered as possible influence factors for the manifestation of Brugada ECG patterns. In this analysis, type 1 ECG pattern is assumed to be detached from the other types, as is the diagnostic pattern. So, we start by analysing the influence of those covariates on the manifestation of

Table 1 Comparing diagnostic with non-diagnostic cases with respect to gender, age and BMI

	Diagnostic (type 1 ECG)		Non-diagnostic (no type 1 ECG)		p-value
Gender	Frequency (% within gender)				Fisher's exact test
(n = 130)	F	6 (9.7 %)	F	56 (90.3 %)	0.782
	M	8 (11.7 %)	M	60 (88.2 %)	
	Mean ± standard deviation				Mann–Whitney test
Age (n = 130)	37.43 ± 11.07		23.74 ± 16.63		0.001
BMI (n = 46)	21.91 ± 2.77		23.33 ± 3.07		0.14

type 1 pattern, that is, on the diagnosis of BS (comparing relatives for whom a type 1 ECG pattern was found with the others). Next, we analyse their influence on the manifestation of one of the Brugada patterns (comparing relatives for whom a Brugada pattern ECG was found with those which have always had normal ECGs). The association between age, BMI and gender with ECG manifestation of Brugada patterns is assessed via an univariable analysis, through logistic regression, Fisher's exact test and Mann–Whitney test, and also via a multivariable analysis through logistic regression. Our data consist of 130 subjects, family relatives of the index case, which were screened with two ECGs within an interval of 6 months. Yet, information for the BMI was collected for only 46 of these 130.

2.1 Diagnostic Versus Non-diagnostic

Table 1 summarizes some informative statistics about age, BMI and gender in the group of subjects for whom BS has been diagnosed (type 1 ECG found) compared with the others.

The results presented in Table 1 give evidence that age has an influence on diagnosis (diagnostic cases are significantly older), but not BMI nor gender. Univariable logistic regression models were adjusted for age, gender and BMI and the results agreed with those in Table 1: only age was identified as a significant predictor. Multivariable analysis, however, showed that both age and BMI are somewhat related to the diagnosis, meaning that, in the presence of age, BMI becomes a useful predictor of the diagnosis. Results for the multivariate model containing the three covariates are shown in Table 2. In fact, age and BMI, but not gender, are identified as significant predictors of diagnosis by both Wald test and likelihood ratio test. Akaike information criterion (AIC) also supports the model with age and BMI but not gender, since it decreases if we remove gender from the model, but it increases if we remove any of the other two covariates (see Burnham and Anderson [4] for an exposition about model comparisons and AIC use).

Interaction between age and BMI was examined in the model with the two covariates, but it was not significant. Indeed, the Wald test statistic for the interaction term is equal to 0.484 with p-value = 0.487 and the likelihood ratio test comparing

Table 2 Multivariable logistic regression models containing age, BMI and gender as covariates ($n = 46$)

Covariate	Wald test (p-value)	Likelihood ratio test[a](p-value)	Change in AIC[b]
Gender	0.000 ($p = 0.991$)	0.000 ($p = 0.991$)	2
Age	5.53 ($p = 0.019$)	6.918 ($p = 0.009$)	−4.92
BMI	4.39 ($p = 0.036$)	5.664 ($p = 0.017$)	−3.66

[a]Comparing the complete model with the model without the covariate
[b]Difference between the AIC of the complete model (49.8) and of the model without the covariate

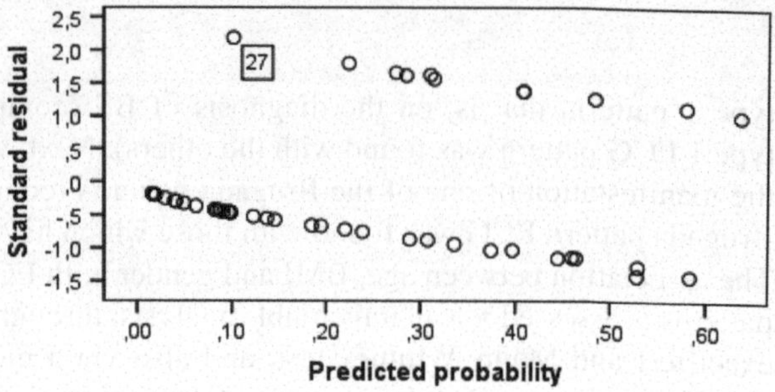

Fig. 1 Standardized pearson residuals against predicted probability for the logistic model with age and BMI

the complete model with the model without the interaction term is equal to 0.51 with p-value = 0.475. Furthermore, the AIC decreases when the interaction term is removed (AIC = 49.29 for the complete model and 47.8 for the model without the interaction term). We also analyse the goodness of fit and other diagnostic statistics for the model with covariates age and BMI. The value of the likelihood ratio test statistic, for overall significance of the two covariates in the model, is 8.811 (p-value = 0.012), so we can conclude that at least one of those covariates is important to predict the diagnosis. Furthermore, the Hosmer and Lemeshow test statistic is equal to 1.808 (p-value = 0.97), which indicates that the model fits reasonably well. The standardized Pearson residuals are plotted in Fig. 1 against the predicted probabilities. As we can see, with the exception of one, all values are less than 2. This again reveals that the model fits quite well, the exception being subject 27 which may be considered less well fitted. However, we note that this subject does not have any abnormal value. Furthermore, this is a diagnostic case, therefore an important case from the clinical point of view, which rules out the possibility of excluding this case from the analysis.

To analyse the influence of each subject on the values of the estimated parameters, we have plotted in Fig. 2 the difference in beta values (dfbeta) for age against the dfbeta for BMI with the size of the symbol proportional to the analog of Cook's influence statistic. All values are small indicating that no subject has a strong influence on the estimated coefficients.

Detailed information about estimated coefficients for the final model is included in Table 3.

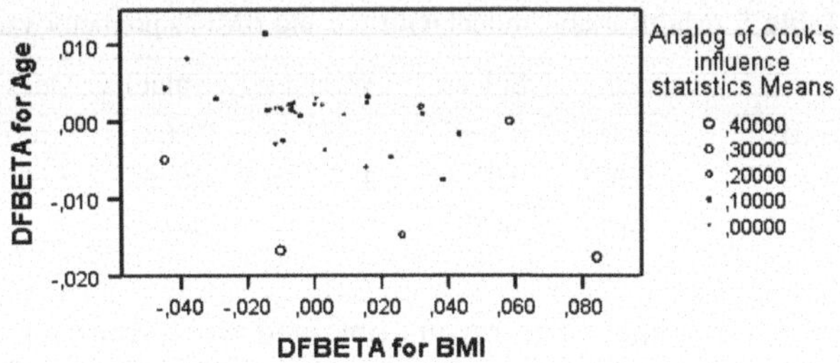

Fig. 2 Dfbeta for age against dfbeta for BMI from the logistic model with age and BMI, with size of the symbols proportional to the analog of Cook's influence statistics

Table 3 Multivariable model with age and BMI as predictor variables ($n = 46$)

	Coeff.	Std. err.	Wald test (p-value)	Odds ratio (OR)	95 % CI for OR	
BMI	−0.384	0.179	4.59 ($p = 0.032$)	0.681	0.479	0.968
Age	0.09	0.038	5.534 ($p = 0.019$)	1.094	1.015	1.179
Constant	4.344	3.206	1.835($p = 0.176$)			

Table 4 Comparing subjects for whom a Brugada pattern was found with the others, with respect to gender, age and BMI

	Diagnostic + suggestive (type 1, 2 or 3 ECG)		Normal (no Brugada pattern)		p-value
Gender	Frequency (% within gender)				Fisher's exact test
($n = 130$)	F	8 (12.9 %)	F	54 (87.1 %)	0.349
	M	14 (20.6 %)	M	54 (79.4 %)	
	Mean ± standard deviation				Mann–Whitney test
Age ($n = 130$)	35.05 ± 11.15		23.21 ± 16.9		0.000
BMI ($n = 46$)	21.34 ± 2.73		23.87 ± 2.86		0.007

We can finally conclude that both age and BMI appear to affect the odds of diagnosis, and we can estimate the effect of 1 year increase in age as increasing the odds of diagnosis by 9.4 % [(1.094–1)100 %] and the effect of one unit increment on the BMI as decreasing the odds of diagnosis by 31.9 % [(0.681–1)100 %]. We also note that the area under the ROC curve generated by the model is 0.78 ($p = 0.006$), which means, according to Hosmer and Lemeshow [5], that the model has an acceptable ability to discriminate between diagnostic and non-diagnostic cases.

2.2 Normal Versus Abnormal ECG

In Table 4 we compare the group of subjects which have had at least one Brugada pattern ECG (type 1, type 2 or type 3) with the other subjects.

The results in Table 4 indicate no significant association between gender and the manifestation of Brugada ECG patterns, but a lower BMI and an older age appear to be significant predictors of abnormal ECG patterns (type 1, 2 or 3). Again, univariable and multivariable logistic regression models were adjusted with these

Table 5 Multivariable model with age and BMI as predictor variables ($n = 46$)

	Coeff.	Std. err.	Wald test (p-value)	Odds ratio (OR)	95 % CI for OR	
BMI	−0.484	0.165	8.547 ($p = 0.003$)	0.617	0.446	0.853
Age	0.071	0.034	4.377 ($p = 0.036$)	1.073	1.004	1.147
Constant	7.871	3.089	6.493 ($p = 0.011$)			

three covariates. The results obtained were similar to those discussed previously when the outcome was the diagnosis of BS. In fact, again age and BMI, but not gender, revealed some association with the outcome, the manifestation of a Brugada pattern (Wald test p-value for gender = 0.537). No significant interaction between age and BMI was found (Wald test p-value for the interaction term = 0.7). As in the previous subsection, we also have AIC supporting the model with only the BMI and age (AIC for the model with BMI and age = 52.65; adding gender AIC = 54.27; adding the interaction term AIC = 54.5). The value of the likelihood ratio test for the model with age and BMI as covariates (12.8, p-value = 0.002) indicates that at least one of these has significant influence on the outcome. Hosmer and Lemeshow test statistic being equal to 4.75 (p-value = 0.69) is indicative of a fairly good fit. Also, standardized Pearson residuals and influence statistics were analysed, to reveal no distinguishable subjects. Table 5 resumes some information about the estimated coefficients for the model with age and BMI. We can again assert that these two covariates appear to have some importance on the manifestation of Brugada patterns, and estimate the odds of this manifestation to increase by 7.3 % for each year older and to decrease by 38.3 % for each BMI unit increase.

3 Analysis of the Intermittency of Brugada Type 1 ECG Pattern

The fluctuations of Brugada ECG patterns demand the need to screen individuals with more than one ECG in the diagnosis of BS. As said before, the 130 individuals were screened with two ECGs separated by 6 months. Of the fourteen to whom BS was diagnosed (type 1 ECG pattern found), 79 % exhibited one non-diagnostic ECG and 43 % had a normal ECG. Furthermore, for these diagnostic subjects, we find no agreement between the first and the second ECG, according to Cohen's Kappa coefficient which is negative (k = −0.45). Note also that for the 22 subjects who exhibited a Brugada pattern ECG, the Cohen's Kappa coefficient equals 0.02 which, according to Landis and Koch [6], may be interpreted as just slight agreement. Also, McNemar's test finds significant the increase in percentage of diagnosis made with two serial ECGs (10.8 %) compared to the percentage made with one ECG (4.6 %) ($p = 0.004$). All these results emphasize the need for more than one ECG on screening for BS.

We now wish to investigate if age, gender and BMI are related to the instability of ECG results in Brugada patients. This could help physicians to assess the importance of doing more than one ECG in the screening of a given individual. Of

Table 6 Comparing stable diagnostic cases (two type 1 ECGs) with diagnostic dynamic cases (one non-diagnostic ECG)

	Diagnostic and stable (2 type 1 ECGs)		Diagnostic and dynamic (one non-diagnostic ECG)		p-value
Gender	Frequency (% within gender)				Fisher's exact test
($n = 130$)	F	1 (33.3 %)	F	5 (45.5 %)	0.6
	M	2 (66.7 %)	M	6 (54.5 %)	
	Mean ± standard deviation				Mann–Whitney test
Age ($n = 130$)	43.7 ± 15.6		35.7 ± 9.8		0.6
BMI ($n = 46$)	24.1 ± 0.5		21.4 ± 2.8		0.4

Table 7 Comparing sensitivities and specificities of ECGs for three different electrodes placements

	ECG Conv	ECG 2IS	ECG 1IS	Cochran's test
Sensitivity	8.3 %	45.8 %	54.2 %	$p = 0.000$
Specificity	100 %	97.1 %	94.3 %	$p = 0.67$

the fourteen subjects with diagnosis of BS, we compare those who have maintained the type 1 ECG pattern with those who have presented fluctuations between diagnostic and non-diagnostic ECG pattern. The results are given in Table 6. In fact, there is no evidence of age, BMI or gender being related to fluctuations in ECG results in patients with BS.

4 Placement of Electrodes in the ECG

It is consensual that raising the positions of the electrodes from the conventional positions increases the number of type 1 ECG patterns, that is, increases the sensitivity of the ECG. Nevertheless, the doubt persists whether it also increases the number of false positives, thus decreasing the specificity. For 59 relatives a genetic test was conducted and ECG recordings were performed with the electrodes in conventional positions (fourth intercostal space—ECG Conv) as well as with upward displacement of the electrodes (second intercostal space—ECG 2IS—and first intercostal space—ECG 1IS). Sensitivity and specificity for these three types of ECG recordings were calculated and the significance of their difference was assessed through Cochran's test. The results are shown in Table 7.

As seen, our data support the consensual idea that raising the electrodes positions leads to a significant increase in the sensitivity. Also, we have registered a decrease on the specificity, yet, no significant difference was found between those.

5 Conclusion

Age and BMI were identified as significant predictors of the manifestation of Brugada patterns. Not only type 1 (diagnostic pattern) but also type 2 and type 3 pattern manifestations tend to occur in older subjects and with low BMI. No significant relation between gender and Brugada pattern manifestations was found. Gender, age and BMI do not seem to be significantly related to fluctuations in ECG results of BS patients. We found a significant increase in the percentage of diagnosis with two serial ECGs (10,8 %) against only one ECG (4,6 %). Besides that, Cohen's Kappa coefficient is indicative of a low degree of agreement between the first and the second ECGs within subjects which have had a Brugada pattern ECG. So, with only one ECG, a BS patient may well not be identified, leading the physician to a misleading prognosis. It then seems important to do more than one ECG on the screening for BS. As for the displacement of the ECG electrodes with respect to the conventional positions, we have found significant increase in the sensitivity for the upward displacements, according to the general consensus, but the accompanying decrease of specificity was not significant.

References

1. Brugada, P., Brugada, J.: Right bundle branch block, persistent ST segment elevation and sudden cardiac death: a distinct clinical and electrocardiographic syndrome. A multicenter report. J. Am. Coll. Cardiol. **20**, 1391–1396 (1992)
2. Brugada, P., Brugada, R., Brugada, J., Priori, S.G., Napolitano, C.: Should patients with an asymptomatic Brugada electrocardiogram undergo pharmacological and electrophysiological testing? Circulation **112**, 279–292 (2005)
3. Brugada, P., Benito, B., Brugada, R., Brugada, J.: Brugada Syndrome: Update 2009. Hellenic J. Cardiol. **50**, 352–372 (2009)
4. Burnham, K.P., Anderson, D.R.: Model Selection and Multimodel Inference: A Pratical Information-Theoretic Approach, 2nd edn. Springer, New York (2002)
5. Hosmer, D.W., Lemeshow, S.: Applied Logistic Regression, 2nd edn. Wiley, New York (2000)
6. Landis, J.R., Koch, G.G.: The measurement of observer agreement for categorical data. Biometrics **33**, 159–174 (1977)

Mortality on Older Portuguese Population due to Circulatory System Diseases and Neoplasms: A Spatio-Temporal Analysis by Age and Sex

Sandra Lagarto, Carla Nunes, Dulce Gomes, and Maria Filomena Mendes

Abstract
There are indicators that suggest that the Portuguese population is aging uneven (Índice de dependência de idosos (N.) por Local de residência; Índice de envelhecimento (N.) por Local de residência; Índice de envelhecimento (N.) por Local de residência). Considering this fact, we propose to identify mortality patterns and regional differences amongst the older Portuguese population (65 or more years). The study of the spatio-temporal distribution of mortality in older people is essential to understand its dynamics and emergent trends as well as to promote health in aging populations. It was used the spatial scan statistic Kulldorff (Commun. Stat. Theor. Meth. 26, 1481–1496, 1997), a method for detecting space-time clusters. This method has a long tradition in spatial epidemiology, particularly, in many applications of public health areas Elliott and Wartenberg (Environ. Health Perspect. 112, 998–1006, 2004) and Nunes et al. (Rev. Port. Sau. Pub. 26, 5–14, 2008). We use stochastic space-time processes

S. Lagarto (✉)
CIMA-UE, Colégio Luís Verney, Rua Romão Ramalho, 59, 7000–671 Évora, Portugal
e-mail: smdl@uevora.pt

C. Nunes
CMDT.LA, Escola Nacional de Saúde Pública, Universidade Nova de Lisboa, Av. Padre Cruz 1660–560 Lisboa, Portugal
e-mail: cnunes@ensp.unl.pt

D. Gomes
CIMA-UE, Department of Mathematics, Colégio Luís Verney – Rua Romão Ramalho, 59, 7000–671 Évora, Portugal
e-mail: dmog@uevora.pt

M.F. Mendes
CIDEHUS-UE, Department of Sociology, Largo dos Colegiais, 2, 7002–554 Évora, Portugal
e-mail: mmendes@uevora.pt

J. Lita da Silva et al. (eds.), *Advances in Regression, Survival Analysis, Extreme Values, Markov Processes and Other Statistical Applications*, Studies in Theoretical and Applied Statistics, DOI 10.1007/978-3-642-34904-1_25,

(according to the level of available geographical disaggregation data) to describe the mortality rates of the older Portuguese population (from 1992 to 2006) associated with diseases of the circulatory system and neoplasms. Results show statistically significant space-time clusters, for different age groups, by sex and cause of death. Those space-time units correspond to simultaneous occurrence of high mortality rates in different regions of the Portuguese mainland. These critical areas were consistent over age groups and sex, concerning diseases of the circulatory system as cause of death; for neoplasm, space-time critical areas presented some variations over age groups for both males and females.

1 Introduction

In Portugal, like in most European countries, the older population doubled in the last four decades of the twentieth century and is still increasing [1]. This phenomenon seems determined to continue and it is expected that the proportion of people aged 65 years, or more, will double again in forty years–reaching 40 % of the population in almost all territory [15]. Population aging is becoming a very pertinent issue, in several and different contexts. The study of mortality in older ages is getting more and more important, especially on population projections and analysis of social–economic impacts resulting from the changes in the classical population structure.

There are many studies on health care in Portugal. Some focus on health care and external death causes [16], others characterize mortality trends without a spatial desegregation [2]. Those which characterize mortality at a subnational level usually apply classical statistical techniques (descriptive or inferential) or some common spatio-temporal approaches based on classical mappings [10, 11]. Also, in recent studies, spatio-temporal clustering analyses identify high-incidence areas of some particular diseases in the Portuguese population [4, 12–14].

Considering that older population is not distributed equally throughout the country and that there are indicators that suggest that the Portuguese population is aging uneven [6], what are the mortality trends and regional differences? The aim of this work is to identify where and when the high mortality rates occur simultaneously in the different regions of the Portuguese mainland for the two major causes of death of Portuguese older population: diseases of the circulatory system (CIRS) and neoplasms (NEO). Also, it is analysed if these critical areas are consistent among age groups, i.e. if critical areas are similar (in a space-time referential) for all ages, per sex and cause of death. For this purpose, spatio-temporal clusters were identified for each cause of death, by age and sex, and a stability indicator was also proposed. The present study can be considered part of the vast area of spatial epidemiology–geography of the causes of death.

2 Methodology

We applied a space-time scanning method, spatial scan statistic, proposed by Martin Kulldorff [9], to identify sets of homogeneous space-time units, clusters, a method widely used in public health [3, 5, 7, 8]. The scan statistic is based on a maximum likelihood ratio for each potential cluster that expresses how much more likely the observed cluster units are, under the hypothesis of clustering, than under the hypothesis of uniformity. Since the exact distribution of the test statistic cannot be determined, Monte Carlo simulation is used to perform the hypothesis test. Scan statistics were applied independently for each age group, per sex and cause of death, identifying critical areas in all sub-levels. The spatio-temporal scanning method was applied using the SaTScan v8.0, developed by M. Kulldorff and available at http://www.satscan.org.

After that, a stability indicator (SI) was built to characterize the regularity of the identified space-time clusters between age groups among older population, by cause of death and by sex. This indicator allows us to understand if the identified clusters (critical areas) were stable in space and time, regarding a reference age group. In this case, we chose the first age group (65–69), because it was the biggest group, in terms of frequencies, providing more robust results. It is also the most import class considering both economic and social impact of lives lost. Cross-tables between the reference age group and other ages groups were built separately for each two age groups, by cause of death and sex, to identify the proportion of areas (considering 420 space-time units = 15 years per 28 regions) which maintain the same classification (belong or not belong to critical areas: cluster or non-clusters) on both age groups. The SI is achieved through the sum of cross-table diagonals, quantifying the clusters proportions which remain constant considering each age pair. Note that for SI definition it is not important if clusters are "the most likely" or secondary clusters, because they are always statistical significant ($p < 0.001$).

3 Case Study

We present a mortality analysis, at a subnational level, focused on two specific causes of death (CIRS and NEO, as mentioned above), only considering Portuguese older population death rates. For that purpose, we identify spatio-temporal clusters for the occurrence of deaths from 1992 to 2006. The space-time referential is defined by NUTIII (Portuguese territorial units for statistical purposes), per year, and the analysis was done, independently, by age group and sex for each specific cause of death. The 28 NUTIII mainland map is available in http://www.ine.pt/.

For individuals aged 65 and over, data was available grouped in age classes: 65–69; 70–74; 75–79; 80–84; 85+ (age 85+ represents the last age group). Annual data–number of deaths–by NUTIII, cause of death, age group and sex, as well as estimates of the average subnational resident population for the selected time period, were provided by INE (Statistical National Institute).

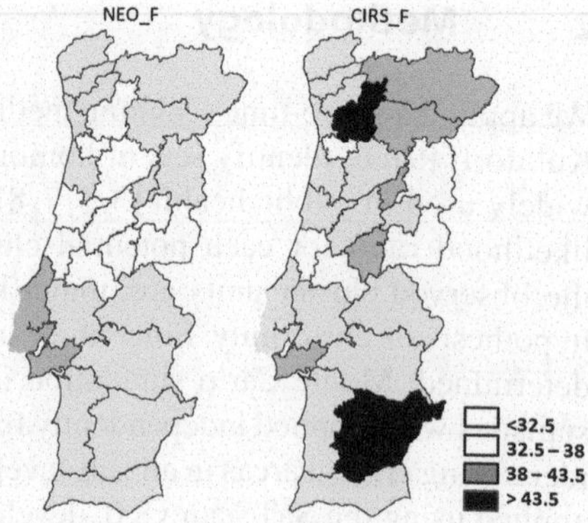

Fig. 1 Average mortality rate by cause of death: females, age 65–69 (1992–2006)

3.1 Main Causes of Death in Portugal

Based on descriptive analysis, we find some regional differences on the selected death causes. These occur only occasionally, especially at ages 65–69 and 85+, by sex. Also, in general, there is an increased range in mortality rates by cause of death in males compared to females. For CIRS, the range in mortality rate values is higher for males, predominantly in the northern and central regions. In the south, on the contrary, the range of NEO deaths rates is wider. The maximum rates are associated to males for both causes of death, with higher dispersion in CIRS. Figure 1 illustrates, as an example, regional average female death rates for age group 65–69. In this particular case, CIRS are the dominant cause, although overlapping (*Minho-Lima, Algarve*) or even occasionally being exceeded in certain regions (*Grande Lisboa* or *Médio Tejo*) by NEO. Still, in some regions, we can only identify a small gap between the two dominant causes (*Tâmega, Pinhal Interior Norte* or *Baixo Alentejo*).

3.2 Identified Clusters by Cause of Death, Age and Sex

Clusters were identified for all age groups, for both males/females, and each cause of death, through retrospective analysis, assuming a Poisson distribution for death occurrences. We used circular windows in the scan (cylinders in the space-time), up to 50 % of the population at risk, looking for high mortality values clusters. The statistical significance of the test was determined by the Monte Carlo method.

For deaths associated with CIRS, three clusters were identified by age group, for all ages and for both males and females ($p < 0.001$). The main (or most likely) cluster–a temporal cluster–is common to all ages, for both males and females, in the period from 1993 to 1999, throughout the territory; the other two space-time clusters (secondary clusters) divide the territory roughly in half, from north to south. The

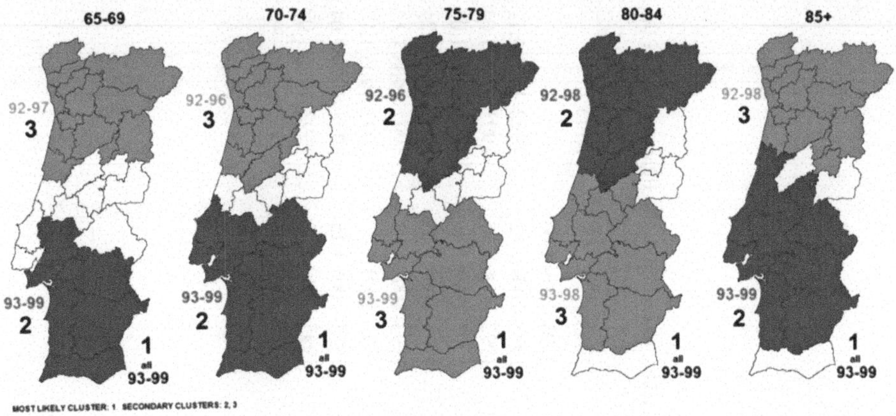

Fig. 2 Identified clusters for CIRS: Males, per age

results point to a certain lack of spatial variation in the Portuguese mainland by age and sex. Figure 2 illustrates identified clusters for CIRS (males, per age).

Figure 2 shows gray scale for the secondary clusters 2 and 3–the most likely as dark gray–while the main cluster 1, in this case, only temporal, is referred in the right bottom corner of each map. Comparing results, there are differences in the ages 75–79 and 80–84 (reversal of primary and secondary clusters, when compared to ages 65–74), however, with no practical impact: test statistic values are similar.

In Table 1 we show the spatio-temporal clustering results. CIRS.2 (M; 65–69), for instance, is an identified cluster on south Portugal (see Fig. 2), with an estimated observed/expected ratio of 1.29, which means that, in that area, 29 % more deaths have occurred compared to the expected ones.

Observing Table 1, one can see that ratios do not present big differences across age and by sex. However, the import issue in discussion, in a space-time referential, is the identification and characterization of critical areas, considering observed/expected ratios. Based on Fig. 2 and Table 1, CIRS deaths have similar patterns by sex. We can only point that the clusters identified for females were more restricted, geographically, than those identified for males, especially aged 75 and over. As for NEO deaths, we identified spatio-temporal clusters, for all ages, for both males and females. The clusters were also located in north and south of the country, occupying a more restricted area, mainly associated with the metropolitan areas of Lisbon and Oporto. For males aged 85+ and females aged 65–69, there was a single spatio-temporal cluster in Lisbon region. Also, for females aged 70–74, it was identified a cluster that covers the entire south of the country.

Table 1 Statistics for identified clusters ($p < 0.001$) by cause of death, age and sex (CIRS.1 to CIRS.3 represent, in each cell, the time period and the observed/expected ratio for clusters 1–3 for females(F) and males(M); the same apply for NEO clusters)

Sex	Age	CIRS.1	CIRS.2	CIRS.3	NEO.1	NEO.2	NEO.3
F	65–69	1993–1999;1.21	1992–1997;1.34	1993–1999;1.25	1993–1999;1.24	1993–1999;1.06	
	70–74	1993–1999;1.22	1992–1998;1.30	1993–1998;1.26	1999–2005;1.07	1998–2004;1.12	1992–1906;1.11
	75–79	1993–1999;1.21	1992–1997;1.38	1993–1998;1.27	1993–1999;1.23	1992–2006;1.15	1993–1999;1.04
	80–84	1993–1999;1.20	1992–1998;1.27	1993–1999;1.24	1993–1999;1.25	1992–1998;1.26	1993–1998;1.07
	85+	1993–1999;1.12	1993–1999;1.18	1992–1996;1.17	1992–2006;1.11	1999–2005;1.05	2000–1905;1.21
M	65–69	1993–1999;1.20	1993–1999;1.29	1992–1997;1.28	1993–1999;1.24	1993–1999;1.07	1992–1906;1.12
	70–74	1993–1999;1.21	1993–1999;1.26	1992–1996;1.33	1994–2000;1.28	1995–2001;1.07	1992–1906;1.17
	75–79	1993–1999;1.19	1992–1996;1.36	1993–1999;1.22	1992–2006;1.15	1992–2006;1.15	1998–1904;1.04
	80–84	1993–1999;1.21	1992–1998;1.27	1993–1998;1.26	1994–2000;1.28	1995–2000;1.06	1992–1906;1.08
	85+	1993–1999;1.17	1993–1999;1.21	1992–1998;1.16	1996–2002;1.23	1996–2002;1.04	

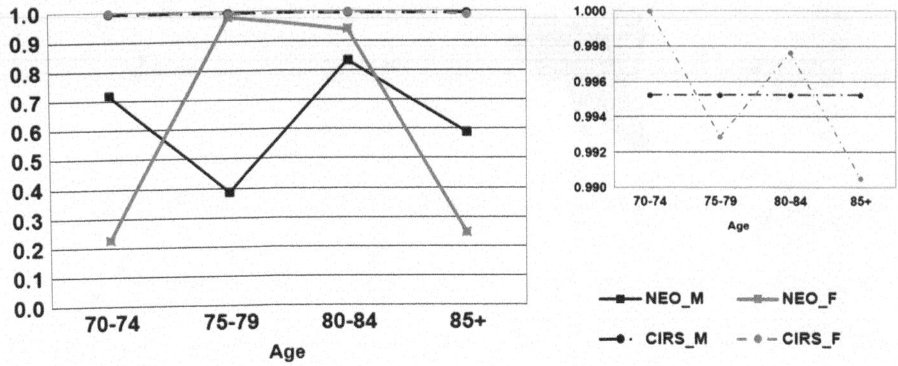

Fig. 3 Spatio-temporal clusters stability by cause of death, age and sex (*left*); "Y axis" detail regarding CIRS deaths (*right*)

3.3 Comparative Analysis by Cause of Death, Age and Sex

As already presented in methodology section, a stability indicator (SI) was built to identify the stability of critical areas, through age groups (using 65–69 as the reference age group), per sex and per cause of death. To this SI (which was built based on contingency tables) we assume that the higher the ratio, the higher the spatio-temporal coincidence of critical areas over studied ages. This SI varies between 0 and 1: 1 means that clusters are precisely the same (space and time) for the two age groups being compared, by sex and cause of death; 0 represents that all space-time units change their status (e.g. areas that belong to clusters in a specific age change to a non-cluster area in other age).

Figure 3 presents stability indicators for the two selected causes of death, across older Portuguese ages, by sex. Note that the age group 65–69 does not appear in the graph because it is the reference comparison group.

For females, the identified clusters of deaths from CIRS remained substantially constant ($SI \simeq 1$), when we compare the age groups 65–69 and 70–74 (we illustrate CIRS deaths stability across ages, in more detail, in Fig. 3 on the right side). It means that the regions maintain the same classification as to their inclusion (or not) in some cluster, within these two age groups. In fact, clusters from CIRS show no significant differences between each age group and the reference group. Under these conditions, stability is maximum, so the space-time overlap of clusters is almost total for all ages. That implies that clusters are almost defined in the same way. Minimal differences occur on females at ages groups 75–79 and 85+.

As for the NEO deaths clusters, the pattern differs by sex and age, with large fluctuations. Differences are larger in ages 70–74 and 85+ for females. For males, the major oscillations occur mainly between age group 75–79. Thus, if one considers the entire older population, there is less stability in clusters related with this cause

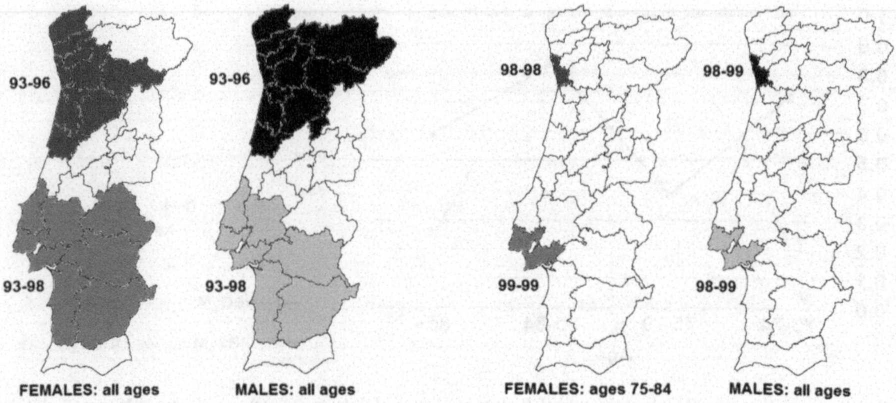

Fig. 4 Homogeneous spatio-temporal units for CIRS (*left*) and NEO (*right*) deaths, by sex

of death. However, when comparing ages 70–74 to the reference group, SI = 0.933, which indicates, in general, a small space-time variability.

Analysing the oscillations by cause of death, age and sex, it is possible to represent the clusters that remained the same in the reviewed time period (Fig. 4). For instance, the second representation on the left (CIRS, males) shows homogeneous areas which are subgroups of the identified clusters in Fig. 2, resulting from clusters overlapping in space and time.

In the identified units of Fig. 4 we highlight the overlapping of large spatio-temporal clusters. For CIRS deaths (Fig. 4, on the left), these critical areas occur for all ages, for both males and females, differentiating north and south regions.

Additionally, considering CIRS deaths, the identified critical areas are mostly common for both male and female (except for some minor regions on central-south, interior centre and northeast of the country). As for clusters from NEO deaths (Fig. 4, on the right), there is also a strong spatio-temporal coincidence by sex, with critical regions located in the Lisbon and Oporto metropolitan areas.

4 Conclusions and Discussion

This work has characterized different spatio-temporal patterns in causes of death among Portuguese older population, at a subnational level. Several spatio-temporal clusters were identified, detecting critical areas with a high number of deaths, in Portuguese mainland, by sex, age and cause of death, from 1992 to 2006.

Identified clusters from neoplasms are associated with metropolitan areas (in this case, on the coast). Considering this cause of death, identified clusters are distinct by sex and age group. As for mortality rates for diseases of the circulatory system, they seem to be mostly connected with the north/south division and the identified clusters are similar by sex and age group.

The dynamics of each cause of death is different throughout the studied period. However, it was possible to identify space-time homogeneous units (in terms of the expected number of mortality rates), but with different temporal patterns.

This is an introductory work to the application of this methodology to Portugal's demographical data. Therefore, further developments are being prepared, like testing other variables or using alternative methods, regarding results assessment.

Acknowledgements This research was supported by the Interdisciplinary Center of History, Cultures and Societies (Project *Forecasting Portuguese Population: the special and spatial contribution of mortality and migration*-FCT/PTDC/SDE/68126/2006) and by the Center of Mathematics and Applications (Program Stochastic and Statistical Applications in Life Sciences Project, FCT)).

References

1. Actualidades do INE, Statistical National Institute. http://alea.ine.pt/html/actual/pdf/act16.pdf. Cited 4 Mar 2009
2. Canudas-Romo, V., Glei, D., Gómez-Redondo, R., Coelho, E., Boe, C.: Mortality changes in the Iberian Peninsula in the last decades of the twentieth century. Population **63**(2), 319–343 (2008). (English edition)
3. Chen, J., Roth, R., Naito, A., Lengerich, E., MacEachren, A.: Geovisual analytics to enhance spatial scan statistic interpretation: an analysis of U.S. cervical cancer mortality. Int. J. Health Geogr. **7**(57) (2008). doi:10.1186/1476-072X-7-57
4. Couceiro, L., Nunes, C., Santana, P.: Pulmonary tuberculosis and risk factors in Portugal: a spatial analysis. Int. J. Tuberc. and Lung Dis. (2011). doi: 10.5588/ijtld.10.0302
5. Elliott, P., Wartenberg, D.: Spatial epidemiology: current approaches and future challenges. Environ. Health Perspect. **112**(9), 998–1006 (2004)
6. Índice de dependência de idosos (N.) por Local de residência; Índice de envelhecimento (N.) por Local de residência; Índice de envelhecimento (N.) por Local de residência. Recenseamento da População e Habitação, Statistical National Institute. http://www.ine.pt. Cited 5 Mar 2009
7. Kadafar, K.(adviser), Kim, A.(RA), Al-Rumaizan, S., Ayalya, C., Been, A., Chen, J., Cook, J., Duarte, G., Durso, C., Gonzales, J., Huynh, C., Huynh, T., Lashuk, I., Wagner, D., Sanabri, S.: A brief evaluation of statistical for detecting disease clusters in time and/or space, vol. 26(MATH 4779/5779). Department of Mathematics, University of Colorado–Denver (2004)
8. Kulldorff, M.: Prospective time periodic geographical disease surveillance using a scan statistic. Journal of the Royal Statistical Society: Series A (Statistics in Society) **164**(1), 61–72 (2001)
9. Kulldorff, M.: A spatial scan statistic. Commun. Stat. Theor. Meth. **26**(6), 1481–1496 (1997)
10. Morais, M.: Causas de Morte no Século XX: Transição e estruturas da Mortalidade em Portugal Continental. Edições Colibri, Lisboa (2002)
11. Nicolau, R., Machado, A., Nunes, B., et al.: Análise da variação concelhia da mortalidade anual média por neoplasias malignas dos órgãos do aparelho respiratório e intra-torácicos em Portugal Continental. Rev. Port. Sal. Pub. **27**(2) 7–16 (2009)
12. Nunes, C., Briz, T., Gomes, D., Dias, C.M.: A Dimensão espácio temporal em saúde pública: da descrição clássica à análise de clustering. Rev. Port. Sau. Pub. **26**(1), 5–14 (2008)
13. Nunes, C., Gomes, D., Matias, C., Briz, T.: Congenital anomalies in Portugal: a spatial and temporal characterization. In: Proceeding of the 56th Session of the International Statistical Institute (2007)

14. Parreira, P., Nunes, C., Miranda, A.: Stomach and lung cancer incidence: a spatiotemporal study in south region of continental Portugal. GRELL 2011 (Groupe des Registres de Langue Latine), Caen, França (2011)
15. Projecções de População Residente, Portugal e Nuts III, 2000–2050, Statistical National Institute. http://www.ine.pt. Cited 5 Mar 2009
16. Santana, P.: A mortalidade evitável em Portugal Continental, 1989 a 1993. Rev. Est. Dem. **32**, 107–146 (2002)

Absolute Diffusion Process: Sensitivity Measures

Manuela Larguinho, José Carlos Dias, and Carlos A. Braumann

Abstract

The constant elasticity of variance (CEV) model of Cox (Notes on Option Pricing I: Constant Elasticity of Variance Diffusions. Working paper, Stanford University (1975)) captures the implied volatility smile that is similar to the volatility curves observed in practice. This diffusion process has been used for pricing several financial option contracts.

In this paper we present the analytical expressions of sensitivity measures for the absolute diffusion process, commonly known as Greeks, and we analyze numerically the behavior of the measures for European options under the CEV model.

1 Introduction

Under the risk-neutral probability measure Q, the constant elasticity of variance (CEV) process of [4] assumes that the asset price $\{S_t; t \geq 0\}$ is described by the following stochastic differential equation:

M. Larguinho (✉)
Centro de Investigação em Matemática e Aplicações, Universidade de Évora and
Department of Mathematics, ISCAC, Quinta Agrícola, Bencata, 3040-316 Coimbra, Portugal
e-mail: mlarguinho@iscac.pt

J.C. Dias
BRU-UNIDE and ISCTE-IUL Business School, Complexo INDEG/ISCTE, Av. Prof. Aníbal Bettencourt, 1600-189 Lisboa, Portugal
e-mail: jose.carlos.dias@iscte.pt

C.A. Braumann
Centro de Investigação em Matemática e Aplicações, Universidade de Évora, Rua Romão Ramalho 59, 7000-671 Évora, Portugal
e-mail: braumann@uevora.pt

J. Lita da Silva et al. (eds.), *Advances in Regression, Survival Analysis, Extreme Values, Markov Processes and Other Statistical Applications*, Studies in Theoretical and Applied Statistics, DOI 10.1007/978-3-642-34904-1_26,

$$dS_t = (r - q)S_t\,dt + \delta\, S_t^{\beta/2}\,dW_t^Q, \tag{1}$$

where W_t^Q is a Wiener process under Q, $r \geq 0$ represents the instantaneous riskless interest rate, which is assumed to be constant, $q \geq 0$ denotes the dividend yield for the underlying asset price, with a local volatility function given by

$$\sigma(S_t) = \delta\, S_t^{\beta/2-1}, \tag{2}$$

where β is a real number, and δ is a positive constant.

The elasticity of return variance with respect to price is equal to $\beta - 2$ given that $dv(S_t)/v(S_t) = (\beta - 2)\,dS_t/S_t$ where $v(S_t) = \delta^2\, S_t^{\beta-2}$ is the instantaneous variance of asset returns. Since volatility is proportional to a power of the underlying asset price, the elasticity of variance is independent of the asset price. The model parameter δ can be interpreted as the scale parameter fixing the initial instantaneous volatility at time $t = t_0$, $\sigma_0 = \sigma(S_{t_0}) = \delta\, S_{t_0}^{\beta/2-1}$.

The CEV specification given by Eq. (1) nests the lognormal assumption of [3, 9] ($\beta = 2$), as well as the square-root diffusion ($\beta = 1$) and the absolute diffusion ($\beta = 0$) models of [5], as special cases. For $\beta < 2$ ($\beta > 2$) the local volatility given by Eq. (2) is a decreasing (increasing) function of the asset price. If $\beta = 2$, the stock price has no influence on the volatility, since the volatility will be a constant over time, $\sigma(S_t) = \delta$, regardless of the underlying asset price.

2 European Options Under the CEV Diffusion

The CEV call option pricing formula for valuing European options has been initially expressed in terms of the standard complementary gamma distribution by [4] for $\beta < 2$, and by [8] for $\beta > 2$. The CEV model was subsequently extended in [10] by expressing the corresponding formulae in terms of the noncentral chi-square distribution as

$$c_t := \begin{cases} S_t\,e^{-q\tau}\,Q(2y; 2 + \frac{2}{2-\beta}, 2x) - X\,e^{-r\tau}\,[1 - Q(2x; \frac{2}{2-\beta}, 2y)] & \text{if } \beta < 2, \\ S_t\,e^{-q\tau}\,Q(2x; \frac{2}{\beta-2}, 2y) - X\,e^{-r\tau}\,[1 - Q(2y; 2 + \frac{2}{\beta-2}, 2x))] & \text{if } \beta > 2, \end{cases} \tag{3}$$

with X being the strike price of option, $Q(w; v, \lambda)$ being the complementary distribution function of a noncentral chi-square law with v degrees of freedom and noncentrality parameter λ, and where

$$k = \frac{2(r-q)}{\delta^2(2-\beta)[e^{(r-q)(2-\beta)\tau} - 1]}, \tag{4}$$

$$x = k S_t^{2-\beta}\, e^{(r-q)(2-\beta)\tau}, \tag{5}$$

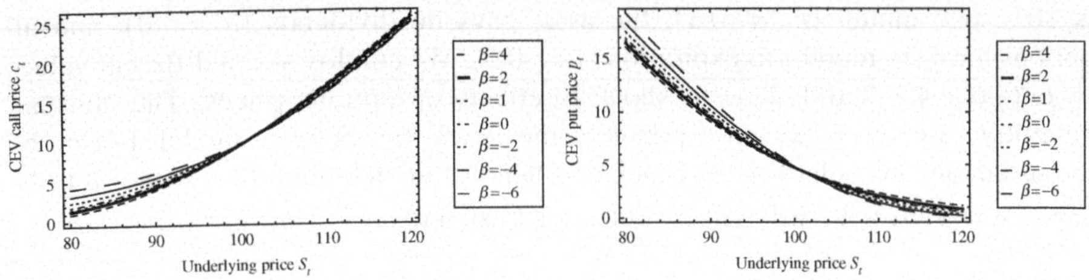

Fig. 1 European call and put option prices under CEV processes and Black–Scholes model as functions of underlying asset price S_t. Parameters: $S_0 = 100$, $X = 100, \sigma_0 = 0.25, \tau = 0.5$, $r = 0.1$, and $q = 0$

$$y = kX^{2-\beta}, \tag{6}$$

$$\delta = \sigma_0 S_0^{1-\beta/2}, \tag{7}$$

$$\tau = T - t. \tag{8}$$

By put–call parity, the CEV put option pricing formulae are

$$p_t := \begin{cases} X\,e^{-r\tau}\,Q(2x;\frac{2}{2-\beta},2y) - S_t\,e^{-q\tau}\,[1 - Q(2y;2+\frac{2}{2-\beta},2x)] & \text{if } \beta < 2, \\ X\,e^{-r\tau}\,Q(2y;2+\frac{2}{\beta-2},2x) - S_t\,e^{-q\tau}\,[1 - Q(2x;\frac{2}{\beta-2},2y))] & \text{if } \beta > 2, \end{cases} \tag{9}$$

In general terms, the underlying asset of the CEV diffusion can be thought of as a stock, a stock index, an exchange rate, or a financial future contract, so long as the parameter q is understood as, respectively, a dividend yield, an average dividend yield, the foreign default-free interest rate, or the domestic risk-free interest rate.

There are several alternative methods for computing the cumulative distribution function of the noncentral chi-square in the literature (see, for instance, [2, 7, 10]). In this work, we use a method based on series of incomplete gamma functions to compute the complementary noncentral chi-square distribution function given by

$$Q(w;v,\lambda) = \sum_{i=0}^{\infty} \frac{(\lambda/2)^i e^{-\lambda/2}}{i!} \frac{\Gamma(v/2+i,w/2)}{\Gamma(v/2+i)}, \tag{10}$$

with $\Gamma(m,n)$ and $\Gamma(m)$ being, respectively, the complementary incomplete gamma function and the Euler gamma function as defined by [1, Eqs. 6.5.3 and 6.1.1].

The next figures show the behavior of European call and put option prices (Fig. 1). We consider the following parameters for our analysis: the initial asset price is $S_0 = 100$, the strike price is $X = 100$, the instantaneous volatility at this price level is 25 % per annum ($\sigma_0 = 0.25$), the risk-free interest rate is 10 % per annum ($r = 0.1$), the asset pays no dividends ($q = 0$), and all options have six months to expiration ($\tau = 0.5$). We employ seven different values to β $(-6, -4, -2, 0, 1, 2, 4)$ to show its effects on options prices. The constant volatility case ($\beta = 2$) corresponds to the Black and Scholes model. Let σ_0 be the instantaneous volatility for Black and Scholes model, then the value of δ to be used for models with different β values is adjusted to be $\delta = \sigma_0\, S_0^{1-\beta/2}$.

3 Sensitivity Measures for the Absolute Diffusion

The absolute diffusion process proposed by [5] is a particular case of the CEV diffusion process in Eq. (1) with $\beta = 0$.

Proposition 1. *Under the CEV diffusion (1) with $\beta = 0$, that is, with a local volatility function given by $\sigma(S_t) = \delta\, S_t^{-1}$, the European call and put option prices are equal to:*[1]

$$c_t = (S_t e^{-q\tau} - X e^{-r\tau})\, N(y_1) + (S_t e^{-q\tau} + X e^{-r\tau})\, N(y_2) + u\,[n(y_1) - n(y_2)], \quad (11)$$

$$p_t = (X e^{-r\tau} - S_t e^{-q\tau})\, N(-y_1) + (S_t e^{-q\tau} + X e^{-r\tau})\, N(y_2) + u\,[n(y_1) - n(y_2)], \quad (12)$$

where $N(x)$ is the cumulative univariate standard normal distribution function, $n(x)$ is the standard normal density function, and

$$u = \delta\left(\frac{e^{-2q\tau} - e^{-2r\tau}}{2(r-q)}\right)^{1/2}, \quad (13)$$

$$y_1 = \frac{S_t e^{-q\tau} - X e^{-r\tau}}{u}, \quad (14)$$

$$y_2 = \frac{-S_t e^{-q\tau} - X e^{-r\tau}}{u}. \quad (15)$$

The sensitivity measures, commonly referred in financial literature as "greek letters" or simply "greeks," are vital tools for risk management and they all represent sensitivity measures of the option price to a small change of a given parameter. The most common greeks are the first-order derivatives: delta, theta, vega, and rho as well as gamma, a second-order derivative of the value function.

[1] Equation (11) of Proposition 1 is standard in the literature and can be found, for example, in [5]. Equation (12) is then easily obtained through the put–call parity relation.

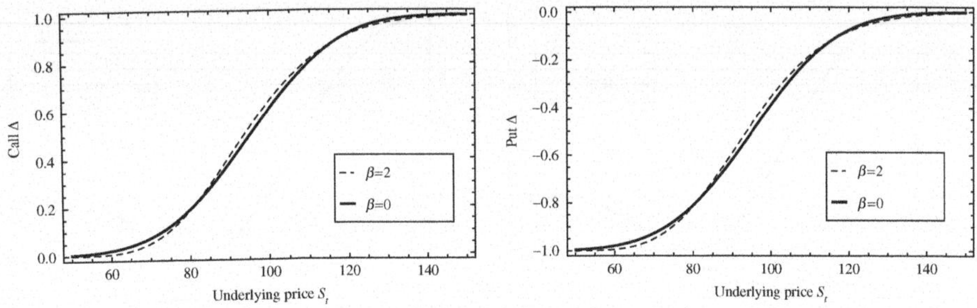

Fig. 2 Variation of delta, Δ, with respect to the underlying asset price S_t. Parameters: $S_0 = 100$, $X = 100, \sigma_0 = 0.25, \tau = 0.5, r = 0.1$, and $q = 0$

In the following we give the analytical expressions for the greek letters under the absolute diffusion process.[2]

3.1 Delta

The delta, Δ, of an option is defined as the rate of change of the option price, V, with respect to the price of the underlying asset, S_t, that is, $\Delta = \partial V/\partial S_t$. It is the slope of the curve that relates the option price to the underlying asset price (Fig. 2). The delta plays a crucial role for hedging portfolios. For European call and put options under the absolute diffusion process on an asset paying a dividend yield q we have

$$\Delta_{call} = e^{-q\tau}\Big(N(y_1) + N(y_2)\Big), \tag{16}$$

$$\Delta_{put} = e^{-q\tau}\Big(- N(-y_1) + N(y_2)\Big), \tag{17}$$

where y_1 and y_2 are defined as in Eqs. (14) and (15).

3.2 Theta

The theta, Θ, of an option is the rate of change of the option price, V, with respect to the passage of time, t, with all else remaining the same, that is, $\Theta = \partial V/\partial t$. Theta is sometimes referred to a time decay effect of the option (Fig. 3). The thetas of European call and put options under the absolute diffusion process, are found, respectively, to be

$$\Theta_{call} = S_t q e^{-q\tau}\Big(N(y_1) + N(y_2)\Big) - X r e^{-r\tau}\Big(N(y_1) - N(y_2)\Big) + A, \tag{18}$$

[2] Due to constraints of space, we have not included proofs of the analytical expressions of sensitivity measures, but they are available upon request.

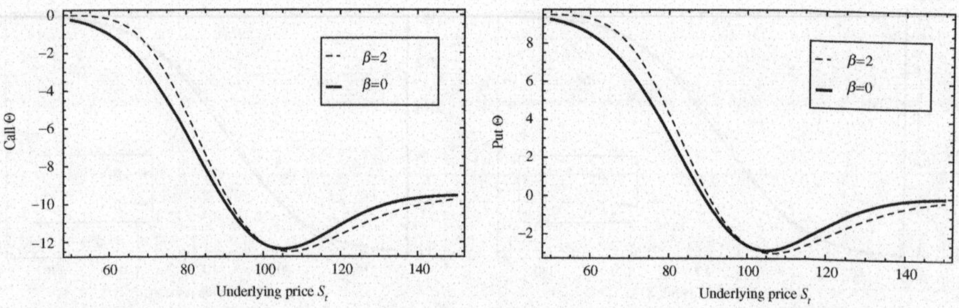

Fig. 3 Variation of theta, Θ, with respect to the underlying asset price S_t. Parameters: $S_0 = 100$, $X = 100$, $\sigma_0 = 0.25$, $\tau = 0.5$, $r = 0.1$, and $q = 0$

Fig. 4 Variation of vega, $\mathcal{V}$, with respect to the underlying asset price S_t. Parameters: $S_0 = 100$, $X = 100$, $\sigma_0 = 0.25$, $\tau = 0.5$, $r = 0.1$, and $q = 0$

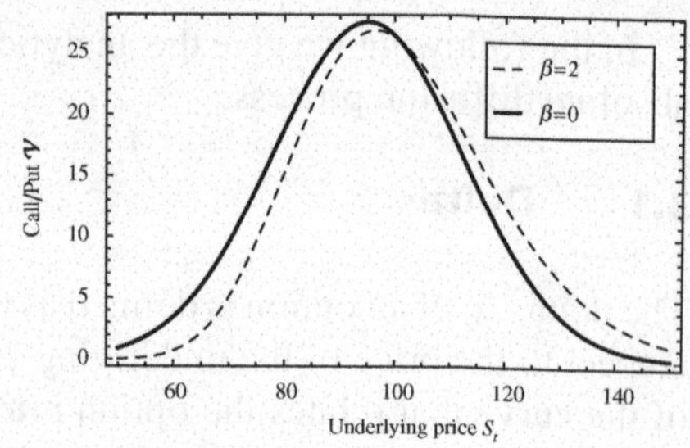

$$\Theta_{put} = -S_t q e^{-q\tau}\Big(N(-y_1) - N(y_2)\Big) + X r e^{-r\tau}\Big(N(-y_1) + N(y_2)\Big) + A, \tag{19}$$

where

$$A = u\Big(n(y_1) - n(y_2)\Big)\frac{q e^{-2q\tau} - r e^{-2r\tau}}{e^{-2q\tau} - e^{-2r\tau}}, \tag{20}$$

with u, y_1, and y_2 being defined as in Eqs. (13), (14), and (15).

3.3 Vega

The vega, $\mathcal{V}$, of an option is defined to be the rate of change of the value of option, V, with respect to asset price volatility, σ, that is, $\mathcal{V} = \partial V/\partial\sigma$ (Fig. 4). For European call and put options under the absolute diffusion process, their vegas are found to be

$$\mathcal{V}_{call} = \mathcal{V}_{put} = \frac{u}{\sigma}\Big(n(y_1) - n(y_2)\Big), \tag{21}$$

where u, y_1, and y_2 are defined as in Eqs. (13)–(15).

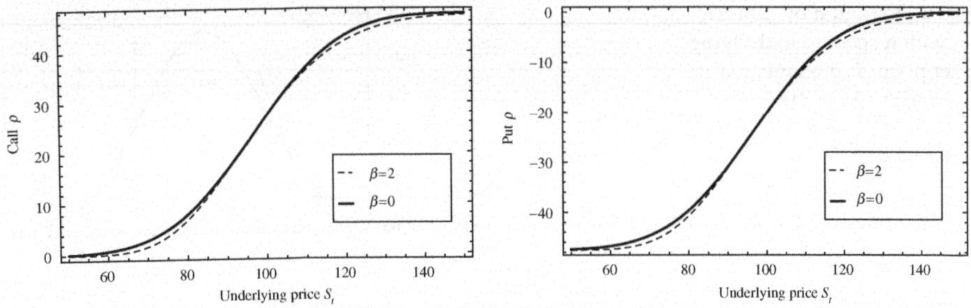

Fig. 5 Variation of rho, ρ, with respect to the underlying asset price S_t. Parameters: $S_0 = 100$, $X = 100, \sigma_0 = 0.25, \tau = 0.5, r = 0.1$, and $q = 0$

3.4 Rho

The rho, ρ, of an option is defined to be the rate of change of the value of an option, V, with respect to the interest rate, r, that is, $\rho = \partial V/\partial r$ (Fig. 5). The rhos of the European call and put option prices under absolute diffusion process are found to be

$$\rho_{call} = X\tau e^{-r\tau}\Big(N(y_1) - N(y_2)\Big) + B, \tag{22}$$

$$\rho_{put} = -X\tau e^{-r\tau}\Big(N(-y_1) + N(y_2)\Big) + B, \tag{23}$$

where

$$B = u\Big(n(y_1) - n(y_2)\Big)\left(\frac{\tau e^{-2r\tau}}{e^{-2q\tau} - e^{-2r\tau}} - \frac{1}{2(r-q)}\right), \tag{24}$$

with u, y_1, and y_2 being defined as in Eqs. (13)–(15).

3.5 Gamma

The gamma, Γ, of an option is defined as the rate of change of delta, Δ, with respect to the asset price, S_t, that is, $\Gamma = \partial^2 V/\partial S_t^2 = \partial\Delta/\partial S_t$ (Fig. 6). For European call and put options under the absolute diffusion process, their gammas are found to be

$$\Gamma_{call} = \Gamma_{put} = \frac{e^{-2q\tau}}{u}\Big(n(y_1) - n(y_2)\Big), \tag{25}$$

where u, y_1, and y_2 are defined as in Eqs. (13)–(15).

The following tables report values of call and put European options under the absolute diffusion process and the Black and Scholes model, as well as their corresponding greeks. Overall, our results show that the misspecification of β may result in significant errors. Thus, similarly to other numerical analysis available in

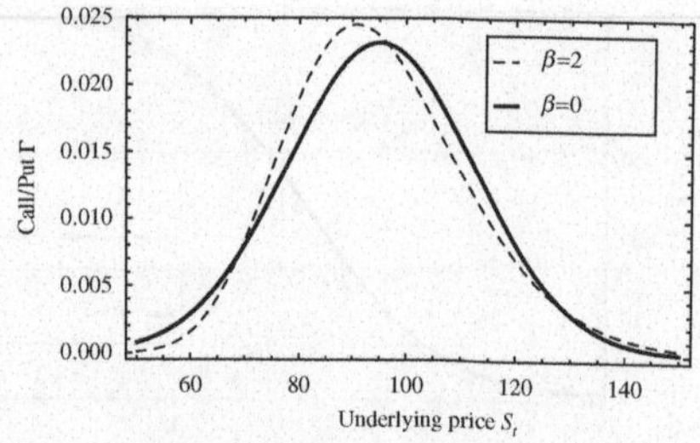

Fig. 6 Variation of gamma, Γ, with respect to underlying asset price S_t. Parameters: $S_0 = 100,\ X = 100,$ $\sigma_0 = 0.25, \tau = 0.5, r = 0.1,$ and $q = 0$

Table 1 Values for call options and greeks under absolute and GBM diffusion processes

X		Call price	Delta	Theta	Vega	Rho	Gamma
95	$\beta = 0$	12.7426	0.7118	−12.0286	23.5433	30.7141	0.0198
	$\beta = 2$	12.5880	0.7458	−11.8663	22.6677	30.9969	0.0181
	% Diff	1.23	4.56	1.37	3.86	0.91	9.14
100	$\beta = 0$	9.5915	0.6113	−12.1002	26.4399	27.4513	0.0222
	$\beta = 2$	9.5822	0.6448	−12.0722	26.3311	27.4472	0.0211
	% Diff	0.10	5.18	0.23	0.41	0.01	5.52
105	$\beta = 0$	6.9403	0.5028	−11.5632	27.5180	23.4182	0.0231
	$\beta = 2$	7.0996	0.5379	−11.6899	28.0819	23.3470	0.0225
	% Diff	2.24	6.53	1.08	2.01	0.31	2.97

Note: Parameters used in calculations: $S_0 = 100, \sigma_0 = 0.25, \tau = 0.5, r = 0.1$, and $q = 0$

Table 2 Values for put options and greeks under absolute and GBM diffusion processes

X		Put price	Delta	Theta	Vega	Rho	Gamma
95	$\beta = 0$	3.1094	−0.2882	−2.9920	23.5433	−14.4693	0.0198
	$\beta = 2$	2.9548	−0.2542	−2.8296	22.6677	−14.1865	0.0181
	% Diff	5.23	13.39	5.74	3.86	1.99	9.14
100	$\beta = 0$	4.7145	−0.3887	−2.5879	26.4399	−20.1102	0.0222
	$\beta = 2$	4.7052	−0.3552	−2.5599	26.3311	−20.1142	0.0211
	% Diff	0.20	9.41	1.09	0.41	0.02	5.52
105	$\beta = 0$	6.8194	−0.4972	−1.5752	27.5180	−26.5213	0.0231
	$\beta = 2$	6.9786	−0.4621	−1.7019	28.0819	−26.5926	0.0225
	% Diff	2.28	7.60	7.44	2.01	0.27	2.97

Note: Parameters used in calculations: $S_0 = 100, \sigma_0 = 0.25, \tau = 0.5, r = 0.1$, and $q = 0$

the literature (e.g., [6]), we conclude that care must be taken when choosing the appropriate diffusion process for pricing and hedging options (Tables 1 and 2).

4 Conclusion

The results of this chapter clearly highlight the importance of the model choice for option pricing and hedging purposes. In fact, we have obtained quite different results when using the Black and Scholes model, the absolute diffusion model, or some other more generalized CEV model that is able to capture both direct ($\beta < 2$) and inverse ($\beta > 2$) leverage effects frequently observed in option market.

Acknowledgements Dias is member of the BRU-UNIDE, and Braumann and Larguinho are members of the Research Center Centro de Investigação em Matemática e Aplicações (CIMA), both centers financed by the Fundação para a Ciência e Tecnologia (FCT). Dias gratefully acknowledges the financial support from the FCTs grant number PTDC/EGE-ECO/099255/2008.

References

1. Abramowitz, M., Stegun, I.A.: Handbook of Mathematical Functions. Dover, New York (1972)
2. Benton, D., Krishnamoorthy, K.: Computing discrete mixtures of continuous distributions: noncentral chisquare, noncentral t and the distribution of the square of the sample multiple correlation coefficient. Comput. Statis. Data Anal. **43**, 249–267 (2003)
3. Black, F., Scholes, M.: The pricing of options and corporate liabilities. J. Polit. Econ. **81**, 637–654 (1973)
4. Cox, J.C.: Notes on Option Pricing I: Constant Elasticity of Variance Diffusions. Working paper, Stanford University (1975)
5. Cox, J.C., Ross, S.A.: The valuation of options for alternative stochastic processes. J. Financ. Econ. **3**, 145–166 (1976)
6. Dias, J.C., Nunes, J.P.: Pricing real options under the constant elasticity of variance diffusion. J. Futures Mark. **31**, 230–250 (2011)
7. Ding, C.G.: Algorithm AS 275: Computing the non-central χ^2 distributions function. Appl. Stat. **41**, 478–482 (1992)
8. Emanuel, D.C., MacBeth, J.D.: Further results on the constant elasticity of variance call option pricing model. J. Financ. Quant. Anal. **17**, 533–554 (1982)
9. Merton, R.C.: Theory of rational option pricing. Bell J. Econ. Manag. Sci. **4**, 141–183 (1973)
10. Schroder, M.: Computing the constant elasticity of variance option pricing formula. J. Financ. **44**, 211–219 (1989)

Scaling Exponents in Heart Rate Variability

Argentina Leite, Maria Eduarda Silva, and Ana Paula Rocha

Abstract

Long recordings of heart rate variability (HRV) display non-stationary characteristics and exhibit long- and short-range correlations. The nonparametric methodology detrended fluctuation analysis (DFA) has become a widely used technique for the detection of long-range correlations in non-stationary HRV data. Recently, we have proposed an alternative approach based on fractional integrated autoregressive moving average (ARFIMA) modelling. These models are an extension of the AR models usual in HRV analysis and have special interest for applications because of their ability for modelling both short- and long-term behaviour of a time series. In this work, DFA is used to assess also short-range scales, further characterizing the data. The methods are applied to 24 h HRV recordings from the Noltisalis database, collected from healthy subjects, patients suffering from congestive heart failure and heart transplanted patients. The analysis of short-range scales leads to a better discrimination between the different groups.

A. Leite (✉)
Departamento de Matemática, Escola de Ciências e Tecnologia, Universidade de Trás-os-Montes e Alto Douro and CM-UTAD, Portugal
e-mail: tinucha@utad.pt

M.E. Silva
Faculdade de Economia, Universidade do Porto and CIDMA, Portugal
e-mail: mesilva@fep.up.pt

A.P. Rocha
Departamento de Matemática, Faculdade de Ciências, Universidade do Porto and CMUP, Portugal
e-mail: aprocha@fc.up.pt

J. Lita da Silva et al. (eds.), *Advances in Regression, Survival Analysis, Extreme Values, Markov Processes and Other Statistical Applications*, Studies in Theoretical and Applied Statistics, DOI 10.1007/978-3-642-34904-1_27,

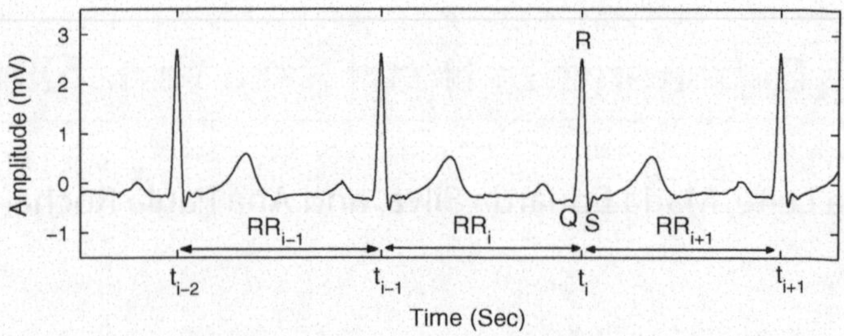

Fig. 1 Schematic representation of electrocardiogram signal and relevant information in each cardiac beat: QRS complexes and RR intervals ($\text{RR}_i = t_i - t_{i-1}$)

1 Introduction

The characterization of the dynamics of a system has become an important and interdisciplinary problem, namely in biomedical applications. Cardiovascular variables such as heart rate, arterial blood pressure and the shape of the QRS complexes in the electrocardiogram, Fig. 1, show variability on a beat to beat basis [14]. This variability reflects the interaction between perturbations to the cardiovascular variables and the corresponding response of the cardiovascular regulatory systems. Therefore, the analysis of such variability can provide a quantitative and non-invasive method to assess the integrity of the cardiovascular system. The discrete series of successive RR intervals, the tachogram, Fig. 2a, is the simplest signal that can be used to characterize heart rate variability (HRV) and has been applied in various clinical situations. The analysis of ambulatory long-term HRV series has become important for clinical diagnosis and risk assessment [14]. These series correspond typically to 24 h recordings and exhibit non-stationary characteristics.

It is well known in the literature that HRV series exhibit not only short but also long-range correlations which were firstly studied with DFA [9]. An alternative parametric approach to describe long-range correlation in HRV data has been proposed by the authors [6], using Fractional integrated autoregressive moving average (ARFIMA) models which are an extension of AR models. The parametric approach has the advantage of allowing the removal of the long-memory component by applying the adequate fractional differencing filter. The remaining short-memory component may then give further insights of the data, as illustrated in Fig. 2.

In this work, ARFIMA models combined with selective adaptive segmentation [6] and DFA scaling exponents are used to describe long- and short-range correlations in 24 h HRV recordings of 30 subjects from the Noltisalis database [13].

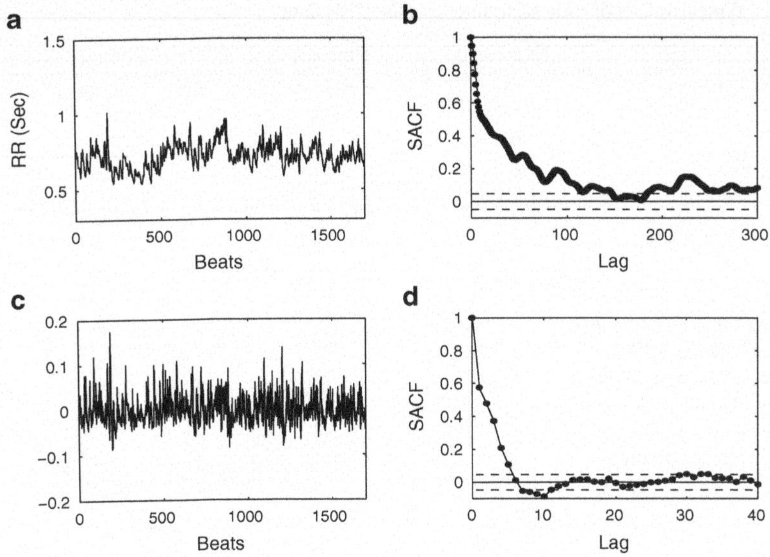

Fig. 2 (**a**) Tachogram of a normal subject; (**c**) same tachogram after removing the long-range correlations with an ARFIMA(0,0.47,0) filter; (**b**) and (**d**) corresponding SACFs

2 Scaling Exponents

DFA [9] has become an important non-parametric tool to assess the correlation properties in non-stationary processes. This methodology was first developed to quantify long-range correlations in non-stationary time series, $x(1), \ldots, x(N)$. The scaling exponent α at time scale k is obtained by fitting a linear model to the log–log relationship

$$F(k) \sim k^{\alpha}, \quad \text{where}$$

$$F(k) = \sqrt{\frac{1}{N}\sum_{i=1}^{N}[y(i) - y_k(i)]^2} \quad \text{with} \quad y(i) = \sum_{t=1}^{i}[x(t) - \bar{x}_i]$$

and $y_k(i)$ is the local linear trend in each segment of length k.

For stationary processes with long-range correlations, random walk theory implies that the scaling behaviour of $F(k)$ is related to the spectral density function, which satisfies $f(\omega) \sim c_f|\omega|^{-\beta}$, $\omega \to 0$, where $\beta \in]0,1[$ and $c_f > 0$. Then, the relation between the exponent β and the mean fluctuation function exponent α is given by $\beta = 2\alpha - 1$ [10].

For uncorrelated data, the scaling exponent is $\alpha = 0.5$. Values of $\alpha > 0.5$ for large scales k indicate long-range correlations in the data. In particular, for 24 h HRV recordings (approximately 100,000 beats) k is taken in the ranges $100 \leq k \leq 100000$ [9] or $128 \leq k \leq 4096$ [3]. This methodology has the disadvantage of

Table 1 Typical values of time scale intervals for HRV data

Length	Exponents	Time scales	References
2 h (∼ 8192 beats)	α_1^{Peng}	$4 \leq k \leq 16$	Peng et al. [9]
	α_2^{Peng}	$16 \leq k \leq 64$	
	α_1^{Pikk}	$4 \leq k \leq 11$	Pikkujämsä et al. [12]
	α_2^{Pikk}	$k > 11$	
	α_2^{Leite}	$64 \leq k \leq 1024$	Leite et al. [7]
10 min (∼ 700 beats)	α_1^{Penzel}	$10 \leq k \leq 40$	Penzel et al. [10]
	α_2^{Penzel}	$70 \leq k \leq 300$	

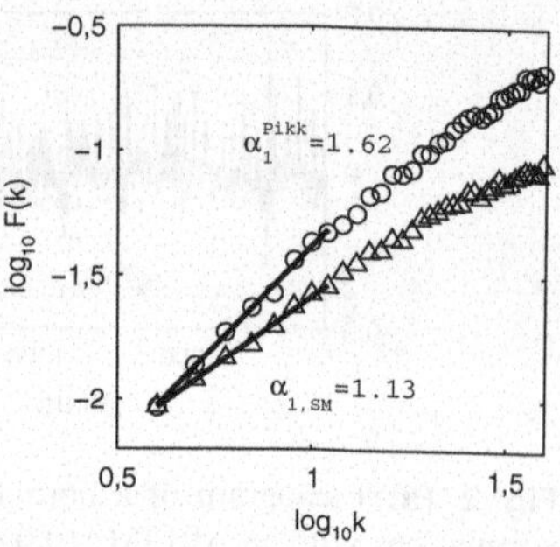

Fig. 3 Results of short-range scaling exponents α_1^{Pikk} (○) and $\alpha_{1,SM}$ (◁) in the range $4 \leq k \leq 11$ with DFA for the data represented in Fig. 2a and Fig. 2c, respectively

requiring large sample sizes for an unbiased estimation of the long memory [9]. Further studies with smaller samples indicated that different ranges of values for k lead to the estimation of other scaling exponents, α_1 and α_2, which may be used to characterize the correlation of the series on small and large time scales [9, 10, 12]. Table 1 summarizes typical values of time scale intervals reported in literature for HRV. For short-range correlated data, α_1 is larger than 0.5 on small scales k (namely, α_1^{Peng}, α_1^{Pikk} and α_1^{Penzel}), and, for long-range correlated data, α_2 is larger than 0.5 on large scales k (namely, α_2^{Leite} and α_2^{Penzel}). In this work α_1 and α_2 are used to estimate short and long range correlations in HRV, respectively.

An alternative approach to describe both long- and short-term correlations is to use ARFIMA$(p, d, 0)$ models, [4], with spectral density function given by $f(\omega) = f_{SM}(\omega)|1 - e^{-i\omega}|^{-2d}$, $-\pi \leq \omega \leq \pi$, where the parameter d characterizes the long-range dependence and $f_{SM}(\omega)$ is the spectral density of the corresponding short-memory AR(p) process. For stationary data with long-range correlations, the parameter d is related to the exponent β and the mean fluctuation function exponent α by $d = 0.5\beta$ and $d = \alpha - 0.5$, respectively [2].

The ARFIMA models have been found adequate to capture and remove long-range correlations in HRV recordings [6]. This suggests studying short-range scaling exponents of the data obtained after removing long memory. Such an exponent is hereafter denoted by $\alpha_{1,SM}$ and is calculated in the range $4 \leq k \leq 11$. In fact, the application of DFA in short-range scales to the data represented in Fig. 2a, c (a tachogram before and after filtering by an ARFIMA(0, 0.47, 0)) is represented in Fig. 3 and suggests that $\alpha_{1,SM}$ may provide different information about the data.

To describe short-range and long-range correlations in the long-term HRV series (24 h, approximately 100,000 beats), ARFIMA modelling combined with selective adaptive segmentation is used [6]: the long record is decomposed into short records of variable length and the break points, which mark the end of consecutive short records, are determined using the AIC criterion for ARFIMA models. The short records thus obtained have a minimum length of 512 and are subsequently modelled using ARFIMA models, to estimate long-range scaling exponent d, and analysed by DFA, to calculate the short-range scaling exponents α_1^{Pikk} and $\alpha_{1,SM}$.

3 Results and Discussion

The methodology presented above is applied to 24 h HRV recordings of 30 subjects from the Noltisalis database [13]: ten healthy subjects (N, 34–56 years), ten patients suffering from congestive heart failure (C, 36–68 years) and ten heart transplanted patients (T, 18–60 years).

Figure 4 illustrates the results for a healthy subject(N6) (a), a patient affected by congestive heart failure(C10) (d) and a heart transplanted patient(T3) (g). The corresponding estimated long-range scaling exponent d in (b), (e) and (h) changes over time and the recordings present multifractality characteristics in concordance with Baillie et al. [1] and Leite et al. [5–7]. Moreover, these estimates present a circadian variation, with lowest values during the night periods. The estimated short-range scaling exponents α_1^{Pikk} and $\alpha_{1,SM}$ for the healthy subject(N6) (c), decrease during the night period, for the heart transplanted patient(T3) (i), increase during this period and for the patients affected by congestive heart failure(C10) (f), are stable during 24 h.

The results, d, α_1^{Pikk} and $\alpha_{1,SM}$, for the three groups of patients during the 24 h, 6 h of night and 6 h of day periods using selective adaptive segmentation (SAS) are summarized in Table 2. For comparison with the results reported in the literature by Peng et al. [9], Table 2 also includes the results for α_2^{Peng}, α_2^{Leite}, α_1^{Peng} and α_1^{Pikk} calculated using segmentation combined with DFA (S), where the long record is decomposed into short records of constant length, $L = 8192$ beats, and the short records are subsequently analysed by DFA. Kruskal–Wallis rank sum test and multiple comparison procedures [11] are used to compare the three groups of patients during 24 h, as well as in the night and day time periods considered.

The long-range scaling exponents, α_2^{Leite} and $d + 0.5$, increase for patients suffering from congestive heart failure and heart transplanted patients, both during night and day periods, with the highest values for the transplanted group. These results are in concordance with the results obtained by Cerutti et al. [3] and Leite et al. [5]. Moreover, the exponent α_2^{Peng} for patients suffering from congestive heart failure is higher than that for the healthy group. This is consistent with previous results reported in literature by Peng et al. [9], indicated in parentheses in Table 2. However, the exponent α_2^{Peng} for the transplanted group is lower than for patients suffering from congestive heart failure. The Kruskal–Wallis test followed by a multiple comparison procedure indicates that both groups N and C differ from group

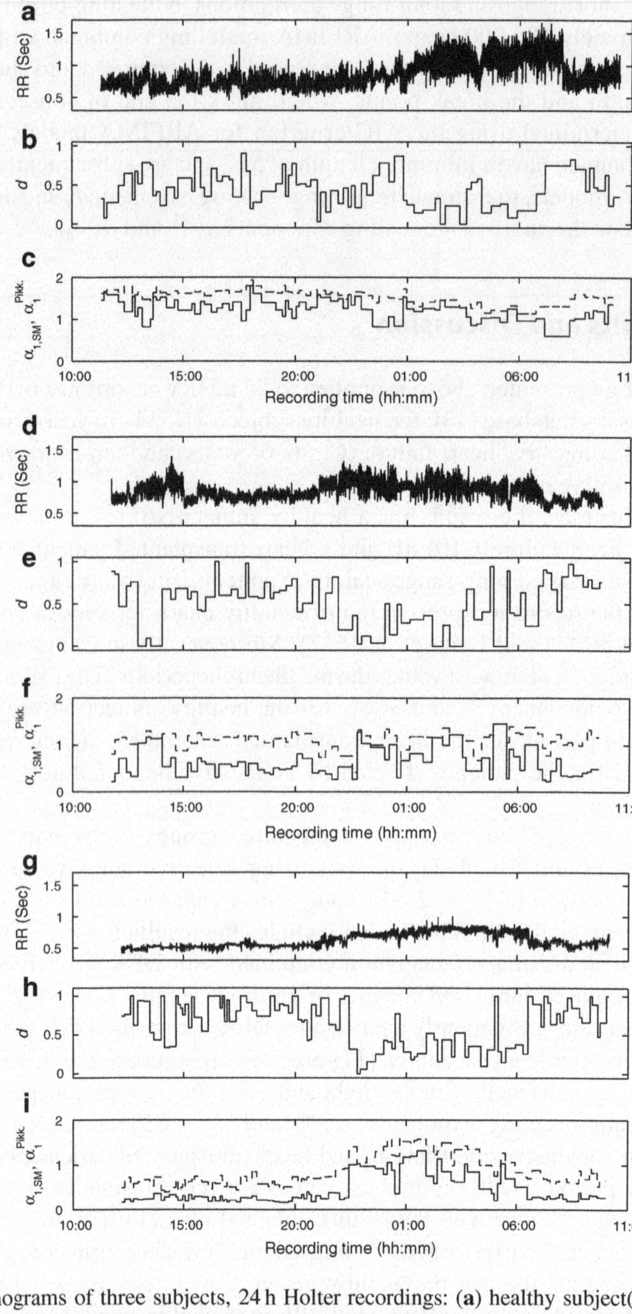

Fig. 4 Tachograms of three subjects, 24 h Holter recordings: (**a**) healthy subject(N6) (**d**) patient affected by congestive heart failure(C10) and (**g**) heart transplanted patient(T3). Corresponding evolution over 24 h of d in (**b**), (**e**) and (**h**) and $\alpha_{1,SM}$ (–) and α_1^{Pikk} (- -) in (**c**), (**f**) and (**i**). d is estimated using ARFIMA models combined with selective adaptive segmentation and $\alpha_{1,SM}$ and α_1^{Pikk} using DFA

Table 2 Scaling exponent values for the three groups of patients from the Noltisalis database: healthy (N), subjects affected by congestive heart failure (C) and transplanted (T), during 24 h, 6 h of night and 6 h of day periods, using segmentation with 8,192 beats (S) and selective adaptive segmentation (SAS)

Method	Exponent	Period	N	C	T	p-value
S	α_2^{Peng}	24 h	0.97 ± 0.13	1.16 ± 0.21	1.03 ± 0.32	0.052
			(1.00 ± 0.12)	(1.13 ± 0.22)	–	
S	α_2^{Leite}	24 h	0.95 ± 0.16	1.04 ± 0.17	1.35 ± 0.23	< 0.001
SAS		24 h	0.94 ± 0.06	1.02 ± 0.14	1.26 ± 0.10	< 0.001
	$d + 0.5$	Night(6 h)	0.84 ± 0.07	0.88 ± 0.16	1.17 ± 0.17	< 0.001
		Day(6 h)	0.96 ± 0.09	1.09 ± 0.16	1.28 ± 0.12	< 0.001
S	α_1^{Peng}	24 h	1.39 ± 0.20	1.22 ± 0.29	0.76 ± 0.32	< 0.001
			(1.20 ± 0.18)	(0.80 ± 0.26)	–	
S	α_1^{Pikk}	24 h	1.46 ± 0.23	1.18 ± 0.30	0.72 ± 0.28	< 0.001
SAS		24 h	1.46 ± 0.18	1.18 ± 0.28	0.70 ± 0.25	< 0.001
	α_1^{Pikk}	Night(6 h)	1.33 ± 0.16	1.15 ± 0.26	0.76 ± 0.29	< 0.001
		Day(6 h)	1.52 ± 0.27	1.19 ± 0.32	0.67 ± 0.28	< 0.001
SAS		24 h	1.16 ± 0.22	0.72 ± 0.24	0.32 ± 0.12	< 0.001
	$\alpha_{1,SM}$	Night(6 h)	1.05 ± 0.19	0.77 ± 0.25	0.40 ± 0.20	< 0.001
		Day(6 h)	1.25 ± 0.30	0.71 ± 0.26	0.29 ± 0.13	< 0.001

Exponents α_1^{Peng} and α_2^{Peng} reported by Peng et al. [9] in parentheses
For each case the average estimates $\pm$ standard deviations are presented
The p-values for the difference between the three groups during the 24 h, night and day time periods, from Kruskal–Wallis rank sum test

T with respect to the long-range scaling exponents α_2^{Leite} and $d + 0.5$ at a 10 % level of significance. However, the exponent α_2^{Peng} differs only between N and C groups at the same significance level.

The short-range scaling exponents α_1^{Peng} and α_1^{Pikk} decrease for patients suffering from congestive heart failure and heart transplanted patients, both during night and day periods, with the lowest values for the transplanted group. For healthy subjects and patients suffering from congestive heart failure, these results are in concordance with the results obtained by Peng et al. [9], reported in parentheses in Table 2. However, the values obtained by Peng et al. are lower than those obtained in this work. This result may be due to the fact that the individuals in the database used by Peng et al. have different characteristics from the individuals in the Noltisalis database regarding age and gender: for example, Platisa and Gal [8] conclude that female subjects had significantly smaller α_1^{Peng} than the male subjects.

The short-range scaling exponent proposed in this chapter, $\alpha_{1,SM}$, also decreases for patients suffering from congestive heart failure and transplanted patients, both during night and day periods, with the lowest values for the transplanted group. The short-range scaling exponents α_1^{Peng}, α_1^{Pikk} and $\alpha_{1,SM}$ differ for groups N and T and C and T. Additionally, the exponent $\alpha_{1,SM}$ differs also for the groups N and C, during 24 h (Fig. 5) and day period.

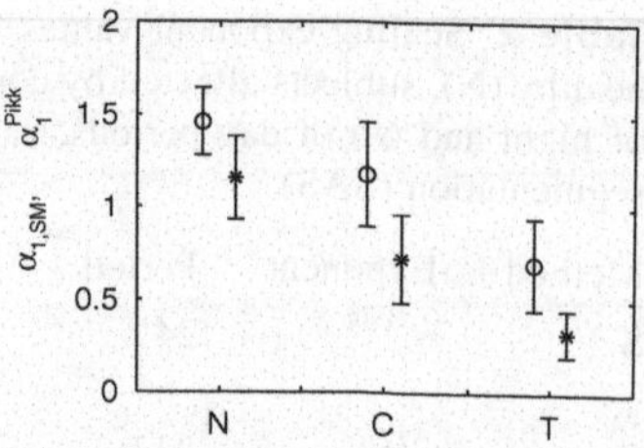

Fig. 5 Average estimates and standard deviations of α_1^{Pikk} (o) and $\alpha_{1,SM}$ (*) for the three groups of patients: healthy (N), subjects affected by congestive heart failure (C) and transplanted (T) during 24 h

4 Final Remarks

It is well know that HRV recordings exhibit long-range correlations. In this work, long memory is removed by fractional differences filtering combined with selective adaptive segmentation. This approach leads to enhanced short-range scaling exponents and a corresponding better discrimination between the different groups of the Noltisalis database.

Acknowledgements Research funded by *FEDER* through the programme COMPETE and by the Portuguese Government through the FCT—Fundação para a Ciência e a Tecnologia—under the projects PEst-C/MAT/UI0144/2011, PEst-C/MAT/UI4106/2011, and PEst-OE/MAT/UI4080/2011.

References

1. Baillie, R.T., Cecen, A.A., Erkal, C.: Normal heartbeat series are nonchaotic, nonlinear, and multifractal: new evidence from semiparametric and parametric tests. Chaos **19**(028503), 1–5 (2009)
2. Boukhan, P., Oppenheim, G., Taqqu, M.S.: Theory and applications of long-range dependence. Birkhäuser, Boston (2003)
3. Cerutti, S., Esposti, F., Ferrario, M., Sassi, R., Signorini, M.G.: Long-term invariant parameters obtained from 24-h Holter recordings: a comparison between different analysis techniques. Chaos **17**(1), 015108–015109 (2007)
4. Hosking, J.R.M.: Fractional differencing. Biometrika **68**, 165–176 (1981)
5. Leite, A., Rocha, A.P., Silva, M.E.: Long memory and volatility in HRV: an ARFIMA-GARCH approach. In: Computers in Cardiology, **36**, 165–168 (2009)
6. Leite, A., Rocha, A.P., Silva, M.E., Costa, O.: Modelling long-term heart rate variability: an ARFIMA approach. Biomed. Technol. **51**, 215–219 (2006)
7. Leite, A., Rocha, A.P., Silva, M.E., Gouveia, S., Carvalho, J., Costa, O.: Long-range dependence in heart rate variability data: ARFIMA modelling vs detrended fluctuation analysis. In: Computers in Cardiology **34**, 21–24 (2007)
8. Platisa, M.M., Gal, V.: Reflection of heart rate regulation on linear and nonlinear heart rate variability measures. Physiol. Meas. **27**, 145–154 (2006)
9. Peng, C.K., Havlin, S., Stanley, H.E., Golberger, A.L.: Quantification of scaling exponents and crossover phenomena in nonstationary heartbeat time series. Chaos **5**, 82–7 (1995)
10. Penzel, T., Kantelhardt, J.W., Grote, L., Peter, J.H., Bunde, A.: Comparison of detrended fluctuation analysis and spectral analysis for heart rate variability in sleep and sleep apnea. IEEE Trans. Biol. Eng. **50**(10), 1143–1151 (2003).
11. pgirmess: Data analysis in ecology. R package version 1.5.1. (2011), http://CRAN.R-project.org/package=pgirmess. Accessed on the 3rd May 2011.

12. Pikkujämsä, S.M., Mäkikallio, T.H., Sourander, L.B., Räihä, I.J., Puukka, P., Skyttä, J., Peng, C.-K., Goldberger, A.L., Huikuri, H.V.: Cardiac interbeat interval dynamics from childhood to senescence: comparison of conventional and new measures based on fractals and chaos theory. Circulation **100**, 393–3991 (1999)
13. Signorini, M.G., Sassi, R., Cerutti, S.: Working on the NOLTISALIS Database: Measurement of nonlinear properties in heart rate variability signals. In: Proceedings of IEEE-EMBS International Conference, Istanbul, Turkey (IEEE, Piscataway), 547–550 (2001)
14. Task Force of the European Society of Cardiology and North American Society of Pacing Electrophysiology: Heart rate variability: Standards of measurement, physiological interpretation and clinical use. Circulation **93**, 1043–1065 (1996)

Prediction of Dementia Patients: A Comparative Approach Using Parametric Versus Nonparametric Classifiers

João Maroco, Dina Silva, Manuela Guerreiro, Alexandre de Mendonça, and Isabel Santana

Abstract

In this chapter, we report a comparison study of seven nonparametric classifiers (multilayer perceptron neural networks, radial basis function neural networks, support vector machines, CART, CHAID and QUEST classification trees, and random forests) as compared to linear discriminant analysis, quadratic discriminant analysis and logistic regression tested in a real data application of mild cognitive impaired elderly patients conversion to dementia. When classification results are compared both on overall accuracy, specificity and sensitivity, linear discriminant analysis and random forests rank first among all the classifiers.

1 Introduction

Traditional parametric statistical classification methods like Fisher's linear discriminant analysis (LDA) and logistic regression (LR) have been extensively used in the past in classification problems for which the criterion variable is dichotomous [1–3]. More recently, attention has been steadily building on the accuracy and efficiency of nonparametric classifiers like neural networks (NN), support vector machines (SVM), classification trees (CART) and random forests (RF) as applied to classification problems [1, 4–6]. Research on the comparative accuracy for both

---

J. Maroco (✉)
Unidade de Psicologia e Saúde, Departamento de Estatística. ISPA-Instituto Universitário, Rua Jardim do Tabaco, 34. 1149-041 Lisboa, Portugal
e-mail: jpmaroco@ispa.pt

D. Silva · M. Guerreiro · A. de Mendonça
Instituto de Medicina Molecular, Universidade de Lisboa, Lisboa, Portugal

I. Santana
Serviço de Neurologia, Hospitais da Universidade de Coimbra, Coimbra, Portugal

J. Lita da Silva et al. (eds.), *Advances in Regression, Survival Analysis, Extreme Values, Markov Processes and Other Statistical Applications*, Studies in Theoretical and Applied Statistics, DOI 10.1007/978-3-642-34904-1_28,

parametric and nonparametric methods has been growing steadily. Some authors defend that nonparametric classifiers have higher accuracy and lower error rates than the traditional parametric methods [7–9]. However, this superiority is not apparent with all data sets, especially with real data [10–14]. Results regarding classification accuracy and stability of the findings are still controversial [6, 15]. Most comparisons are based only on total classification accuracy and/or error rates; they involve human intervention for training and optimization of the nonparametric classifiers vs. out-of-the-box results for the parametric classifiers. According to Duin [16] "(…) a straight forward fair comparison demands automatic classifiers with no user interaction". It also requires a large base comparison taking into account not only total accuracy but also sensitivity, specificity and discriminant power. Having prevented inadequate parametrizations of nonparametric classifiers, we compared total accuracy, sensitivity and specificity of traditional parametric classifiers (LDA, quadratic discriminant analysis (QDA), LR) vs. nonparametric methods derived from data mining and machine learning (NN, SVM, CART, RF). These methods were used to predict the evolution into dementia of 383 elderly people with mild cognitive impairment from several neuropsychological tests with predictive validity. When sensitivity and specificity were taken into account along with total classification accuracy, LDA reveals itself, with random forests, as one of the best classifiers. It is worthwhile to mention that LDA, a classifier devised ca. 100 years ago, still resists the challenges of the new classifiers who required large computing power and user intervention.

2 Classifiers

2.1 Discriminant Analysis

Fisher's LDA estimates discriminant function scores (D) for each of n subjects classified into k groups from p linearly independent predictor variables (X_p) as

$$D_j = w_{j1}X_1 + w_{j2}X_2 + \ldots + w_{jp}X_p \quad (1)$$

where $j = 1,\ldots,\min(k\text{-}1,p)$. Discriminant weights ($w_j$) are estimated by ordinary least squares so that the ratio of the variance within the k groups to the variance between the k groups is minimal. Classification functions of the type

$$C_j = c_{j0} + c_{j1}X_1 + c_{j2}X_2 + \ldots + c_{jp}X_p \quad (2)$$

for each of the $j = 1,\ldots,k$ groups can be constructed. The coefficients of the classification function for the j group are estimated from within sum of squares matrices ($\mathbf{W}$) of the discriminant scores for each group and from the means of the p discriminant predictors in each of the classifying groups ($\mathbf{M}$) as $\mathbf{C}_j = \mathbf{W}^{-1}\mathbf{M}$ with $c_{jo} = \log p - \frac{1}{2}\mathbf{C}_j\mathbf{M}_j$. QDA uses the same within vs. between groups sum of square minimization optimization but on a quadratic form discriminant function:

$$D_j = \sum_{p=1}^{P} w_{jp} X_p + \sum_{p=1}^{P} q_{jp} X_p^2 + \sum_{p=1}^{P-1} r_{jp} X_p X_{p+1} \quad (3)$$

with classification functions

$$C_j = c_{0j} + \sum_{p=1}^{P} c_{jp} X_p + \sum_{p=1}^{P} o_{jp} X_p^2 + \sum_{p=1}^{P-1} m_{jp} X_p X_{p+1} \quad (4)$$

Both on LDA and QDA, a subject is classified into the group for which its classification function score is higher.

2.2 Logistic Regression

Logistic regression (LR) models the probability of occurrence of one (success) of the two classes of a dichotomous criterion. A Logit transformation of the probability of success for each subject (π_i) is iteratively fitted to a linear combination of predictors accordingly to the model

$$Ln[\hat{\pi}_i/(1-\hat{\pi}_i)] = \beta_0 + \beta_1 X_{1i} + \beta_2 X_{2i} + \ldots + \beta_p X_{pi} \quad (5)$$

by means of maximum likelihood estimation. Probability of success for each subject is estimated with the Logit model, and if the estimated probability is greater than 0.5 (or other pre-defined threshold value), the subject is classified in the success group; otherwise, it is classified into the failure group.

2.3 Neural Networks

Neural network (NN) methods have been used in classification problems and this is one of the most active research and application areas in the neural network field. An NN is a multi-stage, multi-unit classifier, with input, hidden or processing, and output layers. For a binary criterion y_k the NN can be described by the general model

$$\hat{y_k} = f_k(\mathbf{x}, \mathbf{w}, o, \mathbf{x}_0, \mathbf{o}_{0k}) = f\left(\sum_{j=1}^{h} o_{kj} \cdot h\left(\sum_{i=1}^{p} w_{ji}\mathbf{x}_i + x_{0j}\right) + o_{0k}\right) \quad (6)$$

where $\mathbf{x}$ is the vector of predictors, $\mathbf{w}$ is the vector of input weights, $\mathbf{o}$ is the vector of hidden weights, $\mathbf{x}_0$ and $\mathbf{o}_{0k}$ are bias constants and $h(.)$ and $f(.)$ are activation functions for the hidden layer and output layer, respectively. Activation functions are one of the general linear, logistic, exponential or Gaussian function families. Several topologies of neural networks (NN) can be used in binary classification problems. Two of the most used NN are the multilayer perceptron (MLP) and the radial basis function (RBF). The main differences between the two NN reside in the activation function of the hidden layer which belongs to the linear family in the MLP and to the Gaussian family in the RBF function. An NN is generally trained in a set of iterations (epochs) for a subset of the data (train set) and tested for the remained subset (test set). Synaptic weights of the NN are upgraded in each iteration in way to maximize the correct classification rate and/or minimize a function of the classification errors (for a detailed description of NN see [17]).

2.4 Support Vector Machines

SVM are machine-learning-derived classifiers which map a vector of predictors into a higher-dimensional linear plane through both linear and non-linear kernel ϕ functions. In a binary classification problem, the two groups, say $\{-1\}$ and $\{+1\}$, are then separated by a higher-dimension hyperplane $\mathbf{w}'\phi(\mathbf{x}) + b = 0$ where $\mathbf{x}$ is the vector of predictors, $\mathbf{w}$ is the weight vector and b is a bias offset. The classification function is then

$$f(\mathbf{x}) = Sign(\mathbf{w}'\phi(\mathbf{x}) + b) \tag{7}$$

To find the optimum plane for both $\{-1\}$ and $\{+1\}$ groups, one strategy is to maximize the distance or margin of separation from the supporting planes, respectively, $\mathbf{w}'\phi(\mathbf{x}) + b \geq +1$ for the $\{+1\}$ group and $\mathbf{w}'\phi(\mathbf{x}) + b \leq -1$ for the $\{-1\}$ group. These support planes are pushed apart until they bum into a small number of observations called "support vectors". This is equivalent to minimize a cost function

$$C(\mathbf{w}) = \frac{\|\mathbf{w}\|^2}{2} + c\sum_{i=1}^{n} \xi_i = \tfrac{1}{2}\mathbf{w}'\phi(\mathbf{w}) + c\sum_{i=1}^{n} \xi_i \tag{8}$$

under the constraints $y_i(\mathbf{w}\prime\phi(\mathbf{x}_i) + b) \geq 1 - \xi_i$ and $\xi_i \geq 0$ where $c > 0$ is penalty parameter for classification errors and ξ_i is the penalty of a misclassified observation. In classification problems the usual kernel functions are the linear kernel $\phi(\mathbf{x}_i, \mathbf{x}_j) = \mathbf{x}_i\prime\mathbf{x}_j$ and the Gaussian $\phi(\mathbf{x}_i, \mathbf{x}_j) = \exp(-\gamma\|\mathbf{x}_i - \mathbf{x}_j\|^2)$ where γ is a kernel parameter (for a complete description of SVM see [7, 18]).

2.5 Classification Trees

Classification trees (CT) are nonparametric classifiers that construct decision trees by splitting a node, accordingly to an "if-then criteria" applied to a set of predictors, into two child nodes repeatedly, from a root node that contains the whole sample.

Thus, CT can select the predictors and its interactions that are most important in determining an outcome for a criterion variable. The development of a CT is supported on three major elements: (1) choosing a sampling–splitting rule that defines the tree branch which connect the classification nodes; (2) the evaluation of the goodness of fit produced by the splitting rule at each node and (3) the criteria used for choosing an optimal or final tree for classification proposes. Accordingly to the features of these major elements, CT can be classified into CART, CHAID and QUEST. In CART trees, the predictors are split (if they are continuous) or classes are separated (if they are qualitative) with the objective of reducing the impurity of the final node produced at each t branch of the tree. The Gini impurity index

$$I_G(t) = 1 - \sum_{c=1}^{C} P(c|t)^2 = \sum_{c=1}^{C} \sum_{c \neq j=1}^{C} P(c|t)P(j|t) \tag{9}$$

is frequently used as a measure of group heterogeneity in CART. $P(c|t)$ is the conditional probability of a class c given the node t:

$$P(c|t) = \frac{P(c,t)}{P(t)} \text{ with } P(c,t) = \frac{\pi(c)n_c(t)}{n_c} \text{ and } P(t) = \sum_{c=1}^{C} P(c,t) \tag{10}$$

where $\pi(c)$ is the probability of observing the group c and $n_c(t)$ is the number of elements in group c at a given node t. The tree grows until no further predictors can be used or the impurity of each group at the final branch of the tree cannot be reduced further. Nonsignificant branches can be pruned from the final tree. In CHAID trees, the homogeneity of the groups is evaluated by a Bonferroni-corrected p-value from the Pearson chi-square statistic applied to two-way classification tables with C classes and K splits. In QUEST, the homogeneity of groups at each branch is evaluated with the ratio of the within group variance and between group variances for continuous predictors or a chi-square like statistic for categorical predictors. Although several other alternative algorithms are also available, in this study we only compared well-established CART, CHAID and QUEST algorithms (see [19]).

2.6 Random Forests

Random forests (RF) construct a series of CART using different bootstrap samples of the original data sample. Each of these CART trees is build from a random subset of the total predictors who maximize the classification criteria at each node. An estimate of the classification error rate can be obtained using each of the CART to predict the data not in the bootstrap sample ("out-of-the bag") used to grow the tree and average the out-of-the bag predictions for the grown forest. These out-of-the bag estimates of the error rate can be quite accurate if enough trees have been grown.

Table 1 Sample demographics[a]

Groups	MCI	Dementia	p-value
Size	262 (68 %)	121 (32 %)	0.001‡
Age (mean ± SD)	68.3±8.5	71.1±8.6	0.003†
Sex (male/female)	157 / 103	75 / 46	0.822‡
Schooling years (mean ± SD)	8.2 ± 4.7	8.6 ± 5.0	0.436†
Time between assessments (year)(mean ± SD)	2.4 ± 1.6	2.4 ± 1.7	0.881†

[a]"MCI"—patients who remained in MCI; and "Dementia"—patients who progressed to dementia p-values for group comparison were obtained from Student's t-test (†) or χ^2test(‡)

Although this classification strategy may lack a perceivable advantage over single CART, accordingly to its creator (Leo Breiman [20]), it has unexcelled accuracy when compared to many classifiers including LDA, NN and SVM.

3 A Classification Application

3.1 Sample

The described classifiers were used to predict the conversion of 383 elderly patients with mild cognitive impairment (MCI) to dementia (see Table 1 for sample demographics).

Thirty-two percent of participants showed dementia (the event to predict). Distributions of sex, schooling years and time between assessments did not differ significantly between the dementia vs. MCI groups. However, mean age was significantly lower for the MCI group ($p \leq 0.05$).

3.2 Criterion and Predictors

The criterion was a dichotomous variable with two groups: MCI and dementia. Predictors used to predict the conversion of MCI into dementia were a set of nine quantitative neuropsychological tests which have previously shown criterion validity (i.e., statistically significant different scores for the MCI vs. dementia groups): Digit Span backward (evaluates working memory), the Logical Memory test (evaluates episodic memory), Verbal Paired Associates Learning (evaluates learning ability), Word Recall (evaluates short-term memory), Orientation (evaluates personal, spatial, and temporal orientation), Semantic Fluency (evaluates verbal initiative), Clock Drawing (evaluates visual constructive abilities), the Raven Progressive Matrices (evaluates non-verbal abstract reasoning), and Proverbs test (evaluates verbal abstract reasoning). Figure 1 shows the scatter plot of these predictors and their frequency histograms. Predictors lack homogeneity of group

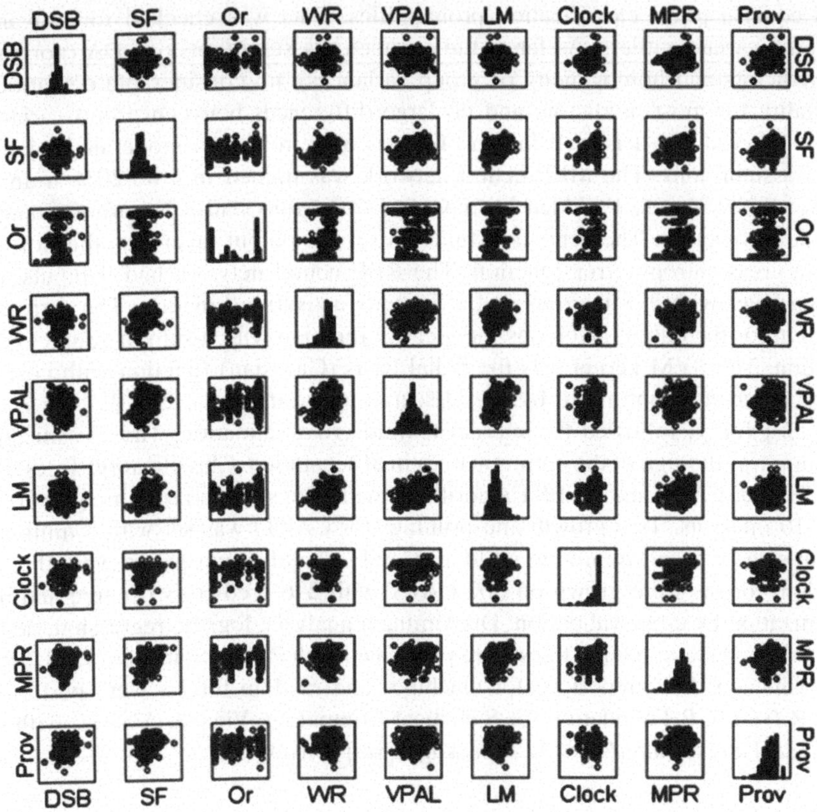

Fig. 1 Scatter plots for MCI (•) and dementia (○) patients in the 9 predictors and its histograms (DSB—Digit Span Backward test; SF—Semantic Fluency; Or—Orientation; WR—Word Recall; VPAL—Verbal Paired Associates Learning; LM—Learning Memory; Clock—Clock Drawing; MPR—Raven Progressive Matrices; Prov—Proverbs)

variances and their histograms show several predictors with a considerable departure from the Gaussian distribution. There were also several outliers.

3.3 Classification Settings

A fivefold cross-validation strategy was followed to train and evaluate all the classifiers. The total sample was divided into 5 proportional subsamples. In each of the 5 steps, 4/5 of the sample was used for training and 1/5 was used for testing. Test results for the 5 runs were then aggregated and the comparative performances of the different classifiers evaluated with Friedman's ANOVA on ranks followed by Dunn's multiple comparisons on mean ranks. Statistical significance was assumed for $p<0.05$. Linear and quadratic discriminant analysis and logistic regression used equal a priori classification probabilities. Data was checked for univariate and multivariate outliers. As far as the parametric assumptions of LDA (normality of predictors and homogeneity of group variances), no considerable deviation of normality for most predictors and no large differences between group variances were observed. As it is well known, LDA is quite robust to moderate violations of its assumptions. The MLP neural network was trained in a 80:20 % train:test setup, with 9 inputs, 1 hidden layer with 4–7 neurons and a hyperbolic tangent activation function. The activation function for the output layer was the Softmax with a cross-entropy error function. The RBF neural network had 9 inputs, one hidden layer with 2–8 neurons and a Softmax activation function. The activation function for the output layer was the identity function with a sum of squares error function. The SVM kernel was the radial basis (Gaussian) function with cost (c) and γ parameters optimized by a grid search in the intervals $[2^{-3}; 2^{15}]$ for c and $[2^{-15}; 2^{3}]$ for γ, followed by internal tenfold cross-validation. The classification function was the sign of the optimum margin of separation. Classification Trees used the CHAID, CART and QUEST algorithms, with α to split and α to merge of 0.05, with 10 intervals. Tree growth and pruning (for CART) was set with a minimum parent size of 5 and minimum child size of 1. Classification priors were 0.5:0.5. Random forests were grown on 500 CART with 2–6 predictors per tree and tree optimization by cross-validation. Discriminant analysis, logistic regression, neural networks and classification trees were performed with PASW Statistics (v. 18, SPSS Inc., Chicago, Il). Support vector machines and random forests were performed with R (v. 2.8, R Foundation for Statistical Computing, Vienna, Austria) with the *e1071*[21] and *randomForest* [22] packages, respectively.

3.4 Results

Classification accuracy, sensitivity and specificity were evaluated in the 5 test sets resulting from the fivefold cross-validation strategy. Data gathered are shown as box plots for the different classifiers. Figure 2 shows the box plots of the total classification accuracy for the 10 classifiers studied. When the distributions of total accuracy are compared with the Friedman test, the observed differences were not statistically significant ($X^2_{Fr}(9) = 13.6$; $p = 0.137$).

The distributions of specificity (that is the proportion of subjects that did not convert into dementia and were correctly predicted by the classifier) are shown in Fig. 3. There were statistical significant differences in the specificity distributions of the different classifiers $X^2_{Fr}(9) = 34.868$; $p < 0.001$). SVM, MLP, LR and RF presented the highest specificity values which were significantly different from a second group composed by LDA, QDA and CART.

Figure 4 shows the distributions of sensitivity (proportion of subjects that were correctly predicted to convert into dementia). There were statistically significant differences in the distributions of sensitivity ($X^2_{Fr}(9) = 37.9$; $p < 0.001$). LDA,

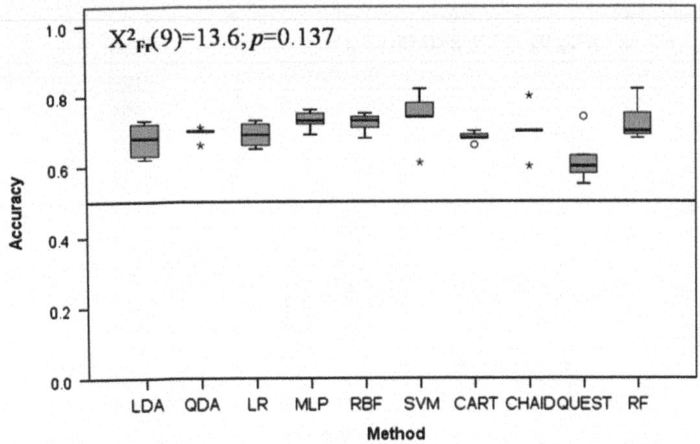

Fig. 2 Box-plot distributions of classification accuracy (number of correct classifications / total sample size) for the 5 test samples resulting from the fivefold cross-validation procedure (see text for abbreviations)

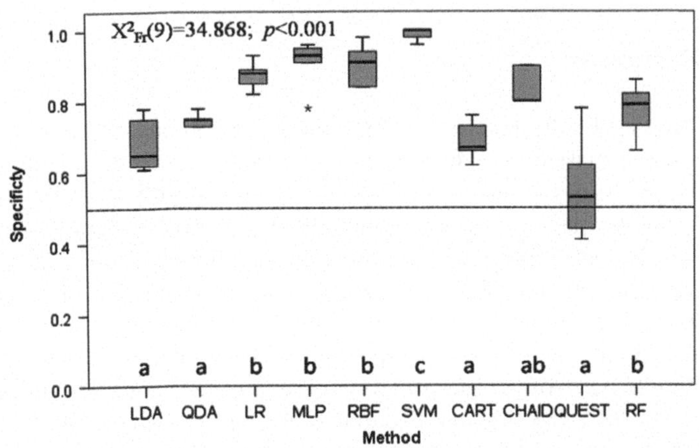

Fig. 3 Box-plot distributions of specificity (number of MCI predicted / number of MCI observed) for the 5 test samples resulting from the fivefold cross-validation procedure (see text for abbreviations). Different letters indicate statistically significant differences between classifiers on a multiple mean rank comparison procedure

CART, QUEST and RF had the highest sensitivity values which were significantly different from a second group composed by LR, MLP, RBF and CHAID. It is worthwhile to mention that this second group had sensitivity lower than 0.5 and that SVM was the classifier with lowest sensitivity.

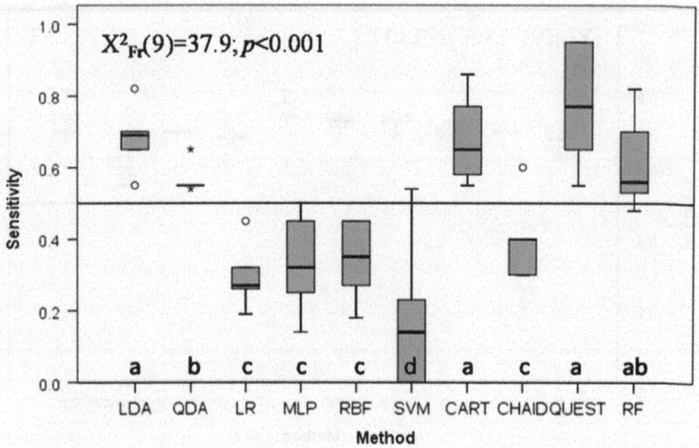

Fig. 4 Box-plot distributions of sensitivity (number of dementia predicted/number of dementia observed) (see text for abbreviations). Different letters indicate statistically significant differences between classifiers on a multiple mean rank comparison procedure

4 Discussion

Although no statistically significant differences were found in total accuracy of the 10 evaluated classifiers (Medians between 0.60 and 0.74), a quite different picture emerges from the analysis of specificity and sensitivity of the classifiers. Median specificity ranged from a minimum of 0.53 (QUEST) to a maximum of 1 (SVM). With the exception of QUEST, all the other classifiers were quite efficient in predicting group membership in the group with larger number of elements (the MCI group corresponding to 68 % of the sample) (median specificity larger than 0.6). However, predictions for the group with lower frequency (the dementia group, corresponding to 32 % of the sample) were quite unsatisfactory. Minimum median sensitivity was 0.14 (SVM) and maximum median sensitivity was 0.7 (LDA). Only five of the ten classifiers tested showed median sensitivity larger than 0.5. Conversion into dementia is the key prediction in this biomedical application, requiring classifiers with high sensitivity. Thus, on this real data example, classifiers like logistic regression, neural networks, support vector machines and CHAID trees are inappropriate for this binary classification task. Also, total accuracy of classifiers is misleading since some classifiers are good only at predicting the larger group membership (high specificity) but quite bad at predicting the smaller group memberships (low sensitivity). Some of the classifiers with the highest specificity (NN and SVM) were also the classifiers with the lowest sensitivity. Unbalance of classification efficiency for small frequency vs. large frequency groups has been found in other real-data studies for logistic regression and neural networks [10, 23–25]. Taking in account both total accuracy, specificity and sensitivity, the oldest Fisher's linear discriminant analysis ranks top with random forests, the newest member of the binary classification family. Similar observations have been made by other authors. For example, Breiman et al. [19] state that LDA does as well as other classifiers in most applications. Meyer et al. [24] point out in their comparison study of data mining classifiers, including NN and SVM, that LDA is a very competitive classifier, "producing good results *out-of-the-box* without the inconvenience of delicate and computationally expensive hyperparameter tuning". For simple binary classification problems, where sample size may compromise training and testing of nonparametric data mining and machine learning classifiers, Fisher's linear discriminant analysis stands up as a simple, efficient and time-proof classifier.

References

1. Goss, E.P., Ramchandani, H.: Comparing classification accuracy of neural networks, binary logit regression and discriminant analysis for insolvency prediction of life insurers. J. Econ. Fin. **19**(3), 1–18 (1995)
2. Lei, P.W., Koehly, L.M.: Linear discriminant analysis versus logistic regression: a comparison of classification errors in the two-group case. J. Exp. Educ. **72**(1), 25–49 (2003)
3. Pohar, M., Blas, M., Turk, S.: Comparison of logistic regression and linear discriminant analysis: A simulation study. Metodološki zvezki **1**(1), 143–161 (2004)
4. Pitarque, A., Roy, J.F., Ruiz, J.C.: Redes neurales vs modelos estadísticos: Simulaciones sobre tareas de predicción y clasificación. Psicológica **19**, 387–400 (1998)
5. Nabney, I.T.: Efficient training of RBF networks for classification. Int. J. Neural Syst. **14**(3), 201–208 (2004)
6. Sommer, M., Olbrich, A., Arendasy, M.: Improvements in personnel selection with neural nets: A pilot study in the field of aviation psychology. Int. J. Aviat. Psychol. **14**(1), 103–115 (2004)
7. Ivanciuc, O.: Applications of support vector machines in chemistry. In: Lipkowitz, K.B., Cundari, T.R. (eds.) Reviews in Computational Chemistry, vol. 23, pp. 291–400. Wiley, Weinheim (2007)
8. Sut, N., Senocak, M: Assessment of the performances of multilayer perceptron neural networks in comparison with recurrent neural networks and two statistical methods for diagnosing coronary artery disease. Expert Syst. **24**(3), 131–142 (2007)
9. Kurt, I., Ture, M., Kurum, A.T.: Comparing performances of logistic regression, classification and regression tree, and neural networks for predicting coronary artery disease. Expert Syst. Appl. **34**(1), 366–374 (2008)
10. Finch, H., Schneider, M.K.: Classification accuracy of neural networks vs. discriminant analysis, logistic regression, and classification and regression trees: Three- and five-group cases. Methodology **3**(2), 47–57 (2007)
11. Gelnarova, E., Safarik, L.: Comparison of three statistical classifiers on a prostate cancer data. Neural Network World **15**(4), 311–318 (2005)
12. Green, M., Björk, J., Forberg, J., Ekelund, U., Edenbrandt, L., Ohlsson, M.: Comparison between neural networks and multiple logistic regression to predict acute coronary syndrome in the emergency room. Artif. Intel. Med. **38**(3), 305–318 (2006)
13. Paulo, J.L.A., Vasconcelos, G.C., Arnaud, A.L., Santos, R.A.F., Cunha, R.C.L.V., Monteiro, D.S.M.P.: Neural Networks vs Logistic Regression: a Comparative Study on a Large Data Set. In: 17th International Conference on Pattern Recognition (ICPR'04), vol. 3, pp. 355–358 (2004)
14. Peter, C.A.: A comparison of regression trees, logistic regression, generalized additive models, and multivariate adaptive regression splines for predicting AMI mortality. Stat. Med. **26**(15), 2937–2957 (2007)

15. Ghaffari, M., Hall, E.L.: Experimental approach for the evaluation of neural network classifier algorithms. Intelligent Robots and Computer Vision XXI: Algorithms, Techniques, and Active Vision. (eds.) Casasent DP, Hall EL, & Röning J (SPIE Bellingham WA) **5267**, 250–256 (2003)
16. Duin, R.P.W.: A note on comparing classifiers. Pattern Recognit. Lett. **17**(5), 529–536 (1996)
17. Bishop, C.: Neural Networks for Pattern Recognition. Oxford University, Oxford (1995)
18. Bennett, K.P., Campbell, C.: Support vector machines: Hype or hallelujah? SIGKDD Expl. **2**(2), 1–13 (2000)
19. Breiman, L., Friedman, J.H., Olshen, R.A., Stone, C.J.: Classification and Regression Trees. Wadsworth, Monterey (1984)
20. Breiman, L.: Random forests. Mach. Learn. **45**(1), 5–32 (2001)
21. Meyer, D.: Support vector machines: the interface to libsvm in package e1071. R News **1/3**, 23–26 (2001)
22. Liaw, A., Wiener, M.: Classification and regression by randomForest. R News **2/3**, 18–22 (2002)
23. Schwarzer, G., Vach, W., Schumacher, M.: On the misuses of artificial neural networks for prognostic and diagnostic classification in oncology. Stat. Med. **19**(4), 541–561 (2000)
24. Meyer, D., Leischa, F., Hornik, K.: The support vector machine under test. Neurocomputing **55**(1–2), 169–186 (2003)
25. Maroco, J., Bartolo-Ribeiro, R.: Métodos de classificação binária no contexto da selecção de pilotos militares. Comparação da precisão classificatória de Redes Neuronais, Regressão Logística e Análise Discriminante Linear. XV Congresso da Sociedade Portuguesa de Estatística, ed Hill M et al (SPE), 289–304 (2008)

Pareto Scale Mixtures

Miguel Martins Felgueiras

Abstract
Pareto scale mixtures are very effective for modelling heavy-tailed data. A new class of models is described, generalizing commonly used slash distributions. Mixture properties and possible applications are discussed.

1 Introduction

Simple models assume a fixed scale parameter. However, in many situations it is advisable to randomize the scale parameter in order to increase variability [4]. For instance, in biostatistical studies, the negative binomial model is sometimes referred to as a "more flexible Poisson" since it is the result of modelling the number of eggs laid by females of certain species, the individual being $Poisson(\lambda)$, but considering that the λ's are values from a $Gamma(\alpha, \delta)$ random variable. This procedure leads to a hierarchical model randomizing the former one and hence more flexible.

In many applications the $Gamma(\alpha, \delta)$ is considered a suitable scale mixing model, because its natural connection with the Laplace transforms brings in a useful toolbox of ready-to-use formulas, and in many cases the resulting mixture is reasonably tractable. But any positive random variables can be used to randomize a scale parameter, although in most cases the resulting mixture is difficult to work with, since usually the corresponding density function is not expressible in a closed form.

The family of Pareto distributions is a suitable randomization subordinator for two main reasons. First, it has a simple analytical form, leading to straightforward

M.M. Felgueiras (✉)
CIIC and ESTG, Polytechnic Institute of Leiria, Portugal and CEAUL Lisbon, Portugal
e-mail: mfelg@estg.ipleiria.pt

J. Lita da Silva et al. (eds.), *Advances in Regression, Survival Analysis, Extreme Values, Markov Processes and Other Statistical Applications*, Studies in Theoretical and Applied Statistics, DOI 10.1007/978-3-642-34904-1_29,

mixture density computation. Second, Pareto's fat tail implies that the resulting densities will have higher kurtosis, useful in heavy-tailed data modelling.

The mixture can be defined, following [6] notation, as

$$Y = \Theta X \tag{1}$$

where Θ and X are independent random variables with X absolutely continuous and $\Theta \sim Pareto\,(\alpha)$,

$$f_{\Theta}\,(\theta) = \alpha\theta^{-\alpha-1}, \quad \theta \geq 1, \alpha > 0. \tag{2}$$

The fact that we use Pareto with left-endpoint $\alpha_{\Theta} = 1$ is in a sense a severe restriction, since it implies that $\mathbb{P}[|Y| > |X|] = 1$. Pareto random variables $\widetilde{\Theta} = \Theta - 1$ with support $\theta \geq 0$ could also be considered, covering all positive values. However, explicit density functions and interesting mixture distributions were not found in that more general setting. On the other hand, as $\theta > 1$, the above mentioned expression has important consequences tied to stochastic ordering.

2 Mixture Densities and Other Properties

The probability density function of the mixture $Y = \Theta X$ can be written as

$$f_Y\,(y) = \int_1^{\infty} \alpha\theta^{-\alpha-2} f_X\left(\tfrac{y}{\theta}\right) d\theta, \tag{3}$$

originating for some usual X distributions the incomplete gamma-based densities (see [2]) presented in Table 1 (see next page) where

$$\gamma\,(a, y) = \int_0^y t^{a-1} e^{-t} dt. \tag{4}$$

Since the support of Θ is $S_{\Theta} = [1, \infty[$, multiplying X by Θ implies expansion of the X values. Clearly, the absolute values of the existing moments of such mixtures are always greater than the corresponding X moments. Further, Y stochastic dominates X, since

$$P\,(Y > t) > P\,(X > t) \Longleftrightarrow \overline{F}_Y\,(t) > \overline{F}_X\,(t)\,, \quad t > 0, \tag{5}$$

a potentially important fact in reliability modelling and in premium computing policies in actuarial applications [1].

When α increases,

$$\lim_{\alpha \to +\infty} \overline{F}_{\Theta_\alpha}\,(\theta) = \lim_{\alpha \to +\infty} \theta^{-\alpha} = \begin{cases} 0, & \theta > 1 \\ 1, & \theta = 1 \end{cases} \tag{6}$$

and Θ_α converges to the degenerate random variable at 1.

Table 1 Some Pareto scale mixture densities

Distribution	Density	Mixture density
$X \sim N(0,1)$	$f_X(x) = \frac{1}{\sqrt{2\pi}} e^{-\frac{x^2}{2}}$	$f_Y(y) = \frac{\alpha 2^{0.5\alpha-1} \gamma\left(\frac{\alpha+1}{2}, \frac{y^2}{2}\right)}{\sqrt{\pi}\,\lvert y\rvert^{\alpha+1}}, \quad y \neq 0$
$\frac{2^{-\frac{3+\beta}{2}} \exp\left[-0.5\,\lvert x\rvert^{\frac{2}{1+\beta}}\right]}{\Gamma\left(\frac{3+\beta}{2}\right)},$	$-1 < \beta \leq 1$[1]	$f_Y(y) = \frac{\alpha(1+\beta)\gamma\left(\frac{\beta+1}{2}(\alpha+1), 0.5\lvert y\rvert^{\frac{2}{1+\beta}}\right)}{2^{-\alpha\frac{\beta+1}{2}} 4\Gamma\left(\frac{3+\beta}{2}\right)\lvert y\rvert^{\alpha+1}}, \quad y \neq 0$
$X \sim Cauchy(0,1)$	$f_X(x) = \frac{1}{\pi}\frac{1}{1+x^2}$	$f_Y(y) = \frac{\alpha y^{-\alpha-1}}{\pi} \int_0^y \frac{z^\alpha}{1+z^2} dz, \quad y \neq 0$
$X \sim Gama(\beta,1)$	$f_X(x) = \frac{1}{\Gamma(\beta)} x^{\beta-1} e^{-x}$	$f_Y(y) = \frac{\alpha y^{-\alpha-1}}{\Gamma(\beta)} \gamma(\alpha+\beta, y), \quad y > 0$
$X \sim Beta(p,q)$	$f_X(x) = \frac{(1-x)^{q-1}}{x^{1-p} B(p,q)}$	$f_Y(y) = \begin{cases} \frac{\alpha B(p+\alpha, q, y)}{y^{\alpha+1} B(p,q)}, & 0 < y < 1 \\ \frac{\alpha B(p+\alpha, q)}{y^{\alpha+1} B(p,q)}, & y \geq 1 \end{cases}$
$X \sim Weibull(\beta,1)$	$f_X(x) = \beta x^{\beta-1} e^{-x^\beta}$	$f_Y(y) = \frac{\alpha \gamma\left(\alpha\beta^{-1}+1, y^\beta\right)}{y^{\alpha+1}}, \quad y > 0$
$X \sim Pareto(\beta)$	$f_X(x) = \beta x^{-\beta-1}$	$f_Y(y) = \begin{cases} \alpha^2 y^{-\alpha-1} \ln y, & \alpha = \beta, \ y > 0 \\ \frac{\alpha\beta\left(y^{-\alpha-1} - y^{-\beta-1}\right)}{\beta-\alpha}, & \alpha \neq \beta, \ y > 0 \end{cases}$

Convergence in distribution to a constant implies convergence in probability, and by convergence in probability properties, when $\alpha \to +\infty$, then

$$Y = \Theta_\alpha X \underset{\alpha\to\infty}{\overset{d}{\Longrightarrow}} X. \tag{7}$$

Thus, the mixture model can be near the original, for large values of α, or more far apart when α is small, leading to a wide range of solutions.

Denoting $E'\left(X^k\right)$ and $E\left(X^k\right)$ as the raw and the central k moments and $\beta_2(X)$ as the kurtosis of X random variable, we can now study the mixture kurtosis as an α function. Assuming that $E(X) = 0$ (otherwise perform a location transformation) and that $\beta_2(X)$ exist, simple calculation show that

[1]Denoted sometimes as the extended Gaussian–Laplace distribution. For $\beta = 0$ we have the Gaussian distribution and for $\beta = 1$ the Laplace distribution.

$$\beta_2(Y) = \frac{E(Y^4)}{E(Y^2)^2} = \frac{E'(Y^4)}{E'(Y^2)^2} = \frac{E'(\Theta^4)\,E'(X^4)}{\left(E'(\Theta^2)\,E'(X^2)\right)^2} = \beta_2(X)\frac{\frac{\alpha}{\alpha-4}}{\left(\frac{\alpha}{\alpha-2}\right)^2} =$$

$$= \beta_2(X)\frac{(\alpha-2)^2}{\alpha(\alpha-4)}, \quad \alpha > 4. \tag{8}$$

Thus, $\beta_2(Y) > \beta_2(X)$ and the mixture can be used to increase the tailweight of the original X distribution.

As for the mixture moments, they can only exist for $k < \alpha$, as we prove in the following theorem.

Theorem 1. *Let $Y = \Theta X$, where Θ and X are independent random variables, X is absolutely continuous and $\Theta \sim$ Pareto (α). Then for $k \geq \alpha$, $E(Y^k)$ does not exist.*

Proof. We prove the result for the border case $k = \alpha$, because it is a well-known result that if $E(Y^\alpha)$ does not exist, then $E(Y^k)$ does not exist, for $k \geq \alpha$. When $E(X^\alpha) = c \neq 0$, then if $E(Y^\alpha)$ exist,

$$E(Y^\alpha) = E(\Theta^\alpha)\,E(X^\alpha) = cE(\Theta^\alpha).$$

Since $E(\Theta^\alpha)$ does not exist for $\Theta \sim$ *Pareto* (α), then it is obvious that also $E(Y^\alpha)$ does not exist. For $E(X^\alpha) = 0$, note that

$$f_Y(y) = \int_1^{+\infty} \alpha\theta^{-\alpha-2} f_X\left(\frac{y}{\theta}\right) d\theta = \int_{-\infty}^{+\infty} \frac{f_X(x)}{|x|} f_\Theta\left(\frac{y}{x}\right) dx =$$

$$= \int_{-\infty}^{+\infty} \frac{f_X(x)}{|x|}\alpha\left(\frac{y}{x}\right)^{-\alpha-1} dx, \quad \frac{y}{x} > 1$$

leading to

$$f_Y(y) = \begin{cases} \frac{\alpha}{y}\int_0^y \left(\frac{x}{y}\right)^\alpha f_X(x)\,dx, & y > x > 0, y > 0, \\ -\frac{\alpha}{y}\int_y^0 \left(\frac{x}{y}\right)^\alpha f_X(x)\,dx, & y < x < 0, y < 0, \end{cases}$$

The expectation of Y^α exists if and only if

$$E(|Y^\alpha|) = \int_{-\infty}^{0} |y^\alpha|\left[-\frac{\alpha}{y}\int_y^0 \left(\frac{x}{y}\right)^\alpha f_X(x)\,dx\right] dy +$$

$$+ \int_0^{+\infty} |y^\alpha|\left[\frac{\alpha}{y}\int_0^y \left(\frac{x}{y}\right)^\alpha f_X(x)\,dx\right] dy$$

is convergent. In what concerns the second integral in the right-hand side of that expression,

$$\int_0^{+\infty} |y^\alpha| \left[\frac{\alpha}{y}\int_0^y \left(\frac{x}{y}\right)^\alpha f_X(x)\,dx\right] dy = \int_0^{+\infty} \frac{\alpha}{y}\left[\int_0^y x^\alpha f_X(x)\,dx\right] dy.$$

If $P(X > b) = 0$ for some $b > 0$ and using straightforward inequalities,

$$\int_0^{+\infty} \frac{\alpha}{y}\left[\int_0^y x^\alpha f_X(x)\,dx\right] dy = \int_0^{+\infty} \frac{\alpha}{y}\left[\int_0^{\min(b,y)} x^\alpha f_X(x)\,dx\right] dy >$$
$$> \int_b^{+\infty} \frac{\alpha}{y}\left[\int_0^b x^\alpha f_X(x)\,dx\right] dy \underset{C>0}{>} \int_b^{+\infty} \frac{\alpha}{y} C\,dy,$$

which is divergent and therefore the expectation of Y^α does not exist. If $P(X > b) \neq 0$ for all $b > 0$ and using straightforward inequalities,

$$\int_0^{+\infty} \frac{\alpha}{y}\left[\int_0^y x^\alpha f_X(x)\,dx\right] dy > \int_1^{+\infty} \frac{\alpha}{y}\left[\int_1^y x^\alpha f_X(x)\,dx\right] dy >$$
$$> \int_1^{+\infty} \frac{\alpha}{y}\left[\int_1^y f_X(x)\,dx\right] dy = \int_1^{+\infty} \frac{\alpha}{y}\left[F_X(y) - F_X(1)\right] dy.$$

As

$$\lim_{y\to+\infty} y \times \frac{\alpha}{y}\left[F_X(y) - F_X(1)\right] = \alpha\left[1 - F_X(1)\right] = C > 0,$$

we conclude that

$$\int_1^{+\infty} \frac{\alpha}{y}\left[F_X(y) - F_X(1)\right] dy$$

is divergent and hence the expectation of Y^α does not exist.

3 Mixture and Slash Distribution Extensions

The mixture can also be regarded as a random variable quotient,

$$Y = \Theta X = \frac{X}{\Theta^{-1}}, \tag{9}$$

where

$$f_{\Theta^{-1}}(\theta) = f_\Theta\left(\theta^{-1}\right)\theta^{-2} = \alpha\theta^{\alpha-1}, \quad 0 < \theta \leq 1, \alpha > 0,$$

and so

$$\Theta^{-1} \sim Beta(\alpha, 1). \tag{10}$$

When $\alpha = 1$, the expressions above simplify, and since $\Theta^{-1} \sim U(0,1)$ we obtain slash distribution family, often used in reliability and robustness studies [3, 5].

In this context, it is obvious that Pareto scale mixtures generalize the class of slash distributions and therefore share their wide range of applications, namely in situations where symmetrical distributions with fat tails are appropriated. For $0 < \alpha < 1$, Pareto scale mixtures have heavier tailweight than the slash distributions, and for $\alpha > 1$ we have the reverse situation.

As a side result, note that for $\alpha = 1$ Theorem 1 implies that slash distributions do not have mean value.

4 Examples

4.1 Pareto Mixtures of Normal Random Variables

Pareto mixtures of normals exhibit some of the important features of Pareto mixtures of a symmetrical population and are potentially the more widely useful. In fact, when $X \sim N(0,1)$, we obtain an infinitely divisible mixture [6] with density

$$f_Y(y) = \alpha 2^{0.5\alpha-1} |y|^{-\alpha-1} \pi^{-0.5} \gamma\left(\frac{\alpha+1}{2}, \frac{y^2}{2}\right), \quad y \neq 0, \tag{11}$$

For instance, for $\alpha = 1$,

$$f_Y(y) = \frac{1 - e^{-y^2/2}}{\sqrt{2\pi}\, y^2}, \quad y \neq 0, \tag{12}$$

and for $\alpha = 3$

$$f_Y(y) = \frac{3\left(2 - (2 + y^2) e^{-y^2/2}\right)}{\sqrt{2\pi}\, y^4}, \quad y \neq 0. \tag{13}$$

As previously stated, $\Theta_\alpha X \xrightarrow[\alpha \to \infty]{d} X$. This can be seen in Fig. 1.

Note that the α parameter works in a similar way as the n parameter in Student's t-distributions. However, in this situation, the Y distribution as heavier tails (for small values of α) and the rate of convergence towards the Gaussian limit are slower than in the t family. The mixture kurtosis can be calculated as (see Eq. 8)

$$\beta_2(Y) = 3\frac{(\alpha - 2)^2}{\alpha(\alpha - 4)}$$

and can assume very large values, as showed in Fig. 2.

Another symmetrical mixture with even heavier tails can be generated for $X \sim Cauchy(0,1)$ and $\alpha = 1$, originating the slash Cauchy density

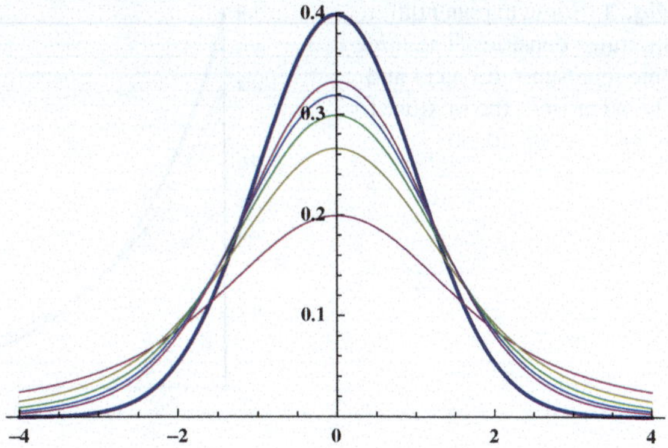

Fig. 1 Some Gaussian mixture densities. The thick line represents $N(0, 1)$ and the other lines the mixture for $\alpha = 1, \ldots, 5, 20, 30$

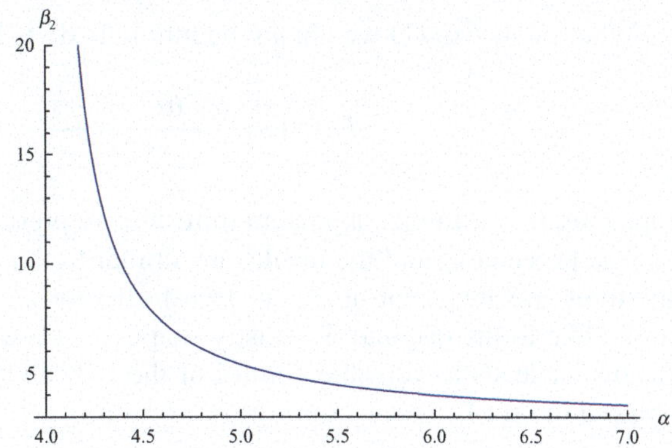

Fig. 2 Mixture kurtosis according with α

Table 2 Probability quantiles for the Cauchy, the slash Gaussian and the slash Cauchy

α	0.5	0.75	0.90	0.95	0.99	0.999
q_α Cauchy	0	1.00	3.08	6.31	31.82	318.31
q_α slash Gaussian	0	1.47	3.99	7.98	39.89	398.94
q_α slash Cauchy	0	2.45	10.75	27.46	200.57	2,850.55

$$f_Y(y) = \frac{\ln\left(y^2 + 1\right)}{2\pi y^2}, \quad y \neq 0. \tag{14}$$

In Table 2, we can observe that Cauchy and slash Gaussian quantiles are not far apart, but the slash Cauchy has very large quantiles and therefore can be useful in modelling very extreme situations.

4.2 Pareto Mixtures of Positive Random Variables

To exemplify Pareto mixtures of positive random variables we choose exponential parent, since it exhibits the more important features of mixtures of a positive support population and it is the more readily useful in applications.

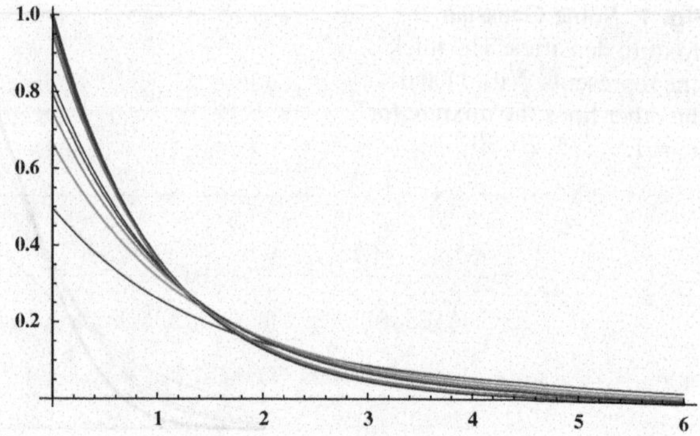

Fig. 3 Some exponential mixtures densities. The *thick line* represents $Exp\,(1)$ and the other lines the mixture for $\alpha = 1, \ldots, 5, 20, 30$

When $X \sim Exp\,(1)$ we obtain the infinitely divisible mixture [7] with density

$$f_Y\,(y) = \frac{\alpha\gamma\,(\alpha+1, y)}{y^{\alpha+1}}, \quad y > 0, \tag{15}$$

replacing $\beta = 1$ in the gamma mixture density presented in Table 1.

The procedures and the results are similar to that we presented for the gaussian mixtures. We are performing scale transformations, so mixture density shape always look alike to the original X density shape. As observed previously, varying the α parameter leads to versatile control of the tailweight of the resulting mixtures, as showed in Fig. 3.

Acknowledgements Research partially sponsored by national funds through the Fundação Nacional para a Ciência e Tecnologia, Portugal FCT under the project (PEst-OE/MAT/UI0006/2011).

References

1. Centeno, M.L., Andrade e Silva, J.: Bonus systems in an open portfolio. Insur. Math. Econ. **28**, 341–350 (2001)
2. Felgueiras, M.: Considerações sobre a distribuição Pareto. Actas do XV Congresso Anual da Sociedade Portuguesa de Estatística, 193–201 (2008)
3. Gómez, H., Quintana, F., Torres, F.: A new family of slash-distributions with elliptical contours. Stat. Probab. Lett. **77**, 717–725 (2007)
4. Johnson, N., Kotz, S., Kemp, A.: Univariate Discrete Distributions. Wiley, New York (1992)
5. Johnson, N., Kotz, S., Balakrishnan, N.: Continuous Univariate Distributions, vol I. Wiley, New York (1994)
6. Kelker, D.: Infinite divisibility and variance mixtures of the normal distribution. Ann. Math. Stat. **42**, 802–808 (1971)
7. Steutel, F.: Preservation of Infinite Divisibility under Mixing and Related Topics. Mathematisch Centrum, Amsterdam (1970)

Fitting Johnson's S_B Distribution to Forest Tree Diameter

Ayana Mateus and Margarida Tomé

Abstract
The simulation of diameter distributions is an essential aid for a more efficient planning of the harvesting operations which usually represents a high percentage in costs associated with production of pulp. In this chapter Johnson's S_B probability density function has been used to model diameter distribution of *Eucalyptus globulus* Labill. in Portugal.

1 Introduction

Eucalyptus globulus Labill. is one of the most important economic forest species in Portugal, occupying an area of 875,000ha of a total forest area of 3,346,000ha. It is a fast-growing species that is mainly used commercially by the pulp industry.

The objective of the research report here is to model the diameter distribution of eucalyptus plantations in Portugal.

To achieve this objective the following partial objectives were needed: to identify the probability density function (*pdf*) that better reproduced the set of observed frequencies based on the estimates obtained for coefficients of skewness (β_1) and kurtosis (β_2) in each plot at each measurement age; to develop a system of equations that relates stand basal area (G) with the noncentral moments of the distribution in

A. Mateus (✉)
CMA, Departamento de Matemática, Faculdade de Ciências e Tecnologia, Universidade Nova de Lisboa, 2829-516 Caparica, Portugal
e-mail: amf@fct.unl.pt

M. Tomé
CEF, Centro de Estudos florestais, ISA-Instituto Superior de Agronomia, UTL-Universidade Técnica de Lisboa, Tapada da Ajuda, 1349-017 Lisboa, Portugal
e-mail: magatome@isa.utl.pt

J. Lita da Silva et al. (eds.), *Advances in Regression, Survival Analysis, Extreme Values, Markov Processes and Other Statistical Applications*, Studies in Theoretical and Applied Statistics, DOI 10.1007/978-3-642-34904-1_30,

Table 1 Characteristics of tree variable used

Tree variable	Minimum	Mean	Maximum
Diameter at breast height (d, cm)	0.1	11.38	45.40

Table 2 Characteristics of stand variables used

Stand variable	Minimum	Mean	Maximum
Basal area (G, m^2ha^{-1})	0.05	16.16	64.55
Number of trees per hectare (N, ha^{-1})	450	1237.34	2811
Productivity (S, m)	10.33	20.20	33.93
Age (t, years)	0.6	9.36	34.70

order to obtain estimates of the *pdf*. The stand basal area is the sum of squared diameter multiplied by a factor to express it on area per hectare, and it is related to the second noncentral moment of the distribution. This variable expresses the competition between trees as it is the area occupied with tree stems. It has a great importance because in the growth of trees, the competition reflects itself mainly by growth in diameter. The algorithm proposed by Parresol [12] was selected as a starting point for the parameter recovery.

2 Methods

2.1 Data

The data used in this study to model diameter distributions of eucalyptus (*Eucalyptus globulus* Labill.) plantations were collected in Portugal in permanent plots installed in first rotation stands.

The information concerning all the trees within a plot includes successive measurements, usually annually, of diameter at breast height (d).

The plots used in the present research have drawn on a very large data set covering stands with different characteristics, namely age (t), stocking (N, number of trees per hectare), and productivity (S) (see Tables 1 and 2).

2.2 Testing the Performance of Johnson S_B Distribution

The analysis of the coefficients of skewness (β_1) and kurtosis (β_2) of the distribution of the diameters could be used to indicate the appropriate pattern followed by a certain population.

For a first identification of the distribution that better reproduced the set of observed frequencies, the estimates of the coefficients of skewness (β_1) and kurtosis (β_2), in each plot at each measurement age, were first analyzed.

The estimators used were, respectively,

$$b_1 = \frac{m_3}{m_2^{3/2}} \quad \textit{and} \quad b_2 = \frac{m_4}{m_2^2}$$

with

$$m_2 = \frac{\sum_{i=1}^{n}(d_i - \overline{d})^2}{n-1} \qquad m_j = \frac{\sum_{i=1}^{n}(d_i - \overline{d})^j}{n} \qquad j = 3, 4.$$

d_i is the diameter at breast height of tree i, $\overline{d}$ is the average diameter of the plot, and n is the number of trees measured in the plot.

The choice of the Johnson S_B distribution as the null hypothesis for the modeling of diameter distributions of eucalyptus has been based on its flexibility to model distributions with different shapes. It has a broader range of the (β_1, β_2) space than other distributions and includes most of the alternative *pdf* [5,6].

Since Hafley and Schreuder [4] introduced the four parameter Johnson's S_B distribution into forest literature, this probability density function has been widely used in forest diameter (and height) distribution modeling by several authors, such as [1,3,7,9,12,13,15,16].

To test the performance of Johnson S_B distribution to model diameter distributions of eucalyptus plantations in Portugal, the b_1 and b_2 estimates were computed in each plot at each measurement age on the fitting data set in order to check if the pairs (b_1, b_2) occur mainly in the parametric space that corresponds to this distribution [12].

In order to complement the methodology used, based on the analysis of coefficients (β_1, β_2) for a first identification of the distribution to be used, the goodness-of-fit Kolmogorov–Smirnov test was also used in order to test the hypothesis that the Johnson S_B distribution fits the diameter distributions on individual plots [11,14]. We used the modified Kolmogorov–Smirnov test because the parameters were unknown and estimated from the data [10]. The test of the qui-square was not used, for being dependent of the grouping of data in classes.

2.3 The Johnson System of Probability Density Functions

The Johnson system corresponds to the distribution of a random variable X, in which a particular transformation is applied, in order to obtain a normal distribution to the random variable processed. This system is composed by three kinds of distributions (Johnson S_L, S_B, and S_U), depending on the transformation applied to the random variable [5].

When the transformation $Z = \gamma + \delta g(X)$ is made on the random variable X, an infinite system of distribution functions (or random variables) is being defined, clearly identified by the transformation $g(X)$, necessary to obtain a transformed with standard normal distribution.

Johnson introduced four parameters γ, δ, ϵ, and λ, with $\gamma, \epsilon \in \mathbb{R}$, $\lambda \in \mathbb{R}^+$, $\delta \in \mathbb{R}\backslash\{0\}$ and expressed the generic transformation defined above in the following way:

$$Z = \gamma + \delta g\left(\frac{X - \epsilon}{\lambda}\right), \tag{1}$$

where γ and δ are shape parameters and ϵ and λ are location and scale parameters, respectively. Although the parameters γ and δ affecting both the skewness and the kurtosis of distribution, the parameter γ is particularly associated with the asymmetry and an increase in the parameter δ corresponds to an increase in the kurtosis [5].

In order to generate distributions with limited support, the transformed chosen is

$$g(Y) = \ln\left(\frac{Y}{1 - Y}\right) \tag{2}$$

that in terms of the variable $Y = \dfrac{X - \epsilon}{\lambda}$ results in

$$Z = \gamma + \delta \ln\left(\frac{X - \epsilon}{\epsilon + \lambda - X}\right), \quad \epsilon < X < \epsilon + \lambda, -\infty < \gamma < \infty, \delta > 0, -\infty < \epsilon < \infty, \lambda > 0 \tag{3}$$

or

$$Z = \gamma + \delta \ln\left(\frac{Y}{1 - Y}\right), 0 < y < 1, -\infty < \gamma < \infty, \delta > 0, -\infty < \epsilon < \infty, \lambda > 0. \tag{4}$$

The system of random variables generated by Eq. (3) or (4) is called the Johnson S_B system of distributions.

2.4 Algorithm to Estimate the Parameters of the Johnson S_B Distribution

The parameters of the Johnson S_B distribution were estimated using the methodology proposed by Parresol [12].

If Eq. (4) is expressed in terms of the variable Y, the following expression is obtained for Y:

$$Y = \left[1 + \exp\left(-\frac{Z - \gamma}{\delta}\right)\right]^{-1}. \tag{5}$$

When the variable Z assumes the null value the median of the variable Y (or X) is obtained:

$$y_{1/2} = \left(1 + e^{\gamma/\delta}\right)^{-1}. \tag{6}$$

Note that the median of Y and X are related, since $y_{1/2} = \frac{x_{1/2} - \epsilon}{\lambda}$.

Equation (6) enables the estimation of the shape parameter γ, according to the median value of the diameter distribution, provided that the shape parameter δ is known:

$$\gamma = \delta \ln \left(\frac{1}{y_{1/2}} - 1 \right) = \delta \ln \left(\frac{\lambda}{x_{1/2} - \epsilon} - 1 \right). \tag{7}$$

However, another equation is needed to estimate the shape parameter δ. As we said in Sect. 1, a variable of great interest in the elaboration of stand models with diameter distribution simulation is the stand basal area (G). This variable is related to the second noncentral moment, $E(X^2)$ of X, through the relation,

$$G = \frac{1}{10000} \sum_{i=1}^{N} \frac{\pi}{4} d_i^2 = c\ N\ E\left(X^2\right) \qquad (m^2 ha^{-1}) \tag{8}$$

with $N =$ number of trees alive, per hectare, $d_i =$ diameter at breast height (cm) measured of tree i, and $c = \frac{\pi}{40000}$ is a conversion constant.

As

$$E\left(X^2\right) = E\left(\epsilon + \lambda\ Y\right)^2 = \epsilon^2 + 2\epsilon\lambda E\left(Y\right) + \lambda^2 E\left(Y^2\right),$$

then

$$G = c\ N\left(\epsilon^2 + 2\epsilon\lambda E\left(Y\right) + \lambda^2 E\left(Y^2\right)\right). \tag{9}$$

The noncentral moments of order r ($E(Y^r)$) $r = 1, 2$ may be determined through the moment-generating function φ of the variable Y

$$\varphi_Y\left(t\right) = \int_{-\infty}^{+\infty} \frac{1}{\sqrt{2\pi}} \exp\left(\frac{t}{1 + e^{-\frac{z-\gamma}{\delta}}}\right) e^{-z^2/2} dz$$

which shows the following relationship:

$$E\left(Y^r\right) = \frac{d^r}{dt^r} \varphi_Y\left(t\right)|_{t=0} = \frac{1}{\sqrt{2\pi}} \int_{-\infty}^{+\infty} \left(1 + e^{-\frac{z-\gamma}{\delta}}\right)^{-r} e^{-z^2/2} dz.$$

The resolution of the system formed by Eqs. (7) and (9), based on known values of the median variable Y(or X), G, and N, allows, by assuming some reasonable values for ϵ and λ, to obtain estimates for the parameters γ and δ. The solution requires the use of numerical iterative methods of numerical integration, as the calculation of moments of the distribution does not have an analytical solution [2,8].

As in any iterative process it is necessary to assign initial values to the parameters. Parresol [12] suggests the attribution of an initial value to δ for a first approach of γ obtained from Eq. (7). Thus the parameter δ was initialized with the estimate obtained for the kurtosis because an increase of δ corresponds to an increase in the kurtosis [5]. The parameter ϵ was fixed as equal to the minimum value of the observed diameter and λ to the difference between the maximum and minimum value of the observed diameter. The values for G and N were obtained from the measurement of each plot in study.

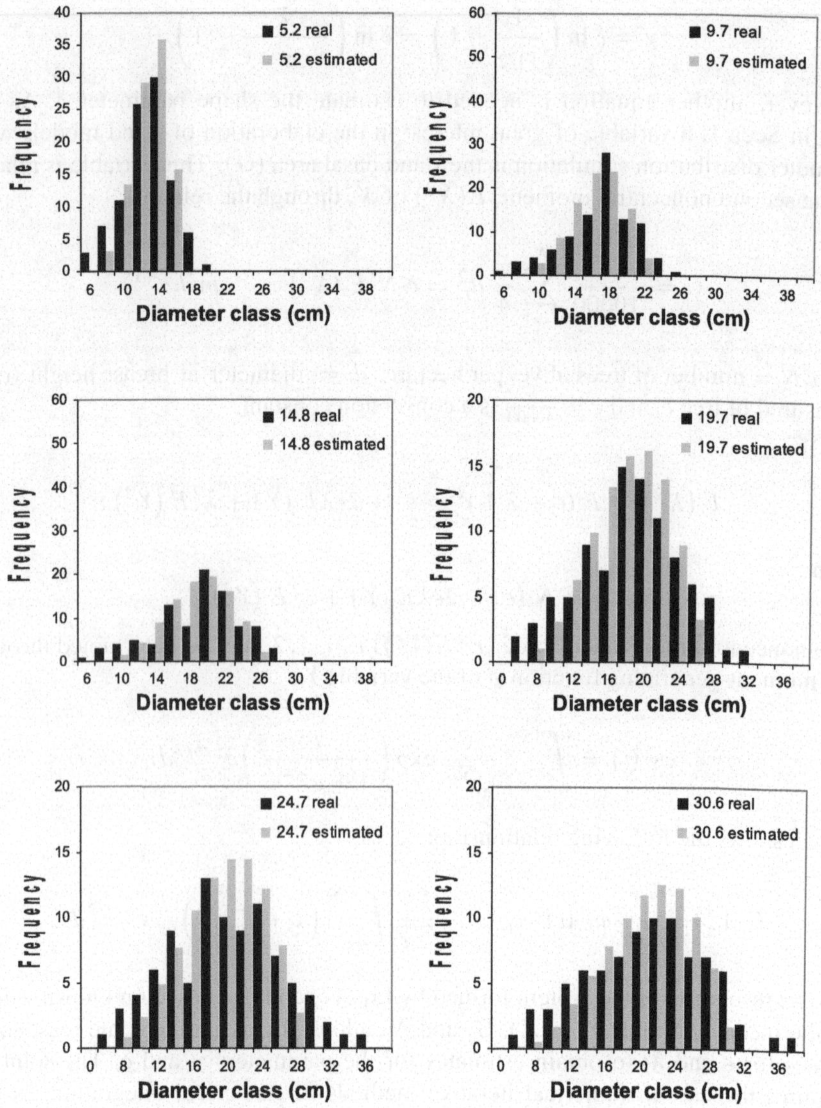

Fig. 1 Comparison of real and estimated diameter distribution from a plot at ages (years) 5.2, 9.7, 14.8, 19.7, 24.7, and 30.6 (*dark* = real values)

3 Results and Conclusions

The ranges for the coefficients (b_1, b_2) estimated with the data set described in Tables 1 and 2 were $-1.3977 \leq b_1 \leq 1.0805$ and $1.8112 \leq b_2 \leq 6.8685$, which indicates the existence of a huge variety of empirical diameter distributions for eucalyptus plantations. This supports the choice of a very flexible distribution. The values observed for the pairs (b_1, b_2) are included in the range of variation for the coefficients of skewness and kurtosis of the Johnson S_B distribution [5].

It was also verified that in the great majority of the plots, the coefficients of skewness assume negative values. In the growth of trees, competition between trees affects growth in tree diameter; this fact explains the negative values for the coefficients of skewness. In other words, the trees that had a higher initial growth in diameter (d) will compete, mainly for light, with the smaller ones making those to continue to have lower growth rates, and the differences between small and large trees tended to increase.

The modified Kolmogorov–Smirnov test with a significance level of 5% showed that the distribution Johnson S_B did not significantly differ from the empirical distribution in 106 out of 111 studied stands, each of them with several remeasurements between 5 and 32 years.

In conclusion, modeling diameter distributions of eucalyptus (*Eucalyptus globulus* Labill.) plantations in Portugal through a probability density function, namely Johnson S_B, using a parameter recovery approach seems to be a good methodology that can be generally applied to the most common values of the pair (b_1, b_2). The main advantage of using parameter recovery models is that the stand variable that was used in the parameter recovery, namely basal area, assures compatibility between the characteristics of the observed population and those obtained through simulation of diameter distribution. This means that basal area computed with the simulated distribution is fairly closed to the one observed.

As an example Fig. 1 shows the evolution of the observed and simulated diameter distribution from 5.2 to 30.6 years of age in one of the permanent plot from the fitting data set when the initialization was made with the measurement at 5.2 years of age. As can be seen the agreement is very good, even for ages far away from the initial one. The results in other long-term series plots were similar.

Acknowledgements This work was supported by the FP6 EFORWOOD IP project (contract 518128) and the FCT project CarbWoodCork (POCI/AGR/57279/2004). This work was partially supported by CMA/FCT/UNL, under the project PEst-OE/MAT/UI0297/2011. The authors gratefully acknowledge Bernard Parresol (USDA Forest Service Southern Research Station) for facilitating the use of a SAS program for Johnson S_B parameter recovery that was used in an initial stage of this research. The authors also thank Celbi and Silvicaima pulp companies that provided a large proportion of the data used in this study.

References

1. Fonseca, T.F., Marques, C.P., Parresol, B.R.: Describing maritime pine diameter distributions with Johnson's S_B distribution using a new all-parameter recovery approach. For. Sci. **55**(4), 367–373 (2009)
2. Gallant, A.: Nonlinear Statistical Models, p. 624. Wiley, New York (1987)
3. Hafley, W.L., Buford, M.A.: A bivariate model for growth and yield prediction. For. Sci. **31**, 237–247 (1985)

4. Hafley, W.L., Schreuder, H.T.: Statistical distributions for fitting diameter and height data in even-ages stands. Can. J. For. Res. **7**, 481–487 (1977)
5. Johnson, N.L.: Systems of frequency curves generated by methods of translation. Biometrika **36**, 147–176 (1949)
6. Johnson, N., Kotz, S.: Continuous Univariate Distribution, vol. 1, p. 756, 2nd edn. Wiley, New York (1994)
7. Kamziah, A.K., Ahmad, M.I., Lapongan, J.: Nonlinear regression approach to estimating Johnson SB parameters for diameter data. Can. J. For. Res. **29**, 310–314 (1999)
8. Law, A.M., Kelton, W.D.: Simulation Modeling and Analysis. McGraw-Hill, New York (1982)
9. Li, F., Zhang, L., Davis, C.J.: Modelling the joint distribution of tree diameters and heights by bivariate generalized beta distribution. For. Sci. **48**(1), 47–58 (2002)
10. Lilliefors, H.W.: On the Kolmogorov-Smirnov test for normality with mean and variance unknown. American Stat. Assoc. J. **62**, 399–402 (1967)
11. Massey, F.J.: The Kolmogorov-Smirnov test for goodness of fit. American Stat. Assoc. J. **46**, 68–78 (1951)
12. Parresol, B.: Recovering parameters of johnson's S_B distribution. Res. pap. SRS-31. Asheville, NC:USDA For. Ser., Southern Research Station, p. 9 (2003)
13. Rennolls, K., Wang, M.: A new parameterization of Johnson's S_B distribution with application to fitting forest tree diameter data. Can. J. For. Res. **35**, 575–579 (2005)
14. Reynolds, M.R., Burk, T.E., Huang, W.: Goodness-of-fit tests and model selection procedures for diameter distributions models. For. Sci. **34**, 377–399 (1988)
15. Zhang, L., Packard, K.C., Liu, C.: A comparison of estimation methods for fitting Weibull and Johnsons SB distribution to mixed spruce-fir stands in northeastern North America. Can. J. For. Res. **33**(7), 1340–1347 (2003)
16. Zhou, B., McTague, J.P.: Comparison and evaluation of five methods of estimation of the Johnson system parameters. Can. J. For. Res. **26**, 928–935 (1996)

On a Continuous-Time Stock Price Model with Two Mean Reverting Regimes

Pedro P. Mota

Abstract

Motivated by the need to describe regime switching in stock prices, we introduce and study a stochastic process in continuous time with two regimes and one threshold driving the change in regimes. When the difference between the regimes is simply given by different sets of real-valued parameters for the drift and diffusion coefficients, we show that there are consistent estimators for the threshold as long as we know how to classify a given observation of the process as belonging to one of the two regimes.

1 Introduction

It seems reasonable to suggest that when stock prices are below a certain threshold they could have a mean reverting dynamics being attracted to a mean value, smaller than the threshold, at a certain velocity and with a certain volatility. In the same way when the stock prices are above the threshold the mean reverting dynamics could be different.

The study of some nonlinear time series models has received renewed attention, namely the threshold models (see [4, 5, 12] or [13]). One of our goals is to study an extension of threshold processes to continuous time and to obtain estimation methods for the parameters of this kind of processes. In [1–3, 6, 7] or [10] some results for threshold continuous-time processes are given when discrete time observations are available.

In this chapter we consider stochastic processes where, in each regime, the process follows the dynamics of a simple continuous-time process and the change

P.P. Mota (✉)
Department of Mathematics and CMA of FCT/UNL, Caparica, Portugal
e-mail: pjpm@fct.unl.pt

J. Lita da Silva et al. (eds.), *Advances in Regression, Survival Analysis, Extreme Values, Markov Processes and Other Statistical Applications*, Studies in Theoretical and Applied Statistics, DOI 10.1007/978-3-642-34904-1_31,

in regime will happen when the process crosses a threshold. We note that the regime switching here proposed is not driven by some external random source and our goal is to define and study some main aspects of the model, namely, parameter estimation and usefulness for practical applications. To build this regime switching process we will define a sequence of stopping times corresponding to the times of the regime changes. The problem that arises is how to ensure the existence of this sequence when we know that by a change of measure, using the Girsanov theorem, the process can be transformed into Brownian motion. Then the problem of defining the threshold hitting times is the same of studying the level sets of the Brownian process, those level sets are perfect sets, that is, closed sets consisting only of limit points of the set itself. One way to overcome this problem is defining the process, in Sect. 2 below, using an auxiliary threshold band $]m-\gamma, m[$ for a fixed (and small) γ, instead of a single threshold m. In Sect. 3 we define consistent estimators for the limits of the threshold band, but the same kind of estimators could be defined, for the single threshold m, if the existence of the process could be ensured with that single threshold.

2 On the Existence of a Solution for the Model

In a context of practical applications, like pricing contingent claims, there is no generality loss if we suppose that $t \in [0,T]$ with T as deterministic finite time horizon. In order to define our process $(X_t)_{t\in[0,T]}$, we will suppose that for $t=0$ a parameter choice, for instance, $\boldsymbol{\theta}_1 = (\boldsymbol{\mu}_1, \sigma_1)$ will give diffusion coefficients $\mu(t, X_t, \boldsymbol{\theta}_1) = \mu(t, X_t, \boldsymbol{\mu}_1)$ and $\sigma(t, X_t, \boldsymbol{\theta}_1) = \sigma(t, X_t, \sigma_1)$ such that the stochastic differential equation

$$dX_t^{\boldsymbol{\theta}_1} = \mu(t, X_t^{\boldsymbol{\theta}_1}, \boldsymbol{\mu}_1)dt + \sigma(t, X_t^{\boldsymbol{\theta}_1}, \sigma_1)dB_t, \quad X_0^{\boldsymbol{\theta}_1} = x_0 \tag{1}$$

has a unique continuous solution $(X_t^{\boldsymbol{\theta}_1, x_0})_{t\in[0,T]}$ for the chosen time horizon $T>0$ (see [8, p. 289] or [9, p. 73], for instance). For simplicity suppose that $x_0 < m$ and let τ_1 be the first stopping time at which the solution process hits the upper limit, m, from below, of the threshold band, that is:

$$\tau_1 := \inf\left\{0 < t < T : X_t^{\boldsymbol{\theta}_1, x_0} = m\right\}. \tag{2}$$

By definition, for $0 \le t \le \tau_1 \wedge T$, we will have that $X_t = X_t^{\boldsymbol{\theta}_1, x_0}$. Now, a regime switch having occurred at time τ_1, consider (if $\tau_1 < T$) the stochastic differential equation given by

$$dX_t^{\boldsymbol{\theta}_2} = \mu(t, X_t^{\boldsymbol{\theta}_2}, \boldsymbol{\mu}_2)dt + \sigma(t, X_t^{\boldsymbol{\theta}_2}, \sigma_2)dB_t, \; \tau_1 \le t \le T, \quad X_{\tau_1}^{\boldsymbol{\theta}_2} = m. \tag{3}$$

Note that this equation should be read, for all $t \in [0, T]$, as

$$X_t^{\theta_2} = m\mathbb{I}_{[\tau_1 \wedge T, T]}(t) + \int_0^t \mu(u, X_u^{\theta_2, x_0}, \mu_2)\mathbb{I}_{[\tau_1 \wedge T, T]}(u)\, du + \\ + \int_0^t \sigma(u, X_u^{\theta_2, x_0}, \sigma_2)\mathbb{I}_{[\tau_1 \wedge T, T]}(u) dB_u \tag{4}$$

and then it is clear that if the diffusion coefficients of the equation

$$dX_t^{\theta_2} = \mu(t, X_t^{\theta_2}, \mu_2)dt + \sigma(t, X_t^{\theta_2}, \sigma_2)dB_t, \ s \le t \le T, \quad X_s^{\theta_2} = m, \tag{5}$$

are such that a unique continuous solution exists for all $s \in [0, T]$, then a continuous unique solution $(X_t^{\theta_2, m)})_{t \in [\tau_1 \wedge T, T]}$ exists also for Eq. (4). In fact, consider a standard theorem, for instance, the one in [8, p. 289] or [11, p. 66]. The sufficient condition for the initial value is verified and it is clear that the integrability, Lipschitz, and sublinear growth conditions verified by the drift and diffusion coefficients of Eq. (5) are still verified by the drift and diffusion coefficients of Eq. (4).

Let now τ_2 be the first stopping time, following τ_1, at which the process hits the lower limit $m - \gamma$, from above, of the threshold band, that is,

$$\tau_2 := \inf\left\{\tau_1 < t < T : X_t^{\theta_2, m} = m - \gamma\right\}. \tag{6}$$

By definition, for $\tau_1 \le t \le \tau_2 \wedge T$ we will have that $X_t = X_t^{\theta_2, m}$. The process may be defined inductively in this way by concatenating together solutions to standard stochastic differential equations defined between stopping times.

3 Consistent Estimators

Under certain assumptions it is possible to define consistent estimators for the limits, m and $m - \gamma$, of the auxiliary threshold band. For each integer n, let $C_n = \{X_{t_1}, X_{t_2}, \dots, X_{t_n}\}$ be the observations of the process at times $t_1, t_2, \dots, t_n$, not necessarily equally spaced, with $\Delta_n^i = t_{i+1} - t_i, i = 1, \dots, n-1$ being such that

$$\lim_{n \to +\infty} \max_{1 \le i \le n} \Delta_n^i = 0.$$

We admit that in the observation protocol, from one step to the next, we keep the observations from the previous step; this implying that for each n, $C_n \subset C_{n+1}$.

We suppose that we observe the random variables $R_1, R_2, \dots, R_n$ where $R_i = 1$ if X_{t_i} belongs to regime 1 or $R_i = 2$ if X_{t_i} belongs to regime 2.

With these hypothesis we can define consistent estimators for m and $m - \gamma$. In fact, knowing the sequence $R_1, R_2, \dots, R_n$, we can split the observations into two sets using the fact that $R_j = 1 \Rightarrow X_{t_j} \le m$ and $R_j = 2 \Rightarrow X_{t_j} \ge m - \gamma$. For that, define the sets

$$C_n^- = \{X_i : R_i = 1, i = 1, \dots, n\} \tag{7}$$

and

$$C_n^+ = \{X_i : R_i = 2, i = d, \ldots, n\}. \tag{8}$$

Finally, we can estimate the threshold band limits in a consistent way.

Theorem 1. *If the process $(X_t)_{t\geq 0}$ has continuous trajectories and if there is at least a change from the first to the second regime, then*

$$\text{with } \hat{m}_n^- = \max C_n^- \text{ we have } \lim_{n\to+\infty} \hat{m}_n^- = m \text{ a.s.} \tag{9}$$

that is, $\hat{m}_n^-$ is a consistent estimator of the threshold m, and if there is at least a change from the second to the first regime, then

$$\text{with } \hat{m}_n^+ = \min C_n^+ \text{ we have } \lim_{n\to+\infty} \hat{m}_n^+ = m - \gamma \text{ a.s.} \tag{10}$$

that is, $\hat{m}_n^+$ is a consistent estimator of $m - \gamma$.

Proof. We will only prove that $\hat{m}_n^- = \max C_n^-$ is a strongly consistent estimator for m, the proof being similar for $\hat{m}_n^+$. Note that as $C_n^- \subset C_{n+1}^-$ we have that $\hat{m}_n^- \leq \hat{m}_{n+1}^-$ and for each n, by definition, $\hat{m}_n^- \leq m$. This implies that $\lim_{n\to\infty} \hat{m}_n^-$ exists and that $\limsup_{n\to+\infty} \hat{m}_n^- \leq m$. Suppose that

$$\lim_{n\to+\infty} \hat{m}_n^- = \lim_{n\to+\infty} \left(\max C_n^-\right) = \limsup_{n\to+\infty} \left(\max C_n^-\right) < m\,. \tag{11}$$

Then, for some $\varepsilon > 0$, there exists $p \geq 1$ such that for all $n \geq p$:

$$\forall i, X_i \in C_n^- \Rightarrow X_i < m - \varepsilon. \tag{12}$$

Let τ be the first random time at which the process has a change in regime from the first to the second. Recall that this implies that $X_\tau = m$ and because the process has continuous trajectories we can choose $\delta = \delta(\omega)$ such that for all t verifying $|\tau - t| < \delta$ we have $|X_t(\omega) - X_\tau(\omega)| < \varepsilon$ and choose $q = q(\varepsilon, \omega)$ such that for $n \geq q$ we have that $\max_{1\leq i\leq n} \Delta_n^i < \delta$. Then, for $n \geq \max(q, p)$ and the observation time $t_i = t_i(\omega) \in \{1, \ldots, n\}$ such that $\tau \in [t_i, t_{i+1}]$ (notice that, $t_i < \tau \Rightarrow X_{t_i} \in C_n^-$), we have that

$$\tau - t_i \leq t_{i+1} - t_i = \Delta_n^i < \delta \tag{13}$$

and so

$$|\, X_{t_i}(\omega) - X_\tau(\omega) \,| = |\, X_{t_i}(\omega) - m \,| < \varepsilon\,, \tag{14}$$

a contradiction with Eq. (12).

Notice that we can use the usual estimators to estimate the other model parameters as if the process has no regimes.

4 Conditional Least Squares Estimators

For practical purposes we do not need the auxiliary threshold band. For instance, when working with stock prices, the smallest unit used is, usually, one cent, so, if in the definition of the threshold band we consider γ smaller than one cent, the band will be indistinguishable from a singular threshold m. In this section our purpose is to implement a practical way of estimating this threshold and the other parameters of the process when we do not know the regime for each observation. The procedure is implemented as follows:

1. For fixed threshold, m, we split the observations in two sets, corresponding to the two regimes. Each observation, X_i, is considered in the first regime if $X_i \leq m$, otherwise is considered as belonging to the second regime, $\forall i = 1, \ldots, n$.
2. Next, we can compute the conditional estimators for the diffusion parameters, namely the drift $\mu(t, X_t, \boldsymbol{\theta})$ and the volatility $\sigma(t, X_t, \boldsymbol{\theta})$, using the observations in each of the regimes and the usual estimators for the parameters in the single regime process context.
3. We define the conditional least squares function,

$$CLS_n(m) = \sum_{j=1}^{n-1} \left(X_{j+1} - \mathbb{E}_{\boldsymbol{\theta},m} \left[X_{j+1} | X_j, \ldots, X_1, \widehat{R}_j, \ldots, \widehat{R}_1 \right] \right)^2 \quad (15)$$

4. Finally we choose as threshold estimate the value that minimizes $CLS_n(m)$, that is,

$$\widehat{m}_n = \operatorname{argmin}_m CLS_n(m). \quad (16)$$

Remark 1. The conditional expectation in Eq. (15) is not explicitly known and we will approximate it by the conditional expectation of the underlying process, in each regime.

4.1 Simulation Study

We are now in conditions to illustrate this procedure with a simulation study for a particular type of process. We define the geometric Ornstein–Uhlenbeck process with regimes (GOUR)

$$dX_t = \mu(t, X_t, \boldsymbol{\theta})dt + \sigma(t, X_t, \boldsymbol{\theta})dB_t, \quad X_0 = x_0, \quad (17)$$

Table 1 Estimates for the GOUR process with $\alpha_1 = 0.05, \beta_1 = 20, \alpha_2 = 0.05, \beta_2 = 35$, and $m = 30$ and for different values of σ

		m	α_1	β_1	σ_1	α_2	β_2	σ_2
$\sigma_1 = 0.04$	Mean	29.157	0.053	20.035	0.041	0.055	35.065	0.021
$\sigma_2 = 0.02$	sd	1.034	0.007	0.280	0.001	0.010	0.357	0.001
$\sigma_1 = 0.08$	Mean	29.600	0.068	19.215	0.082	0.062	35.400	0.031
$\sigma_2 = 0.03$	sd	0.694	0.008	0.514	0.001	0.009	0.394	0.001

where $\boldsymbol{\theta} \in \Theta = \{(\alpha_1, \beta_1, \sigma_1), (\alpha_2, \beta_2, \sigma_2)\} \subset \mathbf{R}^3$ and

$$\mu(t, X_t, \boldsymbol{\theta}) = \alpha_i(\beta_i - \ln(X_t))X_t, \quad \sigma(t, X_t, \boldsymbol{\theta}) = \sigma_i X_t, i = 1, 2, \sigma_i > 0, \quad (18)$$

if at time t the process is in regime $i, i = 1, 2$.

The simulations started with 250 trajectories with 5,000 observations in each one, considering the process parameters, $\alpha_1 = 0.05, \beta_1 = 20, \alpha_2 = 0.05, \beta_2 = 35$, and $m = 30$ for different values of σ (reasonable values as we will see in the next section). For the estimating procedure we introduce the auxiliary conditional least squares contrast function,

$$CLS_n(m) = \sum_{k=1}^{2}\sum_{i\geq 0}^{n-1}\left(X_{i+1} - \exp\left[\ln(X_i)e^{-\widehat{\alpha}_k\Delta} + \left(\widehat{\beta}_k - \frac{\widehat{\sigma}_k^2}{2\widehat{\alpha}_k}\right)\left(1 - e^{-\widehat{\alpha}_k\Delta}\right)\right.\right.$$
$$\left.\left.+\frac{\widehat{\sigma}_k}{4\widehat{\alpha}_k}\left(1 - e^{-2\widehat{\alpha}_k\Delta}\right)\right]\right)^2 1_{\widehat{R}_i(m)=k},$$

where the exponential term corresponds to the conditional expectation of the geometric Ornstein–Uhlenbeck process and where the estimators $\widehat{\alpha}_k$, $\widehat{\beta}_k$, and $\widehat{\sigma}_k$ for $k = 1, 2$ are the usual ones for the considered process and will be computed with the observations in each one of the regimes.

We perform a grid search for m in $[10, 45]$ with grid step of 0.1. The results for the different values of σ under consideration are presented in Table 1 and, as we can see, the results suggest that the procedure works well, getting good approximations for the original values and, as expected, the standard deviation for the estimators gets larger when σ increases.

5 Application with Real-World Data

In this section we apply the procedure to real-world data gathered from Yahoo Finance. The data presented consists of stock daily prices of 15 companies from the Dow Jones Industrial Average index. We have applied the estimation procedure to the complete set of 30 companies from the considered index, but, for 15 of them, the minimum value for the contrast function is obtained when we consider the process with only one regime. In Table 2 we present the results of the

Table 2 Estimated parameters for various stocks; data range from January 2005 to March 2011

Stock	$\widehat{m}$	$\widehat{\alpha}_1$	$\widehat{\beta}_1$	$\widehat{\sigma}_1$	$\widehat{\alpha}_2$	$\widehat{\beta}_2$	$\widehat{\sigma}_2$
Alcoa	29.1	0.0036	17.14	0.0342	0.0494	34.21	0.0243
Cisco	19.0	0.0901	17.35	0.0244	0.0103	24.65	0.0179
City Group	36.4	0.0052	5.85	0.0604	0.0078	43.04	0.0106
Coca-Cola	43.8	0.0218	38.93	0.0143	0.0071	56.45	0.0110
General Electric	26.3	0.0107	15.58	0.0310	0.0116	29.64	0.0115
Hewlett Packard	40.6	0.0060	36.63	0.0220	0.0323	46.40	0.0169
Merck	26.5	0.0835	24.00	0.0253	0.0065	36.49	0.0162
Monsanto	95.3	0.0034	71.37	0.0238	0.1047	113.01	0.0337
Moodys	56.9	0.0050	33.76	0.0307	0.0527	63.12	0.0165
Motorola	91.6	0.0135	45.01	0.0378	0.0050	122.02	0.0166
NY Times	15.5	0.0108	8.79	0.0415	0.0048	19.13	0.0161
Pfizer	18.7	0.0183	15.90	0.0194	0.0270	20.52	0.0119
Pacific Gas and Electricity	36.6	0.0238	33.29	0.0170	0.0173	42.33	0.0118
Philip Morris	41.4	0.0537	36.76	0.0269	0.0059	55.98	0.0140
Walt Disney	25.9	0.0329	22.68	0.0278	0.0074	33.75	0.0142

estimation procedure for the remaining 15 companies. Remark that from the 30 companies analyzed from the Dow Jones Industrial Average index, 15 of them can be characterized with two very distinctive regimes and in all of them (except *M*onsanto) the volatility coefficient is larger in the second regime (higher prices) than in the first regime (lower prices). This could happen because for higher prices we can observe larger oscillations in prices (even if in percentage the oscillation is the same). In Fig. 1 we present the estimated regimes and thresholds for these stock prices. The horizontal lines in each figure are at the levels of the $\hat{\beta}_1$ estimate, the $\hat{m}$ estimates and the $\hat{\beta}_2$ estimate. It is not difficult to accept the existence of regimes and a threshold which makes sense for this stocks. The existence of returning values for financial prices is not a new idea, but it seems reasonable that could exist more than one returning value, one for lower prices and another for higher prices.

6 Conclusions

We introduced and studied a SDE model that can be well fitted for the price evolution of stocks by dividing the phase space in two regions and considering that the solution process follows, in each region, a different diffusion.

We developed a practical and useful procedure for the estimation for all the parameters of the model (the diffusion parameters for the two regimes and threshold) in the particular case in which the SDE defining the diffusion, in each region, corresponds to a geometric Ornstein–Uhlenbeck process. We showed that, for the general case, if we have known to what regime each of the observations belongs, then we can define consistent estimators for the threshold. A simulation study induced us to think that the estimation procedure can give satisfactory results.

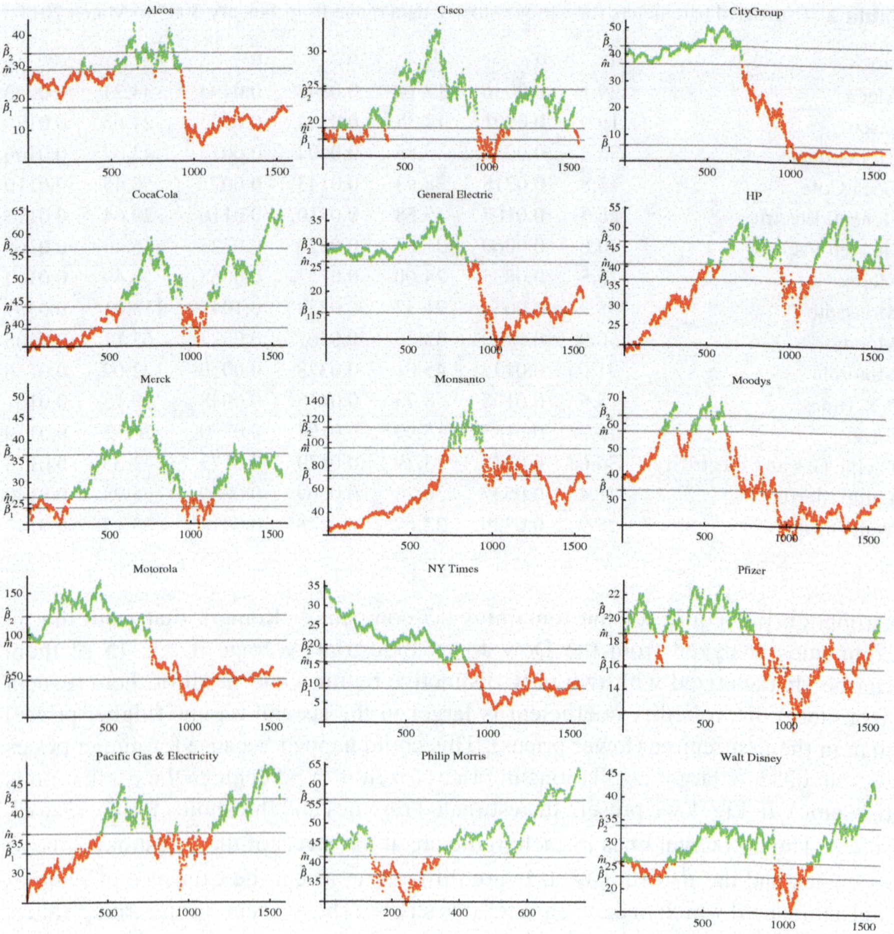

Fig. 1 Estimated regimes and thresholds for various stocks; the two colors differentiate the regimes

We applied the estimation procedure to the 30 components of the Dow Jones Industrial Average index. In half of these companies the procedure enables us to find fairly differentiated regimes for the stock prices. For future work a comparison with alternative linear and nonlinear SDE models could be an interesting line of research.

Acknowledgments This work was partially supported by the Fundação para a Ciência e a Tecnologia (Portuguese Foundation for Science and Technology) through PEst-OE/MAT/UI0297/2011 (CMA).

References

1. Brockwell, P.J.: On continuous time threshold ARMA processes. J. Statist. Plann. Inference **39**, 291–303 (1994)
2. Brockwell, P.J., Stramer, O.: On the approximation of continuous time threshold ARMA processes. Ann. Inst. Statist. Math. **47**, 1–20 (1995)
3. Brockwell, P.J., Williams, R.J.: On the existence and application of continuous-time threshold autoregressions of order two. Adv. Appl. Prob. **29**, 205–227 (1997)
4. Chan, K.S.: Consistency and limiting distribution of the least squares estimator of a threshold autoregressive model. Ann. Stat. **2**(1), 520–533 (1993)
5. Chan, K.S., Tsay, R.S.: Limiting properties of the least squares estimator of a continuous threshold autoregressive model. Biometrica **85**(2), 413–426 (1998)
6. Freidlin, M., Pfeiffer, R.: A threshold estimation problem for processes with hysteresis. Finance. Stochast. **36**, 337–347 (1998)
7. Hansen A.T., Poulsen, R.: A simple regime switching term structure model. Finance. Stochast. **4**(4), 409–429 (2000)
8. Karatzas, I., Shreve, S.: Brownian Motion and Stochastic Calculus, 2nd edn. Springer, Berlin (1991)
9. Lamberton, D., Lapeyre, B.: Introduction to Stochastic Calculus Applied to Finance, 2nd edn. Chapman and Hall/CRC, Boca Raton (2008)
10. Mota, P.P.: Brownian motion with drift threshold model. PhD dissertation, FCT/UNL (2008)
11. Øksendal, B.: Stochastic Differential Equations. Springer, New York (2007)
12. Petrucelli, J.D.: On the consistency of least squares estimators for a threshold AR(1) model. J. Time. Anal. **7**(4), 269–278 (1986)
13. Tong, H.: Non-linear Time Series: A Dynamical System Approach. Oxford University Press, Oxford (1990)

Generalized F Tests in Models with Random Perturbations: The Truncated Normal Case

Célia Nunes, Dário Ferreira, Sandra Ferreira, and João T. Mexia

Abstract

This paper shows how to obtain explicit expressions for non-central generalized F distributions with random non-centrality parameters. We consider the case when these parameters are random variables with truncated Normal distribution, for the usual F distribution and for the generalized F distribution.

1 Introduction

Quotients of linear combinations of chi-squares have relevant applications. For instance, the statistics of the generalized F tests are such quotients. These tests were introduced by [5,6], first for variance components and later for linear combinations of parameters in mixed linear models.

These tests are derived when we have a quadratic unbiased estimator $\widetilde{\theta}$ for a parameter θ and we want to test $H_0 : \theta = 0$ against $H_1 : \theta > 0$. Assuming this quadratic estimator to be a linear combination of statistics we can consider, following [5] and [6], the positive part, where the coefficients are positive, and the negative parte, where the coefficients are negative. Let $\widetilde{\theta}^+$ and $\widetilde{\theta}^-$ be the positive and the negative parts, respectively, of $\widetilde{\theta}$; thus we are led to use the test statistic

$$\mathfrak{I} = \frac{\widetilde{\theta}^+}{\widetilde{\theta}^-}.$$

C. Nunes (✉) · D. Ferreira · S. Ferreira
Department of Mathematics, University of Beira Interior, Covilhã, Portugal
e-mail: celian@ubi.pt; dario@ubi.pt; sandraf@ubi.pt

J.T. Mexia
Department of Mathematics, New University of Lisbon, Caparica, Portugal
e-mail: jtm@fct.unl.pt

J. Lita da Silva et al. (eds.), *Advances in Regression, Survival Analysis, Extreme Values, Markov Processes and Other Statistical Applications*, Studies in Theoretical and Applied Statistics, DOI 10.1007/978-3-642-34904-1_32,

In [2] the exact expressions of generalized F distributions are given when the degrees of freedom in the numerator or the denominator are even. In [7] this result was extended to the non-central case.

When the vector of observations is the sum of a vector corresponding to the theoretical model plus an independent perturbation vector, the distribution of the generalized F statistics has, see [7–9], random non-centrality parameters. This kind of model perturbation is worthwhile to study since it would cover situations in which the collection of the observations was made on non-standardized conditions. Since we have full control of the observations collection, the usual model assumptions would hold.

In this chapter we obtain the expressions of the usual and generalized non-central F distributions, when the non-centrality parameters are random variables with truncated normal distribution.

It is important to refer that our aim is mainly theoretical. We must point out that if practical applications are the main goal, an alternative for our treatment is given, for example, by [1, 4]. This way, the previous approaches such as the one given by [3, 12] may be improved.

This chapter is organized as follows. In Sect. 2 we present the expressions of the central and non-central generalized F distributions. In Sect. 3 we develop the case when the non-centrality parameters are random. This section is divided in two sections. Section 3.1 deals with the expression of the usual F distributions where the non-centrality parameters have truncated normal distribution. Finally, in Sect. 3.2 we present the results for the generalized case.

2 Generalized F and Related Distributions

In this section we present the expressions of the central and non-central generalized F distributions as a preparation for further expansions of random non-centrality parameters. These expressions were obtained in [2, 7].

2.1 Central Generalized F Distributions

Let us consider the independent random variables $U_i \sim \chi^2_{g_{1,i}}$, $i = 1, \ldots, r$, and $V_j \sim \chi^2_{g_{2,j}}$, $j = 1, \ldots, s$, and the vectors a_1^r and a_2^s, with non-negative components and being at least one of them not null. Thus, the distribution of

$$\frac{\sum_{i=1}^{r} a_{1,i} U_i}{\sum_{j=1}^{s} a_{2,j} V_j}$$

will be $F^+(z|a_1^r, a_2^s, g_1^r, g_2^s)$.

With $(v^m)^{-1}$ the vector whose components are the inverses of the components of v^m, the central generalized F distribution will be given by

$$F(z|g_1^r, g_2^s) = F^+(z|(g_1^r)^{-1}, (g_2^s)^{-1}, g_1^r, g_2^s).$$

If a_1^r and a_2^s have all components equal to 1 we will have

$$\overline{F}(z|g_1^r, g_2^s) = F^+(z|1^r, 1^s, g_1^r, g_2^s).$$

Consider that $r = s = 1$, in the first case one will have the usual central F distribution with g_1 and g_2 degrees of freedom, $F(z|g_1, g_2)$, while for the second case one will have the $\overline{F}$ distribution, which is the distribution of the quotient of independent central chi-squares with g_1 and g_2 degrees of freedom, $\overline{F}(z|g_1, g_2)$.

2.2 Non-central Generalized F Distributions

Distributions $\chi^2_{g,\delta}$ are a mixture of the distributions χ^2_{g+2j}, $j = 0, \ldots$. The coefficients in this mixture are the probabilities for non-negative integers of the Poisson distribution with parameter $\frac{\delta}{2}$, $P_{\delta/2}$. Thus, if $U \sim \chi^2_{g,\delta}$, it can be assumed that there is an indicator variable $J \sim P_{\delta/2}$ such that $U \sim \chi^2_{g+2\ell}$, when $J = \ell$, $\ell = 0, \ldots$.

So, if $U_i \sim \chi^2_{g_{1,i},\delta_{1,i}}$, $i = 1, \ldots, r$, and $V_j \sim \chi^2_{g_{2,j},\delta_{2,j}}$, $j = 1, \ldots, s$, are independent, their joint distribution $\chi^2_{g_1^r, g_2^s, \delta_1^r, \delta_2^s} = \prod_{i=1}^{r} \chi^2_{g_{1,i},\delta_{1,i}} \prod_{j=1}^{s} \chi^2_{g_{2,j},\delta_{2,j}}$ will be a mixture with coefficients

$$c(\ell_1^r, \ell_2^s, \delta_1^r, \delta_2^s) = \prod_{i=1}^{r} e^{-\frac{\delta_{1,i}}{2}} \frac{(\frac{\delta_{1,i}}{2})^{\ell_{1,i}}}{\ell_{1,i}!} \prod_{j=1}^{s} e^{-\frac{\delta_{2,j}}{2}} \frac{(\frac{\delta_{2,j}}{2})^{\ell_{2,j}}}{\ell_{2,j}!} \tag{1}$$

of the $\chi^2_{g_1^r+2\ell_1^r, g_2^s+2\ell_2^s} = \prod_{i=1}^{r} \chi^2_{g_{1,i}+2\ell_{1,i}} \prod_{j=1}^{s} \chi^2_{g_{2,j}+2\ell_{2,j}}$.

Moreover, using the mixtures method (see [10] and [11]) the distribution of $\frac{\sum_{i=1}^{r} a_{1,i} U_i}{\sum_{j=1}^{s} a_{2,j} V_j}$ will be

$$F^{+}(z|a_1^r, a_2^s, g_1^r, g_2^s, \delta_1^r, \delta_2^s) =$$

$$= \sum_{\ell_{1,1}=0}^{+\infty} \cdots \sum_{\ell_{1,r}=0}^{+\infty} \sum_{\ell_{2,1}=0}^{+\infty} \cdots \sum_{\ell_{2,s}=0}^{+\infty} c(\ell_1^r, \ell_2^s, \delta_1^r, \delta_2^s) F^{+}(z|a_1^r, a_2^s, g_1^r + 2\ell_1^r, g_2^s + 2\ell_2^s). \tag{2}$$

If, as above, we consider indicator variables, the conditional distribution of

$$\frac{\sum_{i=1}^{r} a_{1,i} U_i}{\sum_{j=1}^{s} a_{2,j} V_j},$$

when $J_{1,i} = \ell_{1,i}$, $i = 1, \ldots, r$ and $J_{2,j} = \ell_{2,j}$, $j = 1, \ldots, s$, will be $F^{+}(z|a_1^r, a_2^s, g_1^r + 2\ell_1^r, g_2^s + 2\ell_2^s)$. Thus, the expression of $F^{+}(z|a_1^r, a_2^s, g_1^r, g_2^s, \delta_1^r, \delta_2^s)$ can be obtained unconditioning with respect to the indicator variables.

3 Random Non-centrality Parameters

Up to now we have considered the indicator variables $J_{1,i}$, $i = 1, \ldots, r$, and $J_{2,j}$, $j = 1, \ldots, s$, to have Poisson distributions with fixed parameters. In this section we assume that these parameters are random variables and we obtain the expressions of the distributions for the truncated normal case.

Consider the random variables L_1^r and L_2^s with components $L_{1,i}$, $i = 1, \ldots, r$, and $L_{2,j}$, $j = 1, \ldots, s$. With $\lambda_{L_1^r, L_2^s}(t_1^r, t_2^s)$ the joint moment-generating function for these variables and

$$\lambda_{L_1^r, L_2^s}^{<\ell_1^r, \ell_2^s>}(t_1^r, t_2^s) = \frac{\partial^{\sum_{i=1}^{r} \ell_{1,i} + \sum_{j=1}^{s} \ell_{2,j}} \lambda_{L_1^r, L_2^s}(t_1^r, t_2^s)}{\prod_{i=1}^{r} \partial t_{1,i}^{\ell_{1,i}} \prod_{j=1}^{s} \partial t_{2,j}^{\ell_{2,j}}}, \tag{3}$$

unconditioning

$$F^{+}(z|a_1^r, a_2^s, g_1^r, g_2^s, l_1^r, l_2^s) =$$

$$= \sum_{\ell_{1,1}=0}^{+\infty} \cdots \sum_{\ell_{1,r}=0}^{+\infty} \sum_{\ell_{2,1}=0}^{+\infty} \cdots \sum_{\ell_{2,s}=0}^{+\infty} c(\ell_1^r, \ell_2^s, l_1^r, l_2^s) F^{+}(z|a_1^r, a_2^s, g_1^r + 2\ell_1^r, g_2^s + 2\ell_2^s) \tag{4}$$

in order to the random parameters vectors L_1^r and L_2^s, there will be

$$F^+(z|a_1^r, a_2^s, g_1^r, g_2^s, \lambda_{L_1^r, L_2^s}) =$$

$$= \sum_{\ell_{1,1}=0}^{+\infty} \cdots \sum_{\ell_{1,r}=0}^{+\infty} \sum_{\ell_{2,1}=0}^{+\infty} \cdots \sum_{\ell_{2,s}=0}^{+\infty} \frac{\lambda_{L_1^r, L_2^s}^{<\ell_1^r, \ell_2^s>}(-\frac{1}{2}1^r, -\frac{1}{2}1^s)}{\prod_{i=1}^{r} \ell_{1,i}!2^{\ell_{1,i}} \prod_{j=1}^{s} \ell_{2,j}!2^{\ell_{2,j}}} F^+(z|a_1^r, a_2^s, g_1^r + 2\ell_1^r, g_2^s + 2\ell_2^s). \tag{5}$$

3.1 Non-centrality Parameters with Truncated Normal Distribution: The Usual Case

As it was previously seen, if $a_1^r = 1^r$ and $a_2^s = 1^s$, with $r = s = 1$, we will have the $\overline{F}$ distribution, which is the distribution of the quotient of independent chi-squares with g_1 and g_2 degrees of freedom. So, (5) can be rewritten as

$$\overline{F}(z|g_1, g_2, \lambda_{X,Y}) = \sum_{i=0}^{+\infty} \sum_{j=0}^{+\infty} \frac{\lambda_{X,Y}^{<i,j>}(-\frac{1}{2}, -\frac{1}{2})}{2^{i+j} i! j!} \overline{F}(z|g_1 + 2i, g_2 + 2j). \tag{6}$$

Let X be a random variable with normal distribution with mean value μ_x and variance σ_x^2, $X \sim N(\mu_x, \sigma_x^2)$, independent of Y, also with normal distribution with mean value μ_y and variance σ_y^2, $Y \sim N(\mu_y, \sigma_y^2)$.

We want to consider the density of those random variables if X and Y take only non-negative values. So X and Y have truncated normal distribution bounded below by zero.

The density of X will be given by

$$f(x|\mu_x, \sigma_x, 0, +\infty) = \frac{\frac{1}{\sqrt{2\pi}\sigma_x} e^{-\frac{(x-\mu_x)^2}{2\sigma_x^2}}}{F(+\infty|\mu_x, \sigma_x) - F(0|\mu_x, \sigma_x)}, \quad x \geq 0,$$

with $F(x|\mu_x, \sigma_x)$ the distribution function of X.

So, the moment-generating function of X will be

$$\begin{aligned} \lambda_X(t_1) &= \frac{1}{1 - F(0|\mu_x, \sigma_x)} \int_0^{+\infty} e^{t_1 x} \frac{1}{\sqrt{2\pi}\sigma} e^{-\frac{(x-\mu_x)^2}{2\sigma_x^2}} dx \\ &= \frac{1}{K_0(\mu_x)} e^{\mu_x t_1} \sum_{p_1=0}^{+\infty} \frac{K_{p_1}(\mu_x)}{p_1!} (t_1 \sigma_x)^{p_1}, \end{aligned} \tag{7}$$

with

$$\begin{cases} K_0(\mu_x) = 1 - F(0|\mu_x, \sigma_x) = \frac{1}{2} + \frac{1}{\sqrt{2\pi}} \sum_{p_1=0}^{+\infty} \frac{(-1)^{p_1}}{p_1! 2^{p_1}} \frac{\left(-\frac{\mu_x}{\sigma_x}\right)^{2p_1+1}}{(2p_1+1)} \\ K_1(\mu_x) = \frac{e^{-\frac{\mu_x^2}{2\sigma_x^2}}}{\sqrt{2\pi}} \\ K_{p_1}(\mu_x) = (p_1-1)K_{p_1-2}(\mu_x) + \frac{\left(-\frac{\mu_x}{\sigma_x}\right)^{p_1-1} e^{-\frac{\mu_x^2}{2\sigma_x^2}}}{\sqrt{2\pi}} \end{cases}.$$

Let us consider $\lambda_X(t_1) = g(t_1) \times h(t_1)$, where

$$g(t_1) = \frac{1}{K_0(\mu_x)} e^{\mu_x t_1}$$

and

$$h(t_1) = \sum_{p_1=0}^{+\infty} \frac{K_{p_1}(\mu_x)}{p_1!} (t_1 \sigma_x)^{p_1}.$$

We will obtain

$$\begin{aligned} \lambda_X^{<i>}(t_1) &= \sum_{h_1=0}^{i} \binom{i}{h_1} g^{<h_1>}(t_1) h^{<i-h_1>}(t_1) \\ &= \sum_{h_1=0}^{i} \binom{i}{h_1} \frac{\mu_x^{h_1} e^{\mu_x t_1}}{K_0(\mu_x)} \sum_{p_1=i-h_1}^{+\infty} \frac{K_{p_1}(\mu_x) \sigma_x^{p_1} t_1^{(p_1-i+h_1)}}{(p_1-i+h_1)!}. \end{aligned} \tag{8}$$

This way, and because the variables are independent,

$$\begin{aligned} \lambda_{X,Y}^{<i,j>}(t_1, t_2) &= \lambda_X^{<i>}(t_1) \lambda_Y^{<j>}(t_2) \\ &= \sum_{h_1=0}^{i} \binom{i}{h_1} \frac{\mu_x^{h_1} e^{\mu_x t_1}}{K_0(\mu_x)} \sum_{p_1=i-h_1}^{+\infty} \frac{K_{p_1}(\mu_x) \sigma_x^{p_1} t_1^{(p_1-i+h_1)}}{(p_1-i+h_1)!} \\ &\times \sum_{h_2=0}^{j} \binom{j}{h_2} \frac{\mu_y^{h_2} e^{\mu_y t_2}}{K_0(\mu_y)} \sum_{p_2=j-h_2}^{+\infty} \frac{K_{p_2}(\mu_y) \sigma_y^{p_2} t_2^{(p_2-j+h_2)}}{(p_2-j+h_2)!} \end{aligned} \tag{9}$$

and (6) will be given by

$$\overline{F}(z|g_1, g_2, \lambda_{X,Y}) = \sum_{i=0}^{+\infty}\sum_{j=0}^{+\infty}\sum_{h_1=0}^{i}\binom{i}{h_1}\frac{\mu_x^{h_1}e^{-\frac{\mu_x}{2}}}{2^i i! K_0(\mu_x)}$$

$$\times \sum_{p_1=h_1}^{+\infty}\frac{K_{p_1}(\mu_x)\sigma_x^{p_1}(-\frac{1}{2})^{(p_1-i+h_1)}}{(p_1-i+h_1)!}\sum_{h_2=0}^{j}\binom{j}{h_2}\frac{\mu_y^{h_2}e^{-\frac{\mu_y}{2}}}{2^j j! K_0(\mu_y)}$$

$$\times \sum_{p_2=h_2}^{+\infty}\frac{K_{p_2}(\mu_y)\sigma_y^{p_2}(-\frac{1}{2})^{(p_2-j+h_2)}}{(p_2-j+h_2)!}\overline{F}(z|g_1+2i, g_2+2j). \quad (10)$$

3.2 Non-centrality Parameters with Truncated Normal Distribution: The Generalized Case

Consider now the generalized case and the independent random variables X^r and Y^s with non-negative components

$$X_i \sim N(\mu_{x_i}; \sigma_{x_i}^2),\ i = 1, \ldots, r,$$

and

$$Y_j \sim N(\mu_{y_j}; \sigma_{y_j}^2),\ j = 1, \ldots, s.$$

The moment-generating function of X^r will be

$$\lambda_{X^r}(t_1^r) = \prod_{i=1}^{r}\lambda_{X_i(t_{1,i})} = \prod_{i=1}^{r}\frac{1}{K_0(\mu_{x_i})}e^{\mu_{x_i}t_{1,i}}\sum_{p_{1,i}=0}^{+\infty}\frac{K_{p_{1,i}}(\mu_{x_i})}{p_{1,i}!}(t_{1,i}\sigma_{x_i})^{p_{1,i}},$$

with

$$\begin{cases} K_0(\mu_{x_i}) = \frac{1}{2} + \frac{1}{\sqrt{2\pi}}\sum_{p_1=0}^{+\infty}\frac{(-1)^{p_1}}{p_1!2^{p_1}}\frac{\left(-\frac{\mu_{x_i}}{\sigma_{x_i}}\right)^{2p_1+1}}{(2p_1+1)} \\ K_1(\mu_{x_i}) = \frac{e^{-\frac{\mu_{x_i}^2}{2\sigma_{x_i}^2}}}{\sqrt{2\pi}} \\ K_{p_{1,i}}(\mu_{x_i}) = (p_{1,i}-1)K_{p_{1,i}-2}(\mu_{x,i}) + \frac{\left(-\frac{\mu_{x,i}}{\sigma_{x,i}}\right)^{p_{1,i}-1}e^{-\frac{\mu_{x_i}^2}{2\sigma_{x,i}^2}}}{\sqrt{2\pi}} \end{cases},$$

and consequently

$$\lambda_{X^r}^{<\ell_1^r>}(t_1^r) = \prod_{i=1}^{r} \lambda_{X_i}^{<\ell_{1,i}>}(t_{1,i})$$

$$= \prod_{i=1}^{r} \sum_{h_1=0}^{\ell_{1,i}} \binom{\ell_{1,i}}{h_1} \frac{\mu_{x_i}^{h_1} e^{\mu_{x_i} t_{1,i}}}{K_0(\mu_{x_i})} \sum_{p_{1,i}=\ell_{1,i}-h_1}^{+\infty} \frac{K_{p_{1,i}}(\mu_{x_i}) \sigma_{x_i}^{p_{1,i}} t_{1,i}^{(p_{1,i}-\ell_{1,i}+h_1)}}{(p_{1,i}-\ell_{1,i}+h_1)!}. \quad (11)$$

Since the variables X^r and Y^s are independent,

$$\lambda_{X^r,Y^s}^{<\ell_1^r,\ell_2^s>}(t_1^r, t_2^s) = \lambda_{X^r}^{<\ell_1^r>}(t_2^s)\lambda_{Y^s}^{<\ell_2^s>}(t_2^s) =$$

$$= \prod_{i=1}^{r} \sum_{h_1=0}^{\ell_{1,i}} \binom{\ell_{1,i}}{h_1} \frac{\mu_{x_i}^{h_1} e^{\mu_{x_i} t_{1,i}}}{K_0(\mu_{x_i})} \sum_{p_{1,i}=\ell_{1,i}-h_1}^{+\infty} \frac{K_{p_{1,i}}(\mu_{x_i}) \sigma_{x_i}^{p_{1,i}} t_{1,i}^{(p_{1,i}-\ell_{1,i}+h_1)}}{(p_{1,i}-\ell_{1,i}+h_1)!}$$

$$\times \prod_{j=1}^{s} \sum_{h_2=0}^{\ell_{2,j}} \binom{\ell_{2,j}}{h_2} \frac{\mu_{y_j}^{h_2} e^{\mu_{y_j} t_{2,j}}}{K_0(\mu_{y_j})} \sum_{p_{2,j}=\ell_{2,j}-h_2}^{+\infty} \frac{K_{p_{2,j}}(\mu_{y_j}) \sigma_{y_j}^{p_{2,j}} t_{2,j}^{(p_{2,j}-\ell_{2,j}+h_2)}}{(p_{2,j}-\ell_{2,j}+h_2)!}. \quad (12)$$

This way, (5) can be rewritten as

$$F^+(z|a_1^r, a_2^s, g_1^r, g_2^s, \lambda_{X^r,Y^s})$$

$$= \sum_{\ell_{1,1}=0}^{+\infty} \cdots \sum_{\ell_{1,r}=0}^{+\infty} \sum_{\ell_{2,1}=0}^{+\infty} \cdots \sum_{\ell_{2,s}=0}^{+\infty} \prod_{i=1}^{r} \sum_{h_1=0}^{\ell_{1,i}} \binom{\ell_{1,i}}{h_1} \frac{\mu_{x_i}^{h_1} e^{-\frac{\mu_{x_i}}{2}}}{\ell_{1,i}! 2^{\ell_{1,i}} K_0(\mu_{x_i})}$$

$$\times \sum_{p_{1,i}=\ell_{1,i}-h_1}^{+\infty} \frac{K_{p_{1,i}}(\mu_{x_i}) \sigma_{x_i}^{p_{1,i}} (-\frac{1}{2})^{(p_{1,i}-\ell_{1,i}+h_1)}}{(p_{1,i}-\ell_{1,i}+h_1)!} \prod_{j=1}^{s} \sum_{h_2=0}^{\ell_{2,j}} \binom{\ell_{2,j}}{h_2} \frac{\mu_{y_j}^{h_2} e^{-\frac{\mu_{y_j}}{2}}}{\ell_{2,j}! 2^{\ell_{2,j}} K_0(\mu_{y_j})}$$

$$\times \sum_{p_{2,j}=\ell_{2,j}-h_2}^{+\infty} \frac{K_{p_{2,j}}(\mu_{y_j}) \sigma_{y_j}^{p_{2,j}} (-\frac{1}{2})^{(p_{2,j}-\ell_{2,j}+h_2)}}{(p_{2,j}-\ell_{2,j}+h_2)!} F^+(z|a_1^r, a_2^s, g_1^r + 2\ell_1^r, g_2^s + 2\ell_2^s). \quad (13)$$

Acknowledgements The authors are grateful to the anonymous referees for their useful comments and remarks. This work was partially supported by the center of Mathematics, University of Beira Interior, under the project PEst-OE/MAT/UI0212/2011.

References

1. Davies, R.B.: Algorithm AS 155: The distribution of a linear combinations of χ^2 random variables. Appl. Stat. **29**, 232–333 (1980)
2. Fonseca, M., Mexia, J.T., Zmyślony, R.: Exact distribution for the generalized F tests. Discuss. Math. Probab. Stat. **22**, 37–51 (2002)
3. Gaylor, D.W., Hopper, F.N.: Estimating the degrees of freedom for linear combinations of mean squares by Satterthwaite's formula. Technometrics **11**, 691–706 (1969)
4. Imhof, J.P.: Computing the distribution of quadratic forms in normal variables. Biometrika **48**, 419–426 (1961)
5. Michalski, A., Zmyślony, R.: Testing hypothesis for variance components in mixed linear models. Statistics **27**, 297–310 (1996)
6. Michalski, A., Zmyślony, R.: Testing hypothesis for linear functions of parameters in mixed linear models. Tatra Mt. Math. Publs. **17**, 103–110 (1999)
7. Nunes, C., Mexia, J.T.: Non-central generalized F distributions. Discuss. Math. Probab. Stat. **26**, 47–61 (2006)
8. Nunes, C., Ferreira, S., Ferreira, D.: Generalized F tests in models with random perturbations: The Gamma case. Discuss. Math. Probab. Stat. **29**(2), 185–198 (2009)
9. Nunes, C., Ferreira, D., Ferreira, S., Mexia, J.T.: Generalized F distributions with random non-centrality parameters: The convolution of Gamma and Beta variables. Far East J. Math. Sci. **62**(1), 1–14 (2012)
10. Robbins, H.: Mixture of distribution. Ann. Math. Statist. **19**, 360–369 (1948)
11. Robbins, H., Pitman, E.J.G.: Application of the method of mixtures to quadratic forms in normal variates. Ann. Math. Statist. **20**, 552–560 (1949)
12. Satterthwaite, F.E.: An approximate distribution of estimates of variance components. Biometrics Bull. **2**, 110–114 (1946)

Generalized Linear Models, Generalized Additive Models and Neural Networks: Comparative Study in Medical Applications

Ana Luisa Papoila, Cristina Rocha, Carlos Geraldes, and Patricia Xufre

Abstract

During the last two decades, evaluating severity of illness and predicting mortality of critical patients became a major concern of all professionals that work in intensive care units all over the world. Due to the binary nature of the response variable, logistic regression models were a natural choice for modelling this kind of data. The objective of this study is to compare the performance of generalized linear models (GLMs) with binary response (McCullagh and Nelder, Generalized Linear Models. Chapman and Hall, London, 1989), with the performance of generalized additive models (GAMs) with binary response (Hastie and Tibshirani, Generalized Additive Models. Chapman and Hall, New York, 1990) and also with the performance of artificial neural networks (ANNs) (Bishop, Neural Networks for Pattern Recognition. Clarendon Press, Oxford, 1995), in what concerns their predictive and discriminative power. A dataset of 996 patients was collected and the entire sample was used for the development

A.L. Papoila (✉)
CEAUL and Faculdade de Ciências Médicas da, Universidade Nova de Lisboa, Campo Mártires da Pátria 130 1169-056 Lisboa, Portugal
e-mail: ana.papoila@fcm.unl.pt

C. Rocha
CEAUL and Faculdade de Ciências, Universidade de Lisboa, Campo Grande 1149-016 Lisboa, Portugal
e-mail: cmrocha@fc.ul.pt

C. Geraldes
Universidade Nova de Lisboa, Campo Mártires da Pátria 130 1169-056 Lisboa, Portugal
e-mail: carlos.geraldes@fcm.unl.pt

P. Xufre
CIOUL and Instituto Superior de Estatística e Gestão de Informação da, Universidade Nova de Lisboa, Campus de Campolide 1070-312 Lisboa, Portugal
e-mail: pxufre@fe.unl.pt

J. Lita da Silva et al. (eds.), *Advances in Regression, Survival Analysis, Extreme Values, Markov Processes and Other Statistical Applications*, Studies in Theoretical and Applied Statistics, DOI 10.1007/978-3-642-34904-1_33,

of the models and also for the validation process, due to the nonexistence of an external, independent dataset. The performance of the proposed methodologies was assessed, not only by the evaluation of the agreement between observed mortality and predicted probabilities of death through the use of calibration plots, but also by their discriminating ability, measured by the area under the receiver operating characteristic (ROC) curve.

1 Introduction

Since 1981 numerical scoring systems and multivariable statistical models have been used to assess the severity of illness of critically ill patients. The former assign, subjectively, weights to variables reflecting the degree of physiologic derangement. In fact, the acute physiology and chronic health evaluation score, referred to as APACHE [4], the simplified acute physiology score, referred to as SAPS [7] and the APACHE II score [5] were built using a panel of experts to select variables and weights. More recently, and because the subjectivity of these procedures may lead to undesired discrepancies, multivariable statistical models were considered. Mortality probability models, referred to as MPM [9–11], the APACHE III score [6] and the SAPS II [8], were then developed, making use of more objective methods such as multiple logistic regression. However, the fact that a non-linear dependence between the binary response variable and the continuous covariates may exist led us to adopt generalized additive models (GAMs) to accomplish the fitting process. In fact, the more recently developed severity of illness scores, SAPS 3 [14, 15] and APACHE IV [17] also make use of more flexible strategies, such as splines and regression trees, to model the data. So, in this chapter, we propose the use of GAMs to estimate the probabilities of death and/or to obtain new adjusted cut-off points with the purpose of categorizing the continuous independent variables, if the main interest is the obtainment of a severity of illness score. SAPS II variables were used because this was the severity of illness score adopted by the clinicians of the Portuguese intensive care unit (ICU) where the dataset analysed in the present study was collected. Since artificial neural networks (ANNs) are an alternative to some statistical methodologies, namely, regression models [16], this study also aims to evaluate the performance of ANNs to predict the outcome under study. Finally, a comparison of the several approaches was carried out.

All statistical analyses were performed using S-PLUS (version 8.0, 2007; Insightful Corporation, Seattle, WA) and, to implement the ANNs, a new software was developed using a standard commercially available mathematics package format (MATLAB R2006b, The Math-Works Inc., 3 Apple Hill Drive, Natick, MA 01760).

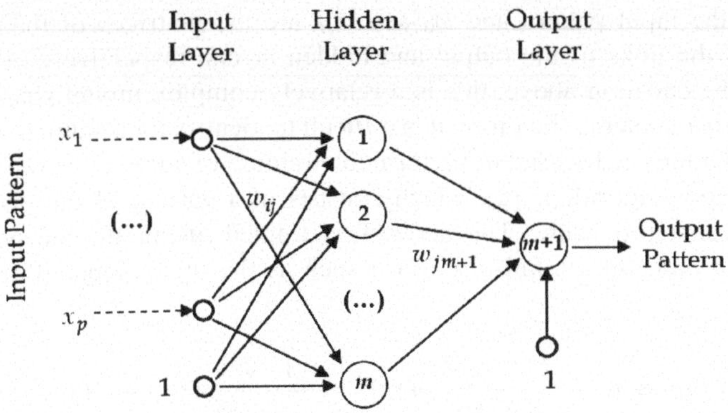

Fig. 1 Multi-layer perceptron architecture

2 Generalized Linear Models and Generalized Additive Models

Let Y be a response variable and $(X_1, \ldots, X_p)$ a vector of p associated covariates that characterize each of n individuals. A GAM is defined by the expression $E(Y|X_1, \ldots, X_p) = h(\beta_0 + \sum_{j=1}^{p} f_j(X_j))$, where Y has a probability mass or density function that belongs to the exponential family, $h(\cdot)$ is the link function and $f_j(X_j)$, $j = 1, \ldots, p$, known as the partial functions, are arbitrary univariable functions that must be estimated from the data and represent the effect of the covariates on the response [3]. A generalized linear model (GLM) is a particular case of a GAM when $f_j(X_j) = \beta_j X_j$ [12].

3 Artificial Neural Networks

An ANN is, fundamentally, a mathematical model composed by a set of units (nodes), where information is processed [2]. These units are connected through unidirectional communication links, which carry numerical data. One of the most studied and used ANN architecture is the multi-layer perceptron (MLP). Fundamentally, one MLP consists of an input–output network, which has the neurons distributed by several layers, fully connected between adjacent layers, and where the flow of information is done in a feed-forward way. The following figure shows an MLP with three layers: an input layer, without neurons, a hidden layer and a layer with one output neuron.

If we have an MLP such as the one represented in Fig. 1 and with the same activation function, f, in all its neurons, then it can be described mathematically as

$$y(x) = f(\omega_0^T f(\omega_H^T x)),$$

where x is the input pattern and ω_0 and ω_H are the matrices of the parameters related with the links of the output and hidden layers, respectively. As it can be seen from the equation above, this is a relatively complex model since it is non-linear in the parameters. Therefore, it is difficult to identify and estimate it correctly. The method traditionally used to perform the training of such networks is the error backpropagation algorithm [2], which consists of a variant of the instantaneous gradient descent procedure. The network is trained, using the steepest descent algorithm, in order to minimize an error such as the mean squared error (MSE) given by

$$MSE \equiv E_N = \frac{1}{2N}\sum_x (e(x))^2 = \frac{1}{2N}\sum_x (y(x) - d(x))^2,$$

where $d(x)$ corresponds to the desired output for the input pattern x and N is the number of individuals of the training dataset. It can be viewed as a sort of non-linear and non-parametric regression. The updating of the synaptic weights is

$$\omega = \omega - \alpha \frac{\partial E_N}{\partial \omega_{ij}},$$

where α is the learning rate. However, this kind of searching methods does not guarantee convergence of the objective function to a global minimum, and the convergence rate is typically very slow during most of the training process. To help in both respects, it is common to consider the inclusion of a momentum term in the weights updates:

$$\Delta\omega_{ij}^{(k)} = -\alpha \frac{\partial E_N}{\partial \omega_{ij}} + \beta \Delta\omega_{ij}^{(k-1)}.$$

4 The New Simplified Acute Physiology Score (SAPS II)

The SAPS II is a severity of illness score, used in ICUs, that has received a lot of attention in Europe for its simplicity and applicability. It includes 17 variables: 12 physiology variables (heart rate, systolic blood pressure, body temperature, the ratio $\frac{PaO_2}{FiO_2}$ for ventilated patients, urinary output, serum urea level, white blood cells count, serum potassium, serum sodium level, serum bicarbonate level, bilirubin level and Glasgow coma score), age, type of admission (scheduled surgical, unscheduled surgical or medical) and three underlying disease variables (acquired immunodeficiency syndrome, metastatic cancer and hematologic malignancy). To develop and validate this score, a large international sample of surgical and medical patients, collected by an European/North American multicentre study, was used [8]. The development phase used 65 % of the available patients, randomly selected, while the remaining 35 % became the validation set. The cut-off points for each of the continuous covariates that revealed to be statistically significant in the univariable analysis were found by using the LOWESS (locally weighted scatterplot smoothing) technique. After the categories were defined, a multiple logistic regression was used and the total severity score was obtained by adding the estimated coefficients of the corresponding design variables multiplied by 10 and rounded off to the nearest integer. Finally, for converting the SAPS II into a probability of hospital mortality, a multiple logistic regression model was fitted with SAPS II and ln(SAPS II + 1) as independent variables. However, when applied to different populations, this model is often unable to adequately predict the outcome, and so, a customization may be done by fitting that model to the new datasets.

Model calibration was evaluated by analysing the agreement between the estimated probabilities of death and the actual observed mortality using the Hosmer–Lemeshow goodness-of-fit test, having obtained a p-value $= 0.104$ for the validation sample. To evaluate the ability of the model to distinguish between patients who live from patients who die, usually referred to as discrimination, receiver operating characteristic (ROC) curves were used and an area under the curve of 0.86 was achieved for the validation sample. Indeed, both results are highly satisfactory; however, when SAPS II was applied to some external databases, the results obtained were far worse (e.g. [1, 13]).

5 Results

Data from 996 patients, consecutively admitted to a Portuguese mixed (medical and surgical) ICU, were analysed. All SAPS data were collected during the first 24 hours after ICU admission. The mean age of the patients was 60.3 (95 % C.I. : 59.3,61.4) years with a median SAPS score of 41 (interquartile range 20–60) and a hospital mortality of 36 %. The original SAPS II scoring system did not produce very good results, namely, in what concerns calibration (p-value < 0.001) (Fig. 2, left), although an area under the ROC curve of 0.82 (95 % C.I. : 0.79, 0.84) was achieved, showing a satisfactory discrimination ability. After customization, by using a logistic regression model with SAPS II and ln(SAPS II + 1) as independent variables, a new equation for the hospital mortality prediction was derived and a better performance was obtained (Fig. 2, right), with a p-value $= 0.517$ attained by the Hosmer–Lemeshow goodness-of-fit test and with the same area under the ROC curve.

The same dataset was used to implement a 3-layered perceptron with 17 input nodes, 5 hidden units, a single output node and a sigmoidal activation function. Firstly, this network was trained using the steepest descent algorithm so to minimize the MSE (Fig. 3, left). The obtained area under the ROC curve was 0.82 (95 % C.I. : 0.79, 0.84). Secondly, the Kullback–Leibler (KL) distance was used instead of the MSE criterium (Fig. 3, right) and the obtained area under the ROC curve was 0.81 (95 % C.I. : 0.78, 0.84).

At last, GAMs were used to analyse the data. Based on the partial functions estimates, we found new cut-off values for each continuous covariate adjusted by the remaining covariates and we fitted a logistic regression model with these new categorical independent variables (Fig. 4, left).

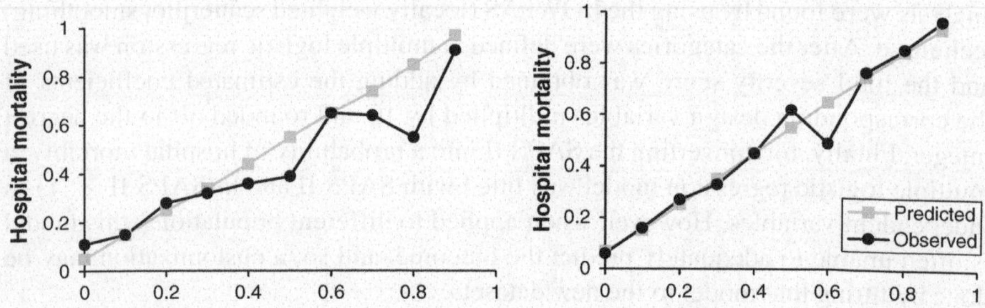

Fig. 2 Predicted versus observed probability of death. Original SAPS II (*left*) and customized SAPS II (*right*)

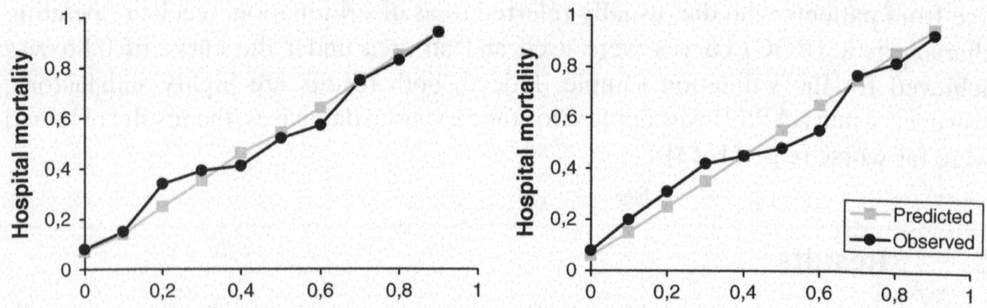

Fig. 3 Predicted versus observed probability of death. Artificial neural network using MSE (*left*) and using the Kullback–Leibler distance (*right*)

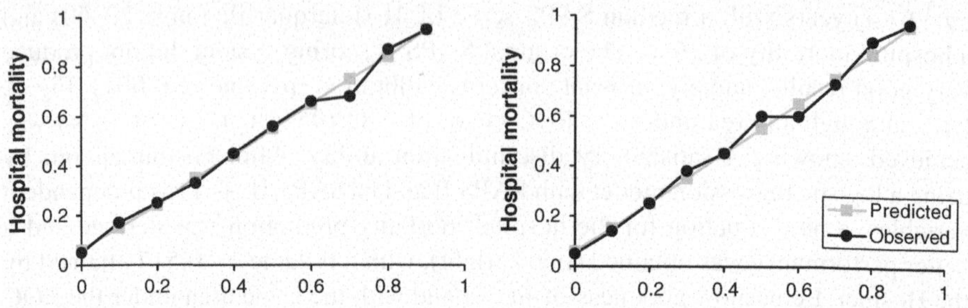

Fig. 4 Predicted versus observed probability of death. Logistic regression with the new categorical covariates (*left*) and a GAM without categorizing the continuous covariates (*right*)

The entire sample was used for model estimation and validation was accomplished by randomly splitting the whole sample into five mutually exclusive groups. Five regression models were then fitted, with each model excluding one group and used to calculate predictions for the excluded group (fivefold cross validation). An area below the ROC curve equal to 0.85 (95 % C.I. : 0.82, 0.87) and a calibration p-value $=$ 0.74 were obtained. The substantial improvements in both calibration and discrimination, even without introducing new prognostic variables, were interesting findings. However, since some information is lost in the categorization process, we also used GAMs to estimate the probabilities of death without categorizing the continuous covariates. After fitting a GAM to our cross-validated sample, good calibration curves (Fig. 4, right) and an area under the ROC curve of 0.87 (95 % C.I. : 0.85, 0.89) were obtained. As it can be seen from Fig. 4, GAMs obtained better results than those presented by the other approaches.

6 Conclusions and Future Work

The performance of GAMs is clearly superior to the GLMs and neural networks used in this study. When comparing these last two approaches, in what concerns their discriminative power, results are according to the ones referred elsewhere (no substantial differences between the areas under the ROC curve). The same did not happen for the predictive power since neural network calibration plots showed a weaker performance, independently of the used criterium (MSE or KL distance). This means that, in our study, there was no relevant advantage in using ANN-MLPs. As future work, other ANN structures, such as Generalized Additive Neural Networks, will be implemented with the purpose of obtaining better results.

Acknowledgements Ana Luisa Papoila's and Cristina Rocha's research is partially sponsored by national funds through the Fundação Nacional para a Ciência e Tecnologia, Portugal—FCT—under the project PEst-OE/MAT/UI0006/2011. The authors would like to thank the medical team of UCIP from the Centro Hospitalar de Lisboa Central (Lisbon), in particular to Dr. Eduardo Silva and Dr. Miguel Robalo, for their collaboration in data collection.

References

1. Apolone, G., Bertolini, G., D'Amico, R., et al.: The performance of SAPS II in a cohort of patients admitted to 99 ICUs: Results from GiViTI. Int. Care Med. **22**, 1368–1378 (1996)
2. Bishop, C.M.: Neural Networks for Pattern Recognition. Clarendon Press, Oxford (1995)
3. Hastie, T., Tibshirani, R.: Generalized Additive Models. Chapman and Hall, New York (1990)
4. Knaus, W.A., Zimmerman, J.E., Wagner, D.P., et al.: APACHE-Acute physiology and chronic health evaluation: A physiologically based classification system. Crit. Care Med. **9**, 591–597 (1981)
5. Knaus, W.A., Draper, E.A., Wagner, D.P., Zimmerman, J.E.: APACHE II: A severity of disease classification system. Crit. Care Med. **13**, 818–829 (1985)
6. Knaus, W.A., Wagner, D.P., Draper, E.A., et al.: The APACHE III prognostic system: Risk prediction of hospital mortality for critically ill hospitalized adults. Chest **100**, 1619–1636 (1991)
7. Le Gall, J.-R., Loirat, P., Alperovitch, A., et al.: A simplified acute physiology score for ICU patients. Crit. Care Med. **12**, 975–977 (1984)
8. Le Gall, J.-R., Lemeshow, S., Saulmier, F.: A New Simplified Acute Physiology Score (SAPS II) based on an European/North American multicenter study. JAMA vol. 270 **24**, 2957–2963 (1993)
9. Lemeshow, S., Teres, D., Pastides, H., et al.: A method for predicting survival and mortality of ICU patients using objectively derived weights. Crit. Care Med. **13**, 519–525 (1985)

10. Lemeshow, S., Teres, D., Avrunin, J.S., Gage, R.W.: Refining intensive care unit outcome prediction by using changing probabilities of mortality. Crit. Care Med. **1**, 470–477 (1988)
11. Lemeshow, S., Teres, D., Klar, J., et al.: Mortality Probability Models (MPM II) based on an international cohort of intensive care unit patients. JAMA vol. 270, **20**, 2478–2486 (1993)
12. McCullagh, P., Nelder, J.: Generalized Linear Models, 2nd edn. Chapman and Hall, London (1989)
13. Metnitz, P.G.H., Valentin, A., Vesely, H., et al.: Prognostic performance and customization of the SAPS II: Results of a multicenter Austrian study. Int. Care Med. **25**, 192–197 (1999)
14. Metnitz, P.G.H., Moreno, R.P., Almeida, E., et al.: SAPS 3: From evaluation of the patient to evaluation of the intensive care unit, Part 1: Objectives, methods and cohort description. Int. Care Med. **31**, 1336–1344 (2005)
15. Moreno, R.P., Metnitz, P.G.H., Almeida, E., et al.: SAPS 3: From evaluation of the patient to evaluation of the intensive care unit, Part 2: Development of a prognostic model for hospital mortality at ICU admission. Int. Care Med. **31**, 1345–1355 (2005)
16. Schumacher, M., Roβner, R., Vach W.: Neural networks and logistic regression: Part I. Elsevier, Comput. Stat. Data An. **21**, 661–682 (1996)
17. Zimmerman, J.E., Kramer, A.A., McNair, D.S., Malila, F.M.: Acute physiology and chronic health evaluation (APACHE) IV: Hospital mortality assessment for today's critically ill patients. Crit. Care Med. **34**, 1297–1310 (2006)

Joint-Regression Analysis and Incorporation of Environmental Variables in Stochastic Frontier Production Function: An Application to Experimental Data of Winter Rye

Dulce Gamito Pereira and Ana Sampaio

Abstract
This chapter joins the main properties of two specific regression techniques, joint-regression analysis (JRA) and stochastic frontier approach (SFA) in the analysis of experimental data sets from a breeding program of winter rye (*Secale cereale* L.), conducted in Poland, *Research Center for Cultivar Testing de Słupia Wielka*, over the period 1997–1998. With JRA, a meta-model, based on several linear regressions, had been estimated in order to analyze multilocation trials of winter rye production and to select the best cultivars (more productive) for a related stratum (locality/genotype). With SFA, another regression model had been investigated to predict production rankings of cultivars, through individual efficiency estimates. These measures resulted from a stochastic production frontier on experimental data of production and different climate conditions. Both techniques show similar dominant cultivars for the same environments.

1 Introduction

Joint-regression analysis (JRA) has been a widely used technique in the analysis of series of experiments designed for cultivar comparison. The series of experiments to be analyzed must cover sufficiently large areas so to make the choice of

D.G. Pereira (✉)
Department of Mathematics and CIMA, University of Évora, R. Romão Ramalho, 59, 7000-671 Évora, Portugal
e-mail: dgsp@uevora.pt

A. Sampaio
Business Research Unit-ISCTE and Department of Mathematics, University of Évora, R. Romão Ramalho, 59, 7000-671 Évora, Portugal
e-mail: sampaio@uevora.pt

J. Lita da Silva et al. (eds.), *Advances in Regression, Survival Analysis, Extreme Values, Markov Processes and Other Statistical Applications*, Studies in Theoretical and Applied Statistics, DOI 10.1007/978-3-642-34904-1_34,

cultivars worthwhile. Formerly, such series consisted of randomized block designs. Nowadays, these have been replaced, in most of Europe, by α-designs, which consist of incomplete blocks. Thus, the usual technique, see Gusmão [16] and [17], of estimating the environmental indexes of the blocks by their average yields, can no longer be used. Mexia et al. [21], introduced L_2 environmental indexes which can be applied to series of experiments using incomplete block.

Stochastic frontier approach (SFA) was originally developed in 1977 by Aigner, Lovell, and Schmidt (ALS) and it has been published, almost simultaneously, by Meusen and van den Broeck (MB), Battese and Corra (BC), and ALS. This technique shares with JRA the linear regression methodology although it considers in the model specification a composed error structure ($\varepsilon = V - U$), where the first error component $V \sim N(0, \sigma_v^2)$ is intended to capture the effects of statistical noise and the second component, U, is intended to capture the effects of technical inefficiency, being $U \geq 0$, as the observation lies on or beneath this stochastic production frontier [12, 14]. Because of the asymmetric component of statistical error, four different distributional assumptions, such as half normal, normal truncated, exponential, and gamma, have been proposed and developed (BC for half normal, ALS for half normal and normal truncated, Greene [13] for gamma, Stevenson [32] for gamma and normal truncated). Distributional assumption on U leads that the composed error is negatively skewed requiring maximum likelihood estimation (MLE). Although the main applications of SFA have been developed in the competitive market context [2, 29], SFA appears also in experimental designs for improving agriculture productivity [3, 6]. Indeed, as the ability of converting inputs into outputs also accounts for environmental variables, the literature allows their inclusion in the production frontier in order to directly [18] or indirectly [8] influence the stochastic component or the nonstochastic component, respectively. In this study we investigate environmental variables' incorporation on the model, by specifying their direct and indirect influence on the dependent variable, through their additional inclusion on the asymmetrical error component [28].

This study seeks to analyze the production of winter rye crop in the year 1997/1998, conducted by *Research Center for Cultivar Testing of Słupia Wielka* in different experimental stations in Poland, from the application of two different regression techniques: JRA and SFA. We examine how certain variables may directly or indirectly affect production, as these variables had been incorporated into the deterministic and stochastic components of the production frontier [9]. By specifying the general model it investigated the direct influence of the environmental index (already obtained with the JRA) and of the two annual environmental factors (average air temperature and average rainfall), on the production of winter rye. The objective is to select the more productive genotype(s) and to identify the impact of some environmental factors in the level of production of winter rye crop (cereal). We also investigate possible relationships between genotypes and levels of fertility.

2 Methodologies

2.1 Joint-Regression Analysis

JRA may be used for the analysis of series of experiments for cultivar comparison. The technique is based on the adjustment of a linear regression, per cultivar, of the yield on a synthetic variable measuring productivity, the environmental index.

The principles of JRA were first developed by Mooers [23] and, after an ephemerous revival by Yates and Cochran [34], were resumed by Finlay and Wilkinson [11] forty-two years later. Important improvements to this technique were given by Eberhart and Russell [10], thus enabling to compare cultivars under a large range of fertility levels.

Despite the criticism still prevailing [19, 33] for not considering specific environmental variables, the technique continues to be largely used as a complement of traditional statistical analysis and is applied, mainly, in the assessment of the genotype x environmental interaction.

Initially the field trials were designed as randomized blocks. Gusmão [15–17] showed that the precision in analyzing series of randomized block experiments was highly increased by considering environmental indexes for individual blocks instead of only one environmental index per experiment. Following Gusmão [16] the (classic) environmental indexes of the blocks were measured by their average yields.

When incomplete blocks are used, such as is the case with α-designs, the classic environmental indexes can no longer be used, since it would lead to highly biased estimates for the environmental indexes corresponding to blocks. To overcome this problem Mexia et al. [21] introduced the L_2 environmental indexes obtained minimizing the sum of sums of squares of residuals, both in order to the coefficients of the regressions and to the environmental indexes.

The upper contour defined by the adjusted regression lines is a convex polygon which, see Mexia et al. [20, 22], can be used to carry out cultivar selection. The cultivars whose regression lines partake of the upper contour are the dominant ones (Fig. 1). The other cultivars are compared with the dominant ones. For each dominant cultivar there is a range of environmental indexes in which it has the highest yields.

This technique was systematically studied by Pereira [25] considering how to use it in cultivar selection.

2.1.1 α-Designs

When using incomplete blocks it is worthwhile to consider designs in which the blocks are grouped in superblocks, each containing any cultivar α times. We thus get, for all cultivars, yields obtained under similar conditions. Such designs will be resolvable in the sense of Shrikhande and Raghavarao [30, 31]. Those with $\alpha = 1$ are more and more used in agriculture and especially in cultivar testing. A very flexible family of resolvable designs is constituted by the α-designs introduced by Patterson and Williams [24]. While we could take $\alpha > 1$, it is usually preferable to

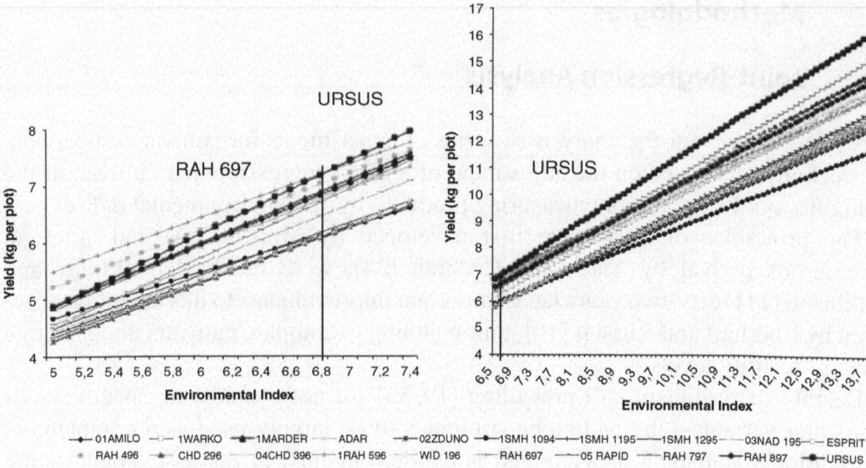

Fig. 1 Adjusted regressions using L_2 environmental indexes for a series of 17 experiments of α-designs of winter rye cultivars, in the years of 1997 and 1998

use $\alpha = 1$ in order to increase the within superblock homogeneity. Per experiment there will be a superblock. Thus, for each cultivar, we will have a replicate per location. As we will see, the choice of $\alpha = 1$ does not raise any problem while using L_2 environmental indexes.

2.1.2 L_2 Environmental Indexes

For convenience, let us consider data arranged in a two-way array with b rows and J columns. Suppose Y_{ij} is a continuous response variate (e.g., yield) for cultivar/genotype j in block i if present. The joint-regression model is

$$Y_{ij} = \alpha_j + \beta_j x_i + e_{ij}, \qquad i = 1, 2, \ldots, b, \;\; j = 1, 2, \ldots, J, \qquad (1)$$

with (α_j, β_j) the regression coefficients, for the J cultivars and the x_i, the block environmental indexes.

The goal function to be minimized will be

$$S(\alpha^J, \beta^J, x^b) = \sum_{i=1}^{b} \sum_{j=1}^{J} p_{ij} (Y_{ij} - \alpha_j - \beta_j x_i)^2. \qquad (2)$$

Usually the weight p_{ij} is 1 [0] when cultivar j is present [absent] from block i. When the cultivar occurs we take $p_{ij} = p_i$. These weights may differ from block to block to express differences in representativity of the blocks.

The main problem in such modeling is how to estimate the parameters. One can observe that the lately proposed and so-called zigzag algorithm is very efficient in finding the estimates of (α_j, β_j) and the x_i (cf. [21, 22, 27]), but it has not been established that it converges to the absolute minimum of the goal function.

We presented an alternative algorithm, double minimization algorithm, for the adjustment of JRA and showed that, in the complete case, it converges to the absolute minimum (cf. [26]). In the incomplete case we are developing a stochastic search algorithm to validate the use of the zigzag algorithm.

2.2 Stochastic Frontier Approach

Generally, SFA approach and efficiency analysis have two objectives [9]. The first purpose is to estimate a stochastic frontier for benchmarking uses through comparisons between productive units based on individual efficiency estimates. The second purpose is concerned with the incorporation of environmental variables, assumed to influence the global context of production, but which are neither inputs to the production process nor outputs of it. The stochastic frontier model is characterized by the utilization of a two-component error term. The symmetric component of the error term, or the statistical noise, captures the random variation of the frontier across observations, measurement errors, or random shocks' external to control. The other component is a one-sided variable which captures the inefficiency of the process or the deviance from the technological frontier. Following the specification proposed by Battese and Coelli [4] and Coelli, Perelman, and Romano [8], this study employs a stochastic production frontier function to measure technical efficiency of winter rye crop production. The theoretical model can be expressed as

$$y_j = f(x_j;\beta)\exp^{\varepsilon_j}, \qquad j = 1,2,\ldots,J, \tag{3}$$

where

y_j is the production of the jth cultivar;
$f(x_j;\beta)$ is a suitable production function;
x_j is a $(1 \times k)$ vector of inputs for the jth cultivar;
β is a $(k \times 1)$ vector of unknown parameters to be estimated;

$\varepsilon_j = v_j - u_j$ represents a stochastic component error, where V is assumed to be independently and identically distributed (i.i.d.) $N(0,\sigma_v)$ random error and independently distributed of the U; and U is a nonnegative random variable, associated with technical inefficiency of production, which is assumed to be independently distributed, such that U is obtained by truncation (at zero) of the normal distribution with mean $Z'\delta$ and variance σ_u^2. The main idea is that the first random variable (V) represents a noise component with an identical role to the error in the classical linear regression model and that the second random variable (U) represents the impact of management (or decisions about experimental design) inefficiencies on the dependent variable. So, when the value of U comes out to be equal to zero, the jth cultivar is located on the frontier, meaning that the greater the magnitude of U far away will be the production unit from the production frontier. In the stochastic frontier model (1), U can be specified as

$$u_j = z_j'\delta + w_j, \tag{4}$$

where Z is a $(1 \times m)$ vector of explanatory variables associated with technical inefficiency of production, δ is an $(m \times 1)$ vector of unknown coefficients, and W is a random variable defined by the truncation of the normal distribution with zero mean and variance σ_u^2, such that the point of truncation is $-z_j\delta$, i.e., $W_j > -z_j\delta$. These assumptions are consistent with U being a nonnegative truncation of the $N(z_j\delta, \sigma_u^2)$ distribution. The method of maximum likelihood is applied for simultaneous estimation of the parameters of the stochastic frontier and of the model for the technical inefficiency effects [5].

2.2.1 The General Model

As Eq. (3) specifies the stochastic frontier production function in terms of the original production values, it follows the logarithmic version as

$$lny_j = \beta_0 + \beta_1 ln(x_{1j}) + \beta_2 ln(x_{2j}) + \beta_3 ln(x_{3j}) + v_j - u_j, \tag{5}$$

where:

$ln(y_j)$ is the natural logarithm (i.e., logarithm to the base e) of rye crop production per acre (kg) and for j cultivar/genotype;
$ln(x_1)$ is the natural logarithm of the block environmental index (used in joint regression);
$ln(x_2)$ is the natural logarithm of average air temperature (air) (°C);
$ln(x_3)$ is the natural logarithm of monthly average amount of precipitation (pre) (in mm/ms).

2.2.2 The Inefficiency Effects Model

Following Eq. (4) the asymmetric component of error (U) is function of some controllable attributes (locality and variety) that can be written as

$$u_j = \delta_0 + \delta_1 z_{1j} + \delta_2 z_{2j} + w_j, \tag{6}$$

where:

z_1 is the natural logarithm of average air temperature (air) (°C);
z_2 is the natural logarithm of monthly average amount of precipitation (pre) (in mm/ms);

W is a random noise assumed to follow a normal distribution. Aigner, Lovell, and Schmidt [1] parameterized the log-likelihood function for the half-normal model in terms of $\sigma^2 = \sigma_v^2 + \sigma_u^2$ and $\lambda^2 = \sigma_u^2/\sigma_v^2 \geq 0$. The main idea is that if $\lambda = 0$, there are no technical inefficiency effects and all the deviations from the frontier are due to noise. The technical efficiency of production for the jth cultivar can be computed as $TE_j = exp(-u_j) = Y_j/Y_j^*$, where Y_j represents the level of observed output and Y_j^* represents the maximum possible output using the given level of inputs.

Table 1 Adjusted and determination coefficients

Cultivar	$\tilde{\alpha}$	$\tilde{\beta}$	R^2
URSUS	−1,59	1,29	0,96
RAH 797	−1,60	1,22	0,97
05RAPID	−0,78	1,12	0,97
1MARDER	−0,73	1,12	0,94
RAH 897	−0,55	1,09	0,95
ESPRIT	−0,22	1,07	0,92
WID 196	−0,38	1,06	0,96
03NAD 195	−0,68	1,05	0,93
02ZDUNO	−0,82	1,02	0,97
1RAH 596	−0,15	1,01	0,95
RAH 496	0,20	1,00	0,95
1WARKO	−0,63	0,99	0,96
CHD 296	−0,55	0,98	0,93
04CHD 396	−0,54	0,98	0,95
1SMH 1195	−0,45	0,96	0,93
ADAR	−0,35	0,96	0,96
RAH 697	0,77	0,95	0,91
01AMILO	−0,27	0,93	0,93
1SMH 1295	−0,16	0,93	0,96
1SMH 1094	0,65	0,80	0,90

3 Results

3.1 Joint-Regression Analysis Results

The data set used in this chapter is from a plant-breeding program of winter rye (*Secale cereale* L.) experiments carried out between 1997 and 1998 by the *Research Center for Cultivar Testing of Słupia Wielka* (Poland). 20 cultivars of winter rye are compared through these experimental designs. By design, there are four superblocks, each with five blocks of four plots. Each cultivar is present in a plot by superblock. Final results of adjustments made by applying the algorithm zigzag are presented in Figure 1 and Table 1.

In order to compare the cultivars that participate in the upper contour (dominant) with the others, we first used unilateral t tests (Table 2).

In addition to the dominant cultivars (URSUS and RAH 697) only the cultivars ESPRIT and RAH 496 are not significantly dominated at the 5 % level of probability, the range of the average superblocks. If we work at the 1 %, we must also include the 1MARDER as nonsignificantly dominated. We also used more robust methods such as the Scheffé and Bonferroni multiple comparison methods (Table 3).

The number of cultivars not dominated increase considerably. These results would point to a high performance of same cultivars.

Table 2 Significantly dominated cultivars at the significance level of 5 %, using the one-sided t test

Dominant cultivars	RAH 697	URSUS
Range of dominance	[5,42 ; 6,84]	[6,84 ; 13,47]
Dominated cultivars at 5 % significance level	RAH 797; 05RAPID; 1MARDER; RAH 897; WID 196; 03NAD 195; 02ZDUNO; 1RAH 596; 1WARKO; CHD 296; 04CHD 396; 1SMH 1195; ADAR; 01AMILO; 1SMH 1295; 1SMH 1094	RAH 797; 05RAPID; 1MARDER; RAH 897; WID 196; 03NAD 195; 02ZDUNO; 1RAH 596; 1WARKO; CHD 296; 04CHD 396; 1SMH 1195; ADAR; 01AMILO; 1SMH 1295; 1SMH 1094

Table 3 Significantly dominated cultivars at the significance level of 5 %, using the Scheffé and Bonferroni multiple comparisons methods

Dominant cultivars	RAH 697	URSUS
Range of dominance	[5,42 ; 6,84]	[6,84 ; 13,47]
Scheffé method	01AMILO; 1SMH 1295; 1SMH 1094; 03NAD 195; 02ZDUNO; 1WARKO; CHD 296; 04CHD 396; 1SMH 1195; ADAR	03NAD 195; 02ZDUNO; 1WARKO; CHD 296; 04CHD 396; 1SMH 1195; ADAR; 01AMILO; 1SMH 1295; 1SMH 1094
Bonferroni method	01AMILO; 1SMH 1295; 1SMH 1094; RAH 797; WID 196; 03NAD 195; 02ZDUNO; 1RAH 596; 1WARKO; CHD 296; 04CHD 396; 1SMH 1195; ADAR	RAH 797; WID 196; 03NAD 195; 02ZDUNO; 1RAH 596; 1WARKO; CHD 296; 04CHD 396; 1SMH 1195; ADAR; 01AMILO; 1SMH 1295; 1SMH 1094

3.2 Stochastic Frontier Approach Results

The general stochastic frontier production model defined by Eq. (5) and the technical inefficiency model defined by Eq. (6) were jointly estimated by the maximum likelihood method, using Frontier 4.1 [7]. For the distributional specification of asymmetric component error, two different distributional assumptions have been assumed, half-normal and normal-truncated distributions. This solution gave rise to alternative prediction ranks of inefficiency. The validity of the model had been investigated through some tests of hypotheses performed using generalized likelihood-ratio statistics, LR. From the maximized log-likelihood values, the first null hypothesis that we tested is concerned with the adequability of the half-normal model ($H_0 : \mu = 0$) against the alternative hypothesis (normal truncated, or $H_1 : \mu \neq 0$). As the likelihood-ratio (LR) statistic $[LR = -2ln[L(H_0/L(H_1) = 12]$ exceeds the 5 % critical value of the chi-square value with one degree of freedom, it leads to the rejection that the half-normal model is adequate. The second null hypothesis tested is $H_0 : \lambda = \delta_0 = \delta_1 = \delta_2 = 0$ which specifies that technical inefficiency effects are not present in the model. This implies that the stochastic frontier function is not appropriate, or not so different from the traditional average

Table 4 Maximum likelihood estimates for SFA

Variables	Parameters	Estimate (t ratio) Normal truncated
Stochastic production frontier		
Constant	β_0	$-8{,}37(-17, 1)^{**}$
$ln(X_1)$	β_1	$1{,}11(9, 85)^{**}$
$ln(X_2)$	β_2	$1{,}90(7, 42)^{**}$
$ln(X_3)$	β_3	$0{,}56(7, 28)^{**}$
Inefficiency effects model		
Constant	δ_0	$-1{,}83(-50, 1)^{**}$
$ln(Z_1)$	δ_1	$1{,}97(8, 89)^{**}$
$ln(Z_2)$	δ_2	0,57(6,46)
	σ^2	$73{,}9(36, 9)^{**}$
	λ	$0{,}9999(43, 2)^{**}$

** Significant at the 0.01 probability level

production function (ordinary least squares—OLS—estimation procedure should be more adequate). This hypothesis was rejected because the LR statistic exceed the 5 % critical value of the chi-square value (9,488) with 4 degrees of freedom (four restrictions). Nine parameters are estimated in the stochastic production frontier model: four in the stochastic frontier model, three in the inefficiency effects model, and two parameters associated with the variances of the component error term ($\sigma^2 = \sigma_v^2 + \sigma_u^2$ and λ). All the seven estimated parameters are statistically significant at five percent level (Table 4).

These results imply that all the selected variables improve wheat productivity significantly. The positive signals of all coefficients (except the constant) are as expected as well as the magnitude of t-ratios either associated with the stochastic production frontier or associated with the inefficiency effects model. Results also show that the average measure of efficiency is 0,70 (sd. 0,11) ranging from 0,43 to 0,999. The major reason for this discrepancy appears to be associated with environmental heterogeneity of experimental data design.

The results indicate that the three most efficient cultivars (URSUS) are located in Rarwino locality and the less productive cultivar (02ZDUNO) is located in Lubliniec Nowy (Table 5). The obtained ranking of efficiency highlights the importance of URSUS in Rarwino local, as this combination (cultivar, location) was the most efficient (0,999). Additionally, and with the same ranking, it was obtained the less efficiency cultivar (0,43) as being 02ZDUNO, growing in the Lubliniec Nowy local.

4 Conclusion

This study is concerned with an experimental design developed in agricultural area. This chapter uses two specific regression techniques for rye crop productivity estimation proposes: JRA and SFA. Both methodologies are concerned with research

Table 5 Spatial distribution of efficiency measures by genotype

Locality	Variety	Efficiency
Most efficiency		
1: Rarwino	URSUS	0,999
2: Glodowo	URSUS	0,997
3: Glodowo	URSUS	0,992
Less efficiency		
1: Lubliniec Nowy	02ZDUNO	0,431
2: Lubliniec Nowy	01AMILO	0,446
3: Dukla	1SMH1195	0,449

on winter rye production level of cultivars obtained from experimental data. With JRA it was adjusted, per cultivar, a linear regression of yield on a summarized productivity variable. Using a new algorithm (zigzag algorithm) two cultivars were selected (RAH 697 and URSUS in Pokoj and Rarwino local, respectively) as the most appropriated for less fertile land and for more fertile land, respectively. URSUS cultivar is the dominant variety on the right of the upper contour and so should be selected to grow in more fertile land. As RAH 697 cultivar is the dominant variety on the left of the upper contour it should be selected to grow in less fertile land.

With SFA approach it had been estimated a stochastic frontier production function with a normal truncated distribution for the asymmetric component of the error term and with the incorporation of an inefficient effects model in the global specification. With this technique, URSUS cultivar, in Rarwino local, was also selected as being the most efficient cultivar (or the most productive) and the 02ZDUNO cultivar, in Lubliniec Nowy local, was selected as the cultivar less productive. We point out that this cultivar was significantly dominated in the range of the average of superblocks, obtained with the JRA technique.

Although both techniques have been widely applied in agricultural field, the use of SFA in order to analyze the impact of genotype x environmental interaction on the production of winter rye represents an innovation as it is a technique mostly applied on economic modeling. Combination of SFA optimization potentialities with the strongness of the JRA technique in cultivar selection task represents an interesting field, especially in the empirical research domain. Using SFA, a rank of cultivars in terms of efficiency was obtained, just applying the MLE method. With JRA, OLS was used in order to select cultivars with the same goal. It is interesting to note the excellent agreement reached with both estimation techniques. JRA and SFA both indicate the URSUS cultivar as being the most efficient and productive one. Nevertheless future research should make an attempt:

1. To cross information between efficiency rankings and two clusters of productivity (left-dominant and right-dominant ones) obtained with JRA
2. To estimate partial efficiency rankings for subsamples of productivity, as suggested by JRA results

Acknowledgments The *Research Center for Cultivar Testing de Słupia Wielka* (Poland) is thanked for allowing the use of their data in this study.

The first author of this work is a member of the CIMA-UE, research center financed by the Science and Technology Foundation, Portugal.

We thank the anonymous referees for their suggestions, which greatly improved this chapter.

References

1. Aigner, D., Lovell, K., Schmidt, P.: Formulation and estimation of stochastic frontier function models. J. Econometrics **6**, 21–37 (1977)
2. Barros, C., Sampaio, A.: Technical and allocative efficiency in airports. Int. J. Transp. Econ. **31**(3), 355–377 (2004)
3. Battese, G.E., Broca, S.S.: Functional forms of stochastic frontier production functions and models for technical inefficiency effects: a comparative study for wheat farmers in Pakistan. J. Prod. Anal. **8**(4), 395–414 (1997)
4. Battese, G.E., Coelli, T.J.: A model for technical inefficiency effects in a stochastic frontier production function for panel data. Empir. Econ. **20**, 325–332 (1995)
5. Battese, G.E., Coelli, T.J.: Frontier production functions, technical efficiency and panel data: with application to paddy farmers in India. J. Prod. Anal. **3**, 153–169 (1992)
6. Battese, G.E., Corra, G.S.: Estimation of a production frontier model: with application to the pastoral zone of Eastern Australia. Aust. J. Agr. Econ. **21**, 169–79 (1977)
7. Coelli, T.J.: A Guide to Frontier Version 4.1: A Computer Program for Stochastic Frontier Production and Cost Function Estimation. CEPA Working Papers, No. 7/96, School of Economics, University of New England, Armidale (1996)
8. Coelli, T., Perelman, S., Romano, E.: Accounting for environmental influences in stochastic frontier models: application to international airlines. J. Prod. Anal. **11**, 251–273 (1999)
9. Coelli, T., Rao, D.S., O'Donnell, C., Battese, G.: An Introduction to Efficiency and Productivity Analysis. Springer, New York (2005)
10. Eberhardt, S.A., Russell, W.A.: Stability parameters for comparing varieties. Crop. Sci. **6**, 36–40 (1966)
11. Finlay, K.W., Wilkinson, G.N.: The analysis of adaptation in a plant-breeding programme. Aust. J. Agric. Res. **14**, 742–754 (1963)
12. Greene, W.: Frontier production functions. In: Pesaran, M., Schmidt, P. (eds.) Handbook of Applied Econometrics, Volume II: Microeconometrics. Blackwell, Oxford (1997)
13. Greene, W.: Maximum likelihood estimation of econometric frontier. J. Econometrics. **13**, 27–56 (1980)
14. Greene, W.: Fixed and random effects in stochastic frontier models. J. Prod. Anal. **23**, 7–32 (2004)
15. Gusmão, L.: A interação genótipo ambiente e a comparação de cultivares de cereais. Ph'd Thesis, Instituto Superior de Agronomia, Universidade Técnica de Lisboa, Lisboa (1986)
16. Gusmão, L.: An adequate design for regression analysis of yield trials. Theor. Appl. Genet. **71**, 314–319 (1985)
17. Gusmão, L.: Inadequacy of blocking in cultivar yield trials. Theor. Appl. Genet. **72**, 98–104 (1986)
18. Kumbhakar, S.C., Ghosh, S., McGuckin, J.T.: A Generalised production frontier approach for estimating determinants of inefficiency in U.S. dairy farms. J. Bus. Econ. Stat. **9**(3), 279–86 (1991)
19. Lin, C.S., Binns, M.R., Lefkovitch, L.P.: Stability analysis: where do we stand? Crop. Sci. **26**, 894–900 (1986)
20. Mexia, J.T., Amaro, A.P., Gusmão, L., Baeta, J.: Upper contour of a joint regression analysis. J. Genet. Breed. **51**, 253–255 (1997)

21. Mexia, J.T., Pereira, D.G., Baeta, J.: L_2 environmental indexes. Biometrical Lett. **36**, 137–143 (1999)
22. Mexia, J.T., Pereira, D.G., Baeta, J.: Weighted linear joint regression analysis. Biometrical Lett. **38**, 33–40 (2001)
23. Mooers, C.A.: The agronomic placement of varieties. J. Amer. Soc. Agron. **13**, 337–352 (1921)
24. Patterson, H.D., Williams, E.R.: A new class of resolvable incomplete block designs. Biometrika **63**, 83–92 (1976)
25. Pereira, D.G.: Análise conjunta pesada de regressões em redes de ensaios. Ph'd Thesis, Universidade de Évora (2004)
26. Pereira, D.G., Mexia, J.T.: Comparing double minimization and zigzag algorithms in joint regression analysis: the complete case. J. Stat. Comput. Sim. **80**, 133–141 (2010)
27. Pereira, D.G., Mexia, J.T.: The use of joint regression analysis in selecting recommended cultivars. Biuletyn Oceny Odmian (Cultivar Testing Bulletin) **31**, 19–25 (2003)
28. Reifschneider, D., Stevenson, R.: Systematic departures from the frontier : a framework for the analysis of firm inefficiency. Int. Econ. Rev. **32**, 715–723 (1991)
29. Sampaio, A.: Medicin de la eficiencia en el servicio pblico de distribucin de agua en Portugal. European Ph'd Thesis, Extremadura University, Spain (2007)
30. Shrikhande, S.S., Raghavarao, D.: A method of construction of incomplete block designs. Sankhya Ser. A. **25**, 399–402 (1963)
31. Shrikhande, S.S., Raghavarao, D.: Affine α-resolvable incomplete block designs. In: Rao, C.R. (ed.) Contributions to Statistic, pp. 471–480. Pregamon Press, Statistical Publishing Society, Calcutta (1964)
32. Stevenson, R.E.: Likelihood functions for generalised stochastic frontier estimation. J. Econometrics **13**, 57–66 (1980)
33. Westcott, B.: Some methods of analysing genotype-environment interaction. Heredity **56**, 243–253 (1986)
34. Yates, F., Cochran, W.G.: The analysis of groups experiments. J. Agric. Sci. (Cambridge) **28**, 556–580 (1938)

On the Maximum and Minimum of a Stationary Random Field

Luísa Pereira

Abstract
We determine the class of nondegenerate joint-limiting distributions for the maximum and minimum of stationary random fields $\mathbf{X} = \{X_{\mathbf{n}}\}_{\mathbf{n}\in\mathbb{N}^2}$ satisfying a long-range dependence restriction for each coordinate direction at a time. Unlike the classical result for i.i.d. random fields the maximum and minimum of $\mathbf{X}$ may be asymptotically dependent. We also give a sufficient condition for the asymptotic independence of the maximum and minimum. Additional conditions are given in order to obtain the asymptotic independence of the locations of maximum and minimum.

1 Introduction

Let $\mathbf{X} = \{X_{\mathbf{n}}\}_{\mathbf{n}\in\mathbb{N}^2}$ be a stationary random field on $\mathbb{N}^2$ with common distribution function F and, for a subset $\mathbf{I}$ of the rectangle of points $\mathbf{R_n} = \{1,\ldots,n_1\} \times \{1,\ldots,n_2\} \subset \mathbb{N}^2$, $M_{\mathbf{n}}(\mathbf{I}) = \max\{X_{\mathbf{i}} : \mathbf{i} \in \mathbf{I}\}$ and $W_{\mathbf{n}}(\mathbf{I}) = \min\{X_{\mathbf{i}} : \mathbf{i} \in \mathbf{I}\}$. When $\mathbf{I} = \mathbf{R_n}$, we simply write $M_{\mathbf{n}}$ and $W_{\mathbf{n}}$.

For $\mathbf{n} = (n_1, n_2)$, the condition $\mathbf{n} \to \infty$ means $n_s \to \infty$, $s = 1, 2$.

We say the pair $\mathbf{I} \subset \mathbb{N}^2$ and $\mathbf{J} \subset \mathbb{N}^2$ is in $S_i(l)$, for each $i = 1, 2$, if the distance between $\Pi_i(\mathbf{I})$ and $\Pi_i(\mathbf{J})$ is greater or equal to l, where $\Pi_i, i = 1, 2$, denote the cartesian projections. The distance, $d(\mathbf{I}, \mathbf{J})$, between sets $\mathbf{I}$ and $\mathbf{J}$ of $\mathbb{N}^2$, is the minimum of distances $d(\mathbf{i}, \mathbf{j}) = \max\{|i_s - j_s| : s \in \{1, 2\}\}$, $\mathbf{i} \in \mathbf{I}$ and $\mathbf{j} \in \mathbf{J}$.

Denote by $\widehat{M}_{\mathbf{n}}$ and $\widehat{W}_{\mathbf{n}}$ the corresponding maximum and minimum of the associated random field, $\widehat{\mathbf{X}} = \{\widehat{X}_{\mathbf{n}}\}_{\mathbf{n}\in\mathbb{N}^2}$, of independent and identically distributed (i.i.d.) random variables having the same distribution function F.

L. Pereira (✉)
University of Beira Interior, Covilhã, Portugal
e-mail: lpereira@ubi.pt

J. Lita da Silva et al. (eds.), *Advances in Regression, Survival Analysis, Extreme Values, Markov Processes and Other Statistical Applications*, Studies in Theoretical and Applied Statistics, DOI 10.1007/978-3-642-34904-1_35,

We shall assume that there are sequences of constants $\{a_{\mathbf{n}} > 0\}_{\mathbf{n}\in\mathbb{N}^2}$, $\{b_{\mathbf{n}}\}_{\mathbf{n}\in\mathbb{N}^2}$, $\{c_{\mathbf{n}} > 0\}_{\mathbf{n}\in\mathbb{N}^2}$ and $\{d_{\mathbf{n}}\}_{\mathbf{n}\in\mathbb{N}^2}$ such that for $\{u_{\mathbf{n}}(x) = a_{\mathbf{n}}x + b_{\mathbf{n}}\}_{\mathbf{n}\in\mathbb{N}^2}$ and $\{v_{\mathbf{n}}(y) = c_{\mathbf{n}}y + d_{\mathbf{n}}\}_{\mathbf{n}\in\mathbb{N}^2}$, $x, y \in \mathbb{R}$,

$$P(\widehat{M}_{\mathbf{n}} \le u_{\mathbf{n}}(x)) \to_{\mathbf{n}\to\infty} \widehat{G}(x) \tag{1}$$

and

$$P(\widehat{W}_{\mathbf{n}} \le v_{\mathbf{n}}(y)) \to_{\mathbf{n}\to\infty} \widehat{H}(y), \tag{2}$$

where $\widehat{H}$ and $\widehat{G}$ are nondegenerate distribution functions.

The classes of the possible nondegenerate distribution functions which may occur as limits in Eqs. (1) and (2) form, respectively, the classes of *max-stable* and *min-stable* distributions.

The asymptotic independence of maximum and minimum, under linear normalizations, holds for $\widehat{\mathbf{X}}$, *id est*,

$$P(\widehat{M}_{\mathbf{n}} \le u_{\mathbf{n}}(x), \widehat{W}_{\mathbf{n}} \le v_{\mathbf{n}}(y)) \to_{\mathbf{n}\to\infty} \widehat{Q}(x, y) = \widehat{G}(x)\widehat{H}(y), \tag{3}$$

and it can be used, for instance, to approximate the distribution of the sample range.

Using the ideas of [2], in Sect. 2, we determine the class of all joint-limiting distributions for the maximum and minimum of stationary random fields, $\mathbf{X}$, under appropriate long-range dependence restrictions. Unlike the result established in Eq. (3) for the i.i.d. case, the maximum and minimum of stationary random fields may be asymptotically dependent, as we shall see here. We also give a necessary and sufficient condition for asymptotic independence also to hold for stationary random fields.

In Sect. 3 we deal with the asymptotic joint distribution of the locations of maximum and minimum, $L_{\mathbf{n}}^{(Max)}$ and $L_{\mathbf{n}}^{(Min)}$, defined in [6] as

$$L_{\mathbf{n}}^{(i)} = \begin{cases} \mathbf{j}^{(1)} & \text{if } \mathcal{P}_i = \{\mathbf{j}^{(1)}\} \\ \mathbf{j}^{(2)} & \text{if } |\mathcal{P}_i| > 1 \wedge \mathcal{Q}_i = \{\mathbf{j}^{(2)}\} \\ \mathbf{j}^{(3)} & \text{if } |\mathcal{Q}_i| > 1 \wedge \mathcal{R}_i = \{\mathbf{j}^{(3)}\} \end{cases} \quad , i \in \{Max, Min\},$$

where $|A|$ denotes the cardinal of a set A, $\Im_{Max} = \{\mathbf{j} \in \mathbb{N}^2 : X_{\mathbf{j}} = M_{\mathbf{n}}\}$, $\Im_{Min} = \{\mathbf{j} \in \mathbb{N}^2 : X_{\mathbf{j}} = W_{\mathbf{n}}\}$, and for each $i \in \{Max, Min\}$,

$$\mathcal{P}_i = \{\mathbf{j} \in \Im_i : \forall \mathbf{j}' \in \Im_i, d(\mathbf{j}, \mathbf{1}) \le d(\mathbf{j}', \mathbf{1})\}, \ \ with\ \mathbf{1} = (1, 1),$$

$$\mathcal{Q}_i = \{\mathbf{j} \in \mathcal{P}_i : \forall \mathbf{j}' \in \mathcal{P}_i, j_1 \le j_1'\} \ \ and\ \ \mathcal{R}_i = \{\mathbf{j} \in \mathcal{Q}_i : \forall \mathbf{j}' \in \mathcal{Q}_i, j_2 \le j_2'\}.$$

The study of the relationship between extreme values and their locations has important practical applications, for instance, when dealing with censored data.

In Pereira [6] it was shown that the normalized location of the maximum of a stationary random field with extremal index $\theta \in (0, 1]$ satisfying a long-range dependence condition for each coordinate at a time converges to a uniform variable on $[0,1]^2$ and is asymptotically independent of the height of the maximum.

Here, by assuming the asymptotic independence of $M_{\mathbf{n}}$ and $W_{\mathbf{n}}$, we establish conditions under which, for each $\epsilon_{11}, \epsilon_{12}, \epsilon_{21}, \epsilon_{22} \in (0,1)$, the random vectors $(M_{\mathbf{n}}([1, n_1\epsilon_{11}] \times [1, n_2\epsilon_{12}] \cap \mathbb{N}^2), M_{\mathbf{n}}(\mathbf{R_n} - ([1, n_1\epsilon_{11}] \times [1, n_2\epsilon_{12}] \cap \mathbb{N}^2)))$ and $(W_{\mathbf{n}}([1, n_1\epsilon_{21}] \times [1, n_2\epsilon_{22}] \cap \mathbb{N}^2), W_{\mathbf{n}}(\mathbf{R_n} - ([1, n_1\epsilon_{21}] \times [1, n_2\epsilon_{22}] \cap \mathbb{N}^2)))$, under linear normalizations, are asymptotically independent, which, in turn, will lead to the asymptotic independence of the locations of maximum and minimum.

2 Asymptotic Independence of Maximum and Minimum

The dependence structure used here is a coordinatewise-mixing condition, which restricts dependence by limiting

$$\begin{aligned}&|P(v_{\mathbf{n}} < W_{\mathbf{n}}(\mathbf{I}_1) < M_{\mathbf{n}}(\mathbf{I}_1) \le u_{\mathbf{n}}, v_{\mathbf{n}} < W_{\mathbf{n}}(\mathbf{I}_2) < M_{\mathbf{n}}(\mathbf{I}_2) \le u_{\mathbf{n}})\\ &- P(v_{\mathbf{n}} < W_{\mathbf{n}}(\mathbf{I}_1) < M_{\mathbf{n}}(\mathbf{I}_1) \le u_{\mathbf{n}})P(v_{\mathbf{n}} < W_{\mathbf{n}}(\mathbf{I}_2) < M_{\mathbf{n}}(\mathbf{I}_2) \le u_{\mathbf{n}})|\end{aligned}$$

with the two index sets $\mathbf{I}_1, \mathbf{I}_2 \subset \mathbf{R_n}$ being "separated" from each other by a certain distance along each direction.

Definition 1. Let $\{u_{\mathbf{n}}\}_{\mathbf{n}\in\mathbb{N}^2}$ and $\{v_{\mathbf{n}}\}_{\mathbf{n}\in\mathbb{N}^2}$ be sequences of real numbers. The random field $\mathbf{X}$ satisfies the condition $\Delta(u_{\mathbf{n}}, v_{\mathbf{n}})$ if there exist integer sequences $\{k_{n_i}\}_{n_i\ge1}$, $\{l_{n_i}\}_{n_i\ge1}$, $i = 1, 2$, such that, as $\mathbf{n} = (n_1, n_2) \to \infty$, we have

$$(k_{n_1}, k_{n_2}) \to \infty \quad \left(\frac{k_{n_1} l_{n_1}}{n_1}, \frac{k_{n_2} l_{n_2}}{n_2}\right) \to \mathbf{0} \quad \left(k_{n_1} \Delta^{(1)}_{\mathbf{n}, l_{n_1}}, k_{n_1} k_{n_2} \Delta^{(2)}_{\mathbf{n}, l_{n_2}}\right) \to \mathbf{0}, \qquad (4)$$

where the coefficients $\Delta^{(i)}_{\mathbf{n}, l_{n_i}}$, $i = 1, 2$, are defined as follows:

$$\begin{aligned}\Delta^{(1)}_{\mathbf{n}, l_{n_1}} &= \sup |P\left(v_{\mathbf{n}} < W_{\mathbf{n}}(\mathbf{I}_1) < M_{\mathbf{n}}(\mathbf{I}_1) \le u_{\mathbf{n}}, v_{\mathbf{n}} < W_{\mathbf{n}}(\mathbf{I}_2) < M_{\mathbf{n}}(\mathbf{I}_2) \le u_{\mathbf{n}}\right)\\ &- P\left(v_{\mathbf{n}} < W_{\mathbf{n}}(\mathbf{I}_1) < M_{\mathbf{n}}(\mathbf{I}_1) \le u_{\mathbf{n}}\right) P\left(v_{\mathbf{n}} < W_{\mathbf{n}}(\mathbf{I}_2) < M_{\mathbf{n}}(\mathbf{I}_2) \le u_{\mathbf{n}}\right)|, \quad (5)\end{aligned}$$

where the supremum is taken over pairs $\mathbf{I}_1$ *and* $\mathbf{I}_2$ *in* $S_1(l_{n_1})$, such that $|\Pi_1(\mathbf{I}_2)| \le \frac{n_1}{k_{n_1}}$,

$$\begin{aligned}\Delta^{(2)}_{\mathbf{n}, l_{n_2}} &= \sup |P\left(v_{\mathbf{n}} < W_{\mathbf{n}}(\mathbf{I}_1) < M_{\mathbf{n}}(\mathbf{I}_1) \le u_{\mathbf{n}}, v_{\mathbf{n}} < W_{\mathbf{n}}(\mathbf{I}_2) < M_{\mathbf{n}}(\mathbf{I}_2) \le u_{\mathbf{n}}\right)\\ &- P\left(v_{\mathbf{n}} < W_{\mathbf{n}}(\mathbf{I}_1) < M_{\mathbf{n}}(\mathbf{I}_1) \le u_{\mathbf{n}}\right) P\left(v_{\mathbf{n}} < W_{\mathbf{n}}(\mathbf{I}_2) < M_{\mathbf{n}}(\mathbf{I}_2) \le u_{\mathbf{n}}\right)|, \quad (6)\end{aligned}$$

where the supremum is taken over pairs $\mathbf{I}_1$ and $\mathbf{I}_2$ in $S_2(l_{n_2})$ such that $\Pi_1(\mathbf{I}_1) = \Pi_1(\mathbf{I}_2)$ and $|\Pi_2(\mathbf{I}_2)| \le \frac{n_2}{k_{n_2}}$.

If in Eqs. (5) and (6) we consider the events $\{M_{\mathbf{n}}(\mathbf{I}_1) \le u_{\mathbf{n}}\}$ and $\{M_{\mathbf{n}}(\mathbf{I}_2) \le u_{\mathbf{n}}\}$ instead of $\{v_{\mathbf{n}} < W_{\mathbf{n}}(\mathbf{I}_1) < M_{\mathbf{n}}(\mathbf{I}_1) \le u_{\mathbf{n}}\}$ and $\{v_{\mathbf{n}} < W_{\mathbf{n}}(\mathbf{I}_2) < M_{\mathbf{n}}(\mathbf{I}_2) \le u_{\mathbf{n}}\}$, respectively, we obtain the coordinatewise-mixing condition of Leadbetter and Rootzén [3], under which the Extremal Types Theorem for stationary random fields is proved.

The extremal index of $\mathbf{X}$, introduced in Choi [1], is the key parameter to relate the limiting distributions for the maxima of $\mathbf{X}$ and $\widehat{\mathbf{X}}$.

Definition 2. The random field $\mathbf{X}$ has extremal index θ, $0 < \theta \le 1$ if for each $\tau > 0$ there exists $\left\{u_{\mathbf{n}}^{(\tau)}\right\}_{\mathbf{n}\in\mathbb{N}^2}$ such that, as $\mathbf{n} \to \infty$, $n_1 n_2 P\left(X_1 > u_{\mathbf{n}}^{(\tau)}\right) \to \tau$ and $P\left(M_{\mathbf{n}} \le u_{\mathbf{n}}^{(\tau)}\right) \to \exp(-\theta\tau)$.

The extremal indexes of $\mathbf{X}$ and $-\mathbf{X} = \{-X_{\mathbf{n}}\}_{\mathbf{n}\in\mathbb{N}^2}$, which are, respectively, measures of clustering of high and low values of the random field $\mathbf{X}$, will be denominated, respectively, by superior and inferior extremal indexes. We denote them by $\overline{\theta}$ and $\underline{\theta}$, respectively.

The existence of the superior and inferior extremal indexes of $\mathbf{X}$ allows us to write

$$\lim_{\mathbf{n}\to\infty} P(M_{\mathbf{n}} \le u_{\mathbf{n}}(x)) = \widehat{G}^{\overline{\theta}}(x), \quad \lim_{\mathbf{n}\to\infty} P(W_{\mathbf{n}} > v_{\mathbf{n}}(y)) = (1 - \widehat{H}(y))^{\underline{\theta}} \tag{7}$$

and $u_{\mathbf{n}}(x) = u_{\mathbf{n}}^{(\tau(x))}$, $v_{\mathbf{n}}(y) = v_{\mathbf{n}}^{(\tau'(y))}$ with $\tau(x) = -\log \widehat{G}(x)$, $\tau'(y) = -\log(1 - \widehat{H}(y))$, $\forall x, y \in \mathbb{R}$.

The condition $\Delta(u_{\mathbf{n}}, v_{\mathbf{n}})$ allows us to obtain the asymptotic independence of certain class of events.

Lemma 1. *Let* $\left\{u_{\mathbf{n}}^{(\tau)}\right\}_{\mathbf{n}\in\mathbb{N}^2}$ *and* $\left\{v_{\mathbf{n}}^{(\tau')}\right\}_{\mathbf{n}\in\mathbb{N}^2}$ *be sequences of real numbers such that*

$$n_1 n_2 P\left(X_1 > u_{\mathbf{n}}^{(\tau)}\right) \xrightarrow[\mathbf{n}\to\infty]{} \tau \ \text{and} \ n_1 n_2 P\left(X_1 < v_{\mathbf{n}}^{(\tau')}\right) \xrightarrow[\mathbf{n}\to\infty]{} \tau', \tag{8}$$

where $\tau, \tau' < +\infty$. *If* $\mathbf{X}$ *verifies* $\Delta(u_{\mathbf{n}}^{(\tau)}, v_{\mathbf{n}}^{(\tau')})$, *for sequences* $\{k_{n_i}\}_{n_i\ge1}$, $\{l_{n_i}\}_{n_i\ge1}$, $i = 1, 2$, $\left\{u_{\mathbf{n}}^{(\tau)}\right\}_{\mathbf{n}\in\mathbb{N}^2}$ *and* $\left\{v_{\mathbf{n}}^{(\tau')}\right\}_{\mathbf{n}\in\mathbb{N}^2}$, *satisfying Eq. (4) and* $\mathbf{V}_{s,t} = I_s \times J_{s,t} \subset \mathbf{R}_{\mathbf{n}}$, $s = 1, \ldots, k_{n_1}$, $t = 1, \ldots, k_{n_2}$, *are disjoint rectangles, then*

$$\left| P\left(\bigcap_{s=1}^{k_{n_1}}\bigcap_{t=1}^{k_{n_2}}\bigcap_{\mathbf{i}\in\mathbf{V}_{s,t}} \left\{v_{\mathbf{n}}^{(\tau')} < X_{\mathbf{i}} \le u_{\mathbf{n}}^{(\tau)}\right\}\right) - \prod_{s=1}^{k_{n_1}}\prod_{t=1}^{k_{n_2}} P\left(\bigcap_{\mathbf{i}\in\mathbf{V}_{s,t}} \left\{v_{\mathbf{n}}^{(\tau')} < X_{\mathbf{i}} \le u_{\mathbf{n}}^{(\tau)}\right\}\right)\right| \xrightarrow[\mathbf{n}\to\infty]{} 0.$$

Proof. Since Eq. (8) holds, from Eq. (4) we can assume, for each $s = 1, \ldots, k_{n_1}$, $t = 1, \ldots, k_{n_2}$, that $\Pi_i(\mathbf{V}_{s,t})$ consists of at least l_{n_i} integers, $i = 1, 2$. If all pairs of rectangles $\mathbf{V}_{s,t}$ are in $S_1(l_{n_1}) \cup S_2(l_{n_2})$ then the result follows inductively from the condition $\Delta(u_{\mathbf{n}}^{(\tau)}, v_{\mathbf{n}}^{(\tau')})$. If some pair of rectangles $\mathbf{V}_{s,t}$ are not in $S_1(l_{n_1}) \cup S_2(l_{n_2})$ we can eliminate l_{n_1} columns or l_{n_2} rows in order to obtain $\mathbf{V}^*_{s,t} \subset \mathbf{V}_{s,t}$, $s = 1, \ldots, k_{n_1}$, $t = 1, \ldots, k_{n_2}$, to which we can apply inductively the condition $\Delta(u_{\mathbf{n}}^{(\tau)}, v_{\mathbf{n}}^{(\tau')})$. □

In the following result we obtain the class of nondegenerate joint-limiting distributions for the maximum and minimum of stationary random fields verifying condition $\Delta(u_{\mathbf{n}}, v_{\mathbf{n}})$.

Proposition 1. *Let* $\mathbf{X}$ *be a stationary random field and* $\{a_{\mathbf{n}} > 0\}_{\mathbf{n}\in\mathbb{N}^2}$, $\{b_{\mathbf{n}}\}_{\mathbf{n}\in\mathbb{N}^2}$, $\{c_{\mathbf{n}} > 0\}_{\mathbf{n}\in\mathbb{N}^2}$ *and* $\{d_{\mathbf{n}}\}_{\mathbf{n}\in\mathbb{N}^2}$ *given sequences of real numbers such that*

$$P(a_{\mathbf{n}}^{-1}(M_{\mathbf{n}} - b_{\mathbf{n}}) \le x, c_{\mathbf{n}}^{-1}(W_{\mathbf{n}} - d_{\mathbf{n}}) \le y) \xrightarrow[\mathbf{n}\to\infty]{} Q(x, y),$$

for all $x, y \in \mathbb{R}$, *where* Q *is a nondegenerate distribution function. Suppose that* $\Delta(u_{\mathbf{n}}(x), v_{\mathbf{n}}(-y))$ *is satisfied for* $u_{\mathbf{n}}(x) = a_{\mathbf{n}}x + b_{\mathbf{n}}$, $v_{\mathbf{n}}(y) = c_{\mathbf{n}}y + d_{\mathbf{n}}$, *for each* $x, y \in \mathbb{R}$. *Then* $Q(x, y) = R(x, \infty) - R(x, -y)$ *where*

$$R(x, y) = \lim_{\mathbf{n}\to\infty} P(M_{\mathbf{n}} \le u_{\mathbf{n}}(x), W_{\mathbf{n}} > v_{\mathbf{n}}(-y))$$

is a bivariate extreme value distribution.

Proof. Since $\mathbf{X}$ verifies $\Delta(u_{\mathbf{n}}(x), v_{\mathbf{n}}(-y))$, by Lemma 1, we have

$$P(M_{\mathbf{n}} \le u_{\mathbf{n}k}(x), W_{\mathbf{n}} > v_{\mathbf{n}k}(-y)) \xrightarrow[\mathbf{n}\to\infty]{} R^{\frac{1}{k}}(x, y), \quad k \in \mathbb{N}.$$

Employing the multivariate analogue of the convergence of types result, there exist constants $A_k > 0$, B_k, $C_k > 0$, D_k such that $R^k(A_k x + B_k, C_k y + D_k) = R(x, y)$. So, the dependence function of R, D_R, verifies $D_R^k(u^{\frac{1}{k}}, v^{\frac{1}{k}}) = D_R(u, v)$, $k \ge 1$, $u, v \in [0, 1]$, and the marginals of R verify $R(\infty, y) = R^k(\infty, C_k y + D_k)$ and $R(x, \infty) = R^k(A_k x + B_k, \infty)$; that is, they are *max-stable* and consequently they are of extreme value type. So, it follows at once, that R is a bivariate extreme value distribution.

Finally,

$$P(a_{\mathbf{n}}^{-1}(M_{\mathbf{n}} - b_{\mathbf{n}}) \le x, c_{\mathbf{n}}^{-1}(W_{\mathbf{n}} - d_{\mathbf{n}}) \le y) = P(M_{\mathbf{n}} \le u_{\mathbf{n}}(x)) - P(M_{\mathbf{n}} \le u_{\mathbf{n}}(x), W_{\mathbf{n}} > v_{\mathbf{n}}(y))$$

converges to $R(x, \infty) - R(x, -y)$, as $\mathbf{n} \to \infty$, concluding the proof. □

We can relate $Q(x, \infty) - Q(x, y)$ with $\widehat{Q}(x, \infty) - \widehat{Q}(x, y)$ through the two-sided extremal index of $\mathbf{X}$, which is a direct extension of the two-sided extremal index of a stationary sequence given in Martins et al. [4].

Definition 3. The random field **X** has two-sided extremal index $\underline{\overline{\theta}} \in [0, 1]$, if there exist sequences $\{u_{\mathbf{n}}\}_{\mathbf{n}\in\mathbb{N}^2}$, $\{v_{\mathbf{n}}\}_{\mathbf{n}\in\mathbb{N}^2}$ such that $n_1 n_2 P(X_1 \leq v_{\mathbf{n}} \vee X_1 > u_{\mathbf{n}}) \xrightarrow[\mathbf{n}\to\infty]{} \nu$ and

$$P\left(M_{\mathbf{n}} \leq u_{\mathbf{n}}, W_{\mathbf{n}} > v_{\mathbf{n}}\right) \xrightarrow[\mathbf{n}\to\infty]{} \exp(\nu\underline{\overline{\theta}}). \tag{9}$$

For sequences $\left\{u_{\mathbf{n}}^{(\tau)}\right\}_{\mathbf{n}\in\mathbb{N}^2}$ and $\left\{v_{\mathbf{n}}^{(\tau')}\right\}_{\mathbf{n}\in\mathbb{N}^2}$, $\tau > 0$, $\tau' > 0$, verifying Eq. (8) we have

$$lim_{\mathbf{n}\to\infty} P(M_{\mathbf{n}} \leq u_{\mathbf{n}}^{(\tau)}, W_{\mathbf{n}} > v_{\mathbf{n}}^{(\tau')}) =$$

$$= \left(lim_{\mathbf{n}\to\infty} P(\widehat{M}_{\mathbf{n}} \leq u_{\mathbf{n}}^{(\tau)}, \widehat{W}_{\mathbf{n}} > v_{\mathbf{n}}^{(\tau')})\right)^{\underline{\overline{\theta}}} = \exp(-\underline{\overline{\theta}}(\tau + \tau')). \tag{10}$$

From Eqs. (7) and (10) we extend to random fields the characterization of the asymptotic independence of the maximum and minimum of a stationary sequence $\mathbf{Y} = \{Y_n\}_{n\geq 1}$ through a linear relation between the extremal indexes, $\overline{\theta}$, $\underline{\theta}$ and $\underline{\overline{\theta}}$ of **Y** (Martins *et al.*,[4]).

Proposition 2. *Let* **X** *be a random field with superior, inferior and two-sided extremal indexes. Then, for* $\left\{u_{\mathbf{n}}^{(\tau)}\right\}_{\mathbf{n}\in\mathbb{N}^2}$ *and* $\left\{v_{\mathbf{n}}^{(\tau')}\right\}_{\mathbf{n}\in\mathbb{N}^2}$, $\tau > 0$, $\tau' > 0$, *verifying Eq. (8), we have*

$$\lim_{\mathbf{n}\to\infty} P\left(M_{\mathbf{n}} \leq u_{\mathbf{n}}^{(\tau)}, W_{\mathbf{n}} > v_{\mathbf{n}}^{(\tau')}\right) = \exp(-\overline{\theta}\tau)\exp(-\underline{\theta}\tau')$$

if and only if

$$\underline{\overline{\theta}} = \overline{\theta}\frac{\tau}{\tau + \tau'} + \underline{\theta}\frac{\tau'}{\tau + \tau'}. \tag{11}$$

We now give an example illustrating Proposition 2.

Example 1. Let $\mathbf{X} = \{X_{\mathbf{n}}\}_{\mathbf{n}\in\mathbb{N}^2}$ be a stationary random field with $E(X_1) = 0$, $E(X_1^2) = 1$ and covariance function $r_{\mathbf{n}} = cov(X_1, X_{\mathbf{n}})$, $\mathbf{n} \geq \mathbf{1}$. It was shown in [1] (see also [5]) that if

$$r_{\mathbf{n}} \log(n_1 n_2) \xrightarrow[\mathbf{n}\to\infty]{} 0, \quad r_{(n_1,0)} \log n_1 \xrightarrow[n_1\to\infty]{} 0, \quad r_{(0,n_2)} \log n_2 \xrightarrow[n_2\to\infty]{} 0, \tag{12}$$

then $P(M_{\mathbf{n}} \leq u_{\mathbf{n}}) \to_{\mathbf{n}\to\infty} \exp(-\tau)$, $\tau < \infty$, if and only if $n_1 n_2 P(X_1 > u_{\mathbf{n}}) \to_{\mathbf{n}\to\infty} \tau$.

Moreover, if $b_{\mathbf{n}} = (2\log(n_1 n_2))^{\frac{1}{2}} - \frac{1}{2}(2\log(n_1 n_2))^{-\frac{1}{2}}(\log\log(n_1 n_2) + \log 4\pi)$ and $a_{\mathbf{n}} = (2\log(n_1 n_2))^{-\frac{1}{2}}$, it follows $P\left(a_{\mathbf{n}}^{-1}(M_{\mathbf{n}} - b_{\mathbf{n}}) \leq x\right) \to_{\mathbf{n}\to\infty} \exp(-\exp(-x))$ and $P\left(a_{\mathbf{n}}^{-1}(W_{\mathbf{n}} + b_{\mathbf{n}}) \leq y\right) \to_{\mathbf{n}\to\infty} 1 - \exp(-\exp(-y))$. Using methods similar to those in [1] (see also [5]) it can be shown that the maximum and minimum are asymptotically independent, so, for $u_{\mathbf{n}}(x) = a_{\mathbf{n}}x + b_{\mathbf{n}}$,

$v_{\mathbf{n}}(y) = -a_{\mathbf{n}} y - b_{\mathbf{n}}$, $\tau(x) = \exp(-x)$ and $\tau(y) = \exp(-y)$ we have $\underline{\theta} = \overline{\theta}\frac{\tau(x)}{\tau(x)+\tau(y)} + \underline{\theta}\frac{\tau(y)}{\tau(x)+\tau(y)} = 1$.

3 Asymptotic Independence Between the Locations of Maximum and Minimum

In the following we generalize the condition $\Delta(u_{\mathbf{n}}, v_{\mathbf{n}})$ to deal with the joint behavior of maxima and minima in disjoint rectangles of indexes and then use this to obtain the asymptotic independence of the locations of maximum and minimum.

Definition 4. Let $\left\{u_{\mathbf{n}}^{(i)}\right\}_{\mathbf{n}\in\mathbb{N}^2}$, $\left\{v_{\mathbf{n}}^{(i)}\right\}_{\mathbf{n}\in\mathbb{N}^2}$, $i = 1, 2$, be sequences of real numbers. The random field $\mathbf{X}$ satisfies the condition $\Delta_2^*((u_{\mathbf{n}}^{(1)}, u_{\mathbf{n}}^{(2)}), (v_{\mathbf{n}}^{(1)}, v_{\mathbf{n}}^{(2)}))$ if in Eqs. (5) and (6) we consider, respectively, the events $\left\{v_{\mathbf{n}}^{(i)*} < W_{\mathbf{n}}(\mathbf{I}_1) < M_{\mathbf{n}}(\mathbf{I}_1) \leq u_{\mathbf{n}}^{(i)*}\right\}$ and $\left\{v_{\mathbf{n}}^{(i)*} < W_{\mathbf{n}}(\mathbf{I}_2) < M_{\mathbf{n}}(\mathbf{I}_2) \leq u_{\mathbf{n}}^{(i)*}\right\}$, where $u_{\mathbf{n}}^{(i)*} \in \left\{u_{\mathbf{n}}^{(1)}, u_{\mathbf{n}}^{(2)}\right\}$, $v_{\mathbf{n}}^{(i)*} \in \left\{v_{\mathbf{n}}^{(1)}, v_{\mathbf{n}}^{(2)}\right\}$.

If $u_{\mathbf{n}}^{(1)} = u_{\mathbf{n}}^{(2)} = u_{\mathbf{n}}$ and $v_{\mathbf{n}}^{(1)} = v_{\mathbf{n}}^{(2)} = v_{\mathbf{n}}$, we obtain the condition $\Delta(u_{\mathbf{n}}, v_{\mathbf{n}})$.

It is worth noting that if in Eqs. (5) and (6) we consider the events $\left\{M_{\mathbf{n}}(\mathbf{I}_1) \leq u_{\mathbf{n}}^{(i)*}\right\}$ and $\left\{M_{\mathbf{n}}(\mathbf{I}_2) \leq u_{\mathbf{n}}^{(i)*}\right\}$ where $u_{\mathbf{n}}^{(i)*} \in \left\{u_{\mathbf{n}}^{(1)}, u_{\mathbf{n}}^{(2)}\right\}$, we obtain the condition $\Delta_2^*(u_{\mathbf{n}}^{(1)}, u_{\mathbf{n}}^{(2)})$ of Pereira [6] under which it is proved that the normalized location of the maximum converges to a uniform variable on $[0, 1]^2$ and is asymptotically independent of the height of the maximum.

As a consequence of condition $\Delta_2^*((u_{\mathbf{n}}^{(\tau_1)}, u_{\mathbf{n}}^{(\tau_2)}), (v_{\mathbf{n}}^{(\tau_1')}, v_{\mathbf{n}}^{(\tau_2')}))$, where $\left\{u_{\mathbf{n}}^{(\tau_i)}\right\}_{\mathbf{n}\in\mathbb{N}^2}$, $\left\{v_{\mathbf{n}}^{(\tau_i')}\right\}_{\mathbf{n}\in\mathbb{N}^2}$ are sequences of real numbers such that

$$n_1 n_2 P(X_1 > u_{\mathbf{n}}^{(\tau_i)}) \xrightarrow[\mathbf{n}\to\infty]{} \tau_i, \quad n_1 n_2 P(X_1 \leq u_{\mathbf{n}}^{(\tau_i')}) \xrightarrow[\mathbf{n}\to\infty]{} \tau_i', \ i = 1, 2, \tag{13}$$

it follows that, if $\mathbf{V}_1, \ldots, \mathbf{V}_k$ are disjoint rectangles of $\mathbf{R}_{\mathbf{n}}$, then

$$\left| P\left(\bigcap_{s=1}^{k} \bigcap_{\mathbf{i}\in\mathbf{V}_s} \{v_{\mathbf{n},s} < X_{\mathbf{i}} \leq u_{\mathbf{n},s}\}\right) - \prod_{s=1}^{k} P\left(\bigcap_{\mathbf{i}\in\mathbf{V}_s} \{v_{\mathbf{n},s} < X_{\mathbf{i}} \leq u_{\mathbf{n},s}\}\right)\right| \xrightarrow[\mathbf{n}\to\infty]{} 0,$$

where, for each $s = 1, \ldots, k$, $u_{\mathbf{n},s} \in \left\{u_{\mathbf{n}}^{(\tau_1)}, u_{\mathbf{n}}^{(\tau_2)}\right\}$ and $v_{\mathbf{n},s} \in \left\{v_{\mathbf{n}}^{(\tau_1')}, v_{\mathbf{n}}^{(\tau_2')}\right\}$.

Lemma 2. *Let $\mathbf{X}$ be a stationary random field with superior, inferior and two-sided extremal indexes verifying Eq. (11). If $\mathbf{X}$ verifies the conditions $\Delta_2^*(u_{\mathbf{n}}^{(\tau_1)}, u_{\mathbf{n}}^{(\tau_2)})$ and $\Delta_2^*((u_{\mathbf{n}}^{(\tau_1)}, u_{\mathbf{n}}^{(\tau_2)}), (v_{\mathbf{n}}^{(\tau_1')}, v_{\mathbf{n}}^{(\tau_2')}))$ and $-\mathbf{X}$ verifies $\Delta_2^*(-v_{\mathbf{n}}^{(\tau_1')}, -v_{\mathbf{n}}^{(\tau'2)})$, where*

$\left\{u_{\mathbf{n}}^{(\tau_i)}\right\}_{\mathbf{n}\in\mathbb{N}^2}$, $\left\{v_{\mathbf{n}}^{(\tau_i')}\right\}_{\mathbf{n}\in\mathbb{N}^2}$ *are sequence of real numbers verifying Eq. (13), then, for each* $\varepsilon_{11}, \varepsilon_{12}, \varepsilon_{21}, \varepsilon_{22} \in (0,1]$, *as* $\mathbf{n} \to \infty$, *we have*

$$P\left(M_{\mathbf{n}}([1,n_1\varepsilon_{11}]\times[1,n_2\varepsilon_{12}]) \le u_{\mathbf{n}}^{(\tau_1)}, M_{\mathbf{n}}(\mathbf{R_n}-[1,n_1\varepsilon_{11}]\times[1,n_2\varepsilon_{12}]) \le u_{\mathbf{n}}^{(\tau_2)},\right.$$
$$\left. W_{\mathbf{n}}([1,n_1\varepsilon_{21}]\times[1,n_2\varepsilon_{22}]) > v_{\mathbf{n}}^{(\tau_1')}, W_{\mathbf{n}}(\mathbf{R_n}-[1,n_1\varepsilon_{21}]\times[1,n_2\varepsilon_{22}]) > v_{\mathbf{n}}^{(\tau_2')}\right)$$

$$-P\left(M_{\mathbf{n}}([1,n_1\varepsilon_{11}]\times[1,n_2\varepsilon_{12}]) \le u_{\mathbf{n}}^{(\tau_1)}, M_{\mathbf{n}}(\mathbf{R_n}-[1,n_1\varepsilon_{11}]\times[1,n_2\varepsilon_{12}]) \le u_{\mathbf{n}}^{(\tau_2)}\right)$$
$$\times P\left(W_{\mathbf{n}}([1,n_1\varepsilon_{21}]\times[1,n_2\varepsilon_{22}]) > v_{\mathbf{n}}^{(\tau_1')}, W_{\mathbf{n}}(\mathbf{R_n}-[1,n_1\varepsilon_{21}]\times[1,n_2\varepsilon_{22}]) > v_{\mathbf{n}}^{(\tau_2')}\right) \to 0.$$

Proof. Let us suppose, for example, that $\varepsilon_{11} < \varepsilon_{21}$ and $\varepsilon_{12} < \varepsilon_{22}$ and consider $\mathbf{I}_{\mathbf{n},1} = [1,n_1\varepsilon_{11}]\times[1,n_2\varepsilon_{12}]$, $\mathbf{I}_{\mathbf{n},2} = \mathbf{R_n} - \mathbf{I}_{\mathbf{n},1}$, $\mathbf{I}_{\mathbf{n},3} = [1,n_1\varepsilon_{21}]\times[1,n_2\varepsilon_{22}] - \mathbf{I}_{\mathbf{n},1}$ and $\mathbf{I}_{\mathbf{n},4} = \mathbf{R_n} - \mathbf{I}_{\mathbf{n},1} \cup \mathbf{I}_{\mathbf{n},3}$. We have

$$\lim_{\mathbf{n}\to\infty} P\left(M_{\mathbf{n}}(\mathbf{I}_{\mathbf{n},1}) \le u_{\mathbf{n}}^{(\tau_1)}, M_{\mathbf{n}}(\mathbf{I}_{\mathbf{n},2}) \le u_{\mathbf{n}}^{(\tau_2)}, W_{\mathbf{n}}(\mathbf{I}_{\mathbf{n},1}) > v_{\mathbf{n}}^{(\tau_1')}, W_{\mathbf{n}}(\mathbf{I}_{\mathbf{n},2}) > v_{\mathbf{n}}^{(\tau_2')}\right)$$
$$= \lim_{\mathbf{n}\to\infty} P\left(M_{\mathbf{n}}(\mathbf{I}_{\mathbf{n},1}) \le u_{\mathbf{n}}^{(\tau_1)}, W_{\mathbf{n}}(\mathbf{I}_{\mathbf{n},1}) > v_{\mathbf{n}}^{(\tau_1')}, M_{\mathbf{n}}(\mathbf{I}_{\mathbf{n},3}) \le u_{\mathbf{n}}^{(\tau_2)}, W_{\mathbf{n}}(\mathbf{I}_{\mathbf{n},3}) > v_{\mathbf{n}}^{(\tau_2')},\right.$$
$$\left. M_{\mathbf{n}}(\mathbf{I}_{\mathbf{n},4}) \le u_{\mathbf{n}}^{(\tau_2)}, W_{\mathbf{n}}(\mathbf{I}_{\mathbf{n},4}) > v_{\mathbf{n}}^{(\tau_2')}\right)$$
$$= \lim_{\mathbf{n}\to\infty} P\left(v_{\mathbf{n}}^{(\tau_1')} < W_{\mathbf{n}}(\mathbf{I}_{\mathbf{n},1}) < M_{\mathbf{n}}(\mathbf{I}_{\mathbf{n},1}) \le u_{\mathbf{n}}^{(\tau_1)}\right) \times$$
$$\prod_{i\in\{3,4\}} \lim_{\mathbf{n}\to\infty} P\left(v_{\mathbf{n}}^{(\tau_2')} < W_{\mathbf{n}}(\mathbf{I}_{\mathbf{n},i}) < M_{\mathbf{n}}(\mathbf{I}_{\mathbf{n},i}) \le u_{\mathbf{n}}^{(\tau_2)}\right)$$

$$= \exp(-\overline{\theta}\tau_1\varepsilon_{11}\varepsilon_{12})\exp(-\underline{\theta}\tau_1'\varepsilon_{11}\varepsilon_{12})\exp(-\overline{\theta}\tau_2(\varepsilon_{21}\varepsilon_{22}-\varepsilon_{11}\varepsilon_{12}))$$
$$\exp(-\underline{\theta}\tau_2'(\varepsilon_{21}\varepsilon_{22}-\varepsilon_{11}\varepsilon_{12}))\exp(-\overline{\theta}\tau_2(1-\varepsilon_{21}\varepsilon_{22}))\exp(-\underline{\theta}\tau_2'(1-\varepsilon_{21}\varepsilon_{22})).$$

By calculating the limit of the second term we obtain the same result. □

As an application of the previous results the asymptotic independence of the locations of maximum and minimum is obtained.

Proposition 3. *Let* $\mathbf{X}$ *be a stationary random field with extremal indexes* $\overline{\theta}$, $\underline{\theta}$ *and* $\overline{\theta}$ *verifying Eq. (11). Let* $\{a_{\mathbf{n}} > 0\}_{\mathbf{n}\in\mathbb{N}^2}$, $\{b_{\mathbf{n}}\}_{\mathbf{n}\in\mathbb{N}^2}$, $\{c_{\mathbf{n}} > 0\}_{\mathbf{n}\in\mathbb{N}^2}$ *and* $\{d_{\mathbf{n}}\}_{\mathbf{n}\in\mathbb{N}^2}$ *be sequences of constants verifying Eqs. (1) and (2). If, for each* $x_1, x_2, y_1, y_2 \in \mathbb{R}$ *and* $u_{\mathbf{n}}(x_i) = u_{\mathbf{n}}^{(\tau_i)} = a_{\mathbf{n}}x_i + b_{\mathbf{n}}$, $v_{\mathbf{n}}(y_i) = v_{\mathbf{n}}^{(\tau_i')} = c_{\mathbf{n}}y_i + d_{\mathbf{n}}$, $i = 1,2$, *the stationary random field* $\mathbf{X}$ *verifies the conditions* $\Delta_2^*((u_{\mathbf{n}}^{(\tau_1)}, u_{\mathbf{n}}^{(\tau_2)}), (v_{\mathbf{n}}^{(\tau_1')}, v_{\mathbf{n}}^{(\tau_2')}))$ *and* $\Delta_2^*(u_{\mathbf{n}}^{(\tau_1)}, u_{\mathbf{n}}^{(\tau_2)})$ *and* $-\mathbf{X}$ *verifies* $\Delta_2^*(-v_{\mathbf{n}}^{(\tau_1')}, -v_{\mathbf{n}}^{(\tau_2')})$, *then, for each* $\varepsilon_{11}, \varepsilon_{12}, \varepsilon_{21}, \varepsilon_{22} \in (0,1]$, *we have*

$$P\left(L_{\mathbf{n}}^{(Max)} \in \left([1, n_1\varepsilon_{11}] \times [1, n_2\varepsilon_{12}] \cap \mathbb{N}^2\right), a_{\mathbf{n}}^{-1}(M_{\mathbf{n}} - b_{\mathbf{n}}) \leq x,\right.$$
$$\left. L_{\mathbf{n}}^{(Min)} \in \left(\mathbf{R_n} - [1, n_1\varepsilon_{21}] \times [1, n_2\varepsilon_{22}] \cap \mathbb{N}^2\right), c_{\mathbf{n}}(W_{\mathbf{n}} - d_{\mathbf{n}}) > y\right)$$
$$\xrightarrow[\mathbf{n}\to\infty]{} \varepsilon_{11}\varepsilon_{12}(1 - \varepsilon_{21}\varepsilon_{22})\widehat{G}^{\overline{\theta}}(x)(1 - \widehat{H}(y))^{\underline{\theta}}$$

We omit the proof since it follows the same line of argument as in the proof of Proposition 3.1. of Pereira and Ferreira [7].

Acknowledgements We would like to thank the referee's careful reading of the manuscript which has resulted in improvements to the final form of this chapter.

This research was supported by the research unit "Centro de Matemática" of the University of Beira Interior through the Foundation for Science and Technology (FCT).

References

1. Choi, H.: Central limit theory and extremes of random fields. Phd Dissertation, Univ. of North Carolina, Chapel Hill (2002)
2. Davis, R.: Limit laws for the maximum and minimum of stationary sequences. Z. Whars. verw. Gebiete **61**, 31–42 (1982)
3. Leadbetter, M.R., Rootzén, H.: On extreme values in stationary random fields, Stochastic processes and related topics. Trends Math, pp. 275–285. Birkhauser, Boston (1998)
4. Martins, A., Pereira, L., Ferreira, H.: Localizaões do máximo e do mínimo em sucessões com índices extremais. Actas do XI Congresso da Sociedade Portuguesa de Estatística (2004)
5. Pereira, L.: On the extremal behavior of a nonstationary normal random field. J. Stat. Plan. Infer. **140**, 3567–3576 (2010)
6. Pereira, L.: The asymptotic location of the maximum of a stationary random field. Stat. Probabil. Lett. **79**, 2166–2169 (2009)
7. Pereira, L., Ferreira, H.: The asymptotic locations of the maximum and minimum of stationary sequences. J. Stat. Plan. Infer. **104**, 287–295 (2002)

Publication Bias and Meta-analytic Syntheses

D. Pestana, M.L. Rocha, R. Vasconcelos, and S. Velosa

Abstract

Aside from more traditional methods of combining p-values, a test based on the geometric mean G_n of a uniform random sample of size n is developed. As $\mathbb{E}(G_n) = \left(\frac{n}{n+1}\right)^n \underset{n\to\infty}{\downarrow} \frac{1}{e}$, it is obvious that publication bias has a bearing on the overall rejection of the null hypothesis and that the recent concepts of random and of generalized p-values deserve full attention.

1 Introduction

Meta-analysis is a successful development of former systematic reviews and is nowadays considered the gold standard of reporting the previous findings by other researchers in medicine (cf. the collection of invited papers by Egger and his co-authors [5–10], published in the *British Medical Journal*), demography [21], epidemiology [31] and pharmacology [25]. The original development has been

D. Pestana (✉)
Universidade de Lisboa, Bloco C6, Piso 4, Campo Grande, 1749-016 Lisboa, and CEAUL, Centro de Estatística e Aplicações da Universidade de Lisboa, Lisboa, Portugal
e-mail: dinis.pestana@fc.ul.pt

M.L. Rocha
Universidade dos Açores (DEG) and CEEAplA, Centro de Estudos de Economia Aplicada do Atlântico, Rua da Mãe de Deus, Apartado 1422, 9501-801 Ponta Delgada, Portugal
e-mail: lrocha@uac.pt

R. Vasconcelos · S. Velosa
Universidade da Madeira (CC CEE), Campus Universitário da Penteada, 9000-390 Funchal, and CEAUL, Centro de Estatística e Aplicações da Universidade de Lisboa, Lisboa, Portugal
e-mail: rita@uma.pt; sfilipe@uma.pt

J. Lita da Silva et al. (eds.), *Advances in Regression, Survival Analysis, Extreme Values, Markov Processes and Other Statistical Applications*, Studies in Theoretical and Applied Statistics, DOI 10.1007/978-3-642-34904-1_36,

made by Glass [12, 13], an expert in education sciences, cf. also his interesting overview [14]. Recent developments appear in [2, 16, 17].

Meta-analysis can be used to build up evidence from several inconclusive studies (namely, when sample size is small and thus the power of tests is scarce) or to resolve conflicting evidence when different studies, eventually conducted with different methodologies, seem to provide antagonistic results. A recent development of meta-analysis, christened cumulative meta-analysis, builds up evidence from costly and eventually ethically challenging studies to draw the line when pooled significant results have been achieved.

Important journals in the area of medicine, such as the *British Medical Journal*, nowadays recommend that substantial papers present a meta-analysis of former results. This is possible because the publishing standards of research in medicine attained some form of standardization in the presentation of evidence, which requires that statistical evidence is clearly reported, namely, providing means and standard deviations, observed values of the test statistics or at least observed p-values. Under those circumstances proper meta-analysis can be performed, either presenting a global estimate of some measured effect or combining p-values to achieve a global decision on some null hypothesis. This is so even when the studies have been conducted with very different precisions (a technique based on funnel plots provides in general interesting evidence), and even when very different treatments are compared, as, for instance, the celebrated studies on pre-eclampsia of pregnant women, where different treatments have been globally compared with a baseline diuretic.

Combining p-values is an important method in meta-analysis (Pestana [23]), since in most systematic reviews the only common reported statistical findings are p-values of tests on the same issue. The rationale is as follows: let us assume that the p-values p_k are known for testing H_{0k} vs. H_{Ak}, $k = 1, \ldots, n$, in n independent studies on some common issue, and our aim is to achieve a decision on the overall question H_0^* : all the H_{0k} are true *vs.* H_A^* : some of the H_{Ak} are true. As there are many different ways in which H_0^* can be false, selecting an appropriate test is in general unfeasible. On the other hand, combining the available p_k's so that $T(p_1, \ldots, p_n)$ is the observed value of a random variable whose sampling distribution under H_0^* is known is a simple issue, since under H_0^*, $\mathbf{p} = (p_1, \ldots, p_n)$ is the observation of a random sample $\mathbf{P} = (P_1, \ldots, P_n)$ from a $Uniform(0, 1)$ population. As usual, the corresponding vector of ascending order statistics is denoted $(P_{1:n}, P_{2:n}, \ldots, P_{n:n})$, and the observed k-th ascending order statistic in a sample of size n is denoted $p_{k:n}$.

In what follows we describe methods that deal directly with the p-values (Tippett, Wilkinson, arithmetic mean) and methods that use transformed p-values (Fisher, Stouffer, logistic).

We also derive a new method using directly the p-values, using the fact that the density function and the distribution function of the geometric mean of a uniform random sample are easily obtained and that the expected value, variance, skewness and kurtosis can easily be computed, either using the independence of the random variables or directly using the density of the geometric mean. Publication bias is an important issue in meta-analysis and thus can be used as a rationale for ranking methods of combining p-values.

2 Methods of Combining p-Values

A rational combined procedure should of course be *monotone*, in the sense that if one set of p-values $\boldsymbol{p} = (p_1, \ldots, p_n)$ leads to rejection of the overall null hypothesis H_0^*, any set of componentwise smaller p-values $\boldsymbol{p}' = (p_1', \ldots, p_n')$, $p_k' \le p_k$, $k = 1, \ldots, n$, must also reject H_0^*.

Tippett [29] used the fact that $P_{1:n} = \min\{P_1, \ldots, P_n\}|_{H_0^*} \frown Beta(1,n)$ to reject H_0^* if the minimum observed p-value $p_{1:n} < 1 - (1-\alpha)^{1/n}$. This *Tippett's minimum method* is a special case of *Wilkinson's method* [30], advising rejection of H_0^* when some low-rank order statistic $p_{k:n} < c$; as $P_{k:n} \frown Beta(k, n+1-k)$, to reject H_0^* at level α the cutoff point c is the solution of $\int_0^c u^{k-1}(1-u)^{n-k} du = \alpha\, B(k, n+1-k)$.

Another way of using directly the observed p-values is to compute their arithmetic mean. However, the exact distribution of $\overline{P}_n = \frac{1}{n}\sum_{k=1}^n P_k$ is cumbersome since under the null hypothesis $P_1 + \cdots + P_n$ is the sum of independent uniforms, with density function

$$f_n(x) = f_{P_1+\cdots+P_n}(x) = \frac{1}{(n-1)!}\left[\sum_{j=0}^{k}(-1)^j\binom{n}{j}(x-j)^{n-j}\,\mathrm{I}_{[k,k+1)}(x)\right]\mathrm{I}_{[0,n)}(x),$$

an expression easily proved by induction using $f_n(x) = \int_{x-1}^k f_{n-1}(x)\,\mathrm{d}x + \int_k^x f_{n-1}(x)\,\mathrm{d}x$, $x \in [k, k+1)$, $k \in \{0, 1, \ldots, n-1\}$, cf. also [24]. For large n an approximation based on the central limit theorem can be used to perform an approximate overall test on H_0^* *vs.* H_A^*. This is the least used method of combining p-values, and rightly so, since the overall test based on the arithmetic mean is not consistent in the sense that it can fail to reject the overall test null hypothesis, although the result of one of the partial tests is extremely significant.

In Sect. 2 we use the geometric mean $G_n = \left(\prod_{j=1}^n P_j\right)^{1/n}$ of n independent uniform random variables, whose distribution function is readily computed, leading to a consistent and more powerful test based on the direct use of all observed p-values; see, also, the discussion on publication bias in Sect. 4.

Alternatively, the construction of combined p-values using additive properties of simple functions of uniform random variables is a popular approach. Fisher [11] used the fact that $P_k \frown Uniform(0,1) \Longrightarrow -2\ln(P_k) \frown \chi_2^2$, and therefore, $-2\sum_{k=1}^n \ln(P_k)|_{H_0^*} \frown \chi_{2n}^2$. Then H_0^* is rejected at the significance level α if the $-2\sum_{k=1}^n \ln(p_k) > \chi_{2n,1-\alpha}^2$. Stouffer *et al.* ([28]) used as test statistic $\sum_{k=1}^n \frac{\Phi^{-1}(P_k)}{\sqrt{n}}\Big|_{H_0^*} \frown Gaussian(0,1)$, where Φ^{-1} denotes the inverse of the distribution function of the standard gaussian, rejecting H_0^* at level α if

$\left|\sum_{k=1}^{n} \frac{\Phi^{-1}(P_k)}{\sqrt{n}}\right| > z_{1-\alpha}$, where $z_{1-\alpha}$ stands for the $1-\alpha$ probability quantile of the standard gaussian. Stouffer's method has been further refined by Liptak [19], using sensible weights.

Another simple transformation of uniform random variables P_k is the logit transformation, $\ln \frac{P_k}{1-P_k} \frown Logistic(0,1)$. As $-\sum_{k=1}^{n} \ln\left(\frac{P_k}{1-P_k}\right) / \sqrt{n\, \frac{\pi^2(5n+2)}{3(5n+4)}} \approx t_{5n+4}$, reject H_0^* at the significance level α if $-\sum_{k=1}^{n} \ln\left(\frac{p_k}{1-p_k}\right) / \sqrt{n\, \frac{\pi^2(5n+2)}{3(5n+4)}} > t_{5n+4,\,1-\alpha}$.

Birnbaum [1] has shown that every monotone combined test procedure is *admissible*, i.e. provides a most powerful test against some alternative hypothesis for combining some collection of tests, and is therefore optimal for some combined testing situation whose goal is to harmonize eventually conflicting evidence or to pool inconclusive evidence. In the context of social sciences Mosteller and Bush [22] recommend Stouffer's method, but Littel and Folks [20] have shown that under mild conditions Fisher's method is optimal for combining independent tests. Observe however that H_A^* states that some of the H_{Ak} are true, and so a meta-decision on H_0^* implicitly assumes that some of the P_k may have non-uniform distribution, cf. [16] (p. 81–84) and [17] (pp. 117–119) and references therein, on the promising concepts of generalized and of random p-values.

3 The Geometric Mean of a Uniform Random Sample

The density function of the product of n independent standard uniform random variables $P_1, \ldots, P_n$ is

$$f_{P_1 \ldots P_n}(x) = \frac{(-\ln(x))^{n-1}}{(n-1)!}\, \mathrm{I}_{[0,1)}(x). \tag{1}$$

(This follows easily from $f_{P_1 P_2}(x) = \int_x^1 \frac{dy}{y}\, \mathrm{I}_{[0,1)}(x) = -\ln(x)\, \mathrm{I}_{[0,1)}(x)$ together with $f_{P_1 \ldots P_{n+1}}(x) = \displaystyle\int_x^1 \frac{(-\ln(x/y))^{n-1}}{(n-1)!} \frac{dy}{y}\, \mathrm{I}_{[0,1)}(x) = \frac{(-\ln(x))^n}{n!} \mathrm{I}_{[0,1)}(x)$ if (1) is assumed. The result also follows easily from the relationship between standard exponential random variables and standard uniforms, recalling that the sum of n iid standard exponentials is a $Gamma(n,1)$.) Although below we use the geometric mean as a suitable statistic to perform a consistent overall test on H_0^* *vs.* H_A^*, any power of the p-value product would provide a useful test statistic, as keenly pointed out by the referee; for more on products and products of powers of uniforms, cf. [3].

Hence the density function of the geometric mean $G_n = \left(\prod_{j=1}^{n} P_j\right)^{1/n}$ of a random sample of size n from the standard uniform population is

$$f_{G_n}(x) = \frac{d}{dx} F_{P_1 \ldots P_n}(x^n) = f_{P_1 \ldots P_n}(x^n)\, n\, x^{n-1}\, \mathrm{I}_{[0,1)}(x) = \frac{n\, [x(-n \ln(x))]^{n-1}}{\Gamma(n)} \mathrm{I}_{[0,1)}(x).$$

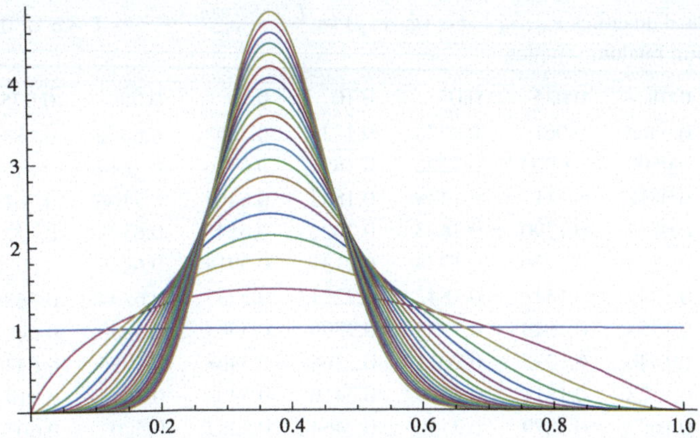

Fig. 1 Probability density functions of G_n, $n = 1, \ldots, 20$ (for $n = 1$, standard uniform; peakedness increases with n)

Table 1 Mean value μ, variance σ^2, skewness γ_1 and kurtosis γ_2 of G_n, $n = 1, \ldots, 20$

n	μ	σ^2	γ_1	γ_2
1	0.500000	0.0833333	0	−1.200000
2	0.444444	0.0524691	0.187180	−0.854118
3	0.421875	0.0380215	0.242030	−0.640618
4	0.409600	0.0297587	0.260104	−0.505923
5	0.401878	0.0244288	0.264457	−0.415154
6	0.396569	0.0207112	0.262968	−0.350538
7	0.392696	0.0179723	0.258854	−0.302510
8	0.389744	0.0158715	0.253580	−0.265570
9	0.387420	0.0142095	0.247859	−0.236365
10	0.385543	0.0128620	0.242051	−0.212747
11	0.383995	0.0117475	0.236342	−0.193284
12	0.382697	0.0108106	0.230825	−0.176991
13	0.381592	0.0100119	0.225543	−0.163163
14	0.380640	0.0093230	0.220513	−0.151291
15	0.379812	0.0087227	0.215736	−0.140993
16	0.379085	0.0081950	0.211205	−0.131980
17	0.378442	0.0077274	0.206908	−0.124029
18	0.377868	0.0073103	0.202834	−0.116965
19	0.377354	0.0069359	0.198968	−0.110650
20	0.376889	0.0065980	0.195296	−0.104971

Figure 1 shows the density functions of G_n for $n = 1, \ldots, 20$. The k-th-order raw moment of G_n is $\mathbb{E}\left(G_n^k\right) = \left(\frac{n}{n+k}\right)^n \underset{n\to\infty}{\longrightarrow} \mathrm{e}^{-k}$, and in particular $\mathbb{E}\left(G_n\right) = \left(\frac{n}{n+1}\right)^n \underset{n\to\infty}{\downarrow} \frac{1}{\mathrm{e}} \approx 0.3679$, the standard deviation decreases to zero, the skewness steadily decreases after a maximum 0.2645 for $n = 5$ and the kurtosis increases from -1.2 ($n = 1$) towards 0. In Table 1 we give the mean,

Table 2 Critical quantiles $g_{n,1-\alpha}$: $F_{G_n}(g_{n,1-\alpha}) = \frac{\Gamma^*(n,-n\ \ln(g_{n,1-\alpha}))}{\Gamma(n)} = 1-\alpha$ of the geometric mean of uniform random samples

$n \setminus 1-\alpha$	0.01	0.025	0.05	0.10	0.90	0.95	0.975	0.99
2	0.0362	0.0617	0.0933	0.1430	0.7665	0.8372	0.8859	0.9284
3	0.0607	0.0900	0.1226	0.1696	0.6926	0.7614	0.8137	0.8647
4	0.0812	0.1117	0.1439	0.1882	0.6465	0.7106	0.7615	0.8140
5	0.0982	0.1290	0.1603	0.2022	0.6148	0.6743	0.7227	0.7743
6	0.1125	0.1430	0.1734	0.2131	0.5914	0.6469	0.6928	0.7426
7	0.1247	0.1548	0.1842	0.2221	0.5733	0.6254	0.6689	0.7169
8	0.1353	0.1648	0.1933	0.2296	0.5588	0.6080	0.6494	0.6954
9	0.1446	0.1735	0.2011	0.2360	0.5468	0.5935	0.6330	0.6772
10	0.1528	0.1811	0.2079	0.2416	0.5368	0.5813	0.6191	0.6616
11	0.1602	0.1879	0.2139	0.2464	0.5282	0.5707	0.6070	0.6481
12	0.1668	0.1939	0.2193	0.2508	0.5208	0.5616	0.5965	0.6361
13	0.1728	0.1994	0.2241	0.2547	0.5142	0.5535	0.5872	0.6255
14	0.1783	0.2044	0.2285	0.2582	0.5084	0.5463	0.5789	0.6160
15	0.1833	0.2089	0.2324	0.2614	0.5033	0.5399	0.5714	0.6075
16	0.1880	0.2130	0.2361	0.2643	0.4986	0.5341	0.5646	0.5997
17	0.1923	0.2169	0.2394	0.2670	0.4944	0.5288	0.5585	0.5926
18	0.1963	0.2204	0.2425	0.2694	0.4905	0.5240	0.5529	0.5861
19	0.2000	0.2237	0.2454	0.2717	0.4870	0.5195	0.5477	0.5801
20	0.2035	0.2268	0.2481	0.2739	0.4837	0.5154	0.5429	0.5746
25	0.2180	0.2397	0.2592	0.2827	0.4706	0.4989	0.5235	0.5520
30	0.2292	0.2495	0.2677	0.2894	0.4610	0.4869	0.5093	0.5354
40	0.2456	0.2637	0.2799	0.2990	0.4478	0.4701	0.4895	0.5121
50	0.2572	0.2737	0.2884	0.3058	0.4389	0.4587	0.4761	0.4963
100	0.2873	0.2996	0.3104	0.3230	0.4172	0.4311	0.4432	0.4574

variance, skewness and kurtosis of the geometric mean G_n of standard uniform random samples of size n, $n = 1,\ldots,20$. The distribution function of G_n is $F_{G_n}(x) = \dfrac{\Gamma^*(n, -n\ \ln(x))}{\Gamma(n)} \mathrm{I}_{[0,1)}(x) + \mathrm{I}_{[1,\infty)}(x)$ where $\Gamma^*(n,z)$ is the incomplete Gamma function $\Gamma^*(n,z) = \int_z^\infty x^{n-1}e^{-x}dx$. The critical quantiles $g_{n,1-\alpha}$ such that $F_{G_n}(g_{n,1-\alpha}) = 1-\alpha$ are easily computed. Table 2 records the quantiles of probability $1-\alpha$, $\alpha = 0.10,\ 0.05,\ 0.025,\ 0.01$ of G_n, $n = 1(1)20$ and $n \in \{25, 30, 40, 50, 100\}$. Further quantiles are easily computed using the built-in function FindRoot in Mathematica or GAMMA.INV in Excel ($g_{n,1-\alpha} = $ *EXP*$(-$*GAMMA.INV*$(\alpha, n, 1)/n)$).

4 Publication Bias

The first step to carry out a meta-analysis is to select properly the evidence. In principle, a clear and fair criterion of inclusion must be adopted. Even so, publication bias must be taken into account, since non-significant results are rarely published in peer-reviewed journals (a general recommendation is to try to include properly chosen unpublished reports). In fact, most (if not all!) available p-values come only from studies considered worth publishing because the observed p-values were small, seeming to point out significant results. Thus the assumption that the p_k's are observations from independent $Uniform(0,1)$ random variables is questionable, since in general they are in fact a set of low-order statistics, given that p-values greater than 0.05, say, have not been recorded.

Observe, for instance, that whenever $p_{n:n}$ falls below the critical rejection point, the geometrical mean test studied in Sect. 3 will lead to the rejection of H_0^*, but $p_{n:n}$ smaller than the critical point (for $n \geq 14$, the expected value of G_n is greater than 0.36 and the standard deviation is smaller than 0.1) is what should be expected as a consequence of publication bias. This obviously enhances one of the ill-resolved problems in meta-analysis: published results have in general significant values, typically less than 0.05. Hence most of the published studies point out that H_0 ought to be rejected and that instead of combining p-values it would be more sensible to combine either generalized p-values [16] or random p-values [17].

A practical way of dealing with publication bias is to compute the number of unpublished studies with non-significant p-values that would be needed to reverse an overall decision of rejection of the null hypothesis; see [26] for details. A meta-analysis on desmoplastic malignant melanoma, using the systematic review [18] and further evidence collected in [27], consultancy for researchers for other areas and extensive simulation, namely, with computationally augmented samples of p-values [4, 15], led to the following ranking of methods of combining p-values, by decreasing power and increasing number of unreported cases needed to reverse the overall conclusion of the meta-analysis:

1. Arithmetic mean (with the *caveat*: inconsistent overall test)
2. Geometric mean
3. Chi-square transformation (Fisher's method)
4. Logistic transformation
5. Gaussian transformation (Stouffer–Liptak's method)
6. Selected order statistics (Wilkinson's method)
7. Minimum (Tippett's method)

Acknowledgements Research partially sponsored by national funds through the Fundação Nacional para a Ciência e Tecnologia, Portugal FCT under the project (PEst-OE/MAT/UI0006/2011). We are thankful to the referee for insightful comments and for sharing his knowledge of the field, and namely for providing the useful references [19, 24].

References

1. Birnbaum, A.: Combining independent tests of significance. J. Amer. Statist. Assoc. **49**, 559–575 (1954)
2. Borenstein, M., Hedges, L.V., Higgins, J.P.T., Rothstein, H.R.: Introduction to Meta-Analysis. Wiley, Chichester (2009)
3. Brilhante, M.F., Mendonça, S., Pestana, D., Sequeira, F.: Using products and powers of products to test uniformity. In: Luzar-Stiffler, V., Jarec, I., Bekic, Z. (eds.) Proceedings of the 32nd International Conference on Information Technology Interfaces, 509–514 (2010)

4. Brilhante, M.F., Pestana, D., Sequeira, F.: Combining p-values and random p-values. In: Luzar-Stiffler, V., Jarec, I., Bekic, Z. (eds.) Proceedings of the 32nd International Conference on Information Technology Interfaces, 515–520 (2010)
5. Davey Smith, G., Egger, M.: Meta-analysis — unresolved issues and future developments. Brit. Med. J. **316**, 221–225 (1998)
6. Egger, M., Davey Smith, G.: Meta-analysis — potentials and promise. Brit. Med. J. **315**, 1371–1374 (1997)
7. Egger, M., Davey Smith, G.: Meta-analysis — principles and procedures. Brit. Med. J. **315**, 1533–1537 (1997a)
8. Egger, M., Davey Smith, G.: Meta-analysis — beyond the grand mean? Brit. Med. J. **315**, 1610–1614 (1997b)
9. Egger, M., Davey Smith, G.: Meta-analysis — bias in location and selection of studies. Brit. Med. J. **316**, 61–66 (1998)
10. Egger, M., Schneider, M., Davey Smith, G.: Meta-analysis — spurious precision? Meta-analysis of observational studies. Brit. Med. J. **316**, 140–144 (1998)
11. Fisher, R.A.: Statistical Methods for Research Workers, 4th edn. Oliver and Boyd, Edinburgh (1932)
12. Glass, G.V.: Primary, secondary, and meta-analysis of research. Edu. Res. **5**, 3–8 (1976)
13. Glass, G.V.: Integrating findings: The meta-analysis of research. Rev. Res. Educ. **5**, 351–379 (1978)
14. Glass, G.V.: Meta-Analysis at 25, http://glass.ed.asu.edu/gene/papers/meta25.html (1999)
15. Gomes, M.I, Pestana, D., Sequeira, F., Mendonça, S., Velosa, S.: Uniformity of offsprings from uniform and non-uniform parents. In: Luzar-Stiffler, V., Jarec, I., Bekic, Z. (eds.) Proceedings of the 31st International Conference on Information Technology Interfaces, pp. 243–248 (2009)
16. Hartung, J., Knapp, G., Sinha, B.K.: Statistical Meta-Analysis with Applications. Wiley, New York (2008)
17. Kulinskaya, E., Morgenthaler, S., Staudte, R.G.: Meta Analysis. A Guide to Calibrating and Combining Statistical Evidence. Wiley, Chichester (2008)
18. Lens, M.B., Newton-Bishop, J.A., Boon, A.P.: Desmoplastic malignant melanoma: A systematic review. British J. Dermatology **152**, 673–678 (2005)
19. Liptak T.: On the combination of independent tests. Magyar Tud. Akad. Mat. Kutato Int. Kozl. **3**, 171–197 (1958)
20. Littel, R.C., Folks, L.J.: Asymptotic optimality of Fisher's method of combining independent tests, I, II. J. Amer. Statist. Assoc. **66**, 802–806 (1971), **68**, 193–194 (1973)
21. Longford, N.T.: Studying Human Populations. Springer, New York (2008)
22. Mosteller, F., Bush, R.: Selected quantitative techniques. In: Lidsey, G. (ed.) Handbook of Social Psychology: Theory and Methods, vol. I, 289–334. Addison-Wesley, Cambridge (1954)
23. Pestana, D.: Combining p-values. In: Lovric, M. (ed.) International Encyclopedia of Statistical Science, 1145–1147. Springer, New York (2011)
24. Sadooghi-Alvandi, S.M, Nematollahi, A.R., Habibi, R: On the distribution of the sum of independent uniform random variables. Statistical Papers **50**, 171–175 (2009)
25. Senn, S.: Statistical Issues in Drug Development, 2nd edn. Wiley, Chichester (2007)
26. Sequeira, F.: Meta-Análise: Harmonização de Testes Usando os Valores de Prova. PhD Thesis, DEIO, Faculdade de Ciências da Universidade de Lisboa (2009)
27. Soares de Almeida, L., Requena, L., Rütten, A., Kutzner, H., Garbe, C., Pestana, D., Marques Gomes, M.: Desmoplastic Malignant Melanoma: A Clinico-Pathologic Analysis of 113 Cases. American J. Dermatopathol. **30**, 207–215 (2008)
28. Stouffer, S.A., Schuman, E.A., DeVinney, L.C., Star, S., Williams, R.M.: The American Soldier, vol. I: Adjustment During Army Life. Princeton University Press, Princeton (1949)
29. Tippett, L.H.C.: The Methods of Statistics. Williams and Norgate, London (1931)
30. Wilkinson, B.: A statistical consideration in psychological research. Psychol. Bull. **48**, 156–158 (1951)
31. Woodward, M.: Epidemiology Study Design and Data Analysis, 2nd edn. Chapman and Hall/CRC (2005)

Self-perception of Health Status and Socio-Economic Differences in the Use of Health Services

Alexandra Pinto, Victor Lobo, Fernando Bação, and Helena Bacelar-Nicolau

Abstract
A problem that Portugal is facing, which needs urgent effective health policies, is the socio-economic differences and inequalities that arise in access to health care. In this study we used data from National Health Survey of 2005/2006 to investigate if socio-economic differences are related both to the frequency which health services are used and to self-perception of health status (SP-HS). We considered all data (Portugal) and also each region of NUTS II (Standard Nomenclature of Territorial Units for Statistics purposes), separately. The study points to a strong association between the SP-HS and the factors: gender, age, education level and income. The number of medical appointments showed weaker results with these factors.

1 Introduction

The social and economic factors have an important role in the development of disease, leading to inequalities not only in health but also in their use of health services [2, 7].

A. Pinto (✉) · H. Bacelar-Nicolau
Laboratório de Biomatemática, 1649-028 Lisbon, Portugal
e-mail: apinto@fm.ul.pt; hbacelar@fpce.ul.pt

V. Lobo
Marinha - Escola Naval, Rua Alfeite, 2810-001 Almada, Portugal
e-mail: vlobo@isegi.unl.pt

F. Bação
ISEGI, Universidade Nova de Lisboa, Campus de Campolide, 1070-312 Lisbon, Portugal
e-mail: bacao@isegi.unl.pt

J. Lita da Silva et al. (eds.), *Advances in Regression, Survival Analysis, Extreme Values, Markov Processes and Other Statistical Applications*, Studies in Theoretical and Applied Statistics, DOI 10.1007/978-3-642-34904-1_37,

In mortality studies, in all social classes, it has been found that the gap between low and high social classes is increasing. Portugal is the EU country with the largest gap between the 20 % richest and the 20 % poorest. According to data from a study conducted in 2006, published by the INE,[1] the yield of the 20 % employees with best income is 6.8 more times the income of the population with lower income [5]. Carlos Farinha Rodrigues argues that the strong discrimination in wages is due to educational level [6]. Despite these differences, the improved access to health care has helped to reduce mortality remarkably [1].

Studies carried out in European countries shows that morbidity is higher in lower socio-economic groups, although the prevalence of some diseases is decreasing [1]. Prevalences among the lower classes have also been identified as an important factor to explain these disparities. Self-perception of health status (SP-HS), i.e. how health status is perceived, physical and mental by the individual, has proved to be an interesting factor to study inequalities. Other aspects such as gender, age, income, education and number of medical appointments are also emphasized and they will be taken into account in this study [2, 7].

Other studies suggest that health care (either treatment or disease prevention) is less used by those who most need it, supporting "the inverse care law", where Hart argues that available medical care tends to vary inversely with the needs of the population [3, 4].

In recent years, health promotion has had a major impact on social classes (more) privileged through the disclosure of information [4].

2 Objective

The aim of this chapter is to investigate whether, in Portugal and in the regions of NUTS II,[2] differences between socio-economic classes may be associated either with disparities in the use of health services or with SP-HS. Another purpose of this study is to establish whether the use of health services increases with poor SP-HS.

3 Material and Methods

This study has been based on data from the 4th National Health Survey (NHS), an instrument that assesses the health of population. This survey was conducted in 2005/2006 and it is the first Portuguese survey on health which includes Açores and Madeira. This survey collected information of some characteristics, among which stand out the most relevant for this study: gender, age, self-perception of health status, education level, family income and number of medical appointments within three months prior to participation in the survey. From the original sample all

[1] National Institute of Statistics of Portugal.

[2] Standard Nomenclature of Territorial Units for statistics purposes.

individuals aged equal or greater than 15 years were selected. It was also necessary to regroup some categories of variables. For example, education level has seven categories which were grouped into two sub-categories: "≤ primary" (individuals with no studies or primary education) and "> primary" (individuals with secondary education or/and university).

Univariate and bivariate statistical analyses were used. Chi-square test and residual analysis were conducted to find significant associations between the relevant response variables (number of medical appointments and SP-HS) and factors (gender, age, education level, income and regions of NUTS II). All significant statistical values showed $p < 0.001$ and the lower significant adjusted residual mentioned in this study was 4.2. We also present a map of response variables for all regions of NUTS II. Statistical analysis was performed in SPSS 16 and the map was constructed in ArcGIS 9.3.

This study was carried out for the whole country and for each of the seven NUTS II regions (Norte, Centro, LVT, [3] Alentejo, Algarve, Açores and Madeira).

4 Results

The sample consists of 35,229 individuals (85 % of the initial sample) of which 52.3 % are women. The age groups have amplitudes of 10 years, varying from 13.5 % to 16.6 % of individuals, until the retirement age (≥ 65). There is a single age group above 65 years old which includes 24.6 % of individuals.

In this sample, there are 15 % (5,282) of individuals with no education and only 10.3 % (3,636) attended the post-secondary education.

Table 1 shows the absolute and relative frequencies observed for response variables.

Chi-square tests were applied and significant associations ($p < 0.001$) were obtained between factors and response variables. We also found significant associations between the two response variables. The results presented (Tables 2 through Table 4) correspond to significant associations and significant relevant residuals. Table 2 shows these results for all sample (Portugal). Tables 3 and 4 show the relevant results to each region of NUTS II. In these tables, an empty cell means no findings of strong patterns of association between two variables.

From the analysis of residuals that led to Table 2 it was concluded that:

- Men are associated with the absence of medical appointments. Women are associated with at least two medical appointments during the same period. Women are also associated with moderate and negative SP-HS.
- Individuals aged less than 45 years are associated with the absence of medical appointments and very good or good self-perception of health status. Ages of 55 and over are associated with the occurrence of medical appointments and poor or very poor SP-HS.

[3] National Institute of Statistics of Portugal.

Table 1 Absolute and relative frequencies of SP-HS and number of medical appointments of all sample

SP-HS	N	%	# Medical appointments	N	%
Very good/good	9,674	27.5	0	16,808	47.7
Moderate	9,483	26.9	1	9,571	27.2
Poor/very poor	4,683	13.3	2	4,019	11.4
			≥3	4,785	13.6
Total	23,840	67.7		35,183	99.9

Table 2 More relevant and significant associations between factors and response variables—Portugal

	Response variables		Factors Sex	Age	Education	Income
Portugal	# Medical appointments	0	M	< 45	> Primary	Higher
		1		≥ 55		
		2	F	≥ 55	≤ Primary	
		≥ 3	F	≥ 55	≤ Primary	Lower
	SP-HS	Very good/good		< 45	> Primary	Higher
		Moderate	F	≥ 45	≤ Primary	
		Poor/very poor	F	≥ 55	≤ Primary	Lower

Table 3 More relevant and significant associations—Açores and Madeira

NUTS II	Response variables		Factors Sex	Age	Education	Income
Açores	# Medical appointments	0	M	< 45	> Primary	
		1	F	≥ 65		
		2	F	≥ 65		
		≥ 3	F	≥ 65	≤ Primary	
	SP-HS	Very good/good	M	< 45	> Primary	Higher
		Moderate	F	55-64	≤ Primary	
		Poor/very poor	F	≥ 65	≤ Primary	Lower
Madeira	# Medical appointments	0	M	< 45	> Primary	Higher
		1	F	≥ 65		
		2	F	≥ 65	≤ Primary	
		≥ 3	F	≥ 65	≤ Primary	Lower
	SP-HS	Very good/good	M	< 45	> Primary	Higher
		Moderate	F		≤ Primary	Lower
		Poor/very poor		≥ 65	≤ Primary	Lower

- Individuals with more than four years of education are associated with the absence of medical appointments and very good or good of SP-HS. Individuals with no studies and with primary education (≤ primary) are associated with the

Table 4 More relevant and significant associations—Norte, Centro, LVT, Alentejo and Algarve

			Factors			
NUTS II	Response variables		Sex	Age	Education	Income
Norte	# Medical appointments	0	M	< 45	> Primary	
		1				
		2	F	≥ 55	≤ Primary	
		≥ 3	F	≥ 55	≤ Primary	Lower
	SP-HS	Very good/good	M	< 45	> Primary	Higher
		Moderate			≤ Primary	
		Poor/very poor	F	≥ 55	≤ Primary	Lower
Centro	# Medical appointments	0	M	< 45	> Primary	Higher
		1				
		2		≥ 65		
		≥ 3	F	≥ 65	≤ Primary	Lower
	SP-HS	Very good/good	M	< 45	> Primary	Higher
		Moderate			≤ Primary	
		Poor/very poor	F	≥ 55	≤ Primary	Lower
LVT	# Medical appointments	0	M	< 45	> Primary	
		1				
		2	F			
		≥ 3	F	≥ 65	≤ Primary	Lower
	SP-HS	Very good/good	M	< 45	> Primary	Higher
		Moderate		≥ 55	≤ Primary	
		Poor/very poor	F	≥ 65	≤ Primary	Lower
Alentejo	# Medical appointments	0	M	< 45	> Primary	Higher
		1				
		2	F			
		≥ 3	F	≥ 65	≤ Primary	Lower
	SP-HS	Very good/good	M	< 45	> Primary	Higher
		Moderate			≤ Primary	
		Poor/very poor	F	≥ 65	≤ Primary	Lower
Algarve	# Medical appointments	0	M	< 45	> Primary	Higher
		1				
		2	F	≥ 65	≤ Primary	
		≥ 3	F	≥ 65	≤ Primary	Lower
	SP-HS	Very good/good	M	< 45	> Primary	Higher
		Moderate		≥ 55	≤ Primary	
		Poor/very poor	F	≥ 65	≤ Primary	Lower

occurrence of two or more medical appointments. These individuals are also associated with moderate and poor or very poor SP-HS.

- More privileged classes appear associated with the absence of medical appointments and very good or good SP-HS, while the lower classes are associated with the occurrence of three or more medical appointments and poor or very poor SP-HS (Fig. 1).

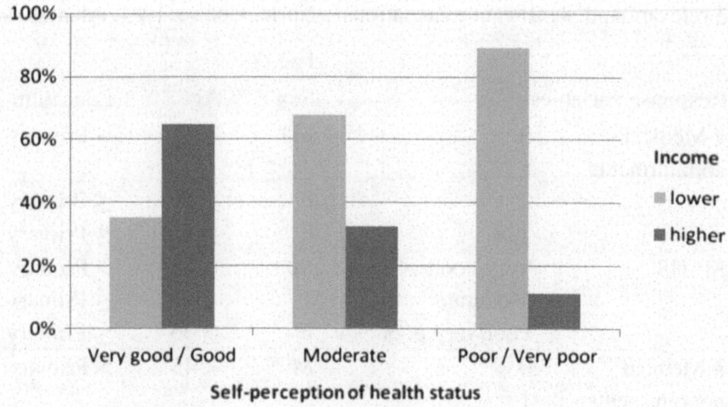

Fig. 1 Distribution of frequencies of SP-HS by income

From the analysis of residuals and the following map (Fig. 2) it can be concluded that:

- Açores and Madeira are associated with the absence of medical appointments and Norte, Centro and LVT tend to be associated with the existence of two or more medical appointments. Algarve and Açores are associated with very good or good SP-HS, while Centro, Alentejo and Madeira appear associated with moderate SP-HS.
- The absence of medical appointments is strongly associated with good or very good SP-HS, while the occurrence of three or more medical appointments is more associated with poor or very poor SP-HS (Fig. 3).

From the analysis of residuals and Tables 3 and 4 it is possible to identify common patterns in regions of NUTS II, namely:

- The trend for the absence of medical appointments and positive SP-HS in men, the opposite to women.
- Younger individuals are associated with the absence of medical appointments and positive SP-HS.
- Individuals with more education go to the doctor less often and have better SP-HS than individuals with few or no qualifications.
- All regions of NUTS II show strong association between SP-HS and family income. Individuals with higher incomes have better SP-HS.

Also from the analysis of Tables 3 and 4 we can conclude that:

- Açores can be distinguished from other regions for not having relevant associations between the number of medical appointments and family income.
- The group of individuals of 45–54 years old appears uncharacterized in this study.

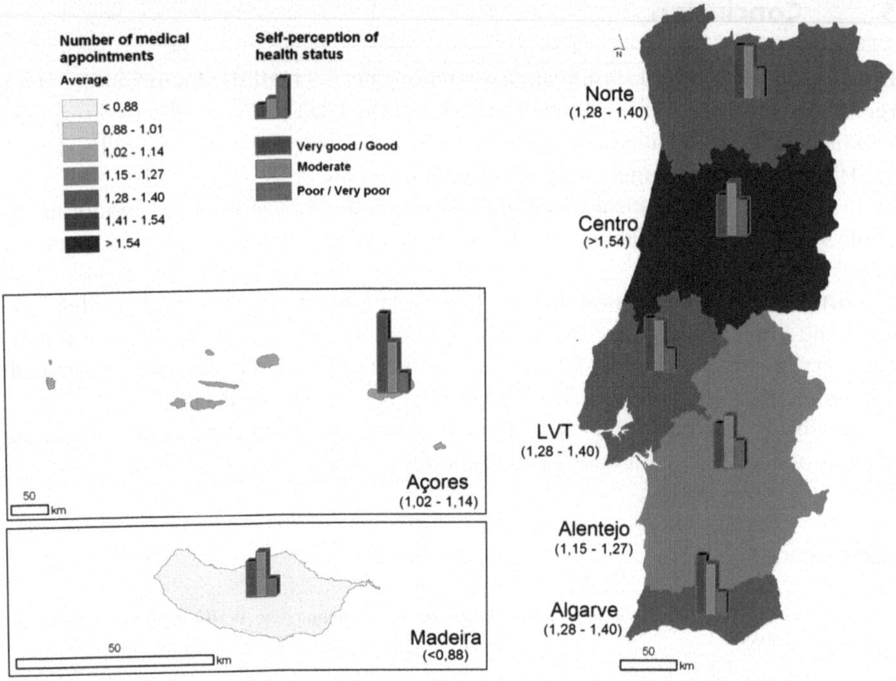

Fig. 2 Map representing the average of medical appointments and SP-HS, at NUTS II

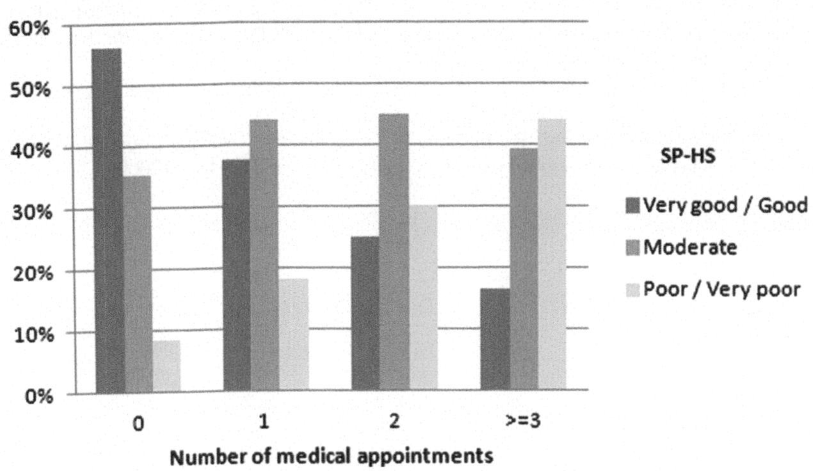

Fig. 3 Distribution of frequencies of number of medical appointments by SP-HS

5 Conclusion

Data from NHS showed significant associations between all the factors under study (gender, age, education level, income and regions NUTS II) and the two response variables (number of medical appointments and self-perception of health status). SP-HS is notoriously higher associated with these factors.

In this study men appear associated with the absence of medical appointments while women are associated with at least two medical appointments and moderate or negative SP-HS.

Furthermore, it was found that for those aged under 45 years, higher education level and higher income are associated with the absence of medical appointments and very good or good SP-HS. It was also found that the absence of medical appointments is strongly associated with good or very good SP-HS.

Finally, it was concluded that there is a pattern of similar behaviour among regions of NUTS II, with only a few exceptions.

References

1. Dahlgren, G., Whitehead, M.: Levelling Up (part 2). Copenhagen: WHO Regional Office for Europe (2006)
2. Fernández, K., Leon, D.: Self-perceived health status and inequalities in use of health services in Spain. Int. J. Epidemiol. **25**, 593–603 (1996)
3. Hart, J.T.: The inverse care law. Lancet **I**, 405–412 (1971)
4. Katz, S.J., Hofer, T.P.: Socioeconomic disparities in preventive care persist despite universal coverage. JAMA **272**, 530–534 (1994)
5. LUSA - Agência de Notícias de Portugal, S.A.: Desigualdades sociais em Portugal são as mais elevadas da Europa e devem-se à educação. RTP (2008) http://tv1.rtp.pt/noticias/?article=92248&visual=3&layout=10.Cited21Fev2008
6. Rodrigues, C.: Distribuição do Rendimento, Desigualdade e Pobreza. Portugal nos Anos 90. In: Colecção "Económicas", II série, **5**. Almedina, Coimbra (2007)
7. Silva, N., Pedroso, G., Puccini, R., Furlani, W.: Desigualdades sociais e uso de serviços de saúde: evidências de análise estratificada. Rev. Saúde Pública, **34**, 44–49 (2000)

Comparison of Modal Variables Using Multivariate Analysis

Isabel Pinto Doria, Áurea Sousa, Helena Bacelar-Nicolau, and Georges Le Calvé

Abstract
Domiciliary palliative care satisfaction and quality were estimated by caregivers via five perception scales with partly ordered answering modalities. The perception scales were codified as symbolic modal variables and analyzed using two multivariate approaches based on complex (symbolic) data to compare modal variables. This study compares the outcomes of previous work by Doria (Representações euclidianas de dados: Uma abordagem para variáveis heterogéneas. Tese de doutoramento, Universidade de Lisboa, Lisboa, 2008), Doria et al.(Livro de Resumos da XI Conferencia Española de Biometria e Primer Encuentro Iberoamericano de Biometria (CEIB2007) 101–102, 2007) and Bacelar-Nicolau et al.(Revista Portuguesa de Filosofia 66(2):427–460, 2010). In particular, it focuses on the differences and similarities of the results obtained with principal component analysis and ascendant hierarchical cluster analysis, directly applied to the similarity matrix S_{LC} and to the generalized affinity matrix, adapted to the comparison of modal variables.

I.P. Doria (✉) · H. Bacelar-Nicolau
FP-UL, LEAD and CEAUL, University of Lisbon, 1649-013-Lisboa, Portugal
e-mail: irpdoria@fp.ul.pt; hbacelar@fp.ul.pt

Á. Sousa
Mathematics Department, University of Azores, 9501-855-Ponta Delgada, and CEAUL, Portugal
e-mail: aurea@uac.pt

G. Le Calvé
Président d'Honneur de la Société Française de Statistique, France
e-mail: g.lecalve@free.fr

J. Lita da Silva et al. (eds.), *Advances in Regression, Survival Analysis, Extreme Values, Markov Processes and Other Statistical Applications*, Studies in Theoretical and Applied Statistics, DOI 10.1007/978-3-642-34904-1_38,

1 Introduction

This study continues the work done by Doria [9], Doria et al. [11], and Bacelar-Nicolau et al. [6]. It deals with the comparison of modal variables using multivariate methods. Its main goal is to make a comparative study of the results obtained from the generalized similarity coefficient s_{LC} and from the generalized affinity coefficient to symbolic data adapted to comparison of symbolic modal variables, when comparing the five perception scales of the Modified SERVQUAL Questionnaire, in the context of multivariate analysis.

A Modified SERVQUAL Questionnaire[1] was used to determine the quality and satisfaction with domiciliary cares performed on oncological patients and to identify their needs. This questionnaire is comprised of five perception scales: A—Tangible Elements, B—Reliability of Treatments and Cares, C—Security/Guarantee (Assurance), D—Interest/Response capability (i.e., capability of inspiring credibility and trust) and E—Empathy Capability, each respectively composed of seven, five, eight, nine and eleven items. These items are measured in a scale with partly ordered modalities (1—Total disagreement, 2—Disagreement, 3—Neither Agreement or Disagreement, 4—Agreement, 5—Total agreement, 6—Does not apply, 9—Doesn't know/Doesn't answer).

The initial classical data matrix (58 x 39) is constituted by 58 palliative caregivers (individuals) and 39 items. In order to compare the scales, the codification of the five scales as symbolic modal variables resulted in a three-dimensional symbolic data matrix $M(58 \times 5 \times 7)$, in which each individual is described according to the profile obtained from their answers to the set of items of each of the scales.

As we know, in a symbolic data matrix, rows correspond to symbolic objects, whereas columns correspond to symbolic variables, which may take values such as subsets of categories, intervals of real axes, or frequency distributions. Each cell can contain just one value, as usual, or several values that can be weighted and linked by logical rules and taxonomies [7].

Formally, a modal variable Y, with domain $\mathcal{Y}$, defined in a set $E = \{a, b, \ldots\}$ of objects, is a mapping $Y(a) = (U(a), \pi_a)$, $a \in E$, where π_a is a non-negative measure in $\mathcal{Y}$, generally a frequency distribution (absolute or relative), a probability or weight distribution on the domain $\mathcal{Y}$ of possible observed values and $U(a) \subseteq \mathcal{Y}$ is the support for π_a in domain $\mathcal{Y}$ [7]. In this study's data matrix the entries correspond to the relative frequency distributions. As an example, the corresponding sub-table for scale A is presented in Table 1. For instance, in this table the value 0.286 means that in the total of answers (one for each item of the scale) the modality 2 (Disagreement) was indicated 2 times by individual 1 (i.e. 28.6 % of the answers given by the individual 1 to the set of seven items of scale A corresponding to the modality 2).

[1] Project SDH. MD/P.I.01.13, subsidized by the Calouste Gulbenkian Foundation and coordinated by Professor Manuel Silvério Marques from IPOFG.

Table 1 Extract from the M(58,5,7) three-dimensional data matrix–Scale A, with the relative frequencies of aggregated answers to the 7 items of the scale, from 1 (Total disagreement) to 5 (Total agreement), plus 6 (Does not apply) and 9 (Doesn't know/Doesn't answer)

	1	2	3	4	5	6	9
1	0.00	0.286	0.143	0.143	0.143	0.286	0.000
2	0.00	0.143	0.000	0.429	0.143	0.286	0.000
⋮	⋮	⋮	⋮	⋮	⋮	⋮	⋮
58	0.00	0.000	0.143	0.000	0.857	0.000	0.000

2 Methods

Five perception scales from Modified SERVQUAL Questionnaire, codified as symbolic/complex modal variables, were compared using principal component analysis (PCA) and ascendant hierarchical cluster analysis (AHCA) directly applied to the generalized similarity matrix S_{LC}. In the case of PCA, the focus is on a representation of lower dimension, so that the scalar products between the vectors are the nearest possible of the similarity table. The classic situation of the principal component analysis on a given data table corresponds to the case in which we use the covariance matrix as a similarity table; thus, when all variables are metric, the PCA of the similarity matrix S_{LC} matches the traditional PCA (e.g., [9, 10, 12]). In addition, the AHCA of the scales–based on the similarity matrix containing the values of the generalized affinity coefficient–was applied to the comparison of modal variables, as defined by formula (2).

In the AHCA, the values from the two similarity coefficients were combined with five aggregation criteria, two of which are classical (single linkage (SL) and complete linkage (CL)), and three of which are probabilistic (AVL, AV1, and AVB—for these aggregation criteria see [3, 16]). The results obtained with the two similarity coefficients were compared.

2.1 The Generalized Similarity Coefficient s_{LC}

The similarity coefficients between variables s, s_{LC}, and P_L were inspired on an idea originally from Daniels [8], later developed by Lerman [14] and generalized by Le Calvé [13]. In this approach, each variable is associated with a score matrix, whose definition depends on the nature of the variable, as well as on the nature of the variable with which it is to be compared. The basic coefficient, s, is defined as the scalar product between the score matrices, the s_{LC} coefficient is the standardized coefficient s, under a certain reference hypothesis, and P_L coefficient corresponds to the probabilistic coefficient. For detailed information on these measures see Doria [9].

Table 2 Three-way data matrix

	$\underline{X}_1$			...		$\underline{X}_p$		
	1	...	m	...		1	...	m
1	$x_{1(1)1}$	...	$x_{1(1)m}$	...	1	$x_{1(p)1}$	...	$x_{1(p)m}$
2	$x_{2(1)1}$	...	$x_{2(1)m}$	...	2	$x_{2(p)1}$	...	$x_{2(p)m}$
⋮	⋮	⋮	⋮		⋮	⋮	⋮	⋮
N	$x_{N(1)1}$	...	$x_{N(1)m}$	...	N	$x_{N(p)1}$	...	$x_{N(p)m}$

Recently, coefficients s, s_{LC}, and P_L were generalized to the comparison of symbolic variables, including modal variables [9, 11]. Given the modal variables $\underline{X}_j = \left(x_{i(j)1}, x_{i(j)2}, \ldots, x_{i(j)m}\right)$, $j = 1, \ldots, p$, all of which with the same number m of modalities, in which $x_{i(j)k}$ $(i = 1, \ldots, N;\ k = 1, \ldots, m)$ is the value taken by the $\underline{X}_j$ variable for statistical unity i (in this case, a relative frequency distribution) (Table 2), the score of the modal variable is defined as follows:

$$x_{ii'} = \mathit{aff}\left(i, i'\right), \text{if } i \neq i'$$
$$x_{ii} = 0, \tag{1}$$

where $\mathit{aff}\,(i, i')$ indicates the basic affinity coefficient (e.g., [2, 3]) between the profiles from the answers given by individuals i and i'. The basic affinity coefficient $\mathit{aff}\,(i, i')$ is defined as

$$\mathit{aff}\left(i, i'\right) = \sum\nolimits_{j=1}^{m} \sqrt{\pi_{ij} \times \pi_{i'j}}$$

in the case of $i = (\pi_{i1}, \ldots, \pi_{im})$ and $i' = (\pi_{i'1}, \ldots, \pi_{i'm})$ being probability distributions or relative frequencies of statistical units i and i', associated to the variable j with m categories.

2.2 The Generalized Affinity Coefficient

In the approach based on the affinity coefficient which is being presented here, let us consider a three-way data matrix (Table 2), containing a sequence of p modal variables, $\underline{X}_j$, $j = 1, \ldots, p$, all of which with the same number of modalities.

For each of the variables $\underline{X}_j$ (briefly, j), i is the index which refers to the data units $(i = 1, \ldots, N)$ and k is the index which corresponds to the columns $(k = 1, \ldots, m)$, where m is the number of answering modalities. Under such conditions, one can compare two variables, j and j', based on the following formula:

$$a(j, j') = \frac{1}{N} \sum_{i=1}^{N} \mathit{aff}\left(j, j'; i\right), \tag{2}$$

where

$$aff\left(j, j'; i\right) = \sum_{k=1}^{m} \sqrt{\frac{x_{i(j)k}}{x_{i(j)\bullet}} \times \frac{x_{i(j')k}}{x_{i(j')\bullet}}} \tag{3}$$

and $x_{i(j)\bullet} = \sum_{k=1}^{m} x_{i(j)k}$ and $x_{i(j')\bullet} = \sum_{k=1}^{m} x_{i(j')k}$. Note that $aff(j, j'; i)$ is the local affinity between variables j and j' on what data unit i is concerned and formula (2) gives the affinity coefficient generalized to the comparison of modal variables [4, 5, 17].

3 Results

The five perception scales, codified as symbolic/complex modal variables, were compared using PCA directly applied to the generalized similarity matrix S_{LC} (e.g., [9]). In addition, AHCA was applied to the scales based on the similarity matrix S_{LC} and on the similarity matrix containing the values of the similarity coefficient corresponding to formula (2). In the AHCA, the values from the two similarity coefficients were combined with five aggregation criteria (SL, CL, AVL, AV1, and AVB).

3.1 Generalized Similarity Coefficient s_{LC}

In conformity with the "level statistics" criterion [1, 2, 15], the best result obtained from the AHCA (s_{LC}+Single Linkage) is given by the partition in two clusters (Fig. 1):

- Cluster 1 = {A.Tangible Elements}. This cluster refers to the equipment and has a different answering profile from all others, as it shows a higher degree of dissatisfaction of caregivers.
- Cluster 2 = {B.Reliability, E.Empathy, C.Security/Guarantee, D.Response capability}. This cluster refers to the ability to correctly apply treatments, to pay personal attention to patients, to promptly conduct the services and to inspire credibility and trust, in which there is a higher number of answers "4—Agreement" and "5—Total agreement".

The PCA directly applied to the similarity matrix S_{LC}, shows that the first principal component explains 42.27 % of the total data variability. The scale "D. Response Capability" stands out (Fig. 2), having a larger proportion of "Total agreement" and "Does not apply" answers. The second principal component, which explains 22.85 % of the variability, corresponds to scale "A.Tangible Elements" showing a higher degree of dissatisfaction (Fig. 2). The third principal component, which explains 17.48 % of the variability, corresponds to the two contrasting scales "D. Response Capability" and "B. Reliability".

When applied to that same matrix, the results of AHCA (s_{LC}+Single Linkage) (Fig. 1) and PCA allow similar interpretations. In the partition obtained from level 3 in two clusters, we recognize the first factorial plane: Cluster 1 = {A.Tangible

Fig. 1 Dendrogram obtained from the Ascendant Hierarchical Cluster Analysis (s_{LC}+Single Linkage). Level 3 is the most important one, followed by level 2, according to the "statistics of levels" [1, 2, 15], STAT(3)=2.13, STAT(2)=2.09

```
                          levels 1 to 4
A.Tangibles          --*--------*
                                |
B.Reliability        --*        |
                       |-----*  |
E.Empathy            --*     |  |
                             |--*
C.Assurance          --*--*  |
                          |--*
D.Responsivenes      --*--*
```

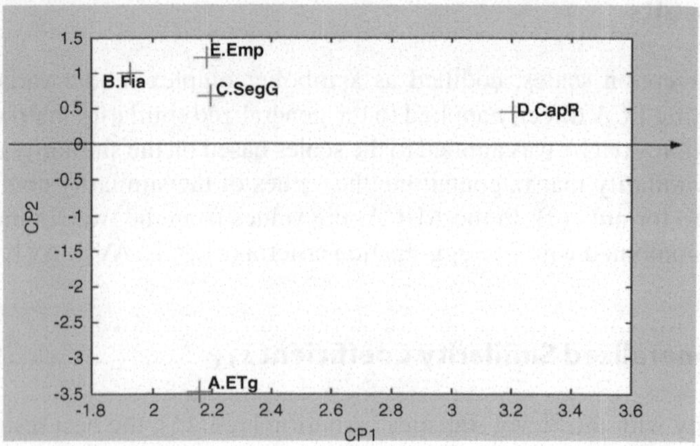

Fig. 2 Representation of the five perception scales from the Modified SERVQUAL Questionnaire in the plane defined by the first two principal components, obtained from the PCA based on the S_{LC} matrix

Elements} and Cluster 2 = {"B.Reliability", "E.Empathy", "C.Security/Guarantee", "D.Response capability"}.

3.2 Generalized Affinity coefficient

The best result obtained from the hierarchical clustering model AHCA ($a(j, j')$+ Single Linkage) is given by the two clusters partition described in the previous section, according to the same "level statistics" criterion. Moreover, in the three clusters partition obtained at level 2, we recognize the three clusters partition shown in the plane defined by the first two principal components of the PCA: Cluster 1 = {A.Tangible Elements}, Cluster 2 = {"B.Reliability", "C.Security/Guarantee", "E.Empathy"}, and Cluster 3 = {"D.Response capability"} (see Figs. 2 and 3): separation between clusters {D} and {B, C, E} explained by (the first principal component of) the PCA is explained by (the second level partition of) the hierarchical clustering model as well.

Fig. 3 Dendrogram obtained from the Ascendant Hierarchical Cluster Analysis ($a(j, j')$+Single Linkage)

```
                          levels 1 to 4
A.Tangibles               --*--------*
                                     |
B.Reliability             --*--*     |
                               |--*  |
C.Assurance               --*  |  |  |
                            |--*  |--*
E.Empathy                 --*     |
                                  |
D.Responsivenes           --*-----*
```

4 Conclusion

The two approaches led to a similar conclusion in what concerns the best partition: Scale "A. Tangible Elements" stands out.

The codification of the perception scales as symbolic/complex modal variables does not imply loss of information and as such presents a clear advantage to traditional multivariate methods. Two approaches were used to compare modal variables. In the first approach the similarities between these scales (modal variables) were evaluated using the generalized similarity coefficient s_{LC}. The second approach is based on the generalized affinity coefficient and provides an alternative way to measure the referred similarities. The results of this study, based on modal symbolic data, support the conclusion that the two approaches lead to similar conclusions and therefore are robust.

Future work on this area may include the application of PCA to the similarity matrix obtained with the generalized affinity coefficient.

Acknowledgments Research is partially sponsored by national funds through the Fundação Nacional para a Ciência e Tecnologia, Portugal—FCT—under the project (PEst-OE/MAT/UI0006/2011).

References

1. Bacelar-Nicolau, H.: Analyse d'un Algorithme de Classification. Thèse de 3 ème cycle, Université Pierre et Marie Curie, Paris (1972)
2. Bacelar-Nicolau, H.: Contribuições ao Estudo dos Coeficientes de Comparação em Análise Classificatória. Tese de doutoramento, Faculdade de Ciências da Universidade de Lisboa, Lisboa (1980)
3. Bacelar-Nicolau, H.: Two probabilistic models for classification of variables in frequency tables. In: Bock, H.-H. (ed.) Classification and Related Methods of Data Analysis, pp. 181–186. North Holland, Amsterdam (1988)
4. Bacelar-Nicolau, H.: The affinity coefficient. In: Bock, H.-H. (ed.) Analysis of Symbolic Data. Exploratory Methods for Extracting Statistical Information from Complex Data, pp. 160–165. Springer, Heidelberg (2000)
5. Bacelar-Nicolau, H., Nicolau, F., Sousa, A.: Measuring similarity of complex and heterogeneous data in clustering of large data sets. Biocybern. Biomed. Eng. **29**(2), 9–18 (2009)
6. Bacelar-Nicolau, H., Sousa, A., Bacelar-Nicolau, L., Marques, M. Silvério: Do univariado ao multivariado: a escala de elementos tangíveis, suas relações com outras escalas e mais além. Revista Portuguesa de Filosofia **66**(2), 427–460 (2010)

7. Bock, H.-H., Diday, E. (eds.): Analysis of Symbolic Data. Exploratory Methods for Extracting Statistical Information from Complex Data. Springer, Heidelberg (2000)
8. Daniels, H.E.: The relation between measures of correlation in the universe of sample permutations. Biometrika **33**, 129–135 (1944)
9. Doria, I.: Representações euclidianas de dados: Uma abordagem para variáveis heterogéneas. Tese de doutoramento, Universidade de Lisboa, Lisboa (2008)
10. Doria, I., Le Calvé, G., Bacelar-Nicolau, H.: Comparision of ultrametrics obtained with real data, using the P_L and VAL_{AW} coefficients. In: Kiers, H.A.L., Rasson, J.-P., Groenen, P.J.F., Schader, M. (eds.) Data Analysis, Classification, and Related Methods, pp. 107–112. Springer, Berlin (2000)
11. Doria, I., Dias, O., Sousa Ferreira, A., Le Calvé, G., Bacelar-Nicolau, H: Comparison of methodologies of multivariate analysis in a palliative care context. In: Livro de Resumos da XI Conferencia Española de Biometria e Primer Encuentro Iberoamericano de Biometria (CEIB2007), pp. 101–102 (2007)
12. Le Calvé, G.: Quelques remarques sur certains aspects de l'analyse factorielle. Cahier 2 du Laboratoire d'Analyse et de Traitement des Données en Sciences Humaines, Universit de Haute-Bretagne, Rennes II (1976)
13. Le Calvé, G.: Un indice de similarité pour des variables de type quelconque. Revue Statistique et Analyse des Données, 01/02, 39–47 (1977)
14. Lerman, I.C.: Étude Distributionnelle de Statistiques de Proximité entre Structures Algébriques Finies de Même Type. Application à la Classification Automatique. Cahiers du Bureau Universitaire de Recherche Oprationnelle, n. 19. Institut de Statistique de l' Université de Paris, Paris (1973)
15. Lerman, I.C.: Sur l'Analyse des Données Préalable à une Classification Automatique. Proposition d'une Nouvelle Mesure de Similarité. MSH, rapport 32 (8e. année), Paris (1970)
16. Nicolau, F., Bacelar-Nicolau, H.: Some Trends in the Classification of Variables. In: Hayashi, C., Ohsumi, N., Yajima, K., Tanaka, Y., Bock, H.-H., Baba, Y. (eds.) Data Science, Classification, and Related Methods, pp. 89–98. Springer (1998)
17. Nicolau, F., Bacelar-Nicolau, H.: Clustering Symbolic Objects Associated to Frequency or Probability Laws by the Weighted Affinity Coefficient. In: Bacelar-Nicolau, H., Nicolau, F.C., Janssen, J. (eds.) Applied Stochastic Models and Data Analysis. Quantitative Methods in Business and Industry Society, pp. 155–158. INE, Lisboa, Portugal (1999)

Disentangling the Relationship Between Entrepreneurship and Job Creation by Poisson Mixture Regressions

Leandro P. Pontes and José G. Dias

Abstract

Entrepreneurship and firm creation are very important drivers of the European policy. We estimate the effect of the characteristics of the new entrepreneurs on job creation, using a sample of 1198 new Portuguese entrepreneurs. The Poisson mixture regression model shows that the population is heterogeneous across three segments. The first segment presents a low growth of employment and the level of education and management experience does not have a significant effect on employment. In the other two segments—moderate and high growth of employment—those variables register a significant effect.

1 Introduction

Entrepreneurship and small- and medium-size business growth are essential in order to foster competition and economic growth in European economies. Entrepreneur characteristics are included in many theoretical and empirical research frameworks as factors that affect the business performance. Most of the research conducted in this field diverges on the causal links between those factors and job creation. This divergence may be due to the underlying hypothesis that population of new entrepreneurs is homogeneous concerning the effect of its characteristics on job creation.

L.P. Pontes (✉)
Statistics Portugal and Instituto Universtário de Lisboa (ISCTE-IUL), BRU, Portugal
e-mail: leandro.pontes@ine.pt

J.G. Dias
Instituto Universtário de Lisboa (ISCTE-IUL), BRU, Portugal
e-mail: jose.dias@iscte.pt

J. Lita da Silva et al. (eds.), *Advances in Regression, Survival Analysis, Extreme Values, Markov Processes and Other Statistical Applications*, Studies in Theoretical and Applied Statistics, DOI 10.1007/978-3-642-34904-1_39,

The hypothesis of heterogeneity is implicit in many empirical investigations on the segmentation of the population of new entrepreneurs [12]. However, those results have been neglected when it comes to assess the effect of the entrepreneur's characteristics on business performance. More recently, Gartner [5] reintroduced this hypothesis, for whom progress in entrepreneurship research depends on assuming the heterogeneity of the population of new entrepreneurs. The main purpose of this study is to discuss the heterogeneity hypothesis in the process of assessing the impact of entrepreneur' socioeconomic profile on job creation. Thus, this chapter models the number of jobs created per firm in the period between the moments of creation of the firm and of the data collection using a Poisson mixture regression model. The structure of the chapter is the following: Sect. 2 presents the methodology; Sect. 3 describes the data used; and Sect. 4 provides the results and its discussion. The chapter ends with concluding remarks and suggestions for further research.

2 Poisson Mixture Regression Model

The Poisson regression model assumes equal mean and variance for the dependent variable y_i. In empirical research, this condition is violated in most economic models as variance tends to exceed the mean. This phenomenon is known as overdispersion and can express the heterogeneity of the sample concerning the mean or the number of occurrences in the period, λ_i, the expected value of y_i. The Poisson mixture regression model [10] extends Poisson regression by allowing model parameters to vary across the mixture components.

The Poisson mixture regression model with concomitant variables contains four types of variables: one independent variable (y); a discrete latent variable or latent class (z), which indicates the segment or latent class; K explanatory variables ($\mathbf{x}$), which explain the mean of y; and L concomitant variables ($\mathbf{w}$), explaining the prior probability that observation i belongs to a given class [7].

The Poisson mixture regression model is defined by

$$P_i(y_i; \boldsymbol{\varphi}, \mathbf{x}_i, \mathbf{w}_i) = \sum_{s=1}^{S} h(z_i; \boldsymbol{\gamma}_s, \mathbf{w}_i) P_{is}(y_i; \boldsymbol{\beta}_s, \mathbf{x}_i),$$

where:

- $P_{is}(y_i; \boldsymbol{\beta}_s, \mathbf{x}_i)$ is the distribution of y_i in segment s. The vector of parameters $\boldsymbol{\beta}_s$ and explanatory variables $\mathbf{x}_i$ represent the systematic component of the GLM for segment s and are linked to the expected value of y_i by the log link function [8]: (1) systematic component: $\eta_{is} = \beta_{0s} + \sum_{j=1}^{K} \beta_{js} x_{ij}$; (2) link function: $\log \lambda_{is} = \eta_{is}$; and $E[y_i | \mathbf{x}_i, z_i] = \lambda_{is}$.
- $h(z_i; \boldsymbol{\gamma}_s, \mathbf{w}_i)$ is the prior probability (π_{is}) that observation i belongs to segment s. Thus, the latent class variable (z_i) is multinomial distributed with categories $s = 1, \ldots, S$ and probabilities π_{is}, with $\pi_{is} > 0$ and $\sum_{s=1}^{S} \pi_{is} = 1$. Thus, the

prior probability $\pi_{is} = h(z_i; \boldsymbol{\gamma}_s, \mathbf{w}_i)$ is regressed on the concomitant variables and varies across individuals. Let L be the number of concomitant variables in vector $\mathbf{w}_i$ and $\boldsymbol{\gamma}_s$ the regression parameters [3, 11]. Then, the logistic submodel is given by

$$\pi_{is} = \frac{\exp(\gamma_{0s} + \sum_{l=1}^{L} \gamma_{ls} w_{il})}{\sum_{r=1}^{S} \exp(\gamma_{0r} + \sum_{l=1}^{L} \gamma_{lr} w_{il})},$$

where for the latent class S, $\boldsymbol{\gamma}_S = 0$ for identification purposes. This submodel allows the profiling of the classes or segments as well as the allocation of new individuals into the classes.

- $P_i(y_i; \boldsymbol{\varphi}, \mathbf{x}_i, \mathbf{w}_i)$ is the marginal distribution of y_i, with parameters $\boldsymbol{\varphi} = (\boldsymbol{\gamma}_1, \ldots, \boldsymbol{\gamma}_{S-1}, \boldsymbol{\beta}_1, \ldots, \boldsymbol{\beta}_S)$.

This model is identified [9]. The estimated posterior probability that observation i belongs to segment s is given by

$$\hat{\alpha}_{is} = \frac{\hat{\pi}_{is} P_{is}(y_i; \hat{\boldsymbol{\beta}}_s, \mathbf{x}_i)}{\sum_{r=1}^{S} \hat{\pi}_{ir} P_{ir}(y_i; \hat{\boldsymbol{\beta}}_r, \mathbf{x}_i)}.$$

Maximum likelihood estimates are obtained by the maximization of the log-likelihood function:

$$\ell_S(\boldsymbol{\varphi}; \mathbf{y}, \mathbf{x}, \mathbf{w}) = \sum_{i=1}^{n} \log P_i(y_i; \boldsymbol{\varphi}, \mathbf{x}_i, \mathbf{w}_i),$$

where $\boldsymbol{\varphi}$ is the set of parameters to be estimated. This estimation can be done by combining the EM algorithm [2] with the Newton-Raphson algorithm [3]. As the log-likelihood surface is extremely complex and with many local maxima, we report the solution with the maximum log-likelihood value out of 500 runs of the algorithm for each S, from 1 to 7, and 10^{-6} as the convergence tolerance.

The number of latent classes, S, is unknown and has to be estimated (model selection). A common model selection strategy based on information criteria is adopted in this chapter. Let $C_S = -2\ell_S(\hat{\boldsymbol{\varphi}}; \mathbf{y}, \mathbf{x}, \mathbf{w}) + dN_S$ be the general information criterion, where $\ell_S(\hat{\boldsymbol{\varphi}}; \mathbf{y}, \mathbf{x}, \mathbf{w})$ is the maximum log-likelihood value, N_S is the number of free parameters, and d is the penalizing constant. For different values of d, one has the AIC—Akaike Information Criterion ($d = 2$), the BIC—Bayesian Information Criterion ($d = \log n$), the AIC3—Modified Akaike Criterion ($d = 3$), and the CAIC—Consistent Akaike Criterion ($d = \log n + 1$). For every criterion, the best solution (S) minimizes the criterion. BIC and CAIC are consistent criteria [1], whereas the AIC is a biased estimate of the true number of latent classes [4, 6]. Thus, in case of lack of agreement, we select the solution suggested by BIC and CAIC.

3 Population, Sample, and Questionnaire

Data were collected by Statistics Portugal between April and October of 2005. The project was co-funded by the Portuguese Government and the European Commission. The target population comprises all Portuguese firms created in 2002, which were still active in 2005. The target population did not include firms created by other firms, reactivated ones, firms purchased by other firms, and changes of the legal status. This is in order to select just the *de novo* creations in 2002. The random sample is stratified by region and industry. A total of 1198 valid responses are used in this research. The survey questionnaire includes eleven nominal variables that describe the entrepreneur and the firm. An additional variable—classes of motivations for start-ups—results from prior research (see Tables 1 and 2).

4 Results

Information criteria BIC and CAIC both register the minimum value for the solution with three segments. The other two criteria, AIC and AIC3, stabilize for the same number of segments. We adopt the solution with three segments based on the consistency properties of the criteria BIC and CAIC [1].

The larger segment represents 58.7 % of the observations and it was labeled as the *low growth* segment. It has an average employment of 5.3 individuals; segments 2 and 3 were labeled as *moderate growth* (33.7 %) and *high growth* (7.6 %), as they present an average employment of 10.4 and 44.5, respectively. Table 1 depicts the profiles of the entrepreneurs in each segment.

The *low growth* segment includes entrepreneurs whose prior occupation was employee, without having much experience in business start-ups and having the lowest level of education. The desire of independence is their main motivation for the creation of a new business. The *moderate growth* segment presents a larger proportion of individuals (50 %) for whom the prior occupation was the management of another firm. They are also motivated mainly by the desire of independence. The *high growth* segment is characterized by the largest proportion of individuals with higher education (30 %) and experience in business start-ups. In this segment, the creation of the firm bases is driven by the desire of expanding their business activity.

Table 2 shows the parameter estimates. Significance tests on individual parameters are performed using the Wald statistic either to test whether a set of parameters is equal to zero – Wald(0) – or whether they are invariant across segments—Wald(=). The coefficients of the variables *Sex*, *Nationality*, *Cooperation with other firms*, and *Professional education in business start-ups* were restricted to be invariant across segments. All the explanatory variables are in nominal scale and consequently, the exponential of the coefficient estimate $exp(\hat{\beta}_{km})$ represents the multiplicative effect of the category m of the variable k on the average of the dependent variable *Employment*.

Table 1 Socioeconomic profile of the entrepreneur

	Segments			
	1	2	3	Aggregate sample
Segment dimension	0.58	0.34	0.08	1.00
Average employment	5.25	10.42	44.52	9.92
Age				
< 30	0.13	0.11	0.11	0.12
30 – 39	0.35	0.33	0.28	0.34
40 – 49	0.30	0.31	0.38	0.31
50 or more	0.22	0.25	0.23	0.23
Education level				
Primary education/lower secondary	0.54	0.55	0.40	0.53
Upper secondary education	0.27	0.22	0.22	0.25
Postsecondary non-tertiary education	0.05	0.08	0.08	0.06
Tertiary education	0.15	0.15	0.30	0.16
Firms started before				
No	0.62	0.52	0.39	0.57
Yes, once	0.27	0.40	0.38	0.32
Yes, more than once	0.10	0.09	0.23	0.11
Sex				
Male	0.84	0.89	0.88	0.86
Female	0.16	0.11	0.12	0.14
Prior occupation				
Other	0.06	0.09	0.11	0.08
Worked as an employee	0.54	0.41	0.43	0.49
Ran another business	0.40	0.50	0.46	0.44
Branch of activity of the new firm				
Manufacturing	0.19	0.22	0.30	0.21
Building	0.16	0.36	0.19	0.23
Trade	0.30	0.19	0.22	0.25
Hotels	0.01	0.03	0.05	0.02
Restaurants	0.07	0.10	0.09	0.08
Cleaning and employment placement agencies	0.01	0.02	0.08	0.02
Other services	0.25	0.10	0.07	0.19
Experience in the branch of activity				
No	0.14	0.17	0.22	0.16
Yes	0.86	0.83	0.78	0.84
Carrying out another paid activity				
No	0.84	0.84	0.61	0.82
Yes	0.16	0.17	0.39	0.18
Classes of motivations for start-ups				
Independence	0.48	0.43	0.31	0.45
Continuity	0.40	0.38	0.36	0.39
Expansion	0.13	0.18	0.34	0.16
Nationality				
Portuguese	0.96	0.97	0.88	0.96
Other	0.04	0.03	0.12	0.04

Table 2 Estimates of the regression parameters

	Segments				
	1	2	3	Wald (=0)	Wald (=)
Intercept	1.27 ***	1.93 ***	3.19 ***	1,380.76 ***	185.33 ***
Prior occupation					
Other	–	–	–	68.85 ***	25.99 ***
Worked as an employee	0.19	0.51 ***	0.16		
Ran another business	0.17	0.47 ***	0.39 ***		
Firms started before					
No	–	–	–	81.47 ***	14.18 **
Yes, once	0.20 ***	0.02	0.07		
Yes, more than once	0.28 ***	0.29 ***	0.41 ***		
Professional education in business start-ups					
No	–	–	–	5.31 *	–
Yes	0.07 *	0.07 *	0.07 *		
Sex					
Male	–	–	–	9.68 **	–
Female	0.12 **	0.12 **	0.12 **		
Nationality					
Portuguese	–	–	–	29.15 ***	–
Other	0.30 ***	0.30 ***	0.30 ***		

Education level					
Primary education/lower secondary	–	–	–	55.56 ***	49.79 ***
Upper secondary education	0.08	0.02	−0.04		
Postsecondary non-tertiary education	−0.09	−0.03	−0.03		
Tertiary education	0.09	0.34 ***	−0.17 **		
Branch of activity of the new firm					
Manufacturing	–	–	–	2,113.53 ***	1,501.03 ***
Building	0.91 ***	−1.25 ***	0.14		
Trade	−0.33 ***	−0.14	−0.01		
Hotels	1.00 ***	−1.33 ***	−0.06		
Restaurants	−0.03	0.41 ***	0.74 ***		
Cleaning and employment placement agencies	0.52 ***	0.81 ***	1.51 ***		
Other services	−0.46 ***	0.14	0.52 ***		
Cooperation with other firms					
No	–	–	–	129.11 ***	–
Yes	0.30 ***	0.30 ***	0.30 ***		
Classes of motivations for start-ups					
Independence	–	–	–	44.03 ***	25.94 ***
Continuity	−0.17 ***	−0.08	−0.17 *		
Expansion	−0.26 **	−0.22 ***	0.12		
Age					
< 30	–	–	–	104.91 ***	31.46 ***
30–39	−0.17 **	−0.06	−0.20 **		
40–49	−0.35 ***	−0.14	−0.52 ***		
50 or more	−0.39 ***	−0.12	−0.23 **		

(continued)

Table 2 (continued)

	Segments				
	1	2	3	Wald (=0)	Wald (=)
Carrying out another paid activity					
No	–	–	–	67.97 ***	16.94 ***
Yes	0.01	−0.33 ***	−0.32 ***		
Experience in the branch of activity					
No	–	–	–	18.06 ***	–
Yes	0.20 ***	0.20 ***	0.20 ***		
Log-likelihood	−3,612.50				
BIC	8,011.81				
Pseudo R^2 (global)	0.95				
Pseudo R^2 (by segment)	0.68	0.79	0.91		

* p-value < 0.05; ** p-value < 0.01; *** p-value < 0.001

The coefficients of *Age* are negative for all segments. It suggests that older entrepreneurs are less prone to expand their firms by contracting new workers than younger ones. This effect is especially strong in the first and third segments. In the *moderate growth* segment coefficients are negative though are not significantly different from the base category.

Variable *Education* suggests important insights. First, there is no significant difference between the base category (primary education) and the two categories immediately above, *i.e.*, entrepreneur education only impacts job creation after reaching the high level of education. Second, the level of education may be important only for the new firms with large dimension; for smaller ones in the *slow growth* segment, the parameter estimates of the three categories are not significantly different from the base category of primary education. Finally, the higher level of education may have opposite impact across segments: while in *moderate growth* segment high education has a positive effect, in the *high growth* segment the signal reverses. Thus, for larger firms, entrepreneurs with higher education are less prone to job creation than their less educated counterparts. The impact of running another business on job creation is positive in *moderate* and *high growth* segments. For the *low growth* segment, the effects are not different across the prior occupations.

Prior experience in starting a new business is highly significant for those entrepreneurs who have started a new business more than once. Although we do not have information about the success or failure of the prior start-ups, these results suggest that society and public policies should support those entrepreneurs who insist in the path of entrepreneurship. They have more chances of paying back through the creation of jobs. Another type of experience, branch experience, has a highly significant impact on job creation as well, although invariant across segments.

5 Conclusion

Job creation is one of the main goals of economic policy and this research suggests directions for an efficient promotion of entrepreneurship. Most of prior research assumes that the dependent variable (*Employment*) is either normal distributed or its log transformation is normally distributed. This research focused on the Poisson distribution. Using a mixture of Poisson distributions, we showed that the population of new entrepreneurs is heterogeneous concerning the effects of their socioeconomic characteristics on job creation, as the impact of the covariates tends to be different across segments. We concluded that entrepreneurs with previous knowledge of the branch of activity and experience on starting up new businesses should be encouraged in order to foster employment growth. Further research can replicate this methodology for other measures of firm performance.

References

1. Bozdogan, H.: Model selection and Akaike's information criterion (AIC): The general theory and its analytical extensions. Psychometrika **52**(3), 345–370 (1987)
2. Dempster, A.P., Laird, N.M., Rubin, D.B.: Maximum likelihood from incomplete data via the EM algorithm. J. Roy. Stat. Soc. B **39**(1), 1–38 (1977)
3. Dias, J.G.: Finite Mixture Models. Review, Applications, and Computer-intensive Methods (PhD Thesis). Ridderprint, The Netherlands (2004)
4. Dias, J.G.: Latent class analysis and model selection. In: Spiliopoulou, M., Kruse, R., Borgelt, C., Nrnberger, A., Gaul, W. (eds.) From Data and Information Analysis to Knowledge Engineering, pp. 95–102. Springer, Berlin (2006)
5. Gartner, W.B.: A conceptual framework for describing new venture formation. Acad. Manage. Rev. **10**(4), 696–706 (1985)
6. Hurvich, C.M., Tsai, C.-L.: Regression and time series model selection in small samples. Biometrika **76**(2), 25–43 (1989)
7. McLachlan, G., Peel, D.: Finite Mixture Models. Wiley, New York (2000)
8. McCullagh, P., Nelder, J.A.: Generalized Linear Models London. Chapman and Hall, London (1989)
9. Teicher, H.: Identifiability of mixtures. Ann. Mathematic. Stat. **32**(1), 244–248 (1961)
10. Wedel, M., Desarbo, W.S., Bult, J.R., Ramaswamy, V.: A latent class Poisson regression model for heterogeneous count data. J. Appl. Econom. **8**(4), 397–411 (1993)
11. Wedel, M.: Concomitant variables in finite mixture models. Stat. Neerl. **56**(3), 362–375 (2002)
12. Woo, C.Y., Cooper, A.C., Dunkelberg, W.C.: Entrepreneurial typologies: Definitions and implications. In: Zacharakis, A. (ed.) Frontiers of Entrepreneurship Research, pp. 165–172. Center for Entrepreneurial Studies, Babson College, Wellesley, Mass (1988)

Simulation Study of the Calibration Technique in the Extremal Index Estimation

D. Prata Gomes, João T. Mexia, and M. Manuela Neves

Abstract
Classical extreme value methods were first derived when the underlying process is assumed to be a sequence of independent and identically distributed random variables. However, when observations are taken along the time and/or the space, the independence is an unrealistic assumption. A relevant parameter that arises in this situation is the extremal index, θ, characterizing the degree of local dependence in the extremes of a stationary series. Most of the semi-parametric estimators of this parameter show a strong dependence on the threshold u_n, with an increasing bias and a decreasing variance as such a threshold decreases. A procedure based on the calibration methodology is here considered as a way of controlling the bias of an estimator. Point and interval estimates for the extremal index are obtained. A simulation study has been performed to illustrate the procedure.

1 Introduction

Classical extreme value theory gives conditions for the existence of normalizing sequences $\{a_n > 0\}$ and $\{b_n\}$ such that for $u_n = a_n x + b_n$, $P\{M_n \leq u_n\} \to G(x)$ as $n \to \infty$, where $G(\cdot)$ is a non-degenerate distribution function that necessarily

D.P. Gomes (✉) · J.T. Mexia
CMA and Mathematics Department, Faculdade de Ciências e Tecnologia, Universidade Nova de Lisboa, Lisboa, Portugal
e-mail: dsrp@fct.unl.pt; jtm@fct.unl.pt

M.M. Neves
CEAUL and Mathematics Department, Instituto Superior de Agronomia, Universidade Técnica de Lisboa, Lisboa, Portugal
e-mail: manela@isa.utl.pt

J. Lita da Silva et al. (eds.), *Advances in Regression, Survival Analysis, Extreme Values, Markov Processes and Other Statistical Applications*, Studies in Theoretical and Applied Statistics, DOI 10.1007/978-3-642-34904-1_40,

belongs to one of the Gumbel, Fréchet and Weibull families that are usually termed as the extreme value distributions and M_n is the maximum of a sequence of independent and identically distributed (i.i.d.) random variables. However, in real situations, extreme events often occur in clusters of large values. So, for a dependent structure, the exceedances over a high level tend to occur in clusters instead of happening in an isolated way. The characterization of extremes of stationary processes, the most natural generalization of a sequence of i.i.d random variables, appeared then.

Let $\{X_n\}$ be a stationary sequence, where M_n is the maximum and $\{\widehat{X}_n\}$ the associated i.i.d. sequence, with the same marginal distribution F. Let $\widehat{M}_n$ be the maximum of the i.i.d. sequence. If the distribution of the maximum $\widehat{M}_n$ suitably normalized by constants $\{a_n > 0\}$ and $\{b_n\}$ converges to a non-degenerate law, i.e., $P[\widehat{M}_n \le a_n x + b_n] \to G(x)$, where $G(\cdot)$ is the extreme value distribution, denoted by $GEV(\lambda, \delta, \gamma)$, then the distribution of M_n also converges with the same set of normalizing constants to $G^\theta(\cdot)$, where θ is the extremal index, see [14]. $G^\theta(\cdot)$ is a $GEV(\lambda_\theta, \delta_\theta, \gamma_\theta)$ distribution with

$$\lambda_\theta = \lambda + \delta \frac{\theta^\gamma - 1}{\gamma}, \qquad \delta_\theta = \delta\theta^\gamma, \qquad \gamma_\theta = \gamma.$$

In a dependent setup the estimation of θ is then important not only by itself but also because of its influence in the estimation of other parameters of extreme events. Most of the semi-parametric estimators of θ show the same type of behaviour: nice asymptotic properties, but a strong dependence on the level u_n, with an increasing bias and a decreasing variance as the level decreases.

In this chapter we shall illustrate the behaviour of one of the most well-known estimators, the *up-crossing estimator*, given in Sect. 2. Extremal properties of two stationary sequences will be reviewed and conditions for the existence of the extremal index will be presented. The stationary models considered in this chapter are:

Model I. The Moving-Maximum Process of order q, see [5], or MM(q) in short, where $q \ge 1$ is a fixed integer, defined by

$$X_t = \max_{0 \le i \le q} Z_{t-i}, \qquad t > q, \tag{1}$$

where the $\{Z_i\}_{i\ge 1}$ are independent standard Fréchet random variables.

Model II. The Max-Autoregressive Process of order one, or ARMAX, see [2], defined by

$$X_1 = Z_1, \quad X_i = \max\{\beta X_{i-1}, (1-\beta)Z_i\}, \qquad i \ge 2, \tag{2}$$

where the $\{Z_i\}_{i\ge 1}$ are independent standard Fréchet random variables and $\beta \in [0, 1)$.

In Sect. 3 a brief description of the calibration technique will be given and used as a suggestion for dealing with the bias–variance trade-off in the extremal index estimation. A simulation study using the two aforementioned stationary sequences will be carried out. Some concluding remarks will end the chapter.

2 Extremal Index: Definition and Estimation

The extremal index, θ, measures the relationship between the dependence structure of the data and the behaviour of the exceedances over a high threshold u_n. This threshold u_n is such that, with τ fixed, the underlying distribution function F verifies

$$F(u_n) = 1 - \tau/n + o(1/n) \quad \text{as} \quad n \to \infty. \tag{3}$$

Definition 1 (See [14]). Let $\{X_n\}_{n\geq 1}$ be a strictly stationary sequence with marginal distribution function F and $M_n = \max_{1\leq k\leq n} X_k$. We say that the process has an **extremal index,** $\theta \in [0,1]$, if for every $\tau > 0$, there exists a sequence of thresholds $\{u_n(\tau)\}_{n\geq 1}$ such that $nP(X_n > u_n(\tau)) \to \tau$ and $P(M_n \leq u_n(\tau)) \to \exp\{-\theta\tau\}$ as $n \to \infty$.

This definition does not involve any dependence restriction on the sequence $\{X_n\}$. However, some results do exist characterizing the extremal behaviour of a stationary sequence under some dependence conditions. One form of short range dependence is formulated in $D(u_n)$ *condition* of Leadbetter et al., see [14], that makes precise the notion of extreme events near independent if they are sufficiently distant.

Definition 2 (See [14]). A stationary series $\{X_n\}_{n\geq 1}$ is said to satisfy the $D(u_n)$ *condition* if, for all $i_1 < \cdots < i_p < j_1 < \cdots < j_q$, with $j_1 - i_p > l$,

$$|P\{X_{i_1} \leq u_n, \ldots, X_{i_p} \leq u_n, X_{j_1} \leq u_n, \ldots, X_{j_q} \leq u_n\}$$

$$-P\{X_{i_1} \leq u_n, \ldots, X_{i_p} \leq u_n\} P\{X_{j_1} \leq u_n, \ldots, X_{j_q} \leq u_n\}| \leq \alpha(n, l_n),$$

where $\alpha(n, l_n) \to 0$ for some sequence l_n such that $l_n/n \to 0$, as $n \to \infty$.

If $\{X_n\}_{n\geq 1}$ is a stationary sequence with $D(u_n(\tau))$ holding for each $\tau > 0$ ($u_n(\tau)$ satisfying (3)) it may be shown that if $P(M_n \leq u_n(\tau))$ converges for some $\tau > 0$, then $P(M_n \leq u_n(\tau)) \to \exp\{-\theta\tau\}$ for all $\tau > 0$ and $\{X_n\}$ has an **extremal index,** $\theta \in [0,1]$, see [13]. Leadbetter [12] gave a condition restricting the clustering of high level exceedances, known as $D'(u_n)$ *condition*. If this condition holds, then $\theta = 1$.

Apart from some few models for which $D(u_n)$ and $D'(u_n)$ *conditions* are easy to verify and the extremal index value can be obtained, those conditions are, in general, difficult to study.

For **Model I** it is easy to show that in the limit, clusters of high exceedances are of size $q + 1$ (with probability 1) and the extremal index of the MM(q) process is $\theta = 1/(q + 1)$, see [19].

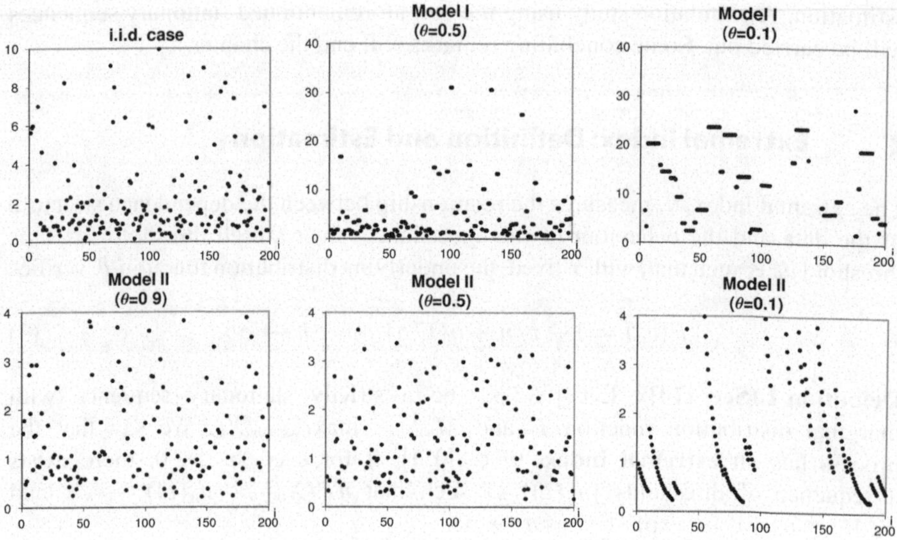

Fig. 1 Simulated data of $n = 1000$ observations of a i.i.d sequence (*top-left*) and MM(q) process with $q = 9$ and $q = 1$ (*top*); ARMAX process with $\beta = 0.1,\ 0.5,\ 0.9$ (*bottom*)

For **Model II** it is easy to show that the marginal distribution of $\{X_n\}$ is Fréchet and $D(u_n)$ *condition* holds. $D'(u_n)$ *condition* fails if $\beta > 0$, see [3]. For $0 < x < \infty$ and $u_n = nx$,

$$P[M_n \leq u_n] \to \exp\{-(1-\beta)/x\} \quad \text{as} \quad n \to \infty.$$

The extremal index of the ARMAX process is $\theta = 1-\beta$ with $0 < \theta \leq 1$, see [2–4].

Although for some pathological cases where $D(u_n)$ holds there is no extremal index, most practical interesting situations are those for which the extremal index θ does exist and is nonzero.

An illustration of a partial realization of i.i.d Fréchet variables and variables following Model I ($q = 9$ and 1, which corresponds to $\theta = 0.1$ and 0.5, respectively) and Model II ($\beta = 0.9,\ 0.5$ and 0.1, which gives $\theta = 0.1,\ 0.5$ and 0.9, respectively) is shown in Fig. 1.

As we can see the high values behave differently in the process, depending on the degree of dependence.

One way of interpreting the extremal index of a stationary sequence is in terms of the tendency of the process to cluster at extreme levels. A rough interpretation of θ is to consider it as the inverse of the limiting mean cluster size, where the limiting is in the sense of clusters of exceedances of increasingly high thresholds. If $\theta = 1$ exceedances occur singly at the limit, while if $\theta < 1$ they tend to cluster at the limit.

The clusters of exceedances may be identified asymptotically as runs of consecutive exceedances and cluster sizes as run lengths. Under regularity conditions, the conditional expected run length is approximately equal to $1/\theta$, see [15].

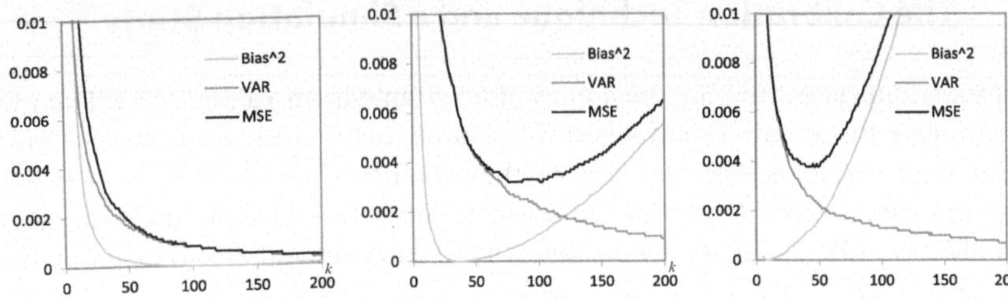

Fig. 2 Estimated mean squared error, squared bias and variance of $\hat{\Theta}^{UC}$ in Model II, for $\theta = 0.1, 0.5$ and 0.9 (from *left* to *right*)

A suggestion was then to estimate θ by the reciprocal of the sample average run length.

Given a sequence of random variables, $X_1, X_2, \ldots, X_n$, from a process which satisfies the $D(u_n)$ *condition*, where n is large and u_n is a high threshold, the most basic form of cluster identification (that does not require any knowledge of clustering characteristics of the process) leads to a naive nonparametric estimator of θ, the **up-crossing estimator**, $\widehat{\Theta}_n^{UC}(u_n)$, see [6–8, 15], defined as

$$\widehat{\Theta}_n^{UC} := \frac{\sum_{i=1}^{n-1} I\left(X_i \leq u_n < X_{i+1}\right)}{\sum_{i=1}^{n} I(X_i > u_n)}.$$

The asymptotic properties of the up-crossing estimator were established in [15, 18, 20], under several different conditions. Nandagopalan, see [15], showed that, for random levels u_n, $\widehat{\Theta}_n^{UC}(u_n)$ is a weakly consistent estimator of θ.

The asymptotic normality of $\widehat{\Theta}_n^{UC}(u_n)$ was derived in [10, 20]. The first moments of the estimator $\widehat{\Theta}_n^{UC}(u_n)$, the variance and the bias were obtained in [10].

Figure 2 illustrates the estimates of mean squared error, squared bias and variance of the estimator for **Model II** and for finite sample size with $n = 1000$, using 1000 replicates and some values of θ. The estimates are plotted at a range of thresholds chosen up to 20 % of the sample length, where $u_n := X_{k:n}$, $(5 \leq k \leq 0.2 \times n)$ and $X_{1:n} \geq X_{2:n} \geq \ldots \geq X_{n:n}$ are the descending ordered statistics associated to the sample.

A problem that arises is how to choose the level u_n (or k) for obtaining the estimates. Intensive computational methods such as *Bootstrap*, *Jackknife* and *subsampling*, see [9, 11], have been considered to obtain estimates of the level. However, this is out of the scope of this chapter. The objective of this study is to show how calibration methodology can be used for reducing the bias of the estimator as well as for providing confidence intervals for the extremal index. This is a preliminary study; some simulation results already obtained are encouraging, but more work is needed.

3 The Calibration Technique and a Simulation Study

Calibration aims at estimating the values of a variable from values of a related variable. We have linear calibration when we assume that there is a linear relationship between both variables. For our case we then shall have $\widehat{\Theta}^{UC} = \beta_1 + \beta_2\theta$ where we use the values of $\widehat{\Theta}^{UC}$ in order to estimate the values of θ. In the general case we would have $\widehat{\Theta}^{UC} = g(\theta)$, with g known. To carry out calibration we obtain the values of $\widehat{\Theta}^{UC}, \widehat{\theta}^{UC}$, for given values of θ and adjust the function g.

In the case of linear calibration we are led to fit a linear regression of $\widehat{\Theta}^{UC}$ on θ, see [1, 17, 21].

In our case, known values of θ, $(\theta_1, \cdots, \theta_{n_\theta})$ are considered, the up-crossing estimates $\widehat{\theta}^{UC}$ are obtained for each value of k $(u_n := X_{k:n})$ and the linear regression is estimated

$$\widehat{\theta}^{UC} = \widehat{\beta}_1(k) + \widehat{\beta}_2(k)\theta, \tag{4}$$

where $\widehat{\beta}_1(k)$ and $\widehat{\beta}_2(k)$ are the least squares estimates for the coefficients. Besides adjusting the linear regression we can obtain the corresponding confidence band.

The α level confidence band is bounded by

$$\widehat{\beta}_1(k) + \widehat{\beta}_2(k)\theta(-1)^h\widehat{\sigma}\left(c_1 + c_2\left(n_\theta^{-1} + Q(\theta - \overline{\theta})^2\right)^{1/2}\right), \tag{5}$$

where $h = 1$ (for lower), 2 (for upper) and $\widehat{\sigma}$ is the estimate for the variance error. Constants c_1 and c_2 are calculated as follows. Let us define $S_1 = n_\theta^{-1/2}$ and $S_2 = \left(n_\theta^{-1} + QM^2\right)^{1/2}$, where $M = \max\left\{\overline{\theta} - \theta^{(1)}, \theta^{(2)} - \overline{\theta}\right\}$, $Q = 1/\sum_{i=1}^{n_\theta}(\theta_i - \overline{\theta})^2$, $\overline{\theta} = \sum_{i=1}^{n_\theta}\theta_i/n_\theta$ and $\theta^{(1)}$ and $\theta^{(2)}$ are the minimum and the maximum of θ_i, respectively. After c has been obtained by entering tables in [17] with $s_1 = S_1/z_\alpha$ and $s_2 = S_2/z_\alpha$ where z_α is the upper $\alpha/2$-point of the standard normal distribution, c_1 and c_2 are given by $c_1 = cz_\alpha\nu^{1/2}\left(\chi_{1-\delta}^{\chi_\nu^2}\right)^{-1/2}$ and $c_2 = c\left(p\,\chi_\delta^{F_{p,\nu}}\right)^{1/2}$, with $p = 2$ where $\chi_\delta^{F_{p,\nu}}$ is the upper δ-point of the F-distribution with p and ν df and $\chi_{1-\delta}^{\chi_\nu^2}$ is the lower δ-point of the chi-square distribution with ν df. We can now invert the Eq. (4)

$$\theta = a(k)\,\widehat{\theta}^{UC} + b(k) \tag{6}$$

and obtain the limits (5) as

$$\theta_{UP} = \overline{\theta} + C^{-1}\left(\widehat{\beta}_2(k)D_1 + \widehat{\sigma}c_2\left(n_\theta^{-1}C + QD_1^2\right)^{1/2}\right), \tag{7}$$

$$\theta_{LOW} = \overline{\theta} + C^{-1}\left(\widehat{\beta}_2(k)D_2 - \widehat{\sigma}c_2\left(n_\theta^{-1}C + QD_2^2\right)^{1/2}\right), \tag{8}$$

with $C = \widehat{\beta}_2^2(k) - (\widehat{\sigma}c_2)^2 b$, $D_1 = D_1(\widehat{\theta}^{UC}) = \widehat{\theta}^{UC} - \widehat{\beta}_1(k) - \widehat{\beta}_2(k)\overline{\theta} + \widehat{\sigma}c_1$,

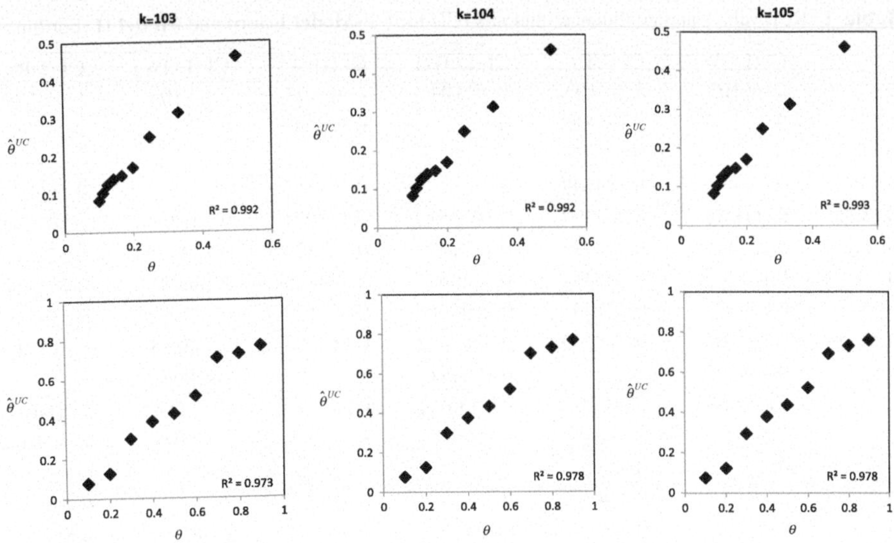

Fig. 3 $\hat{\theta}^{UC}$ against the true value of θ, for Model I (*top*) and Model II (*bottom*) for three values of k chosen within a "stability" region of the trajectories of $(k, \hat{\theta}^{UC}(k))$

$$D_2 = D_2(\widehat{\theta}^{UC}) = \widehat{\theta}^{UC} - \widehat{\beta}_1(k) - \widehat{\beta}_2(k)\overline{\theta} - \widehat{\sigma}c_1.$$

These expressions give the confidence bands of θ for the α level, once $\widehat{\theta}^{UC}$ is obtained. To use Eqs. (6), (7) and (8) we need the endpoints of the three calibration intervals:

- For $v = 1, 2, \widehat{\theta}^{UC(v)} = \beta_1 + \beta_2\theta^{(v)}$.
- $\widehat{\theta}^{UC(I1)}$ ($\widehat{\theta}^{UC(I2)}$) is found by putting $\theta = \theta^{(1)}$ ($\theta^{(2)}$) in Eq. (5) with $h = 2(1)$.
- $\widehat{\theta}^{UC(01)}$ ($\widehat{\theta}^{UC(02)}$) was found by putting $\theta = \theta^{(1)}$ ($\theta^{(2)}$) in Eq. (5) with $h = 1(2)$.

Once $\widehat{\theta}^{UC}$ is obtained:

- For $\widehat{\theta}^{UC(1)} \leq \widehat{\theta}^{UC} \leq \widehat{\theta}^{UC(2)}$, the point estimate of θ is given by putting $\widehat{\theta}^{UC}$ in Eq. (6).
- For $\widehat{\theta}^{UC(01)} \leq \widehat{\theta}^{UC} \leq \widehat{\theta}^{UC(I2)}$ ($\widehat{\theta}^{UC(I1)} \leq \widehat{\theta}^{UC} \leq \widehat{\theta}^{UC(02)}$), the upper (lower) endpoint of the interval estimate for θ is given by putting $\widehat{\theta}^{UC}$ in Eqs. (7) and (8).

In Prata Gomes, see [16], a simulation study considering several stationary processes was carried out. Here we shall show results for **Model I** and **Model II**. For several values of q and β, a sample of size $n = 1000$ was obtained. For the same values of k used in Fig. 2, i.e., $5 \leq k \leq 0.2 \times n$, estimates $\widehat{\theta}^{UC}$ were plotted to look for a stability region. Three values of k for both regions were chosen for illustrating the application of the method. Figure 3 represents the n_θ pairs, $(\theta, \widehat{\theta}^{UC})$. The good results for R^2 led us to use the linear calibration technique, described above.

To illustrate the application of the calibration technique, Table 1 shows the upper and lower confidence limits for a set of θ values. Given a sample of observed values

Table 1 Real values and confidence intervals (CI) for θ—**Model I** (top) and **Model II** (bottom)

θ	CI (LOW) $k = 103$	CI (UP) $k = 103$	CI (LOW) $k = 104$	CI (UP) $k = 104$	CI (LOW) $k = 105$	CI (UP) $k = 105$
0.1	0	0.158898	0	0.15864	0	0.157282
0.1111	0	0.170274	0	0.17	0	0.168642
0.125	0	0.184648	0	0.184352	0	0.182989
0.1429	0.100246	0.203397	0.100651	0.203071	0.101763	0.201694
0.1667	0.127416	0.228792	0.127762	0.22842	0.128815	0.227012
0.2	0.164797	0.265251	0.165067	0.264806	0.166048	0.263331
0.25	0.219072	0.321707	0.219245	0.321139	0.220158	0.319536
0.3333	0.304702	0.418379	0.30476	0.417593	0.305688	0.415743
0.5	0.468591	1	0.468491	1	0.469676	1
θ	CI (LOW) $k = 103$	CI (UP) $k = 103$	CI (LOW) $k = 104$	CI (UP) $k = 104$	CI (LOW) $k = 105$	CI (UP) $k = 105$
0.1	0	0.336	0	0.320	0	0.318
0.2	0	0.434	0	0.418	0	0.417
0.3	0.103	0.539	0.131	0.522	0.130	0.521
0.4	0.229	0.653	0.254	0.633	0.254	0.633
0.5	0.350	0.774	0.372	0.752	0.373	0.752
0.6	0.464	1	0.483	0.875	0.484	0.876
0.7	0.568	1	0.587	1	0.588	1
0.8	0.666	1	0.685	1	0.686	1
0.9	0.761	1	0.779	1	0.781	1

and once fitted a model for which the extremal index exists, after constructing the table, the estimate of θ is obtained through the inverse reading of the table.

Although this is a preliminary study, we think that it deserves some more attention, mainly because it is based on a very popular and well-studied technique.

Acknowledgements Research is partially supported by CMA and National Funds through **FCT**—Fundação para a Ciência e a Tecnologia, project PEst-OE/MAT/UI0006/2011 and PTDC/FEDER. We thank the referees for carefully reading the manuscript and for their constructive remarks that greatly improved this chapter.

References

1. Andrews, F.: Calibration and statistical inference. Int. J Am. Stat. Assoc. **65**, 1233–1242 (1970)
2. Alpuim, M.T.: Contribuições à teoria de extremos em sucessões dependentes.. Ph. D. Thesis, FCUL (1989)
3. Beirlant, J., Goegebeur, Y., Teugels, J., Segers, J., Waal, D., Ferro, C.: Statistics of Extremes: Theory and Applications. Wiley, Chichester (2004)
4. Canto e Castro, L.: Sobre a Teoria Assintótica de Extremos. Ph. D. Thesis, FCUL (1992)
5. Deheuvels, P.: Point processes and multivariate extreme values. J. Multivar. Anal. **13**, 257–272 (1983)

6. Gomes, M.I.: Statistical inference in an extremal markovian model. In: Proceedings in Computational Statistics (COMPSTAT 1990), pp. 257–262 (1990)
7. Gomes, M.I.: Modelos extremais em esquemas de dependência. I Congresso Ibero-Americano de Esdadistica e Investigacion Operativa, pp. 209–220 (1992)
8. Gomes, M.I.: On the estimation of parameters of rare events in environmental time series. Statistic for the Environmental, pp. 226–241 (1993)
9. Hall, P., Horowitz, J.L., Jing, B.-Y.: On blocking rules for the bootstrap with dependent data. Biometrika **50**, 561–574 (1995)
10. Hsing, T.: Extremal index estimation for weakly dependent stationary sequence. Ann. Stat. **21**, 2043–2071 (1993)
11. Lahiri, S., Furukawa, K., Lee, Y.-D.: Nonparametric plug-in method for selecting the optimal block lengths. Stat. Methodol. **4**, 292–321 (2007)
12. Leadbetter, M.: On extreme values in stationary sequences. Z. Wahrsch. Verw. Gebiete **28**, 289–303 (1974)
13. Leadbetter, M., Rootzen, H.: Extremal theory for stochastic processes. Ann. Probab. **16**, 431–478 (1988)
14. Leadbetter, M.R., Lindgren, G., Rootzen, H.: Extremes and Related Properties of Random Sequences and Series. Springer, New York (1983)
15. Nandagopalan, S.: Multivariate Extremes and Estimation of the Extremal Index. PhD Thesis. Techn.Report 315, Center for Stochastic Processes, Univ North-Caroline (1990)
16. Prata Gomes, D.: Métodos computacionais na estimação pontual e intervalar do índice extremal. Tese de Doutoramento, Universidade Nova de Lisboa, Faculdade de Ciências e Tecnologia (2008)
17. Scheffé, H.: A statistical theory of calibration. Ann. Stat. **1**, 1–37 (1973)
18. Smith, R., Weissman, I.: Estimating the extremal index. J. Roy. Stat. Soc. B **56**, 515–528 (1994)
19. Weissman, I., Cohen, U.: The extremal index and clustering of high values for derived stationary sequences. J. Appl. Probab. **32**, 972–981 (1995)
20. Weissman, I., Novak, S.: On blocks and runs estimators of the extremal index. J. Stat. Plan. Infer. **66**, 281–288 (1998)
21. Williams, E.J.: Regression methods in calibration problems. In: Proc. 37th Session, Bull. Int Stat Inst 43 book, pp. 17–28 (1969)

Semi-parametric Building of the Optimal Screening Region in Supervised Classification

Sandra Ramos, Maria Antónia Amaral Turkman, and Marília Antunes

Abstract

In the screening problem, in addition to the classification of a new individual according to the possible outcomes of a categorical variable Y, it is possible to calculate a set of predictive probabilities, called operational characteristics, which constitutes an advantage over the known and well-established classification methods. The procedure consists on the determination a specification region $C_{\mathbf{X}}$ based on a feature vector $\mathbf{X}$ from each individual. In general, a multivariate normal distribution was admitted for $\mathbf{X}$ conditional to the category. In this work, we describe a semi-parametric Bayesian approach that relaxes the distributional assumptions (possibly invalid) using kernel methods to estimate the predictive densities of $\mathbf{X}$ in each group. We demonstrate its usefulness when applied to pairs of gene expression levels for binary classification purposes.

1 Introduction

Supervised classification methods have important applications in a wide variety of contexts such as engineering, medicine, and biology; see, for example, [4, 9]. Ramos et al. [12] proposed a Bayesian optimal screening method (BOSc) based

S. Ramos (✉)
ISEP, School of Engineering, Polytechnic of Porto, Portugal

CEAUL, Lisboa, Portugal
e-mail: sfr@isep.ipp.pt

M.A.A. Turkman · M. Antunes
Faculty of Sciences of University of Lisbon, Lisboa, Portugal

CEAUL, Lisboa, Portugal
e-mail: antonia.turkman@fc.ul.pt; marilia.antunes@fc.ul.pt

J. Lita da Silva et al. (eds.), *Advances in Regression, Survival Analysis, Extreme Values, Markov Processes and Other Statistical Applications*, Studies in Theoretical and Applied Statistics, DOI 10.1007/978-3-642-34904-1_41,

on the observation of pairs of covariates, **X**, for binary classification purposes. This method gives a simple parametric and flexible decision boundary and allows calculating a set of operating characteristics, which is an advantage over all the other classification methods. It allows also to incorporate prior information on the prevalence of population success in the model, improving the performance of the classifier.

The screening procedure may be described by a specification region $C_{\mathbf{X}}$ for the feature vector **X**, where an individual passes the screen only if $\mathbf{X} \in C_{\mathbf{X}}$. Therefore, and assuming a Bayesian framework, an optimal region $C_{\mathbf{X}}$ contains the values **x** for which the Bayesian predictive probability of success conditional on **x** is above a certain threshold [14]. The procedures are derived assuming certain parametric models, namely, a bivariate normal for **X** conditional to the group. However, in many practical situations, a parametric model cannot be expected to describe in an appropriate manner the mechanism which generates the observed dataset, and unrealistic modelling can lead to unsatisfactory classifiers. In such cases, it is important that parametric assumptions are relaxed in order to gain modelling flexibility. The aim of this work is to develop optimal screening methods for classification purposes within a Bayesian framework but without assuming parametric models for $\mathbf{X}|Y$. A semi-parametric solution was obtained using bivariate kernel estimates (see [7, 15]) for predictive densities of **X** in each group and then the predictive probability of success conditional on **x** is calculated. Note that the proposed method is particularly useful when the sampling is done separately from the two populations.

We illustrate the usefulness of this methodology on three public microarrays data sets. The results are compared with those obtained from the parametric approach, assuming a bivariate normal model for $\mathbf{X}|Y$ (BOSc method). The rest of this chapter is organized as follows. A semi-parametric method for classification based on the observation of pairs of covariates is described in Sect. 2. In Sect. 3, we test our method and the results, including comparisons with the parametric solution, are presented. Finally, we draw some conclusions and make final remarks in Sect. 4.

2 Method

We introduce a semi-parametric solution for classification based on the observation of pairs of covariates by using kernel smoothing techniques. In this section we explain the main theoretical derivations that are necessary to fully understand the methodology and its application.

2.1 Model

Let $\mathbf{X} = (X_1, X_2)$ be a pair of covariates, with each **X** having a true class label in $\{0, 1\}$. Let Y be a binary random variable that assumes value 1 (success) if **X** has class 1 and assumes value 0 otherwise. Suppose that the data consist of a random sample of n individuals, $\mathscr{D} = \{(y_1, x_{11}, x_{21}), (y_2, x_{12}, x_{22}) \cdots, (y_n, x_{1n}, x_{2n})\}$, from the unscreened population and that the binary classifications are known with certainty. The optimal screening problem has been stated by Turkman and Amaral Turkman [14] and, accordingly, in this particular case, the optimal classification region of size α is

$$C_{\mathbf{X}} = \left\{ \mathbf{x} \in \mathbb{R}^2 : P\left(Y = 1|\mathbf{x}, \mathscr{D}\right) = \frac{P\left(Y = 1|\mathscr{D}\right) p\left(\mathbf{x}|Y = 1, \mathscr{D}\right)}{\sum_{i=0,1} P\left(Y = i|\mathscr{D}\right) p\left(\mathbf{x}|Y = i, \mathscr{D}\right)} \geq k \right\}, \tag{1}$$

where k is such that $P\left(\mathbf{X} \in C_{\mathbf{X}}|\mathscr{D}\right) = \alpha$.

We consider the case where Y follows a Bernoulli distribution with parameter θ. If a beta distribution is considered as a prior distribution for the θ ($\theta \sim \text{Beta}(a, b)$, $a > 0$, $b > 0$), the predictive probability of a future individual to be a success, and the predictive probability of a future individual not being a success are (see [12]), respectively, $P\left(Y = 1|\mathscr{D}\right) = \dfrac{n_1 + a}{n + a + b}$ and $P\left(Y = 0|\mathscr{D}\right) = \dfrac{n_0 + b}{n + a + b}$, where n_i is the number of individuals in the sample for which $Y = i$, $i = 0, 1$ ($n = n_0 + n_1$).

The predictive densities of a future observation in class $Y = i$, $p(\mathbf{x}|Y = i, \mathscr{D})$, are estimated by using kernel smoothing techniques, namely,

$$\widehat{p}\left(\mathbf{x}; \mathbf{H}_i|Y = i, \mathscr{D}\right) = n_i^{-1} \sum_{j=1}^{n_i} K_{\mathbf{H}_i}\left(\mathbf{x} - \mathbf{X}_j\right), \tag{2}$$

where $\mathbf{x} = (x_1, x_2)^t$ and $\mathbf{X}_j = \left(X_{j1}, X_{j2}\right)^t$, $j = 1, 2, \cdots, n_i$, $i = 0, 1$. Here $K\left(\mathbf{x}\right)$ is the bivariate kernel (which we assume to be a probability density function); $\mathbf{H}_i$ is the bandwidth matrix of group i which is symmetric and positive-definite; and $K_{\mathbf{H}_i}\left(\mathbf{x}\right) = |\mathbf{H}_i|^{-1/2} K\left(\mathbf{H}_i^{-1/2}\,\mathbf{x}\right)$. The choice of K is not crucial: we take the standard normal. In contrast, the choice of $\mathbf{H}_i$ is very important in determining the performance of $\widehat{p}\left(\mathbf{x}; \mathbf{H}_i|Y = i, \mathscr{D}\right)$. Bivariate bandwidth selection is a difficult problem. Duong [1] introduced a new R package **ks**—available form the Comprehensive R Archive Network at http://CRAN.R-project.org/—which implements diagonal and unconstrained data-driven bandwidth matrices for kernel density estimation based on cross-validation (least squares, biased and smoothed), bootstrap, and plug-in methods. We recommend [2, 3] and references therein for more details on the multivariate bandwidth selection problem. In this chapter, we make use of **ks** package and the $p(\mathbf{x}|Y = i, \mathscr{D})$ estimates are from the unconstrained smoothed cross-validation selectors. Hence our semi-parametric solution is

$$\widehat{C}_{\mathbf{X}} = \left\{ \mathbf{x} \in \mathbb{R}^2 : \frac{P\left(Y = 1|\mathscr{D}\right) \widehat{p}\left(\mathbf{x}; \mathbf{H}_1|Y = 1, \mathscr{D}\right)}{\sum_{i=0,1} P\left(Y = i|\mathscr{D}\right) \widehat{p}\left(\mathbf{x}; \mathbf{H}_i|Y = i, \mathscr{D}\right)} \geq k \right\}. \tag{3}$$

The following posterior predictive probabilities are called operating characteristics (OC) of the screening problem:

1. Predictive probability of a randomly selected individual to be a success

$$\gamma = P(Y = 1|\mathscr{D}). \tag{4}$$

2. Size of the screening region

$$\begin{aligned}\alpha &= P(\mathbf{X} \in C_{\mathbf{X}}|\mathscr{D}) \\ &\approx \gamma \int_{\widehat{C}_{\mathbf{X}}} \widehat{p}(\mathbf{x}; \mathbf{H}_1|Y = 1, \mathscr{D})\, d\mathbf{x} + (1-\gamma) \int_{\widehat{C}_{\mathbf{X}}} \widehat{p}(\mathbf{x}; \mathbf{H}_0|Y = 0, \mathscr{D})\, d\mathbf{x}. \end{aligned} \tag{5}$$

3. Predictive probability of an individual selected by the screening procedure to be a success

$$\delta = P(Y = 1|\mathbf{X} \in C_{\mathbf{X}}, \mathscr{D}) \approx \frac{\gamma}{\alpha} \int_{\widehat{C}_{\mathbf{X}}} \widehat{p}(\mathbf{x}; \mathbf{H}_1|Y = 1, \mathscr{D})\, d\mathbf{x}. \tag{6}$$

4. Predictive probability of an individual excluded by the screening procedure to be a success

$$\epsilon = P(Y = 1|\mathbf{X} \notin C_{\mathbf{X}}, \mathscr{D}) = (\gamma - \delta\alpha)/(1-\alpha). \tag{7}$$

The region $\widehat{C}_{\mathbf{X}}$ does not have a closed form, so the screening region boundaries have to be approximated. The procedure implemented was the following:

1. Build a fine grid $G = \left\{(x_1, x_2) \in \mathbb{R}^2\right\}$ such that $P[(X_1, X_2) \in G|\mathscr{D}] \approx 1$.
2. For each $(x_1, x_2) \in G$ calculate

$$\widehat{P}[Y = 1|(x_1, x_2), \mathscr{D}] = \frac{P(Y = 1|\mathscr{D})\, \widehat{p}(x_1, x_2; \mathbf{H}_1|Y = 1, \mathscr{D})}{\sum_{i=0,1} P(Y = i|\mathscr{D})\, \widehat{p}(x_1, x_2; \mathbf{H}_i|Y = i, \mathscr{D})}.$$

3. For several values of k ($\gamma < k < 1$), form the sets $\widehat{C}_{\mathbf{X},k}$ indexed by k,

$$\widehat{C}_{\mathbf{X},k} = \left\{(x_1, x_2) \in G : \widehat{P}[Y = 1|(x_1, x_2), \mathscr{D}] \geq k\right\}.$$

4. Fit a smooth function l_k to the boundaries of each of these sets, to approximate the optimal region $\widehat{C}_{\mathbf{X},k}$ by $\{(x_1, x_2) : x_1 \in \mathbb{R}, x_2 \in I_{l_k(x_1)}\}$, where $I_{l_k(x_1)}$ is an interval of the form $]-\infty, l_k(x_1)]$ or $[l_k(x_1), +\infty[$, depending on the shape of the screening region. The functions $l_k(x_1)$ can be polynomials in x_1 or polynomial splines in x_1 with a small number of knots, depending on the shape of the region. To avoid problems of overfitting, as well as for simplicity of the whole process, low-order polynomials should be considered.

The construction of the boundaries for $\widehat{C}_{\mathbf{X},k}$ can be made more efficient if solution $\left\{(x_1, x_2) : x_2 \in \mathbb{R}, x_1 \in I_{f_k(x_2)}\right\}$ $\left(I_{f_k(x_2)}\right.$ is an interval of the form $]-\infty, f_k(x_2)]$ or $[f_k(x_2), +\infty[)$ is also considered and the most efficient among the two is chosen.

2.2 Classification Procedure

Let $\mathscr{P} = \{\mathbf{X}_j = (X_{j1}, X_{j2}), j = 1, \ldots, m\}$ be a family of m distinct, independent pairs of covariates. For each pair in $\mathscr{P}$, the optimal screening region estimate, $\widehat{C}_{\mathbf{X}_j,k}$, and the OC are obtained for several values of k. The optimal k, k_{opt}, is the one which renders the best collection of OC and gives the smallest number of individuals incorrectly classified (see [12] for details of the algorithm involved in the k_{opt} selection scheme).

Consider a new individual, with family $\mathscr{P}$ with m pairs. Based on the jth pair, the individual is classified in $C_j = 1$ if the observed $\mathbf{x}_j$ belongs to $\hat{C}_{\mathbf{X}_j,k_{opt}}$. Otherwise the individual is classified in $C_j = 0$. Let $\delta_{j,k_{opt}}$ be the predictive probability of success given that the $\mathbf{x}_j$ belongs to $\widehat{C}_{\mathbf{X}_j,k_{opt}}$. The classifications based on each of the m pairs, C_j, $j = 1, \ldots, m$, are then combined to produce the final classification of the individual [12], given by

$$C = I_{[0.5,1]}\left(\frac{C_1\delta_{1,k_{opt}} + C_2\delta_{2,k_{opt}} + \cdots + C_m\delta_{m,k_{opt}}}{\delta_{1,k_{opt}} + \delta_{2,k_{opt}} + \cdots + \delta_{m,k_{opt}}}\right), \qquad (8)$$

where $I_A(.)$ represents the indicator function of the set A.

3 Application

Classification methods have several important applications in the field of microarray data analysis, namely, to classify individuals into one of two or more categories of a disease, particularly cancer; see, for example, [4, 6, 10]. One feature of microarray studies is the fact that the number of samples collected tends to be much smaller than the number of genes per chip. The small-sample dilemma in statistical methods for classification is well documented in the literature (see [4]), with some type of regularization or variable reduction appearing as necessary. Geman et al. [5] propose the use of marker gene pairs (pairs of genes with expression levels that allow class separation) for classification purposes. Once the family of gene pairs is chosen, a profile is classified based on a rule which aggregates the results involving each gene pair in the family.

3.1 The Data

We considered three real microarray data sets (prostate, leukemia, and breast) to illustrate the application of the proposed methodology. Here $\mathbf{X} = (X_1, X_2)$ are expression levels of a gene pair (measured using DNA microarrays). The prostate study (see [13]) assigns profiles to either tumor or normal tissue classes. There are $n_1 = 49$ prostate tumor samples and $n_0 = 43$ non-tumor samples. The leukemia study (see [6]) compares two different types of leukemia with 7129 probes from

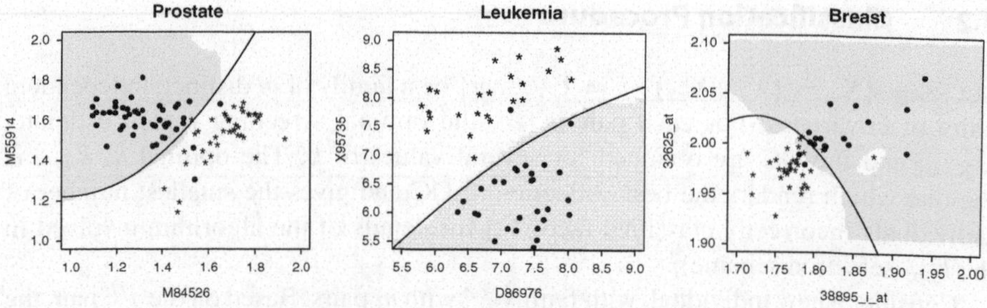

Fig. 1 Scatterplot for a top scoring pair of genes for each study. Classes are represented using *dots* (C_1) and *stars* (C_0). The *curves* represent the decision boundary (quadratic decision rule)

Table 1 Performance of semi-parametric screening region defined by a quadratic decision rule

Study	k_{opt}	$\gamma = P(Y = 1 \mid \mathscr{D})$	$\alpha = P(\mathbf{X} \in C_{\mathbf{X}} \mid \mathscr{D})$	$\delta = P(Y = 1 \mid \mathbf{X} \in C_{\mathbf{X}}; \mathscr{D})$	$\epsilon = P(Y = 1 \mid \mathbf{X} \notin C_{\mathbf{X}}; \mathscr{D})$
Prostate	0.67	0.5319	0.5186	0.9420	0.0921
Leukemia	0.79	0.6486	0.5796	0.9896	0.1786
Breast	0.56	0.3519	0.3848	0.7787	0.0849

47 samples of ALL and 25 of AML. The breast data set (see [8]) consists of gene expression profiles measured on 52 women with breast cancer. Of these, $n_0 = 34$ women did not experience recurrence of the tumor during a three-year time period and $n_1 = 18$ experienced the recurrence.

In the leukemia study, three pairs are considered for $\mathscr{P}$ while both in the prostate and breast examples, there is only one such pair.

3.2 Classification Results

In this section, we present the results of classification based on the proposed methodology. For each study, and for each gene pair in the corresponding $\mathscr{P}$ family, the approximate screening region is computed together with the OC (computed numerically, [11]) for several values of k. The approximate regions were computed over the same 100×100 grid for the different values of k. The results were obtained by considering non-informative prior distributions. An intuitive appreciation of the nature of the decision boundaries defined by semi-parametric solution can be achieved in Fig. 1. For each data set and for the k_{opt}, the figure displays the scatterplot of the log expression levels for gene pair—the unique pair for prostate and breast data and one of the three pairs for the leukemia data. The optimal regions and decision boundaries are also plotted for each case.

Table 1 displays the OC of the semi-parametric optimal screening region defined by a quadratic function for each study reported in Fig. 1.
In predicting the presence of disease in the prostate study, the results obtained show good discriminative power of the semi-parametric solution. It can be seen in Table 1

Table 2 Performance of parametric screening region defined by a quadratic decision rule

Study	k_{opt}	$\gamma = P(Y = 1\|\mathscr{D})$	$\alpha = P(\mathbf{X} \in C_{\mathbf{X}}\|\mathscr{D})$	$\delta = P(Y = 1\|\mathbf{X} \in C_{\mathbf{X}}; \mathscr{D})$	$\epsilon = P(Y = 1\|\mathbf{X} \notin C_{\mathbf{X}}; \mathscr{D})$
Prostate	0.63	0.5319	0.5267	0.8956	0.0931
Leukemia	0.86	0.6486	0.5675	0.9918	0.1985
Breast	0.42	0.3519	0.3128	0.8458	0.1269

that the predictive power of the semi-parametric screening classifier is very high. The predictive probability of success is raised from 0.5319 to 0.9420 when $\mathbf{X} \in \mathbf{C_X}$ is considered and the predictive probability of an excluded individual to be a success is 0.0921. In order to distinguish AML from ALL leukemias the semi-parametric screening classifier produces very satisfactory results. The predictive probability of success is raised from 0.6486 to 0.9896 (Table 1). For the classification of nodal metastatic states and relapse for breast cancer patients, the semi-parametric solution also gives good results (Fig. 1). In this study the predictive probability of an excluded individual to be a success is 0.0849 and the predictive probability of success is raised from 0.3518 to 0.3848.

3.3 Comparison with the Parametric Approach

Table 2 summarizes the OC of the parametric optimal screening region (assuming a bivariate normal model for $\log(\mathbf{X}|\mathbf{Y})$ [12]) defined by a quadratic function for each pair reported in Fig. 1. It is useful to note that, for all studies, the bivariate normal distribution was considered adequate to model the $\log(\mathbf{X}|\mathbf{Y})$.

The values of OC from the two approaches, parametric and semi-parametric, are similar (Table 2 and Table 1). For the breast and leukemia studies, the semi-parametric solution tends to give smaller classification regions (α) than the parametric approach, and, consequently, the predictive probabilities δ and ϵ decrease.

3.4 Comparison with Other Methods

For performance comparison purposes between the proposed method and traditional classification procedures (LDA—linear discriminant analysis, QDA—quadratic discriminant analysis, and SVMs—support vector machines), the estimation prediction errors of the classifiers for each study, based on 0.632+ bootstrap rule, were calculated. All methods gave good results but, in general, the proposed method produced the best or second best prediction error estimates (in terms of average and dispersion). The proposed method gives a simple parametric and flexible decision rule, which is an advantage over both LDA and QDA which produce non-flexible decision rules in terms of shape.

4 Conclusions

We have derived a semi-parametric methodology for classification purposes based entirely on the observation of pairs of covariates. The effectiveness of this solution is illustrated in three microarray data sets, with the method proving to be able to distinguish different classes with high accuracy. We compared this approach with the parametric approach. All solutions gave good results, which means that the semi-parametric solution is an adequate alternative to the parametric approach when assumptions about parametric models are not acceptable. Our approach produced the best or second best prediction error estimate when compared with some well-known classification methods. This method also allows the calculation of OC, which is an advantage over all the other classification methods.

Acknowledgements This research has been partially supported by National Funds through FCT Fundação para a Ciência e a Tecnologia, project PEst-OE/MAT/UI0006/2011 and PTDC/MAT/118335/2010.

References

1. Duong, T.: ks: Kernel density estimation and kernel discriminant analysis for multivariate data in R. J. Stat. Software **21** (2007) http://www.jstatsoft.org/v21/i07/paper (accessed in 2008)
2. Duong, T., Hazelton, M.L.: Plug-in bandwidth matrices for bivariate kernel density estimation. J. Nonparametr. Stat. **15**, 17–30 (2003)
3. Duong, T., Hazelton, M.L.: Cross-validation bandwidth matrices for multivariate kernel density estimation. Scandinavian J. Stat. **32**, 485–506 (2005)
4. Dudoit, S., Fridlyand, J.: Classification in microarrays experiments. In T. Speed, editor, Statistical Analysis of Gene Expression Microarray Data. Chapman and Hall, New York (2003)
5. Geman, D., d'Avignon, C., Naiman, D.: Classification gene expression profiles from pairwise mRNA comparisons. Stat. Appl. Genet. Mol. Biol. **3**(1) (2004) http://www.ncbi.nlm.nih.gov/pmc/articles/PMC1989150/ (accessed in 2007)
6. Golub, T.R., Slomin, D.K., et al.: Molecular classification of cancer: Class discovery and class prediction by gene expression monitoring. Science **286**, 531–537 (1999)
7. Ḧardle, W., Müller, M., Sperlich, S., Werwatz, A.: Nonparametric and Semiparametric Models. Springer, New York (2004)
8. Huang, E., Cheng, S.H., et al.: Gene expression predictors of breast cancer outcomes. The Lancet **361**(9369), 1590–1596 (2003)
9. Kotsiantis, S.: Supervised machine learning: A review of classification techniques. Informatica J. **31**, 249–268 (2007)
10. Li, J., Tang X.: A new classification model with simple decision rule for discovering optimal feature gene pairs. Comput. Biol. Med. **37**, 1637–1646 (2007)
11. Piessens, R., deDoncker-Kapenga, E., berhuber, C., Kahaner, D.: QUADPACK: A Subroutine Package for Automatic Integration. Springer, New York (1983)
12. Ramos, S., Amaral Turkman, M.A., Antunes, M.: Bayesian classification for bivariate normal gene expression. Comput. Stat. Data An. **54**, 2012–2020 (2010)

13. Singh, D., Febbo, P.G. et al.: Gene expression correlates of clinical prostate cancer behaviour. Cancer Cell **1**(2), 203–209 (2002)
14. Turkman, K.F., Amaral Turkman, M.A.: Optimal screening methods. J. Royal Stat. Soc. B **51**, 287–295 (1989)
15. Wand, M.P., Jones, M.C.: Multivariate plug-in bandwidth selection. Comput. Stat. **9**, 97–116 (1994)

An Application of Statistical Methods of Indirect Estimation and Projection of Internal Migration Flows Within the Portuguese Mainland

Maria Filomena Mendes, Antonio Caleiro, Sandra Lagarto, and Filipe Ribeiro

Abstract
The study of migration flows is always problematic, essentially because there is not a systematic process of collecting background information. In the case of internal migration, restrictions on the available data are even more problematic and make it totally unfeasible to measure directly those movements. When the data on (regional) migration is incomplete, inadequate, or unavailable, the estimation or quantification of regional migration flows is made possible by the application of indirect methods of estimation. Andrei Rogers, along with several others, developed and tested over several years methodologies that allow us to analyze and quantify indirectly the different behaviors of regional migration. These methodologies are applied in the chapter, considering the case of Portugal.

1 Introduction

Mainly due to its temporary nature, migrations are always difficult to study, and the demographic lethargy that characterizes the mortality does not happen in migrations. Furthermore, the lack of register in changes of residence, especially at internal level, does not allow knowing the behavior of the migrants that could help to support the demographic projections, especially at regional level.

M.F. Mendes (✉)
Departamento de Sociologia, Universidade de Évora, Largo dos Colegiais n^o2, 7000 – 803 Évora, Portugal
e-mail: mmendes@uevora.pt

A. Caleiro · S. Lagarto · F. Ribeiro
Universidade de Évora, Évora, Portugal
e-mail: caleiro@uevora.pt; smdl@uevora.pt; flipjribeiro@hotmail.com

J. Lita da Silva et al. (eds.), *Advances in Regression, Survival Analysis, Extreme Values, Markov Processes and Other Statistical Applications*, Studies in Theoretical and Applied Statistics, DOI 10.1007/978-3-642-34904-1_42,

We can make projections for the country and separate at regional level, or project for the regions and group for the country. However, in one case or another, the estimation of internal migration is essential not only in demographic terms but also in terms of planning, because information about the attractiveness of regions must be taken into account in the design for a balanced future, and for a sustainable development of regions and country. Not taking the demographic lethargy as a scale of the scenario of the next 20 years (mortality without great variation and fertility with a great break down of 1980 for 2009), we still have to realize in which measure we can find a standard of internal migratory behaviors in function of gender and age that could provide us a reasonable future visibility.

Also, the exchanges between regions are extremely important for the demographic projections, since in terms of migratory flows it is important to know the origin of the migrants that settle down in a different region or even know if the relation of exchanges between two regions has remaining stable along the last decades.

In addition to the volume of migration flows and its movements (origin/destiny), it is still crucial to know the causes leading to the decision of migrating from a region to another one, as well as defining a profile for ages that allow to identify the ages where migration trends are more evident.

2 Context and Data

Data for this chapter was collected through the Census of Portuguese Population in 1991 and 2001 and made available by the IPUMS International Database (Integrated Public Use Microdata Series)/INE (National Statistics Institute Portugal). Such as in most countries, the questionnaire used for the Portuguese census contained a question about the location (region) of residence at the present time and in a given period of time in the past (in this case, 5 years before the census).

We used the resident population by age group and gender, at the level of the geographical area of residence in agreement with the mainland Portuguese NUTS II (North, Center, Lisbon and Tejo Valley, Alentejo and Algarve) in 31 of December of 1985 and 1995, and resident population, by age groups and gender, according to the NUT II of residence of the mother at date of the birth. This analysis excluded those who living or born abroad (the latter were not considered relevant to the analysis of individuals in the age group 0–4 years old).

3 Method

A close observation of migration flows allows us to detect the existence of some specific features of its own, such as the differences in the odds of migrating associated with age [2], and a linear relationship between those at ages 0–4 and their parents.

Furthermore, given the existence of real observed values, the associated error predictions can be measured.

3.1 Age-Specific Regularities

We found that the higher odds occur early in adult life, when individuals are leaving home to attend a degree of higher education, join the military service, to raise a family, or simply get to work. This is an often result known to be called as "labor peak" [7]. Moreover, the odds of migrating lower rates occur in late adolescence and usually after entering the labor market until the beginning of retirement.

As for the probability of migrating children, it appears that this reflects the migration of parents, usually young adults. Despite migration childhood reaches values higher than in adolescence, the retirement age, especially in developed countries, leads to an increased migration probability resulting in a "return peak" close to 65 years old.

The complete model migration schedule has four components [1,6]: (1) the pre-labor force stage (children), (2) the labor force (adults), (3) the post-labor force stage (elderly), and (4) a constant curve, which can be translated by the following expression:

$$m(x) = N_1(x) + N_2(x) + N_3(x) + c$$

$$m(x) = a_1 exp(-\alpha_1 x + a_2 exp{-}\alpha_2(x-\mu_2) - exp[-\lambda_2(x-\mu_2)]$$

$$+a_3 exp{-}\alpha_3(x-\mu_3) - exp[-\lambda_3(x-\mu_3)] + c \qquad (1)$$

where $m(x)$ is the migration probability at age x; N_1 the pre-labor force stage (children); N_2 the labor force (adults); N_3 the post-labor force stage (elderly); c the constant curve; α and μ the parameters; and x the age.

Its implementation is based on the assumption that migrations of adults are linearly linked to migration of children between 0 and 4 years old , which provides the identification of three key assumptions:

1. Regardless of the size or intensity of migration flows and regions of origin/destination, the rates associated with migration have a very similar pattern when analyzed by age groups.
2. The question on region of birth is present in almost all censuses of population for children from the age group of 0–4 years old [3,4], and because it is a five-year period, it is representative of a recent pattern in relation to migration.
3. As children migrate always (or mostly) with parents, young adults, their migration reflect, in a larger scale, the migration of other age groups.

3.2 The Linear Relationship

This method uses the technique of linear regression between the proportions of children aged 0–4 years old, which were born in region i and living in region j at the time of census, and the proportions of people in each age group that lived in region i five years before the census, and at time census is living in region j. This relation can explain the specific probability of migration, by age, from a parameter estimate based on information from the child migration [5].

The above-mentioned assumptions, that in similarity with the indirect estimates of mortality are based on a single infant mortality rate to estimate the mortality curve at all ages, result in estimates of the "survival rates" of migrants in a given age x, represented by $S_{ij}(x)$:

$$S_{ij}(x) = \frac{Migrants_{ij}(x)}{Total\ Population_i(x)},\ x = age \tag{2}$$

In this sense, $Migrants_{ij}(x)$ denotes the number of individuals that at the time of the census are living in a location j, but 5 years before were living in i, and $Total\ Population_i(x)$ represents the sum of all individuals, aged x, who lived in i, 5 years before the census date. In this way, $S_{ij}(x)$ is a measure that translates, for a person with age x, and lives in i, the probability of survival in j, t years later (in this case $t = 5$).

To estimate the specific survival rates for migration, we first take a ratio of child migration—$r_{ij}(x, -5)$—also known as ATI (age infant-to-migration ratio), which reflects the ratio of the probability of migrating at any age and probability of migration for children aged 0 to 4 years (i.e., all those born to 5 years before):

$$r_{ij}(x, -5) = \frac{S_{ij}(x)}{S_{ij}(-5)},\ x = 0,\ 5,\ 10,\ \ldots,\ 70+ \tag{3}$$

This ratio allows us to obtain estimates $S_{ij}(x)$ for 10 years later (the usual interval between census):

$$\hat{S}^t_{ij}(x) = r^{t-10}_{ij} S^t_{ij}(-5) \tag{4}$$

That results in an approach to a simple linear relationship type: $\hat{S}_{ij}(x) = \mathrm{a} + \mathrm{b}\, S_{ij}(-5) + \varepsilon$, where the estimated values $S_{ij}(x)$ are explained in terms of $S_{ij}(-5)$ through the line of regression line and its associated error (ε).

3.3 Measuring Error

Once all estimates are associated with a certain level of error and that we had access to the data from two censuses, it is convenient to use a measure of goodness of fit, like mean absolute percentage error (MAPE), to evaluate the results:

$$MAPE = \frac{\sum_x \frac{\|\hat{S}_{ij}(x) - S_{ij}(x)\|}{S_{ij}(x)}}{N} \times 100\ (\textit{for a particular flow}) \tag{5}$$

$$MAPE_{ij} = \frac{\sum_{i=1}^{n} \sum_{j \neq i}^{n} \sum_x \frac{\|\hat{S}_{ij}(x) - S_{ij}(x)\|}{S_{ij}(x)}}{n(n-1)N} \times 100\ (\textit{for all the flows}) \tag{6}$$

Table 1 Regression statistics for full samples by ages

Age group	α	β	S.E.	R^2	MAPE(%)
0–4	–	–	–	–	–
5–9	−0.00005	1.57279	0.06102	0.97	34.02
10–14	0.00030	1.40662	0.06579	0.96	31.35
15–19	−0.00033	1.57709	0.05386	0.98	29.88
20–24	−0.00145	1.73924	0.06062	0.98	30.15
25–29	−0.00158	1.73977	0.07231	0.97	32.21
30–34	−0.00066	1.78629	0.09220	0.95	38.85
35–39	0.00007	1.67198	0.06608	0.97	39.54
40–44	0.00025	1.53799	0.10456	0.92	37.51
45–49	0.00011	1.41712	0.08149	0.94	30.88
50–54	0.00014	1.40917	0.10944	0.90	35.27
55–59	0.00028	1.33062	0.12396	0.86	37.15
60–64	0.00040	1.11801	0.05143	0.96	36.20
65–69	0.00045	0.95372	0.07001	0.91	28.76
70+	0.00018	1.23467	0.03602	0.98	33.48
Totals	−0.00039	1.64892	0.01892	0.96	31.68

4 Analysis

Turning to examine the explanatory capacity of the estimates made by age groups (Table 1), results revealed that in all groups the recorded R^2 values are very high. On the whole, the estimates have an explanatory power higher than 90.0 %, except for the age group 55–59, which only explains about 86.0 % of the values actually observed.

However, these numbers only indicate the suitability of the model to each age group, and so, it is also essential to evaluate the associated error, which varies between 28.8 % and 38.9 %.

Note also that the method used has an explanatory power of 96.0 % and an associated error of 31.7 % for all the movements, including the analysis by region or by age group in Portugal (except islands).

A similar situation can be seen in the observation of Table 2, where there is no overall explanatory power below 70.0 %. Of relevance also are some differences with respect to the associated error, which varies between 13.3 % and 49.6 %.

5 Results

Taking into account the total estimated and observed migration flows (Fig. 1), we observe that they are mainly from the region of Lisbon and Tejo Valley (L.V.T.), which contributes with approximately 40.0 % of the total. Of relevance are the North and Center areas, with approximately 20.0 % each, and finally the regions

Table 2 Observed and predicted flows, R^2 and MAPE

Reg. 91	Reg. 01	Predicted	Observed	R^2	MAPE(%)
North	Center	8, 627	11, 669	0.96	21.22
North	L.V.T.	9, 990	14, 376	0.98	27.21
North	Alentejo	596	1, 015	0.94	38.50
North	Algarve	1, 427	2, 446	0.99	46.83
Center	North	6, 826	10, 247	0.98	29.06
Center	L.V.T.	12, 568	19, 933	0.98	32.28
Center	Alentejo	1, 037	1, 335	0.95	21.70
Center	Algarve	1, 236	2, 112	0.94	38.02
L.V.T.	North	10, 466	13, 788	0.82	23.59
L.V.T.	Center	18, 056	20, 644	0.88	13.32
L.V.T.	Alentejo	6, 666	11, 032	0.96	38.84
L.V.T.	Algarve	6, 201	8, 832	0.94	30.27
Alentejo	North	971	985	0.95	18.03
Alentejo	Center	902	1, 400	0.96	35.93
Alentejo	L.V.T.	7, 362	11, 421	0.99	30.63
Alentejo	Algarve	19, 22	3, 380	0.91	39.66
Algarve	North	689	1, 334	0.78	49.64
Algarve	Center	1, 025	1, 132	0.71	24.87
Algarve	L.V.T.	3, 802	5, 660	0.98	29.14
Algarve	Alentejo	945	1, 688	0.84	44.91

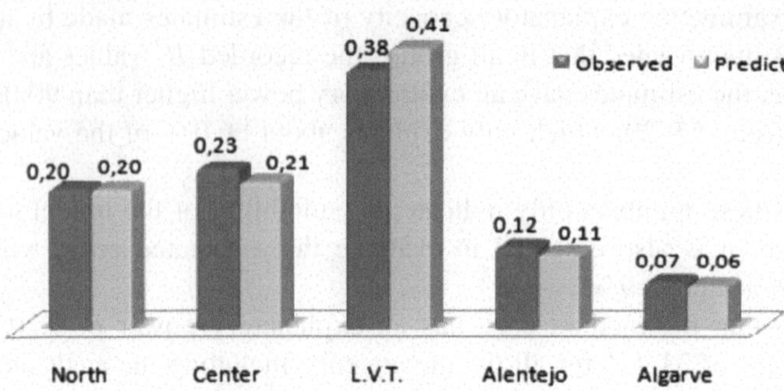

Fig. 1 Migrations flows by the outcoming region

of Alentejo and Algarve with 12.0 % and 7.0 %, respectively. These values are obviously related to the size of populations resident in each of the Portuguese NUTS II.

Considering now, the observed and predicted migration flows between all regions, it was found that, from all NUTS II, the region of Lisbon and Tejo Valley was the one that attracted more migrants, registering very close to 50.0 % or more of the total (Fig. 2).

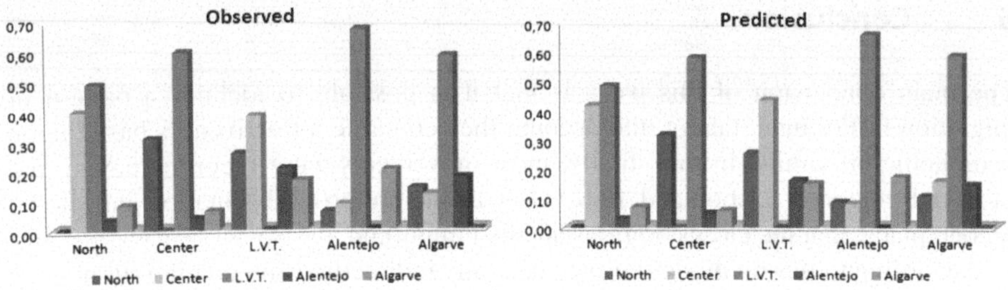

Fig. 2 Migration flows between all regions

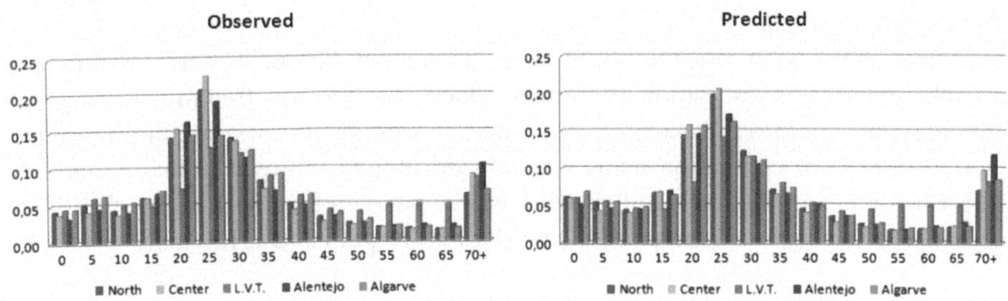

Fig. 3 Migrations flows by age

In the case of Lisbon and Tejo Valley, we found that although about 40.0 % of the migrants moving to the center region, the distribution of these migrations flows occurred more evenly.

At this point we tend to identify from the outset a migratory pattern, the geographical proximity, in that, firstly, the majority of registered movements have always had in common the same fate as the preferred region of Lisbon and Tejo Valley, and moreover, the second option, even in terms of preferential movement, was to the regions that are geographically closer.

By age, was we can see in Fig. 3, the results allowed to identify three distinct phases, where the first corresponds to the children had ages up to 9 years old; the second identifies individuals aged between 20 and 34 years old; and finally, the third, consisting of those aged 70 years old or even more.

The analysis of the differences between the estimates made and the observed values shows that, despite the existence of a lag, it denotes a good approximation to the behavior patterns actually recorded.

In any of the presented situations, the difference between the estimates and the actual values results in an underestimation of the proportion of migration flows by age and regions.

6 Conclusion

The main conclusion of this work is that it is possible to identify a pattern of migration in Portugal, taking into account the economic attractiveness based upon geographic proximity, in that, firstly, most movements were recorded having the most rich region of Lisbon and Tejo Valley as the preferred destination and, in the second place, that the closest regions are also important.

According to this methodology, one can only indirectly estimate migration when the data is regular, which we assumed by considering that the migration observed from 2001 did not suffer from structural changes. Only based on this assumption was possible to determine the standards in relation to age structure of internal migration in Portugal.

Spatial analysis of the migration flows is one of the next steps, within the possible lines of investigation to follow. Once that we are treating an original approach that fits into a broader effort aimed to identify the best methodology for internal migration estimation using recently developed indirect methods, another via for further examination in future work will depend on the application of indirect methods, including those linked to the work of A. Rogers and J. Raymer.

Acknowledgements The authors gratefully acknowledge the helpful comments and suggestions from the referee.

References

1. Little, J.S., Rogers, A.: What can the age composition of a population tell us about the age composition of its out-migrants? Popul. Space Place **13**(1), 23–39 (2007)
2. Raymer, J., Rogers, A.: Using age and spatial flow structures in the indirect estimation of migration streams. Demography. **44**(2), 199–223 (2007)
3. Rogers, A.: Demographic modeling of the geography of migration and population: A multiregional perspective. Geogr. Anal. **40**(3), 276–296 (2008)
4. Rogers, A., Jordan, L.: Estimating migration flows from birthplace-specific population stocks of infants. Geogr. Anal. **36**(1), 38–53 (2004)
5. Rogers, A., Raymer, J., Jordan, L.: Inferring migration flows from birthplace-specific population stocks. In: Population Program, Institute of Behavioral Science, University of Colorado, Working Paper POP 2003–0002, p. 55, Colorado (2003)
6. Rogers, A., Castro, L.J., Lea, M.: Model migration schedules: Three alternative linear parameter estimation methods. Math. Pop. Stud. **12**(1), 17–38 (2005)
7. Rogers, A., Jones, B., Partida, V., Muhidin, S.: Inferring migration flows from the migration propensities of infants: Mexico and Indonesia. The Ann. Reg. Sci. **41**, 443–465 (2007)

Robust Clustering Method for the Detection of Outliers: Using AIC to Select the Number of Clusters

Carla M. Santos-Pereira and Ana M. Pires

Abstract
In Santos-Pereira and Pires (Computational Statistics, pp. 291–296. Physica, Heidelberg, 2002) we proposed a method to detect outliers in multivariate data based on clustering and robust estimators. To implement this method in practice it is necessary to choose a clustering method, a pair of location and scatter estimators, and the number of clusters, k. After several simulation experiments it was possible to give a number of guidelines regarding the first two choices. However, the choice of the number of clusters depends entirely on the structure of the particular data set under study. Our suggestion is to try several values of k (e.g., from 1 to a maximum reasonable k which depends on the number of observations and on the number of variables) and select k minimizing an adapted AIC. In this chapter we analyze this AIC-based criterion for choosing the number of clusters k (and also the clustering method and the location and scatter estimators) by applying it to several simulated data sets with and without outliers.

C.M. Santos-Pereira (✉)
CEMAT, IST and Departamento de Engenharia Civil, Faculdade de Engenharia da Universidade do Porto, Rua Roberto Frias, 4200-465 Porto, Portugal
e-mail: carlasp@fe.up.pt

A.M. Pires
Departamento de Matemática and CEMAT, Instituto Superior Técnico, Av. Rovisco Pais 1, 1049-001, Lisboa, Portugal
e-mail: apires@math.ist.utl.pt

J. Lita da Silva et al. (eds.), *Advances in Regression, Survival Analysis, Extreme Values, Markov Processes and Other Statistical Applications*, Studies in Theoretical and Applied Statistics, DOI 10.1007/978-3-642-34904-1_43,

1 Methodology

The procedure most commonly used to detect outliers in multivariate data sets is based on the Mahalanobis distances, $(\mathbf{x}_i - \hat{\boldsymbol{\mu}})^T \hat{\boldsymbol{\Sigma}}^{-1} (\mathbf{x}_i - \hat{\boldsymbol{\mu}}), i = 1, \ldots, n$. To avoid the masking effect it is recommended to use robust estimates, $\hat{\boldsymbol{\mu}}$ and $\hat{\boldsymbol{\Sigma}}$, instead of the classical estimates, i.e., the sample mean vector and the sample covariance matrix (see, e.g., [5, 12]). However the performance of that procedure is still highly dependent of multivariate normality of the bulk of the data [2], or on the data being elliptically contoured. To avoid this dependency, a method to detect outliers in multivariate data based on clustering and robust estimators was introduced in [14]. A somehow similar method designed to work with nonoverlapping clusters was proposed later in [4]. Both [14] and [4] have been referenced recently in relation to robust clustering [3, 8].

Consider a multivariate data set with n observations in p variables. The basic ideas of the method proposed in [14] are described in the following steps:

1. Segment the n point cloud (of perhaps complicated shape) in k smaller subclouds using a partitioning clustering method with the hope that each subcloud (cluster) looks "more normal" than the original cloud.
2. Then apply a simultaneous multivariate outlier detection rule to each cluster by computing Mahalanobis-type distances from all the observations to all the clusters. An observation is considered an outlier if it is an outlier for every cluster. All the observations in a cluster may also be considered outliers if the size of that cluster is small taking into account the number of variables (our proposal is less than $2p + 2$, since in that case the covariance matrix estimates are very unreliable).
3. Remove the observations detected in 2 and repeat 1 and 2 until no more observations are detected.
4. The final decision on whether all the observations belonging to a given cluster (not previously removed, i.e., with size at least $2p + 2$) are outliers is based on a table of between clusters Mahalanobis-type distances.

In [14] we presented results from a simulation study with several distributional situations, three clustering methods (k-means, *pam*, and *mclust*) and three pairs of location and scatter estimators (classical and two robust), from which it was possible to conclude that for normal data all the methods behave well, whereas for non-normal data the best performance is usually achieved by *mclust*, without large differences between the classical and the robust estimators of location and scatter. A general conclusion from [14] is that the exploratory method proposed for outlier detection works well both under elliptical and non-elliptical data configurations.

The aim of this chapter is to propose a criterion for selecting an appropriate number of clusters, k, to use in the above algorithm, and to assess the robustness of that criterion. In the next section we introduce the new criterion; in Sect. 3 we present the results of a simulation study and in Sect. 4 we state some conclusions.

2 AIC-Based Criterion

One of the difficulties encountered in the implementation of the method was the choice of the number of clusters, k, as well as the clustering method and the location and scatter estimators. In [14] it is suggested to try several values of k (e.g., from 1 to a maximum possible k which depends on the number of observations and on the number of variables) and decide after a careful analysis of the results. A less subjective way for choosing k (and also the clustering method and the location and scatter estimators) is to minimize an adapted AIC (see [13]):

$$AIC = -2\sum_{i=1}^{n} \log \hat{f}(\mathbf{x}_i) + 2k\left(p + \frac{p(p+1)}{2}\right). \tag{1}$$

The full specification of AIC needs $\hat{f}$. This can be either a nonparametric estimate or the density from a parametric model with estimated parameters. The model we consider in this chapter is a finite mixture of multivariate normal densities:

$$\hat{f}(\mathbf{x}) = \sum_{j=1}^{k} \frac{n_j}{n_T} f_N(\mathbf{x}; \hat{\boldsymbol{\mu}}_j, \hat{\boldsymbol{\Sigma}}_j), \text{ and } n_T = \sum_{j=1}^{k} n_j, \tag{2}$$

where

$$f_N(\mathbf{x}; \hat{\boldsymbol{\mu}}, \hat{\boldsymbol{\Sigma}}) \text{ is the density of } \mathbf{N}_p(\hat{\boldsymbol{\mu}}, \hat{\boldsymbol{\Sigma}}). \tag{3}$$

The number of components of the mixture (i.e., the number of clusters), k, is limited in practice ($K_{\max}$). As a generic guidance we can take the advice given in [6], that one should have at least 5–10 observations per variable. This means to choose $k_{\max}$ somewhere between $0.1n/p$ and $0.2n/p$.

In this chapter we assess the robustness of the AIC-based criterion (1) for choosing the number of clusters, k. This is done by comparing results of simulations with and without outliers, for some non-normal distributional situations described in [14].

3 Simulation Study

In order to evaluate the robustness of this AIC-based criterion (1) for choosing the number of clusters, k, we conducted a simulation study with:

- Three clustering methods, k-means, *pam* (partitioning around medoids [7]), and *mclust* (model-based clustering for gaussian distributions [1]), each of them with $k = 2, 3, 4, 5, 6$. The case $k = 1$, for which the clustering method is irrelevant, was also considered.
- Three pairs of location and scatter estimators: classical $(\bar{\mathbf{x}}, \mathbf{S})$ with asymptotic detection limits, RMCD25 [11], and $\text{OGK}_{(2)}(0.9)$ [9] with detection limits determined previously by simulation with 10,000 normal data sets.

Table 1 Proportion of simulations for which each k was chosen within each clustering × estimator combination (the proportion corresponding to the more often chosen K is represented in Bold)

	k	MCD	Classical	OGK
(a) *distributional situation 1, theoretical* $k = 3$				
k-means	1	0.00	0.00	0.00
	2	0.01	0.00	0.02
	3	0.28	0.01	0.32
	4	0.26	0.28	0.14
	5	0.15	0.26	0.19
	6	**0.30**	**0.45**	**0.33**
pam	1	0.00	0.00	0.00
	2	0.00	0.00	0.00
	3	0.29	0.02	0.27
	4	0.20	0.23	0.19
	5	0.14	0.23	0.11
	6	**0.37**	**0.52**	**0.43**
mclust	1	0.00	0.00	0.00
	2	0.00	0.00	0.00
	3	**0.61**	**0.48**	**0.66**
	4	0.30	0.28	0.24
	5	0.06	0.15	0.08
	6	0.03	0.09	0.02
(b) *distributional situation 2, theoretical* $k = 4$				
k-means	1	0.00	0.00	0.00
	2	0.00	0.00	0.00
	3	0.03	0.00	0.00
	4	0.17	0.18	0.12
	5	0.31	0.31	0.33
	6	**0.49**	**0.51**	**0.55**
pam	1	0.00	0.00	0.00
	2	0.00	0.00	0.00
	3	0.00	0.00	0.00
	4	0.27	0.03	0.31
	5	**0.43**	0.44	0.30
	6	0.30	**0.53**	**0.39**
mclust	1	0.00	0.00	0.00
	2	0.00	0.00	0.00
	3	0.13	0.07	0.14
	4	**0.46**	**0.40**	**0.56**
	5	0.27	0.21	0.14
	6	0.14	0.32	0.16

- Four distributional situations:
  1. Non-normal ($p = 2$) without outliers, 50 observations from $N_2(\boldsymbol{\mu}_1, \Sigma_1)$, 50 observations from $N_2(\boldsymbol{\mu}_2, \Sigma_2)$, and 50 observations from $N_2(\mathbf{0}, \Sigma_1)$, with $\boldsymbol{\mu}_1 = (0, 12)^T$, $\Sigma_1 = \text{diag}(1,0.3)$, $\boldsymbol{\mu}_2 = (1.5, 6)^T$, and $\Sigma_2 = \text{diag}(0.2,9)$
  2. Non-normal ($p = 2$) with outliers, 150 observations as in the previous case plus ten outlying observations from $N_2((-2, 6)^T, 0.01\mathbf{I})$

Table 2 Proportion of simulations for which each k was chosen within each clustering × estimator combination (the proportion corresponding to the more often chosen K is represented in Bold)

	k	MCD	Classical	OGK
(a) *distributional situation 3, theoretical* $k = 2$				
k-means	1	0.00	0.00	0.00
	2	0.00	0.00	0.00
	3	0.04	0.00	0.01
	4	0.16	0.09	0.10
	5	**0.41**	**0.47**	0.38
	6	0.39	0.44	**0.51**
pam	1	0.00	0.00	0.00
	2	0.00	0.00	0.00
	3	0.14	0.02	0.03
	4	0.13	0.04	0.02
	5	0.30	0.36	0.47
	6	**0.43**	**0.58**	**0.48**
mclust	1	0.00	0.00	0.00
	2	**0.68**	**0.46**	**0.56**
	3	0.12	0.12	0.18
	4	0.06	0.16	0.12
	5	0.09	0.11	0.07
	6	0.05	0.15	0.07
(b) *distributional situation 4, theoretical* $k = 3$				
k-means	1	0.00	0.00	0.00
	2	0.01	0.00	0.03
	3	0.07	0.00	0.02
	4	0.05	0.03	0.04
	5	0.19	0.25	0.25
	6	**0.68**	**0.72**	**0.66**
pam	1	0.00	0.00	0.00
	2	0.00	0.00	0.00
	3	0.02	0.00	0.01
	4	0.02	0.00	0.00
	5	0.16	0.05	0.07
	6	**0.80**	**0.95**	**0.92**
mclust	1	0.00	0.00	0.00
	2	0.02	0.02	0.01
	3	**0.68**	**0.47**	**0.60**
	4	0.17	0.21	0.21
	5	0.08	0.15	0.12
	6	0.05	0.15	0.06

3. Non-normal ($p = 2$) without outliers, 75 observations from $N_2(\mathbf{0}, \Sigma_3)$ and 75 observations from $N_2(\mathbf{0}, \Sigma_4)$, with $\Sigma_3 = \text{diag}(1,81)$ and $\Sigma_4 = \text{diag}(81,1)$
4. Non-normal ($p = 2$) with outliers, 150 observations as in the previous case plus 20 outlying observations from $N_2(\mathbf{10}, 0.1\mathbf{I})$

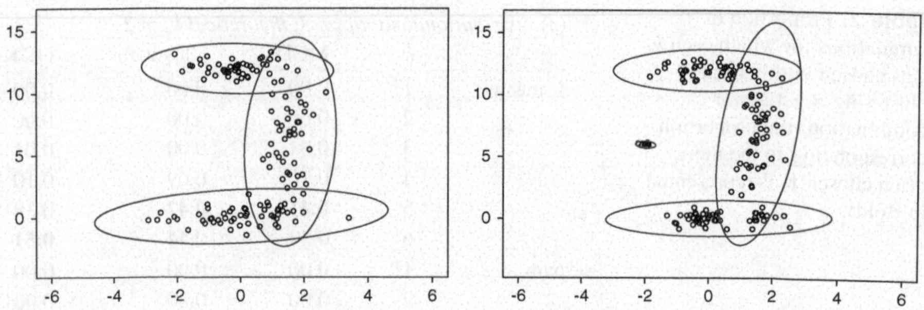

Fig. 1 Distributional situations 1 and 2 with contours (theoretical $k = 3, 4$, respectively)

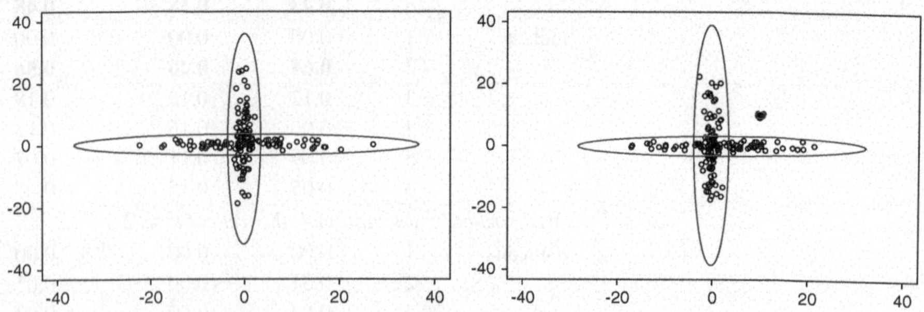

Fig. 2 Distributional situations 3 and 4 with contours (theoretical $k = 2, 3$, respectively)

We have not considered normal data in this simulation study because we have concluded in [14] that in that case the choice of k is not critical. For each distributional situation one hundred data sets were generated.

In each distributional situation we recorded (in each simulation) the chosen k for each clustering $\times$ estimator combination (i.e., the value of k minimizing AIC), and also the overall minimizing combination (i.e., the specific values of (clustering, estimator, k) which minimizes AIC, at each simulation). Tables 1 and 2 give, for the four distributional situations, the proportion of simulations for which each k was chosen (within each clustering $\times$ estimator combination).

The overall minimizing combination was always the *mclust* $\times$ classical, which agrees with the simulations in [14] and shows that this choice can be recommended irrespective of the characteristics of the data sets. This conclusion, which may look unexpected, can be justified as follows: the algorithm either isolates or removes the outliers, leaving almost exclusively "good" observations, and it is well known that in this case, the classical estimators are more efficient.

For the *mclust* cases, the value of k chosen more often is the expected according to the distributional situation (see Figs. 1 and 2). Note that k must be increased by 1 when the outliers are introduced and this is captured by the AIC criterion.

4 Conclusions

The results of the limited simulation study presented in Sect. 3 how that the adapted AIC criterion (1) for selecting k and the clustering method is a useful tool. Moreover, we can also conclude that this criterion is, in association with the present algorithm, robust, since it works well both with and without outliers. An explanation for this robust behavior is that the outliers are either deleted or isolated in their own clusters, before computing the AIC. We thus conclude that in this setup there is no need to consider other more complicated criteria such as the adapted AIC with M-estimators, introduced in [10].

In spite of the good results of this promising technique, one shall not forget that outlier detection in multivariate data is a very difficult task and will always remain an open problem.

Acknowledgments The authors wish to thank the referees for many valuable comments and suggestions.

References

1. Banfield, J., Raftery, A.: Model-based Gaussian and non-Gaussian clustering. Biometrics **49**, 803–822 (1992)
2. Cerioli, A.: Multivariate outlier detection with high breakdown estimators. J. Am. Stat. Assoc. **105**, 147–156 (2010)
3. Garcia-Escudero, L., Gordaliza, A., Matrn, C., Mayo-Iscar, A.: A review of robust clustering methods. Adv. Data Anal. Classif. **4**, 89–109 (2010)
4. Hardin, J., Rocke, D.: Outlier detection in multiple cluster setting using the minimum covariance determinant estimator. Comput. Stat. Data Anal. **44**, 625–638 (2004)
5. Hubert, M., Rousseeuw, P.J., Van Aelst, S.: High-breakdown robust multivariate methods. Stat. Sci. **23**, 92–119 (2008)
6. Jain, A., Chandrasekaran, B.: Dimensionality and sample size considerations in pattern recognition practice. In: Krishnainh, P., Kanal, L. (eds.) Handbook of Statistics, vol. 2, pp. 835–855. North Holland, Amsterdam (1982)
7. Kaufman, L., Rousseeuw, P.: Finding Groups in Data: An Introduction to Cluster Analysis. Wiley, New York (1990)
8. Kumar, M., Orlin, J.B.: Scale-invariant clustering with minimum volume ellipsoids. Comput. Oper. Res. **35**, 1017–1029 (2008)
9. Maronna, R., Zamar, R.: Robust estimates of location and dispersion for high dimensional data sets. Technometrics **44**, 307–317 (2002)
10. Ronchetti, E.: Robustness aspects of model choice. Stat. Sin. **7**, 327–338 (1997)
11. Rousseeuw, P.J.: Multivariate estimation with high breakdown point. In: Grossman, W., Pflug, G., Vincze, I., Werz, W. (eds.) Multivariate Estimation with High Breakdown Point, vol. B, pp. 283–297. Reidel, Dordrecht (1985)
12. Rousseeuw, P.J., von Zomeren, B.C.: Unmasking multivariate outliers and leverage points. J. Am. Stat. Assoc. **85**, 633–639 (1990)
13. Sakamoto, Y., Ishiguro, M., Kitagawa, G.: Akaike Information Criterion Statistics. Kluwer, New York (1988)
14. Santos-Pereira, C., Pires, A.: Detection of outliers in multivariate data: A method based on clustering and robust estimators. In: Härdle, W., Rönz, B. (eds.) Computational Statistics, pp. 291–296. Physica, Heidelberg (2002)

HLA Allele and Haplotype Frequencies of the Portuguese Bone Marrow Donors Registry

Ricardo São João, Ana Luisa Papoila, Dário Ligeiro, and Hélder Trindade

Abstract

Genes in human leucocyte antigens (HLA) System are important in the study of autoimmune diseases and responsible for the rejection of transplants of organs and tissues. HLA genes are part of the human major histocompatibility complex (MHC) which is characterized by the presence of several multigene families, extensive polymorphism at many loci and significant linkage disequilibrium between alleles at particular loci. We analysed HLA-A,-B,-DRB1 locus phenotypes through a sample of 1,021 subjects that were randomly selected among the volunteers recruited by the Portuguese Bone Marrow Donors Registry (Cedace) in order to evaluate allele, gene, haplotype and phenotype frequencies. Allelic frequencies in each of the studied locus were obtained by direct counting. Maximum-likelihood haplotype frequencies were estimated using an expectation-maximization (EM) algorithm [2]. Locus phenotype and gene relative frequencies were estimated according to Baur and Danilov [1]. Hardy–Weinberg equilibrium were tested. The data presented is a definition of HLA genetic repertoire of Cedace with relevance on the strategic management for the increase of a more diverse register with clinical utility.

R.S. João (✉)
Polytechnic Institute of Santarém-Business School, Complexo Andaluz Apt 295 2001-904 Santarém, Portugal
e-mail: ricardo.sjoao@esg.ipsantarem.pt

A.L. Papoila
CEAUL and Faculdade de Ciências Médicas-Universidade Nova de Lisboa, Campo Mártires da Pátria 130 1169-056 Lisboa, Portugal
e-mail: ana.papoila@fcm.unl.pt

D. Ligeiro · H. Trindade
Centro de Histocompatibilidade do Sul, Alameda das Linhas de Torres 117, 1769-001 Lisboa, Portugal
e-mail: dario@chsul.pt; helder.trindade@chsul.pt

J. Lita da Silva et al. (eds.), *Advances in Regression, Survival Analysis, Extreme Values, Markov Processes and Other Statistical Applications*, Studies in Theoretical and Applied Statistics, DOI 10.1007/978-3-642-34904-1_44,

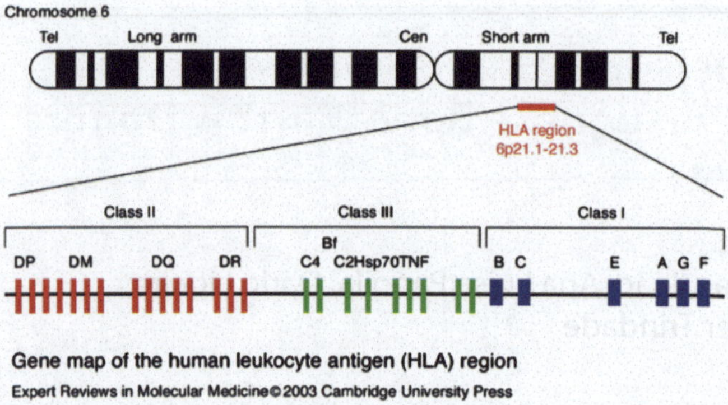

Fig. 1 Gene map of the human leucocyte antigen (HLA) region the HLA region spans 4 × 10^6 nucleotides on chromosome 6p21.1 to p21.3, with class II, class III and class I genes located from the centromeric (Cen) to the telomeric (Tel) end. Figure from http://www.expertreviews.org/ with authors', Narinder K. Mehra and Gurvinder Kaur, consent

1 Introduction

As a species, man has had his development supported in the capacity to generate human leucocyte antigens (HLA) diversity, as T cell restriction molecules. This evolution results in great antigen diversity that renders it virtually impossible to find two identical individuals, with the exception of twins. HLA antigens are also responsible for tissue compatibility and for that reason they are target for allogeneic immunological response, which means they are a biological barrier to cell, tissue and organ transplantation. In organ transplantation new imunossupressor therapies allow transplant between donor–recipient pairs without full HLA identity. In haematopoietic stem cell transplantation, on the contrary, a high degree of HLA compatibility is necessary in order to achieve better patient survival. HLA identity is firstly sought by a low-resolution technique, looking for three main loci, HLA A, B and DRB1, and only after this first successful match, is another technical approach for allelic resolution run. In fact, due to intensive polymorphism of HLA genes, the selection of a non-related donor with the necessary degree of HLA gene compatibility to a patient is a difficult task, only possible at large databases of volunteers haematopoietic stem cell donors genotypes. A large database was created, the National Donor Registry, known as CEDACE (Centro Nacional de Dadores de Células de Medula Óssea, Estaminais ou de Sangue do Cordão) typed for more than 95 % at HLA main loci A,B and DRB1.

The HLA system is located in the short arm of chromosome 6 (see Fig. 1).

Within the HLA system, three constituent regions are distinguished. Near the centromere (Cen) of chromosome 6 is the class II region that contains the class II genes, while nearest the telomere (Tel) of the short arm of chromosome 6 is the class I region that contains the class I genes.

2 Data

The human major histocompatibility complex, of which the HLA class I and class II genes are part, is characterized by the presence of several multigene families, extensive polymorphism at many loci and significant linkage disequilibrium between alleles at particular loci. In most populations, a few alleles are frequent (gene frequency greater than 10 %) but most occur at low frequency (gene frequency lower than 10 %) and a number of the latter may be rare (gene frequency lower than 1 %). As is the case for other genetic polymorphisms, the frequency of HLA alleles differs among populations. An allele that is common in one population may be rare in another. Some alleles are limited to particular ethnic populations, while others are widely shared among ethnically distinct populations. We analysed HLA-A,-B,-DRB1 locus phenotypes through a sample of 1,021 subjects that were randomly selected among the volunteers recruited by Cedace in order to evaluate allele, gene, haplotype and phenotype frequencies. These data represent an important resource for investigators in the fields of transplantation and population genetics.The key limiting factor in the use of bone marrow transplantation (BMT) is the lack of donors. Because only 25–30 % of patients have an HLA-identical sibling, alternative donors are often required. Marrow can be procured from unrelated living donors; marrow donation is a simple, safe procedure. National and international registries of prospective volunteer donors are being expanded to increase the likelihood of finding an exact HLA match for any given recipient. The gene and haplotype frequencies of a registry can be used in advice clinicians and patients about the probability of finding an HLA match for BMT.

3 Methodology

3.1 Hardy–Weinberg Equilibrium

In population genetics, it is very important to study the relationship between allele and genotype frequencies. Godfrey Harold Hardy [4] and Wilhelm Weinberg [7], in 1908, detected, independently, a principle that describes the referred relationship and is known as the Hardy–Weinberg law (HWL). It says that, in a large random-mating population with no selection, mutation or migration, the genotype and allele frequencies remain stable from generation to generation and that there is a fixed relationship between allele and genotype frequencies.

If, for an m-allele autosomal locus with alleles $A_1, A_2, \cdots, A_m$, the genotypic array is given by

$$\sum_i p_i^2 A_i A_i + \sum_{i<j} 2 p_i p_j A_i A_j,$$

where p_i is the allelic frequency of A_i, it is said that the population with these genotype frequencies, known as Hardy–Weinberg proportions (HWP), is in Hardy–Weinberg equilibrium at that locus.

For a sample of size n, data may be organized as an array $\mathbf{f} = (f_{11}, f_{21}, f_{22}, \cdots, f_{mm})$, where $f_{ij} (1 \leq j \leq i \leq m)$ is the observed number of genotype $A_i A_j$. If we consider $f_i = f_{ii} + \sum_{j=1}^{k} f_{ij}$ (where $f_{ij} = f_{ji}$ if $j > i$), then f_i represents the number of A_i alleles in the sample. Assuming Hardy–Weinberg equilibrium, the probability of obtaining **f**, conditional on $\{f_i\}$ is [6]:

$$Pr(\mathbf{f}) = \frac{n! \prod_{i=1}^{m} f_i!}{(2n)! \prod_{j>i} f_{ij}!} 2^{\sum_{j>i} f_{ij}}.$$

In order to compare the observed genotype counts to those expected under HWL, the exact test of Guo–Thompson was used [3]. Given the observed sample **f**, this test has to evaluate

$$P = \sum_{\mathbf{g} \in \varphi} Pr(\mathbf{g}),$$

where $\varphi = \{\mathbf{g} : Pr(\mathbf{g} \leq Pr(\mathbf{f}), \mathbf{g} \in \Gamma_0\}$, and $\Gamma_0 = \Gamma(\mathbf{f}) = \{\mathbf{g} : \mathbf{g}$ has the same allele counts$\{g_i\}$ as $\mathbf{f}\}$. Rejection of the null hypothesis occurs when the value of P is lower than the considered significance level α.

3.2 Allele, Haplotype and Genotype Frequencies

HLA haplotypes are specific sets of HLA-A,-B,-DR locus alleles inherited together from a parent. Haplotypes are usually determined by genotyping a sufficient number of family members to establish a gametic assignment of the detected alleles. It is possible to estimate population haplotype frequencies by genotyping a sufficient number of unrelated individuals to estimate the allele associations that are consistent with the observed genotype data. In fact, this estimation is based on a maximum-likelihood approach and haplotype frequencies are estimated using an expectation-maximization (EM) algorithm [2]. For large populations that are in Hardy–Weinberg equilibrium, it is possible to estimate even relatively rare haplotypes (e.g. frequency 0.01 %) with reasonably accuracy. Locus phenotype and gene relative frequencies were estimated according to Baur and Danilov [1]. Allelic frequencies in each of the studied locus were obtained by direct counting.

4 Results

4.1 Single-Locus Analysis

HLA-A,-B,-DRB1 locus phenotypes were analysed through a sample of 1,021 subjects, selected randomly among the volunteers recruited by Cedace, in order to evaluate allele, phenotype and genotype frequencies. Table 1 presents the three locus phenotype and gene relative frequencies calculated according to the method

Table 1 Counts, phenotype, genotype and standard deviations (SD) frequencies of HLA-A,-B,-DRB1 loci allele groups

Allele N#	# of alleles	Phenotype frequency (4.802)	Genotype frequency (4.802)	Std deviations
A*01	1.021	0,2126	0,1127	0,00343
A*02	2.730	0,5686	0,3432	0,00598
A*03	899	0,1872	0,0985	0,00320
A*11	627	0,1306	0,0676	0,00265
A*23	412	0,0858	0,0439	0,00214
A*24	1.006	0,2095	0,1109	0,00340
A*25	137	0,0285	0,0144	0,00122
A*26	346	0,0721	0,0367	0,00196
A*29	491	0,1023	0,0525	0,00234
A*30	334	0,0696	0,0354	0,00192
A*31	233	0,0485	0,0246	0,00160
A*32	359	0,0748	0,0381	0,00199
A*33	344	0,0716	0,0365	0,00195
A*34	43	0,0090	0,0045	0,00068
A*36	12	0,0025	0,0013	0,00036
A*43	1	0,0002	0,0001	0,00010
A*66	67	0,0140	0,0070	0,00085
A*68	479	0,0998	0,0512	0,00231
A*69	27	0,0056	0,0028	0,00054
A*74	24	0,0050	0,0025	0,00051
A*80	11	0,0023	0,0011	0,00035
B*07	606	0,1239	0,0640	0,00258
B*08	611	0,1250	0,0646	0,00259
B*13	139	0,0284	0,0143	0,00122
B*14	744	0,1522	0,0792	0,00287
B*15	500	0,1023	0,0525	0,00234
B*18	563	0,1151	0,0593	0,00249
B*27	287	0,0587	0,0298	0,00176
B*35	1.250	0,2556	0,1372	0,00378
B*37	115	0,0235	0,0118	0,00111
B*38	271	0,0554	0,0281	0,00171
B*39	143	0,0292	0,0147	0,00124
B*40	332	0,0679	0,0345	0,00190
B*41	112	0,0229	0,0115	0,00110
B*42	18	0,0037	0,0018	0,00044
B*44	1.439	0,2943	0,1599	0,00408
B*45	132	0,0270	0,0136	0,00119
B*46	1	0,0002	0,0001	0,00010
B*47	29	0,0059	0,0030	0,00056
B*48	5	0,0010	0,0005	0,00023

(continued)

Table 1 (continued)

Allele N#	# of alleles	Phenotype frequency (4.802)	Genotype frequency (4.802)	Std deviations
B*49	344	0,0704	0,0358	0,00193
B*50	288	0,0589	0,0299	0,00176
B*51	975	0,1994	0,1052	0,00331
B*52	92	0,0188	0,0095	0,00099
B*53	127	0,0260	0,0131	0,00117
B*54	0	0	0	–
B*55	117	0,0239	0,0120	0,00112
B*56	41	0,0084	0,0042	0,00066
B*57	241	0,0493	0,0250	0,00161
B*58	223	0,0456	0,0231	0,00155
B*59	0	0	0	–
B*67	2	0,0004	0,0002	0,00015
B*73	11	0,0022	0,0011	0,00034
B*78	19	0,0039	0,0019	0,00045
B*81	1	0,0002	0,0001	0,00010
B*82	1	0,0002	0,0001	0,00010
B*83	0	0	0	–
DRB1*01	1.352	0,2377	0,1269	0,00363
DRB1*03	1.302	0,2289	0,1219	0,00356
DRB1*04	1.618	0,2844	0,1541	0,00401
DRB1*07	1.661	0,2920	0,1586	0,00406
DRB1*08	432	0,0759	0,0387	0,00201
DRB1*09	77	0,0135	0,0068	0,00084
DRB1*10	165	0,0290	0,0146	0,00123
DRB1*11	1.332	0,2341	0,1249	0,00361
DRB1*12	176	0,0309	0,0156	0,00127
DRB1*13	1.893	0,3327	0,1831	0,00437
DRB1*14	336	0,0591	0,0300	0,00177
DRB1*15	928	0,1631	0,0852	0,00298
DRB1*16	320	0,0562	0,0285	0,00172

described by Baur and Danilov [1]. For **locus A** the most frequent specificities are A⋆02 (34,3 %), A⋆01 (11,3 %), A⋆24 (11,1 %), A⋆03 (9,8 %) and A⋆11 (6,8 %), which are the classical alleles of European Caucasoid populations. The rare alleles are A⋆43 (0,01 %), A⋆80 (0,11 %), A⋆36 (0,13 %) and A⋆74 (0,25 %), specificities normally described in anthropological different populations. At **locus B** the most frequent specificities are B⋆44 (16 %), B⋆35 (13,7 %), B⋆51 (10,5 %), B⋆14 (7,9 %) and B⋆08 (6,5 %), all alleles typical of Caucasians. This locus has groups of alleles extremely rare such as B⋆81(0,01 %), B⋆82 (0,01 %), B⋆67 (0,02 %), B⋆73 (0,11 %) and others completely absent – B⋆54, B⋆59 and B⋆83. The fact that we can detect with one or two examples of extremely rare HLA-B specificities even in closed and distant ethnic groups is significant to describe the degree of genetic heterogeneity of the Portuguese population. **DRB1 locus** has only 13 allele groups, all of them represented in the probed population. The most common are DRB1⋆13 (18,3 %), DRB1⋆07 (15,9 %),DRB1⋆04 (15,4 %), DRB1⋆01 (12,7 %) and DRB1⋆11 (12,5 %). Less frequent are DRB1⋆09 (0,68 %), DRB1⋆10 (1,4 %), DRB1⋆12 (1,6 %) and DRB1⋆16(2,8 %).

4.2 Multi-locus Analysis

HLA haplotypes are specific sets of HLA-A,-B,-DR locus alleles inherited together from a parent. Haplotypes are usually determined by genotyping a sufficient number of family members to establish a gametic assignment of the alleles detected. Because of the extraordinarily large number of possible HLA haplotypes, it is impractical to determine anything but the most common haplotype frequencies by doing family studies. Nevertheless, it is possible to estimate population haplotype frequencies by genotyping a sufficient number of unrelated individuals and using a computer algorithm, to estimate the allele associations that are consistent with the observed genotype data. For large populations that are in Hardy–Weinberg equilibrium, it is possible to estimate even relatively rare haplotypes (e.g. frequency <0.01 %) with reasonably accuracy. In this study we used the Lencaster and Nelson [5] population genetics analysis package, PyPop (http://allele5.biol.berkeley.edu/pypop/). This program implements an iterative expectation-maximization (EM) [2] algorithm on the genotyping data of a maximum of 1,021 randomly selected samples leading to the maximum-likelihood estimate of haplotype frequency for loci: A:B:DRB1. From the sample of 1,021 individuals it was reported 996 unique phenotypes, 3,381 genotypes and 2,082 haplotypes with an estimated frequency above 0.00001 and a log likelihood obtained via the EM algorithm $ln(L_1) = -11296.3$. The exact test of Guo and Thompson [3] was performed for deviations of HWP. The *p*-value provided describes how probable the observed set of genotypes is, with respect to a large sample of other genotypic configurations (conditioned on the same allele frequencies and 2n). *p*-values lower than 0.05 can be interpreted as evidence that the sample does not fit HWP. Table 2 presents the HLA-A:B:DRB1 haplotypes with an estimated frequency greater than or equal to 0.5 %. The well-known Caucasoid haplotype A⋆01-B⋆08-DRB1⋆03, due to hard disequilibrium linkage, comes out as the most frequent in the probed population. The five most frequent HLA HLA-A:B:DRB1 haplotypes are 01:08:03 (3,1 %), 02:44:07 (2,3 %), 02:44:04 (2,1 %), 02:51:11 (1,9 %) and 29:44:07 (1,6 %) which are all typical haplotypes of European Caucasian populations. In fact from this analysis we can detect only 11 haplotypes with frequencies greater than or equal to 1 %.

The Guo and Thompson exact test for HWP (see Table 3) reveals that the population submitted to haplotype estimation does fit the Hardy–Weinberg equilibrium. To be in HW equilibrium means that the sampled of individuals have random mating and does not suffer of evolutive pressures, which turns possible to apply the frequency data to a larger population.

Table 2 Sample output of HLA-A-B-DRB1 haplotype frequency estimation

Haplotypes sorted by frequency			
Haplotype	# Copies	Frequency	SD
01:08:03	63.2	0.03097	0,0038
02:44:07	47.6	0.02332	0,0033
02:44:04	42.5	0.02079	0,0031
02:51:11	39.2	0.01918	0,0030
29:44:07	32.0	0.01569	0,0027
33:14:01	30.8	0.01510	0,0027
03:07:15	27.2	0.01331	0,0025
03:35:01	23.5	0.01150	0,0023
02:44:13	22.9	0.01124	0,0023
02:18:03	20.6	0.01007	0,0022
23:44:07	20.1	0.00982	0,0022
02:51:13	19.3	0.00946	0,0021
02:50:07	17.8	0.00871	0,0020
11:35:01	16.3	0.00800	0,0020
02:51:08	16.3	0.00797	0,0019
30:18:03	15.9	0.00778	0,0019
68:51:13	14.9	0.00730	0,0019
24:35:11	14.4	0.00706	0,0018
02:14:01	14.0	0.00687	0,0018
26:38:13	13.8	0.00678	0,0018
02:18:11	13.8	0.00673	0,0018
24:35:07	13.3	0.00653	0,0018
02:15:04	12.7	0.00622	0,0017
01:57:07	12.3	0.00602	0,0017
24:35:13	11.2	0.00548	0,0016
02:07:01	10.9	0.00534	0,0016
03:14:01	10.4	0.00511	0,0016
02:51:04	9.9	0.00485	0,0015
33:44:13	9.9	0.00483	0,0015

Table 3 Guo and Thompson exact test for Hardy–Weinberg proportions

Guo and Thompson exact test for HWP		
	p-value	SD
HLA-A	0.6110	0.01086
HLA-B	0.6383	0.01185
HLA-DRB1	0.6557	0.008386

4.3 Most Common Phenotypes at PBMRD

In a sample of 20.000 individuals of Cedace it was detected 17,055 different HLA-A,-B-DRB1 phenotypes. The 2,945 that remains are repeated phenotypes at

Table 4 Output of HLA-A-B-DRB1 phenotype frequency in a random sample of 20,000 individuals

HLA-A	HLA-B	HLA-DRB1	Count	Frequency × 10^4
A⋆01,29	B⋆08,44	DRB1⋆03,07	28	14,13
A⋆01,02	B⋆08,44	DRB1⋆03,04	17	8,58
A⋆02,29	B⋆44,51	DRB1⋆07,11	16	8,07
A⋆01,02	B⋆08,44	DRB1⋆03,07	15	7,57
A⋆01,03	B⋆08,51	DRB1⋆03,08	13	6,56
A⋆01,11	B⋆08,35	DRB1⋆01,03	12	6,06
A⋆02,33	B⋆14,44	DRB1⋆01,04	11	5,55
A⋆01,33	B⋆08,14	DRB1⋆01,03	10	5,05
A⋆02,03	B⋆35,44	DRB1⋆01,04	10	5,05
A⋆01,02	B⋆08,18	DRB1⋆03,11	10	5,05
A⋆01,23	B⋆08,44	DRB1⋆03,07	10	5,05
A⋆01,68	B⋆08,53	DRB1⋆03,13	10	5,05
A⋆02,29	B⋆44	DRB1⋆07,13	10	5,05
A⋆03,29	B⋆07,44	DRB1⋆07,15	10	5,05
A⋆01,02	B⋆08,50	DRB1⋆03,07	9	4,54
A⋆01	B⋆08	DRB1⋆03	9	4,54
A⋆01,02	B⋆44,57	DRB1⋆04,07	9	4,54
A⋆02,03	B⋆07,44	DRB1⋆04,15	9	4,54
A⋆02,29	B⋆44	DRB1⋆04,07	9	4,54

different proportions. Table 4 identifies the most common phenotypes, the absolute counts and relative frequency in the probed population. As expected, it is noted the influence of HLA-A,-B-DRB1 haplotypes frequencies on phenotype frequencies, the A⋆01-B⋆08-DRB1⋆03 appears on 63 % of the 19 most frequent phenotypes. In fact the 100 most frequent ABDRB1 phenotypes at the south PBMDR, are direct combinations of the haplotypes with a frequency greater than 0.5 % (28 haplotypes).

5 Conclusions

From the obtained results, the population under study revealed an anthropologic proximity with the European Caucasian* populations [1]. The characterization of HLA gene and haplotype frequencies of a Bone Marrow Donors Registry, is a valuable resource not only in the prediction of the probability of finding a matched haematopoietic stem cell donor, considering the receptor HLA phenotype, but also to determine donor recruitment goals and strategies.

Furthermore, the extend of the populations at the registry represent an important source of information for investigators interested in population genetics and HLA-disease association studies.

References

1. Baur, M.P., Danilov, J.A.: Population Analysis of HLA-A,B,C,DR, and other genetic markers. In: Terasaki, P.I. (ed.) Histocompatability Testing, pp. 955–957, P.I. UCLA Tissue Typing Lab., Los Angeles (1980)
2. Excoffier, L., Slatkin, M.: Maximum-likelihood estimation of molecular haplotype frequencies in a diploid population. Mol. Biol. Evol. **12**, 921–927 (1995)
3. Guo, S., Thompson, E.: Performing the exact test of Hardy-Weinberg proportion for multiple alleles. Biometrics **48**, 361–72 (1992)
4. Hardy, G.H.: Mendelian proportions in mixed populations. Science **28**, 49–50 (1908)
5. Lancaster, A., Nelson, M., Single, R., Meyer, D., Thomson, G.: PyPop a software framework for population genomics: analyzing large-scale multi-locus genotype data. In: Altman, R.B. et al. (eds.) Pacific Symposium on Biocomputing, vol. 8, pp. 514–525 (2003)
6. Levene, H.: On a matching problem arising in genetics. Annals of Math. Stat. **20**, 91–94 (1949)
7. Weinberg, W.: über den Nachweis der Vererbung beim Menschen. Jahreshefte des Vereins für vaterländische Naturkunde in Württemberg. **64**, 368–383 (1908)

Independent Component Analysis for Extended Time Series in Climate Data

Fernando Sebastião and Irene Oliveira

Abstract
Various techniques of multivariate data analysis have been proposed to study time series, including the multi-channel singular spectrum analysis (MSSA). This technique is a principal component analysis (PCA) of the extended matrix of initial lagged series, also called extended empirical orthogonal function (EEOF) analysis in a climatological context. This work uses independent component analysis (ICA) as an alternative to the MSSA method, when studying the extended time series matrix. Often, ICA is more appropriate than PCA to analyse time series, since the extraction of independent components (ICs) involves higher-order statistics whereas PCA only uses the second-order statistics to obtain the principal components (PCs), which are not correlated and are not necessarily independent. An example of time series for meteorological data and some comparative results between the techniques under study are given. Different methods of ordering ICs are also presented, including a new one, which may influence the quality of the reconstruction of the original data.

F. Sebastião (✉)
Department of Mathematics, School of Technology and Management, Polytechnic Institute of Leiria, and CM-UTAD, Campus 2, Morro do Lena - Alto do Vieiro, Apartado 4163, 2411-901 Leiria, Portugal
e-mail: fsebast@ipleiria.pt

I. Oliveira
Department of Mathematics, University of Trás-os-Montes and Alto Douro, and CM-UTAD, Apartado 1013, 5001-801 Vila Real, Portugal
e-mail: ioliveir@utad.pt

J. Lita da Silva et al. (eds.), *Advances in Regression, Survival Analysis, Extreme Values, Markov Processes and Other Statistical Applications*, Studies in Theoretical and Applied Statistics, DOI 10.1007/978-3-642-34904-1_45,

1 Introduction

Independent component analysis (ICA) is a technique widely used in areas such as image processing, biomedical signals, telecommunications and econometric time series among others [10]. ICA is beginning to be applied in climatology (e.g., [1,5]) as an alternative to the classical principal component analysis (PCA) [11], which does not extract all the essential information underlying a data set in space and time, since it uses only second-order statistics to obtain the principal components (PCs). The series can be analysed using the multi-channel singular spectrum analysis (MSSA) which uses the matrix of the extended series of original data, which can be also employed for the ICA implementation.

This work presents a brief description of the classical linear ICA model. The objectives of using MSSA as well as some of its aspects are also considered. A large number of existing algorithms to extract independent components (ICs) do not rank the ICs according to any criterion. We present some existing methods of ordering ICs and suggest a new one, involving the comparison of correlations between PCs and ICs. In this work we consider a data set on monthly mean pressure, at sea level in the North Pacific Ocean. We present some comparisons of the coordinates of the first five components between the three techniques described in the study. We also analysed the quality of the reconstructions of the original data between techniques through the sum of squared errors, taking into account the different methods of ordering ICs. The new method will produce some good results and may be considered as a good alternative for ordering ICs.

2 Description of ICA and MSSA

ICA is a statistical and computational technique introduced in the early 1980s by Hérault and Ans [6] and Hérault et al. [7] and presented in a clear way by Comon [4]. In the last decade some books have appeared showing the development and the applicability of this technique in several areas of science (e.g. [10, 14, 15]). The main objective of ICA is to find hidden components or factors that relate sets of random variables. In the model, we assume that variables of observed data are linear mixtures (combinations) of latent variables, which cannot be observed directly and are independent. ICA is distinct from other similar methods, since we must assume that components are statistically independent and have nongaussian distributions. This technique is related to PCA, but has greater affinity with factor analysis (although the latter does not take into account the nongaussianity of the data). Usually ICA is a more powerful technique for finding hidden factors when classical methods fail completely.

To apply ICA, we consider the classical linear model (without noise) for a sample consisting of p multivariate time series of n observations, which can be modelled in a matrix form by $\mathbf{X} = \mathbf{S}\mathbf{A}$, where:

- $\mathbf{X}$ is the $n \times p$ matrix of observed data (the matrix of p mixtures).

- **S** is the $n \times k$ matrix of k independent components.
- **A** is the $k \times p$ matrix of coefficients of the mixtures (unknown parameters), whose columns must be linearly independent.

To estimate the model, we must admit that components s_i $(i = 1, \ldots, k)$ are statistically independent, that at least $k - 1$ components of s_i $(i = 1, \ldots, k)$ have nongaussian distributions and, for simplicity, that $p \geq k$ [10].

Before applying an algorithm to implement ICA in a data set, some pre-processing techniques are generally used, which help in estimating the model parameters, such as centering and whitening. Whitening is used to estimate ICs and consists in linearly transforming the matrix of observed data and multiplying it by a certain matrix with the goal to obtain a new whitened matrix consisting of uncorrelated components and with variances equal to unity. The new matrix of coefficients of the mixtures will be orthogonal, which is useful due to its algebraic properties.

There are many algorithms that allow the extraction of ICs [10]. In this work we used the FastICA [9] to extract the ICs simultaneously (in parallel), since it is considered one of the most efficient algorithms and has a fast convergence. This algorithm uses the classic method of approximating negentropy as a measure of optimization of nongaussianity to estimate the sample components that should be close to independence.

MSSA, a generalization of singular spectrum analysis (SSA), is also an extension of PCA applied to multivariate time series lagged in time. In the climatological context this is called extended empirical orthogonal function (EEOF) analysis [17]. Its aim is to identify spatio-temporal patterns from a sequential series of maps over a given timescale.

MSSA analyses the periodicity, trend and oscillatory behaviour of multivariate time series, but the main objective is the extraction of joint temporal information of the interrelations between observations of the variables and the interrelations between lagged original variables. One problem in studies with lagged time series is to decide which size to choose for the lagged vector called lag or window length m. Based on some empirical results, Plaut and Vautard [13] suggest that the use of a window length m allows the distinction of oscillations with periods in the range $(m/5, m)$. Consequently, the $n \times p$ data matrix **X** becomes a $(n - m + 1) \times (mp)$ matrix, known as "augmented matrix of lagged data", where we can extract information of covariances between variables (time series) in each lag up to lag $m - 1$.

After retaining the most important components, we obtain the series reconstructed with a subset of k components. Since k depends on the choice of m, there is no consensus on the number of components to retain in the reconstruction. So, a possible solution is to retain the components whose eigenvalues have a value greater than a obvious break point. It is assumed that the first k components will retain the dominant information in the data, while the remaining $m - k$ components represent some external noise [16].

3 Methods of Ordering Independent Components

In many practical applications it is necessary to order ICs with a view to comparing or identifying the components that extract the most meaningful information in the study. In some ICA algorithms, such as FastICA, ICs are not sorted out according to any criterion, in contrast to PCA where PCs are ranked by decreasing order of variances. We must take into account that different methods may rank components differently, according to their statistical properties.

We consider a brief description of some of the methods in the literature on ordering ICs. We consider also the introduction of a new criterion (M5), which can be applied when using ICA and PCA on the same data set.

1. *(M1) — Maximization of kurtosis of ICs:*

This method ranks the ICs according to the decrease of the absolute value of the difference between its kurtosis and the kurtosis of a normal random variable [3, 12].

2. *(M2) — Maximization of the vector norms of the estimated matrix of mixtures, A:*

ICs are ordered according to decreasing values of the vector norms of the rows of the estimated matrix of mixtures, **A**. This is reminiscent of the PCA ordering, since the vector norms of the rows of **A** provide the contributions of the corresponding ICs to the variances of observed variables [8].

3. *(M3) — Minimization of the sum of squared errors when reconstructing the original data matrix:*

ICs are ranked by an algorithm that is based on the increase in the sum of squared errors, which is similar to a method proposed by Cheung and Xu [2]. Residuals were obtained by the differences between the original data and the reconstructions made from subsets of components.

4. *(M4) — Canonical correlation analysis (CCA):*

The canonical correlations between the reconstructions of the original data using the ICs and the original data can be used to order ICs. Sorting is done according to the decrease in the value of canonical correlations between the original data and each of the subsets formed by ICs. This process starts with the IC that is most correlated with the original data. After fixing the first IC and using multiple correlations, we choose the subset of two ICs most correlated with the original data and that includes the first fixed IC. A third IC is then added to the subset of two ICs obtained in the previous step, again using the largest multiple correlation as a criterion. The process is repeated until the final ordering. A similar method of CCA application can be found in Youssef et al. [18], which presents an algorithm to analyse functional magnetic resonance imaging (fMRI).

5. *(M5) — Correlations between principal components and ICs:*

When PCA can be applied in parallel with ICA to a given data set, the method of comparing the correlations between PCs and ICs can be presented as a new alternative criterion for ordering ICs. Since the PCs are naturally ranked by decreasing order of their variance, for a fixed order of PCs, the ICs are ranked by maximizing their correlation with each PC. This is a natural way of ordering ICs, since in general, each IC is highly correlated with a distinct single PC.

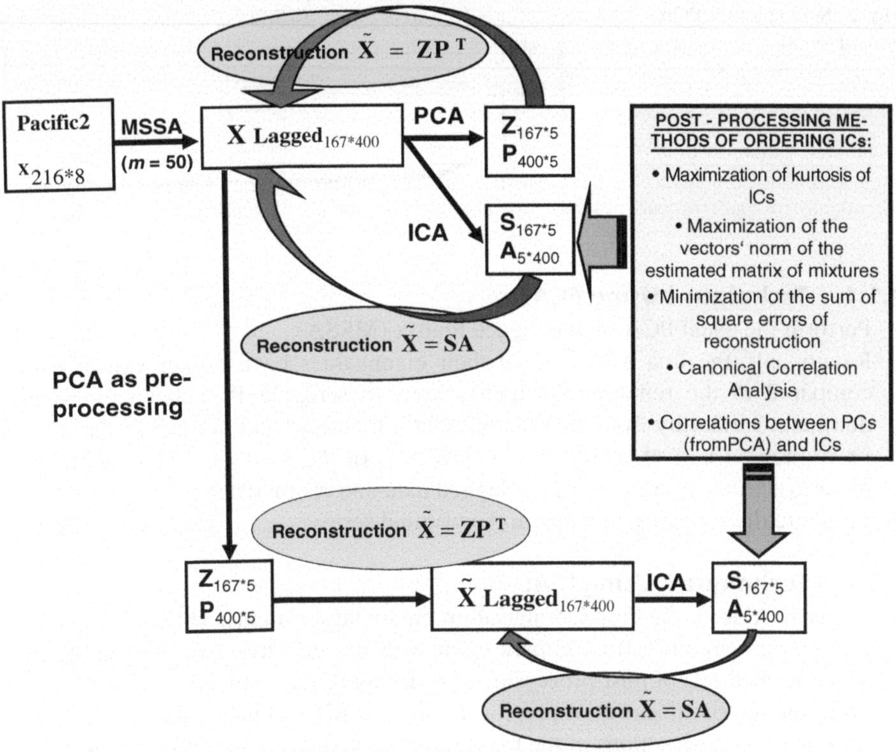

Fig. 1 Methodology for techniques I, II and III with $m = 50$. **Z** is the matrix of scores, **P** is the matrix of eigenvectors, **S** is the matrix of independent components and **A** is the matrix of coefficients of the mixtures

4 Case Study

4.1 Data and Methodology

Consider 216 values of monthly mean-sea-level pressure (from January 1979 to December 1996), for eight weather stations (1 — Crescent City; 2 — San Diego; 3 — San Francisco; 4 — Hilo; 5 — Honolulu; 6 — NeahBay; 7 — Seldovia and 8 — Sitka) in the North Pacific Ocean in the states of Alaska, California, Hawaii and Washington.

The matrix of original data (Pacific2) was transformed into an extended matrix of original time series with a lag $m = 50$, since this value is close to $n/4$ as Vautard et al. [16] suggested. In terms of methodology, we applied three different techniques that are described below by steps and illustrated in Fig. 1 for a lag $m = 50$.

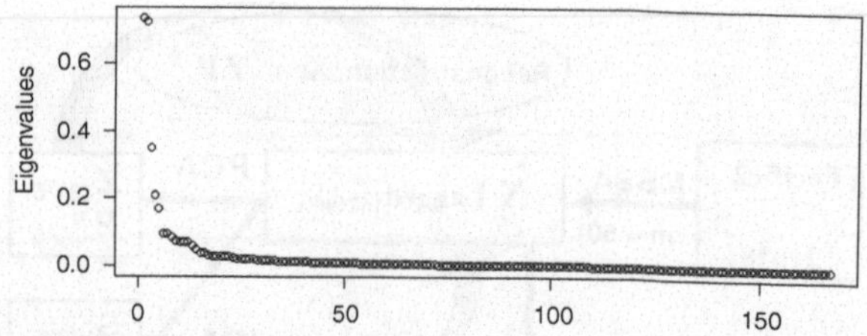

Fig. 2 Scree plot for PCA with $m = 50$

4.1.1 Technique I Using PCA:

- Perform the usual PCA on the lagged matrix (MSSA).
- Retain only the first 5 PCs since their eigenvalues have higher values when compared to the remaining, which allows to separate PCs that capture the dominant variability from the remaining PCs that represent noise (Fig. 2).
- From the matrices of scores and eigenvectors of the 5 retained PCs, obtain the reconstructions of the matrices of lagged data and original data.
- Compare the reconstructions with the original series.

4.1.2 Technique II Using ICA:

- Apply ICA using the FastICA algorithm on the lagged matrix (MSSA).
- Extract only five ICs (to compare them with the first five PCs) and apply the different methods of post-processing of ordering ICs, as defined in Sect. 3.
- From the matrix of scores of previously ordered ICs and the matrix of mixtures, obtain the reconstructions of the matrices of lagged data and original data.
- Compare the reconstructions with the original series.

4.1.3 Technique III Using ICA After PCA:

- Perform the usual PCA on the lagged matrix (MSSA) as pre-processing for ICA, as a way to retain the higher variability of the data in a small number of PCs and ignoring much of the noise.
- From the matrices of scores and eigenvectors of 5 retained PCs, obtain the reconstruction of the lagged matrix.
- Apply ICA using the FastICA algorithm on the reconstructed lagged matrix.
- Extract only 5 ICs (in order to compare them with the 5 components extracted in the other two techniques) and apply the different methods of post-processing of ordering ICs, as defined in Sect. 3.
- From the matrix of scores of previously ordered ICs and the matrix of mixtures, obtain the reconstructions of the matrices of lagged data and original data.
- Compare the reconstructions with the original series.

4.2 Results

For the three presented techniques, we obtained some comparative results associated with the components that are shown in the next figures. We also present some

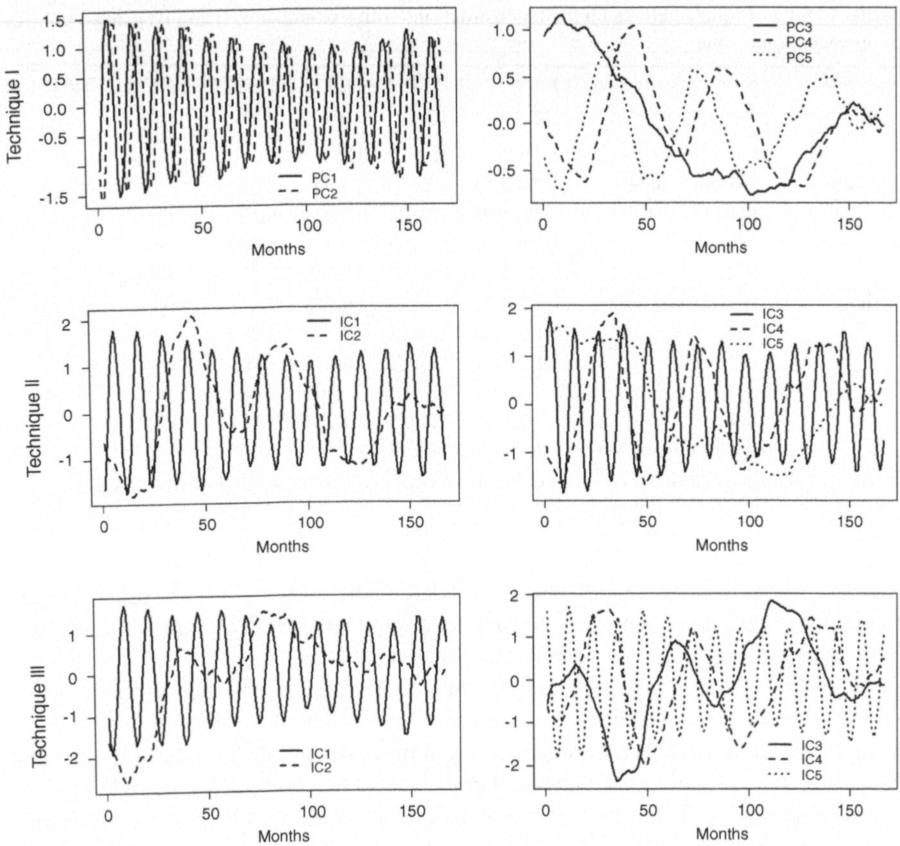

Fig. 3 Scores for $k = 5$ extracted components obtained by techniques I, II and III, for a lag $m = 50$

tables to describe the sum of squared errors in order to analyse the quality of the reconstructions of the original data according to the ordering methods of ICs. In this case study, when we apply PCA to the lagged matrix with lag $m = 50$, we can consider two relevant eigenvalues before the noise floor (Fig. 2), although we retain the first five eigenvalues since they are clearly higher than the others. The percentage of variance explained by the first five (in a total of 167) PCs is close to 55 %, while the percentage explained by the first two PCs is only 36.8 %.

Fig. 3 compares the coordinates (scores) of the first five components for techniques I, II and III for $m = 50$. The PCs are ranked as usual, while the ICs are ranked in an arbitrary order. A visual inspection of the temporal behaviour, it is obvious that ICs are not ranked as the PCs, and therefore we consider the various methods of ordering described in Sect. 3.

To decide what are the most appropriate methods of ordering ICs we compared the reconstructions with the original data. We calculated the sum of squared errors

Table 1 Sum of squared errors for the reconstructions with techniques I, II and III using a number of distinct components

		Order	PC1	PC12	PC123	PC1234	PC12345
Technique I	Natural order	(1,2,3,4,5)	12.103	10.433	8.666	7.751	7.344
	Ordering methods	Order	IC1	IC12	IC123	IC1234	IC12345
Technique II	Arbitrary order	(1,2,3,4,5)	12.120	10.941	9.264	8.610	7.344
	M1	(1,3,5,4,2)	12.120	10.430	8.683	7.887	7.344
	M2; M3; M5	(1,3,5,2,4)	12.120	10.430	8.683	7.749	7.344
	M4	(1,5,3,4,2)	12.120	10.460	8.683	7.887	7.344
Technique III	Arbitrary order	(1,2,3,4,5)	12.251	10.238	9.455	9.057	7.344
	M1	(5,1,4,2,3)	12.118	10.415	9.387	7.722	7.344
	M2; M5	(5,1,2,3,4)	12.118	10.415	8.467	7.718	7.344
	M3	(5,2,1,3,4)	12.118	10.261	8.467	7.718	7.344
	M4	(1,2,5,4,3)	12.251	10.238	8.467	7.722	7.344
	Alternative of M4	(5,2,1,4,3)	12.118	10.261	8.467	7.722	7.344

Techniques II and III presents each of the orders in the different methods of ordering ICs
Techniques used lagged series with lag $m = 50$ and $k = 5$ extracted components

(by time series) and we synthesized in Table 1 the sums of these values for all weather stations, using a number of distinct components for techniques I, II and III in the reconstruction of the original data for $m = 50$.

In Table 1, in techniques II and III, the row that represents the arbitrary order $(1, 2, 3, 4, 5)$ of ICs is obtained by application of the FastICA algorithm. From the arbitrary order, ICs are reordered according to the corresponding methods to obtain a new order. For example, in technique II in the row of M1 (maximization of kurtosis), the order $(1, 3, 5, 4, 2)$ means that the first IC is the same in the arbitrary order, the new second IC is the third IC in the arbitrary order, the new third IC is the fifth IC in the arbitrary order, the new fourth IC is the fourth IC in the arbitrary order and finally the new fifth IC is the second IC in the arbitrary order.

In each of the techniques, the last five columns represent the sum of squared errors for reconstructions from one component until five components according to the order in the corresponding method. For example, in technique II in the row of M1, the value 8.683 represents the sum of squared errors for the reconstruction by first three ICs (IC1, IC3 and IC5) in the respective order $(1, 3, 5, 4, 2)$.

After comparing the various methods of ordering ICs used in the technique II, three (M2, M3 and M5) of the five methods provide the same order $(1, 3, 5, 2, 4)$ to minimize the sums of sum of squared errors for all weather stations. Moreover, technique III suggests the algorithm of minimization of sum of squared errors (M3) with the order $(5, 2, 1, 3, 4)$ as the best method to minimize the sums of sum of squared errors of the weather stations, but an alternative to maximize the correlations of CCA (alternative of M4) with the order $(5, 2, 1, 4, 3)$ shows similar results.

A comparative analysis between the sum of squared errors of the reconstructions for the three techniques emphasizes very similar results between technique I and the order $(1, 3, 5, 2, 4)$ that minimizes the sum of squared errors in technique II, for three methods including the new method M5. Results seem to be relatively better in technique III when compared with techniques I and II.

5 Conclusions

Three techniques involving PCA and ICA were proposed to analyse extended time series, emphasizing the application of ICA to the augmented lagged matrix in two of them. The five dominant components that are naturally ranked in PCA, do not appear in the same order in ICA. We examined five methods of ordering ICs. In the three techniques, similar results were obtained in the scores of components for a lag $m = 50$, after reordering the ICs. The quality of the reconstructions of the original data was analysed through tables of sum of squared errors, with similar results for techniques I and II, while technique III shows slightly better results.

The new method M5 appears to be one of the best, mainly for technique I and therefore can be considered as a good method of ordering ICs. One data set is not sufficient to ensure a conclusive comparative assessment and other data sets must be assessed in similar ways in order to compare the new method's performance with that of the remaining methods.

Acknowledgements The authors would like to thank CM-UTAD and to the research funded by the Portuguese Government through the FCT (Fundação para a Ciência e Tecnologia) under the project PEst-OE/MAT/UI4080/2011 for the financial support for this study.

References

1. Basak, J., Sudarshan, A., Trivedi, D., Santhanam, M.S.: Weather data mining using independent component analysis. J. Mach. Learn. Res. **5**, 239–253 (2004)
2. Cheung, Y., Xu, L.: Independent component ordering in ICA time series analysis. Neurocomputing **41**, 145–152 (2001)
3. Cichocki, A., Thawonmas, R., Amari, S.: Sequential blind signal extraction in order specified by stochastic properties. Electron. Lett. **33**(1), 64–65 (1997)
4. Comon, P.: Independent component analysis, a new concept? Signal Process. **36**, 287–314 (1994)
5. Hannachi, A., Unkel, S., Trendafilov, N.T., Jolliffe, I.T.: Independent component analysis of climate data: a new look at EOF rotation. J. Clim. **22**, 2797–2812 (2009)
6. Hérault, J., Ans, B.: Circuits neuronaux à synapses modifiables: décodage de messages composites par apprentissage non supervisé. Comptes Rendus de l'Académie des Sciences **299**(III-13), 525–528 (1984)
7. Hérault, J., Jutten, C., Ans, B.: Détection de grandeurs primitives dans un message composite par une architecture de calcul neuromim étique en apprentissage non supervisé. In Actes du Xème colloque GRETSI, pp. 1017–1022, Nice, France (1985)
8. Hyvärinen, A.: Survey on independent component analysis. Neural Comput. Surv. **2**, 94–128 (1999)
9. Hyvärinen, A., Oja, E.: A fast fixed-point algorithm for independent component analysis. Neural Comput. **9**(7), 1483–1492 (1997)

10. Hyvärinen, A., Karhunen, J., Oja, E.: Independent Component Analysis. Wiley, New York (2001)
11. Jolliffe, I.T.: Principal Component Analysis, 2nd edn. Springer, New York (2002)
12. Lu, W., Rajapakse, J.C.: Eliminating indeterminacy in ICA. Neurocomputing **50**, 271–290 (2003)
13. Plaut, G., Vautard, R.: Spells of low-frequency oscillations and weather regimes in the Northern Hemisphere. J. Atmos. Sci. **51**(2), 210–236 (1994)
14. Roberts, S., Everson, R.: Independent Component Analysis: Principles and Practice. Cambridge University Press, Cambridge (2001)
15. Stone, J.V.: Independent Component Analysis: A Tutorial Introduction. MIT, Cambridge (2004)
16. Vautard, R., Yiou, P., Ghil, M.: Singular spectrum analysis: A toolkit for short, noisy chaotic signals. Physica D **58**, 95–126 (1992)
17. von Storch, H., Zwiers, F.W.: Statistical Analysis in Climate Research. Springer, New York (1999)
18. Youssef, T., Youssef, A.M., LaConte, S.M., Hu, X.P., Kadah, Y.M.: Robust ordering of independent components in functional magnetic resonance imaging time series data using canonical correlation analysis. Proc. SPIE **2**, 5031–5037 (2003)

Life Satisfaction: A MIMIC Approach with a Discrete Latent Variable

Patrícia Serra, José G. Dias, and Maria de Fátima Salgueiro

Abstract
This chapter proposes modeling a battery of items concerning life satisfaction using a *Multiple Indicator Multiple Cause* (MIMIC) model with a discrete latent variable. Portuguese data from year 2001 of the *European Community Household Panel* (ECHP) are used. Life satisfaction variables include satisfaction with work or main activity, financial situation, housing situation, and the amount of leisure time. Some personal characteristics are considered as explanatory variables of the latent life satisfaction variable. Four classes of individuals are obtained, with distinct patterns of association between dependent and concomitant variables.

1 Introduction

The popularity of finite mixture models has recently increased, mainly due to the availability of fast computing technology able to support this type of models. In particular, the social sciences area, with a big tradition in latent class models, has contributed for the popularity of finite mixture models [3].

A finite mixture model for discrete data is known as a latent class model. The units of a latent class model are assumed to belong to some discrete class ($s = 1, \ldots, S$) and class membership is unknown. Moreover, the classes can be viewed as the categories of a categorical latent variable.

P. Serra (✉)
Instituto Universitário de Lisboa (ISCTE-IUL), BRU-IUL, Lisbon, Portugal
e-mail: patricia.serra@iscte.pt

J.G. Dias · M. de Fátima Salgueiro
Instituto Universitário de Lisboa (ISCTE-IUL), Business Research Unit, Lisbon Portugal
e-mail: jose.dias@iscte.pt; fatima.salgueiro@iscte.pt

J. Lita da Silva et al. (eds.), *Advances in Regression, Survival Analysis, Extreme Values, Markov Processes and Other Statistical Applications*, Studies in Theoretical and Applied Statistics, DOI 10.1007/978-3-642-34904-1_46,

In order to investigate heterogeneity in both measurement and structural relationships, a latent class model with latent and manifest variables called *Multiple Indicator Multiple Cause* (MIMIC) model is used in this chapter to model life satisfaction. Indeed, life satisfaction is a very common proxy to measure subjective well-being (SWB). SWB reflects the extent to which a person thinks or feels that his/her life is going well. Thus, life satisfaction is supposed to reflect a personal assessment of the general living conditions, taking into account the background of individual aspirations, expectations, and guidance [1].

SWB is associated with several indicators either at an individual level or at a contextual level. As far as sex is concerned, there is some evidence that males are less satisfied than females. According to [4] and [5], elder people report higher levels of satisfaction, and so do married people. A good health status increases life satisfaction and housing has a positive relationship with life satisfaction. The more hours people work per week, the less satisfied they are. Individuals with higher income tend to be more satisfied and promotion opportunities increase life satisfaction. Clark et al. [2] proposed a latent class approach to model the relationship between income and self-reported well-being, for twelve European countries. Four classes of individuals were identified, in which individual characteristics and country of residence were found to be strong predictors of class membership.

This chapter uses data from the *European Community Household Panel* (ECHP). Four life satisfaction variables are used and ten possible determinants of life satisfaction are considered. The statistical package Latent Gold 3.0 is used. The structure of the chapter is the following: Sect. 2 presents the sample under analysis. Section 3 describes the proposed MIMIC model. Section 4 summarizes the main results of the statistical modeling undertaken and Sect. 5 provides a discussion of the results.

2 The Data

The ECHP is a longitudinal household survey conducted between 1994 and 2001 and representative of several European Union countries [6]. The ECHP collects data on perceptions of life satisfaction and on demographic characteristics, employment, income, health, education and training, housing, among others. The current study uses the 2001 Portuguese data. The four following questions have been considered as indicators of life satisfaction: satisfaction with work or main activity; satisfaction with financial situation; Satisfaction with housing situation; and satisfaction with amount of leisure time. Respondents have been asked to rate their satisfaction level using a *Likert*-type scale from 1—not satisfied to 6—totally satisfied.

A subsample of 5,742 individuals is used. Individuals are aged 17 years old or more, work more than 15 h per week, and gave valid answers to all four life satisfaction variables under analysis. Figure 1 summarizes the distribution of the responses concerning the life satisfaction variables. It is possible to conclude that

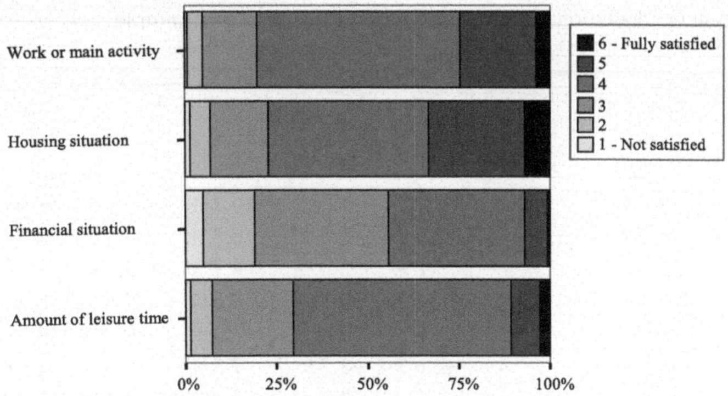

Fig. 1 Distribution of life satisfaction variables

the majority of the responses are at the level 4 of satisfaction for all items. Housing situation and work itself are the dimensions with the highest mean satisfaction levels, whereas satisfaction with financial situation has the lowest mean levels.

Explanatory variables used in this study as possible determinants of life satisfaction include sex, age, marital status, existence of children under twelve years old in the household, education, personal income, health status, degree of urbanization, number of hours worked per week, and job status.

In terms of sex, 55.3 % of the 5,742 respondents are male. The average age of the sample is 38 years old. As far as marital status is concerned, 64.6 % of the respondents are married. Regarding the existence of children under 12 years old in the household, 39 % have children and 61 % do not have. In terms of education, only 13.4 % have the recognized third level of education (ISCED 5–7), 16.4 % have the second stage of secondary level education (ISCED 3), and 70.2 % have less than the second stage of secondary education (ISCED 0–2). The distribution of personal income is the following: earnings under 500 €, 48.3 %; 500 to 1,000 €, 37.9 %; 1,000 to 1,750 €, 10.6 %; the remaining 3.2 % of the sample earns 1,750 €—or more. In terms of perceived health status, 65.1 % of the respondents consider it very good or good, 30.8 % fair, and only 4.2 % rate it as bad or very bad. Regarding the degree of urbanization, almost half of the respondents (48 %) live in a densely populated area, 28.2 % live in an intermediate area, and the remaining 23.8 % live in a thinly populated area. In terms of the number of hours worked per week, 1.8 % work less than 20 h; 86.7 % work between 21 and 45 h, 9.8 % work between 46 and 60 h, and the remaining 1.8 % work 61 h or more per week. The large majority of the respondents are nonsupervisory employees (90.6 %), 5.2 % are intermediate, and 4.2 % are supervisors. This aggregate sample descriptive statistics are displayed in the last column of Table 1.

Table 1 Percentage deviations in each class against the aggregate sample

Variables	Categories	Classes 1	2	3	4	Aggregate sample
Sex	Male	−3.6 %	1.0 %	0.8 %	3.1 %	55.3 %
	Female	3.6 %	−1.0 %	−0.8 %	−3.1 %	44.7 %
Age	17–24	−0.1 %	3.6 %	−10.2 %	−5.2 %	15.9 %
	25–34	−2.7 %	−1.6 %	8.5 %	4.7 %	33.7 %
	35–44	3.1 %	−2.0 %	4.7 %	−3.2 %	23.4 %
	45–54	0.1 %	−0.8 %	0.6 %	2.8 %	18.2 %
	55–65	−0.4 %	0.8 %	−3.6 %	0.9 %	8.8 %
Marital status	Married	−2.6 %	−2.3 %	7.5 %	8.3 %	64.6 %
	Not married	2.6 %	2.3 %	−7.5 %	−8.3 %	35.4 %
Children < 12 years old	Yes	5.0 %	−2.8 %	5.2 %	−4.5 %	39.0 %
	No	−5.0 %	2.8 %	−5.2 %	4.5 %	61.0 %
Education	ISCED 5–7	−7.5 %	−5.0 %	25.1 %	14.0 %	13.4 %
	ISCED 3	−2.5 %	−1.6 %	7.4 %	5.3 %	16.4 %
	ISCED 0–2	9.9 %	6.7 %	−32.5 %	−19.4 %	70.2 %
Personal income	< 500 €	14.6 %	5.0 %	−39.3 %	−16.4 %	48.3 %
	500–1000 €	−4.0 %	1.6 %	2.4 %	0.1 %	37.9 %
	1000–1750 €	−7.5 %	−4.7 %	26.0 %	11.8 %	10.6 %
	> 1750 €	−3.1 %	−1.9 %	10.9 %	4.5 %	3.2 %
Health status	Very good	−11.0 %	−0.6 %	12.9 %	15.2 %	65.1 %
	Fair	8.2 %	0.9 %	−8.7 %	−14.1 %	30.8 %
	Very bad	2.8 %	−0.3 %	−4.1 %	−1.1 %	4.2 %
Degree of urbanization	Densely populated	2.5 %	−5.3 %	13.2 %	4.4 %	48.0 %
	Intermediate	1.3 %	0.5 %	−4.2 %	−0.8 %	28.2 %
	Thinly populated	−3.7 %	4.9 %	−9.0 %	−3.6 %	23.8 %
Number of hours worked per week	< 20	0.7 %	−0.2 %	−1.1 %	0.2 %	1.8 %
	21–45	−2.7 %	4.8 %	−19.3 %	4.3 %	86.7 %
	46–60	1.3 %	−3.6 %	16.6 %	−3.4 %	9.8 %
	> 61	0.7 %	−1.0 %	3.8 %	−1.1 %	1.8 %
Job status	Supervisory	−2.3 %	−1.5 %	8.8 %	2.9 %	4.2 %
	Intermediate	−2.4 %	−1.7 %	12.3 %	0.7 %	5.2 %
	Nonsupervisory	4.7 %	3.1 %	−21.1 %	−3.6 %	90.6 %

3 The Proposed MIMIC Model

In latent class models individuals are assumed to belong to some discrete class ($s = 1, \ldots, S$), and class membership is unknown. Moreover, classes can be viewed as the categories of a categorical latent variable.

A latent class MIMIC model is a one-factor model where the factor (a categorical latent variable) is measured by multiple indicators and regressed on several observed covariates. The MI component is the measurement component and models the relationship between the observed items (in this case the four life satisfaction variables). The MIC component is the structural component of the model (in this case the impacts of the ten determinants of life satisfaction on the categories of the life satisfaction latent variable). Consider a sample of n observations. An observation is denoted by i ($i = 1, \ldots, n$) and is characterized by J attributes. Let $\mathbf{y}_i$ be the vector of the J dependent variables. The MIMIC model with S latent classes for $\mathbf{y}_i$ is defined by the composite density

$$f(\mathbf{y}_i; \boldsymbol{\varphi}, \mathbf{w}_i) = \sum_{s=1}^{S} \pi_{is}(\mathbf{w}_i, \boldsymbol{\gamma}_s) f_s(\mathbf{y}_i; \boldsymbol{\theta}_s), \tag{1}$$

where the discrete latent variable, z_i, has a multinominal distribution, such that $z_i \sim Multi_{S-1}(\boldsymbol{\pi}_i)$, with $\boldsymbol{\pi}_i = (\pi_{i1}(\mathbf{w}_i, \boldsymbol{\gamma}_1), \cdots, \pi_{i,S-1}(\mathbf{w}_i, \boldsymbol{\gamma}_{S-1}), \pi_{is}(\mathbf{w}_i, \boldsymbol{\gamma}_s) > 0$, and $\sum_{s=1}^{S} \pi_{is}(\mathbf{w}_i, \boldsymbol{\gamma}_s) = 1$. The vector of the J dependent variables is defined by $\mathbf{y}_i = (y_{i1}, \ldots, y_{iJ})$ and the vector of the K concomitant variables is $\mathbf{w}_i = (w_{i1}, \ldots, w_{iK})$. The conditional probability function (of class s) is $f_s(\mathbf{y}_i; \boldsymbol{\theta}_s)$, and the parameters of the model are defined by $\boldsymbol{\varphi} = (\boldsymbol{\gamma}_1, \ldots, \boldsymbol{\gamma}_{S-1}, \boldsymbol{\theta}_1, \ldots, \boldsymbol{\theta}_S)$, where $\boldsymbol{\gamma}_s$ and $\boldsymbol{\theta}_s$ are the vector of parameters to estimate in each class s.

An important issue in finite mixture modeling is estimation. The maximum likelihood estimate of a set of independent observations can only be obtained by iterative procedures, such as the *expectation-maximization* (EM) algorithm. This algorithm is a method for maximum likelihood estimation with incomplete data that reintroduces the additivity of the log-likelihood function, using data augmentation. This algorithm is divided into two steps: first, the E-step that consists on associating each individual observation with its conditional expectation of class membership, given the observed values. The next, the M-step that consists in maximizing the full data log-likelihood function ($\ell_s(\hat{\varphi}; \mathbf{y}, \mathbf{w})$) using the complete data as the observed data.

To find the optimal number of classes for the latent variable in the model, one has to select the model that minimizes

$$C_s = -2\ell_s(\hat{\varphi}; \mathbf{y}, \mathbf{w}) + d \cdot N_s, \tag{2}$$

where N_s represents the number of parameters in the model. According to different criteria we can give different values to d. For the *Akaike information criterion* (AIC) $d = 2$, for the *modified Akaike information criterion* (AIC3) $d = 3$, and for the *Bayesian information criterion* $d = \log n$.

4 Results

The diagram of the MIMIC model proposed to model life satisfaction and its determinants is shown in Fig. 2. The dependent variables in the model are the four life satisfaction variables. Ten explanatory variables are considered: sex, age, marital status, existence of children under twelve years old in the household, education, personal income, health status, degree of urbanization, number of hours worked per

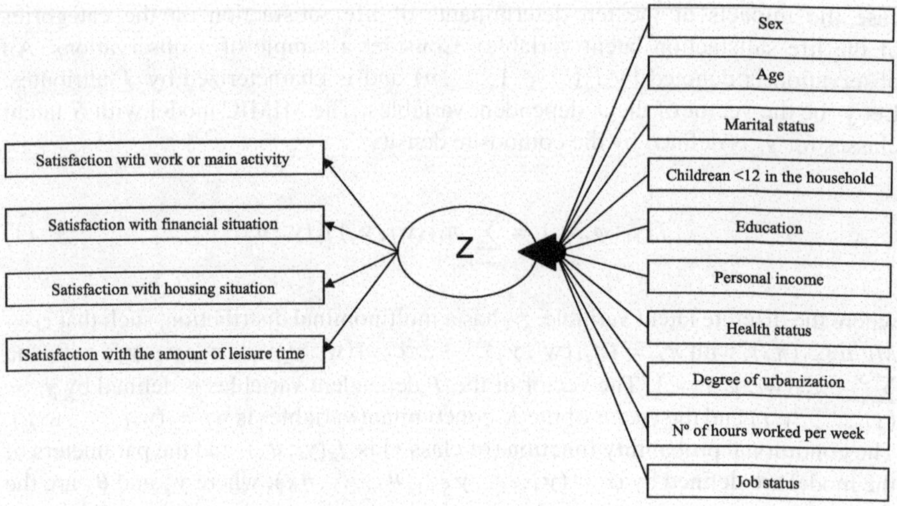

Fig. 2 Diagram of the proposed MIMIC model

week, and job status. The latent class MIMIC model implies that (1) items on satisfaction are conditionally independent given the latent class (local independence) (2) dependent items are conditionally independent of the concomitant variables, given the latent class.

Latent class models with different number of classes, ranging from $S = 1$ to $S = 8$, were estimated. For the EM algorithm random initialization was considered and the convergence tolerance level equals 10^{-6}. The decision concerning the choice of the number of classes of the latent variable was made based on the BIC. A (global) minimum for BIC was reached when $S = 4$ and therefore a solution with four latent classes was chosen. The characterization of the four classes in terms of size and distribution of the four life satisfaction variables follows.

Class 1 corresponds to 26.6 % of the sample and includes the least satisfied respondents with all four components of life satisfaction. In terms of opinion about work or main activity, 43.4 % present a level of satisfaction lower than 3, in contrast with the 16.8 % for the aggregate sample. Responses about satisfaction with financial situation are dominated by the levels 1 and 2 of satisfaction (50.74 %). In terms of satisfaction with housing situation, levels 1, 2, and 3 are dominant (for 54.7 % of respondents), in contrast with the 15.7 % of the aggregate sample.

Class 2 is the largest, representing 50.2 % of the sample. It corresponds to respondents with an intermediate level of life satisfaction, namely, level 4, although in the case of satisfaction with the financial situation, level 3 prevails (45.1 % of respondents). 70.6 % of the individuals chose level 4 for satisfaction with work or main activity, in contrast with the 57.2 % in the sample. As far as the amount of leisure time is concerned, 83.3 % of the individuals answered level 4, compared to 60.41 % of the aggregate sample.

Class 3 corresponds to a group of respondents (11.3 % of the sample) with high levels of satisfaction with work, financial, and housing situation, but very low levels of satisfaction with amount of leisure time. In terms of the financial situation, 67.9 % of the participants respond level 4 of satisfaction (*versus* 39.2 % in the aggregate). Regarding the amount of leisure time, levels 1, 2, and 3 of satisfaction are dominant (67.24 %), contrasting with the 28.2 % in the aggregate sample.

Class 4 corresponds to 11.9 % of the respondents and includes the individuals most satisfied with life. In fact, for this class, 86.5 % of the respondents have high levels (5 and 6) of satisfaction with work or main activity, while in the aggregate sample such percentage equals 26 %. In terms of satisfaction with financial situation, 41.2 % state levels 5 and 6 of satisfaction, contrasting with the 7.1 % in the aggregate sample. Levels 5 and 6 of satisfaction with the housing situation and the amount of leisure time are responded by 85.3 % and 61 %, respectively, *versus* 33.4 % and 11.4 % for the aggregate sample.

The four classes are now characterized in terms of the ten explanatory variables: the determinants of life satisfaction. Since concomitant variables are categorical, the probabilities associated with each category of each explanatory variable were estimated for each of the four categories of the latent variable. Table 1 displays such probabilities in terms of percentage deviations, in each class, against the sample aggregate.

The variables that best distinguish the four classes are education, personal income, perception of health status, degree of urbanization, and number of hours worked per week, since they lead to the highest deviations against the aggregate sample. In class 1, and in comparison to the sample aggregate, there are 9.9 % more individuals with less than the second stage of education; 14.6 % more individuals earning less than 500 €; and 11 % less individuals with a good or very good perception of health status. In class 2 there are 6.7 % more individuals with less than the second stage of education than in the sample aggregate and 5 % more individuals earning less than 500 €. In comparison with the aggregate sample, 4.8 % more individuals work 21–45 hours per week and 3.1 % more individuals have a nonsupervisory position. Class 3 has 25.1 % more individuals with the third level of education and 32.5 % less individuals with the lowest education level, when compared to the aggregate sample. The percentage of individuals with income 1,000–1,750 €is 26 % higher than in the aggregate sample; 12.9 % more individuals perceive their health status as very good, 13.2 % more individuals live in densely populated areas, 16.6 % more respondents work 46–60 h per week, and 21.1 % less individuals have nonsupervisory positions. In class 4, and in comparison to the aggregate sample, there are 8.3 % more individuals married, 4.5 % more individuals with no children under 12 years old in the household, 11.8 % more individuals with income 1,000–1,750 €, 15.2 % more respondents with perceptions of good or very good health status, and 4.3 % more individuals working 21–45 h per week.

5 Discussion

In this chapter we have proposed modeling four life satisfaction items, using a MIMIC model: a discrete latent variable model where the latent variable is explained by a set of characterization variables at the individual level. Four distinct classes of individuals were obtained. Class 1 includes individuals less satisfied with life, with low levels of education, low income, worse perceptions of health status, and nonsupervisory positions. Class 2 has intermediate levels of satisfaction and differs from class 1 mostly in terms of a better perception of health status and a higher income. Class 3 includes individuals that are very happy with their job and financial situation, but very dissatisfied with their amount of leisure time. They work long hours per week, have supervisory positions and high earnings. Class 4 has high levels of life satisfaction and includes individuals with high levels of education, earning more than 1,000 €, with a good perception of health status, living in a densely populated area, and working 21 to 45 h per week.

Future research could address dealing with missing values in life satisfaction variables and taking into account the complex survey design.

Acknowledgments This research was supported by the Fundação para a Ciência e a Tecnologia, grant PTDC/GES/72784/2006.

References

1. Christoph, B., Noll, H.H.: Subjective well-being in the European Union during the 90s. Soc. Indic. Res. **64**, 521–546 (2003)
2. Clark, A., Etil, F., Postel-Vinay, F., Senik, C., Straeten, K.: Heterogeneity in reported well-being: Evidence from twelve European countries. Econ. J. **115**, C118–C132 (2005)
3. Dias, J.G.: Finite Mixture Models. Review, Applications, and Computer-intensive Methods (PhD Thesis). Ridderprint, The Netherlands (2004)
4. Judge, T.A., Watanabe, S.: Another look at the job satisfaction - Life satisfaction relationship. J. Appl. Psychol. **78**(6), 939–948 (1993)
5. Khattab, N., Fenton, S.: What makes adults happy? Employment and non-work as determinants of life satisfaction. Sociology **43**, 11–26 (2009)
6. Peracchi, F.: The European community household panel: A review. Empir. Econ. **27**, 63–90 (2002)

An Application of MRMC ROC Curves on Radiology

Carina Silva-Fortes, Maria Antónia Amaral Turkman, Luis Lança, Ricardo Silva, and Gonçalo Marques

Abstract

The scientific area of radiology of the Higher School of Health Technology of Lisbon, conducted an experimental study with the goal of investigating the influence of the tube potential (kV) on the detection of simulate chest lesions in a chest phantom. Exposure parameters influence the quality and quantity of a X-ray beam and consequently image quality, therefore influencing the observer capacity to detect lesions. To produce images with high quality, readers' performances were compared as well as the accuracy of lesions detection associated with different tube potential and ROC (receiver operating characteristic) methodology was used to select the best ones. The proper binormal ROC curve model was used to select the reader with best performance. However, the conventional ROC curve is not adequate to select the best tube potential, because the evaluation of the images also depends on the reader's interpretation. So, the MRMC (multiple readers multiple cases) ROC curves were proposed and, for their estimation, Dorfman–Berbaum–Metz method was used. All calculations were performed on the `PROPROC`, `DBM-MRMC 2.2` and `R` free softwares.

C. Silva-Fortes (✉)
Higher School of Health Technology of Lisbon and CEAUL, Lisbon, Portugal
e-mail: carina.silva@estesl.ipl.pt

M.A.A. Turkman
Faculty of Sciences of University of Lisbon and CEAUL, Lisbon, Portugal
e-mail: antonia.turkman@fc.ul.pt

L. Lança · R. Silva
Higher School of Health Technology of Lisbon, Lisbon, Portugal
e-mail: luis.lanca@estesl.ipl.pt; ricardo.silva@estesl.ipl.pt

G. Marques
SAMS Hospital and CUF Hospital, Lisbon, Portugal
e-mail: gnpmarques@gmail.com

J. Lita da Silva et al. (eds.), *Advances in Regression, Survival Analysis, Extreme Values, Markov Processes and Other Statistical Applications*, Studies in Theoretical and Applied Statistics, DOI 10.1007/978-3-642-34904-1_47,

1 Introduction

Chest radiography is the most often performed exam in Portugal on the radiology area. Due to the fact that in thoracic region there exist areas with different densities and contrasts, it is important to adopt the best techniques and procedures in the execution of chest X-ray, in order to produce images with high diagnostic quality.

Radiographs were obtained in a chest phantom. The phantom includes the heart, the lungs, the liver and the thorax skeleton. The images were acquired using a chest phantom for two reasons: ethical (avoiding unnecessary exposures to patients) technical (due to the need to control the variations in lesion localization and tube potential intensities).

The adjustment of the energy of the X-ray beam is an important practical action to be taken by the radiographer to contribute for a valuable clinical image [9]. So, this study aims to investigate the readers' performances and the accuracy on the detection of simulated chest lesions in a computed radiography (CR) system associated with different tube potential (kV).

2 Materials and Methods

The radiological images were obtained using a chest phantom at the radiology skill lab of the Higher School of Health Technology of Lisbon. This phantom has regions with different densities corresponding to the structures which constitute a human chest. To achieve an overall representativeness of the chest, we selected six different regions where the lesions were placed. We also considered five tube potential intensities (81 kV, 90 kV, 109 kV, 125 kV and 141 kV). These values agree with the ones used elsewhere [8–11].

Seven trials were considered, six corresponding to the phantom with each lesion at a time and one with the phantom without lesions. Each trial contained five images, corresponding to each tube potential intensity, resulting the experiment in a total of 35 radiographs. The images were randomly observed by six radiologists (readers) who did not know where the lesion positions were and if the images had lesion or not. The readers gave to each image a score according to a five points scale (1—very confident case is normal, 2—confident case is normal, 3—somewhat confident case is abnormal, 4—confident case is abnormal and 5—very confident case is abnormal). All readers evaluated the images obtained using all tube potential intensities. This study is namely fully crossed design [16] (Table 1).

A receiver operating characteristics (ROC) graph is a technique for visualizing, organizing and selecting classifiers based on their performance. ROC graphs have long been used in signal detection theory to depict the trade off between hit rates and false alarm rates of classifiers [19]. ROC analysis has been extended for use in visualizing and analysing the behaviour of diagnostic systems [20].

Metz [12] proposed a binormal model for estimating ROC curves when we have rating data and a latent variable under the binormal assumption is used to construct

Table 1 Fully crossed experimental design

	81 kV			i kV			141 kV		
	Rdr. 1	Rdr. j	Rdr. 6	Rdr. 1	Rdr. j	Rdr. 6	Rdr. 1	Rdr. j	Rdr. 6
No lesion	x_{110}	x_{1j0}	x_{160}	x_{i10}	x_{ij0}	x_{i60}	x_{510}	x_{5j0}	x_{560}
Lesion 1	x_{111}	x_{1j1}	x_{161}	x_{i11}	x_{ij1}	x_{i61}	x_{511}	x_{5j1}	x_{561}
Lesion k	x_{11k}	x_{1jk}	x_{16k}	x_{i1k}	x_{ijk}	x_{i6k}	x_{51k}	x_{5jk}	x_{56k}
Lesion 6	x_{116}	x_{1j6}	x_{166}	x_{i16}	x_{ij6}	x_{i66}	x_{516}	x_{5j6}	x_{566}

x_{ijk} is the score given by reader (Rdr.) j to an image with a lesion on position k and obtained with a potential intensity i

a smooth curve. When degenerated ROC curves are produced (curves that cross the 45° chance line), Metz and Pan [13] developed a proper binormal model and a software entitled PROPROC which uses that model to fit convex ROC curves by maximum likelihood estimation. This method is called proper because it forces the curve shape always to be convex. Accordingly, a ROC curve cannot drop below the 45° chance line.

One way to quantify the diagnostic accuracy of a classifier is to express his performance by a single number. The most common global measure is the area under the ROC curve (AUC). By convention, the AUC ranges between 0.5 (no apparent distributional difference between the two groups of test values) and 1 (perfect separation). To compare the performances between classifiers it is usual to select the one that corresponds to the maximum value of AUC.

The decision of which tube potential intensity produces X-rays with better quality, depends on the comparison of the accuracy with which the tube potential produces the images. However, identification of the lesions depends not only on the tube potential, but also on the readers. Therefore, we will have to consider variability between observers, variability inherent to lesions positions, correlation between intensities and correlation between readers for the same intensity.

ROC curves are the most commonly used tool to compare diagnostic systems in their ability to discriminate between two mutually exclusive populations (in this case presence vs no presence of a lesion). The MRMC (multiple cases multiple readers) [2–6,15] ROC curve is the most appropriate for this experiment, because we have to include in the analysis the variability between readers. MRMC methodology accounts for multiple readers, each one reading multiple cases. In general, we will assume that a reader analyses an image (case) and produces a value that reflects his confidence about the existence of a lesion.

Several statistical methods have been developed for analysing data using MRMC ROC curve methodology. Obuchowski et al. [15] compared five methods, namely, Dorfman–Berbaum–Metz (DBM) method [2], Obuchowski–Rockette (OR) method [14], Beiden–Wagner–Campbell (BWC) method [1], multivariate WMW statistic [18] and hierarchical ordinal regression for ROC curves (HROC) [7].

In this study, we used the DBM method. Dorfman et al. [2] proposed an ANOVA of pseudovalues to analyse multireader ROC data. Their basic idea is to compute jackknife pseudovalues. The jackknife pseudovalue of the k-th case is simply the weighted difference in the accuracy, estimated from all cases, minus the accuracy estimated without the kth case (1). These pseudovalues are transformations of the original data:

$$\hat{y_{ijk}} = c\hat{\theta}_{ij} - (c-1)\hat{\theta}_{ij(k)}, \tag{1}$$

where $\hat{y_{ijk}}$ denotes the estimated AUC pseudovalue for treatment (intensity) i, reader j and case (lesion) k; $\hat{\theta}_{ij}$ denotes the estimated AUC based on the ith treatment and j-th reader, for all cases; $\hat{\theta}_{ij(k)}$ denotes the estimated AUC based on the same data after removing the kth case; c is the total number of cases.

AUC pseudovalues are computed using the jackknife separately for each reader/treatment combination (2):

$$\widehat{AUC}_{ij} = \frac{1}{c}\Sigma_{k=1}^{c}\hat{y}_{ijk}. \tag{2}$$

A mixed-effects ANOVA model is performed (3) on the pseudovalues to test the null hypothesis that the mean accuracy of readers is the same for all intensities. The model is

$$y_{ijk} = \mu + \alpha_i + B_j + C_k + (\alpha B)_{ij} + (\alpha C)_{ik} + (BC)_{jk} + (\alpha BC)_{ijk} + \varepsilon_{ijk}, \tag{3}$$

where μ denotes the global mean; α_i denotes the fixed effect of treatment i; B_j denotes the random effect of reader j; C_k denotes the random effect of case k; the multiple symbols in parentheses denote interactions; and ε_{ijk} is the error term. The interaction terms are all random effects and they are assumed to be mutually independent and normally distributed with zero means and variances corresponding to each random effect.

In this study, accuracy is characterized by AUC, but any summary measure (e.g., sensitivity, specificity, partial area under the ROC curve and sensitivity at a fixed false-positive rate) can also be used. Furthermore, these measures of accuracy can be estimated parametrically or nonparametrically.

The software `MRMC DBM 2.2` from Medical Image Perception Laboratory [21] and Kurt Rossman Laboratories for Radiologic Image Research [22] implements this method.

3 Application

3.1 Comparison of Readers' Performances

To estimate the correspondent ROC curve for each reader (Fig. 1) we used the proper binormal ROC model [13], since the binormal model [12] produced degenerated ROC curves. The AUC for each reader was calculated under the proper binormal ROC model (Table 2).

Analysing Fig. 1 and Table 2 we verified that reader four had the best performance.

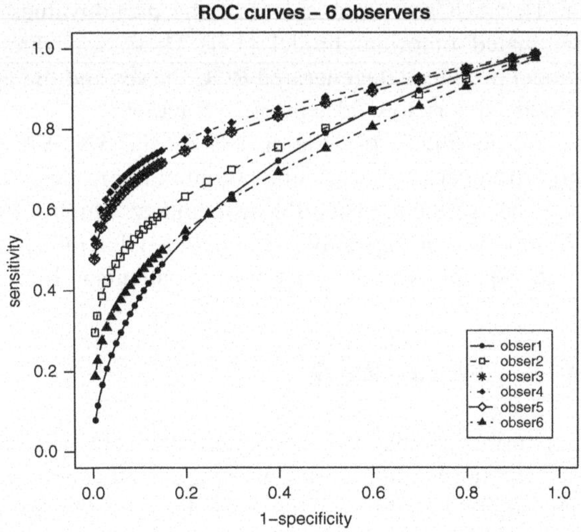

Fig. 1 ROC curves corresponding to each reader

Table 2 AUCs under proper binormal assumption for the six readers

Reader	AUC
1	0.7289
2	0.7731
3	0.8432
4	0.8593
5	0.8424
6	0.7207

3.2 Comparison of Tube Potential Accuracies

Typically, MRMC ROC curves involve c patients (with or without disease), $t \geq 2$ treatments and r readers. In our case we have $c = 7$ (phantom with no lesions and phantom with each lesion position), $t = 5$ tube potential intensities and $r = 6$ radiologists. In this experiment the readers and the lesions were not selected randomly, so, in order to test the hypothesis that the mean value of the AUC pseudovalues values did not differ significantly, ($H_0 : \alpha_1 = \alpha_2 = \ldots = \alpha_5$), we used the factorial model with fixed effects without replicates:

$$y_{ijk} = \mu + \alpha_i + \beta_j + \gamma_k + (\alpha\beta)_{ij} + (\alpha\gamma)_{ik} + (\beta\gamma)_{jk} + \varepsilon_{ijk}. \qquad (4)$$

The `DBM MRMC 2.2` software only considers the following types of analysis: (i) both readers and cases as random samples (the traditional MRMC analysis), (ii) only cases as random samples and (iii) treating readers as random samples. In our experiment we do not have any of these situations. Hence, we used `DMB MRMC 2.2` to estimate the AUC pseudovalues and `R` software to make the variance analysis of factorial fixed effects.

The AUCs used to calculate the pseudovalues were estimated using the contaminated binormal model [17]. These was due to the fact that the binormal model produced degenerated ROC curves and the application of PROPROC model produced very low false positive fractions.

We obtained a p-value of 0.89 for the ANOVA and we concluded that there are no differences between tube potential intensities. Naturally, we chose the one with less risk for the patients. Because higher values of the tube potential correspond to a lower radiation exposure, we concluded that tube potential intensity which produces higher quality with less risk for the patients is 141 kV.

4 Future Work

For further studies we recommend the selection and randomization of a previous defined number of readers and images so that a desired power may be achieved [20]. The model should take into account not only the lesions but also their locations because it is important that a reader correctly identifies a lesion as well as its location.

Acknowledgements Research was partially sponsored by national funds through the Fundação Nacional para a Ciência e Tecnologia, Portugal FCT under the project (PEst-OE/MAT/UI0006/2011), and by the FCT PhD scholarship SFRH/BD/45938/2008. The authors wish to thank an anonymous referee for all valuable suggestions on this manuscript.

References

1. Beiden, S.V., Wagner R.F., Campbell, G.: Components-of-variance models and multiple-bootstrap experiments: an alternative method for random-effects, receiver operating characteristic analysis. Acad. Radiol. **7**, 341–349 (2000)
2. Dorfman, D.D., Berbaum, K.S., Metz, C.E.: Receiver operating characteristic rating analysis. Generalization to the population of readers and patients with the jackknife method. Invest. Radiol. **27**, 723–731 (1992)
3. Hillis, S.L.: Monte Carlo validation of the Dorfman-Berbaum-Metz method using normalized pseudovalues and less data-based model simplification. Acad. Radiol. **12**, 1534–1541 (2005)
4. Hillis, S.L.: A comparison of denominator degrees of freedom for multiple observer ROC analysis. Stat. Med. **26**, 596–619 (2007)
5. Hillis, S.L., Berbaum, K.S.: Power estimation for the Dorfman-Berbaum-Metz method. Acad. Radiol. **11**, 1260–1273 (2004)
6. Hillis, S.L., Obuchowski, N.A., Schartz, K.M., Berbaum, K.S.: A comparison of the Dorfman-Berbaum-Metz and Obuchowski-Rockette methods for receiver operating characteristic (ROC) data. Stat. Med. **24**, 1579–1607 (2005)
7. Ishawaran, H., Gatsonis, C.A.: A general class of hierarchical ordinal regression models with applications to correlated ROC analysis. Can. J. Stat. **28**, 731–750 (2000)
8. Kimme-Smith, C., Aberle, D.R., Sayre, J.W., Hart, E.M., Greaves, S.M., Brown, K., Young, D.A., Deseran, M.D., Johnson, T., Johnson, S.L.: Effects of reduced exposure on computed radiography: comparison of nodule detection accuracy with conventional and asymmetric screen-film radiographs of a chest phantom. Am. J. Roentgenol. **165**, 269–273 (1995)

9. Lança, L., Silva, A., Alves, E., Serranheira, F., Correia, M.: Evaluation of Exposure Parameters in Plain Radiography: A Comparative Study with European Guidelines. Radiat. Prot. Dosim. (2007) doi:10.1093/rpd/ncn144
10. Lança, L., Silva, R., Marques, G., Silva-Fortes, C.: The influence of the tube potential (kV) on the detection of simulated chest lesions in a CR system. Eur. Radiol. **19**(S1):S292 (2009)
11. McEntee, M.F., Brennan, P.C., Connor, G.O.: The effect of X-ray tube potential on the image quality of PA chest radiographs when using digital image acquisition devices. Radiography **10**, 287–292 (2004)
12. Metz, C.E., Herman, B.A., Shen, J.H.: Maximum likelihood estimation of receiver operating characteristic (roc) curves from continuously-distributed data. Stat. Med. **17**, 1033–1053 (1998)
13. Metz, C.E., Pan, X.: Proper binormal ROC curves: Theory and maximum-likelihood estimation. J. Math. Psycho. **43**, 1–33 (1999)
14. Obuchowski, N.A., Rockette, H.E.: Hypothesis testing of the diagnostic accuracy for multiple diagnostic tests: an ANOVA approach with dependent observations. Commun. Stat. Simulat. **24**, 285–308 (1995)
15. Obuchowski, N.A., Beiden, S., Berbaum, K.S., Hillis, S.L., Ishwaran, H., Song, H.H., Wagner, R.F.: Multireader, multicase receiver operating characteristic analysis. Acad. Radiol. **11**, 980–995 (2004)
16. Roe, C.A., Metz, C.E.: Dorfman-Berbaum-Metz method for statistical analysis of multireader, multimodality receiver operating characteristic data: Validation with computer simulation. Acad. Radiol. **4**, 298–303 (1997)
17. Roe, C.A., Metz, C.E.: Variance-component modeling of receiver operating characteristic index estimates. Acad. Radiol. **4**, 587–600 (1997)
18. Song, H.H.: Analysis of correlated ROC area in diagnostic testing. Biometrics **53**, 370–382 (1997)
19. Sweets, J.A.: ROC curve analysis applied to the evaluation of medical imaging techniques. Invest. Radiol. **14**, 109–121 (1979)
20. Sweets, J.A.: Measuring the accuracy of diagnostic systems. Science **240**, 1285–1293 (1988)
21. http://perception.radiology.uiowa.edu. Cited 1 Set 2009
22. http://xray.bsd.uchicago.edu/krl. Cited 1 Set 2009

Some Remarks About Gibbs Variable Selection Performance

Júlia Teles and Maria Antónia Amaral Turkman

Abstract
Gibbs variable selection is one of the Bayesian approaches to the variable selection problem in generalized linear models and, in particular, in linear regression. One of the advantages of this method is that it can be easily implemented in WinBUGS. The results obtained after Gibbs sampling convergence enable us to estimate, in a straightforward manner, the posterior model probabilities and the posterior variable inclusion probabilities. These probabilities allow us to identify the maximum a posteriori model and, if it exists, the median probability model, respectively. A simulation study was performed to study the importance of sample dimension and the number of predictors in the Gibbs variable selection performance in the scope of linear regression models. The results attained suggest that Gibbs variable selection is more demanding in terms of minimum sample sizes requirements than other well-known techniques.

1 Introduction

Model selection and, in particular, variable selection in regression problems is a much studied and discussed matter. In the classical approach, several solutions are often used. Usually they combine a search method, like forward selection,

J. Teles (✉)
CIPER and Mathematics Unit, Technical University of Lisbon, Estrada da Costa, 1495-688 Cruz Quebrada-Dafundo, Portugal
e-mail: jteles@fmh.utl.pt

M.A.A. Turkman
CEAUL and DEIO, University of Lisbon, Edifício C6, Piso 4, Campo Grande, 1749-016 Lisboa, Portugal
e-mail: maturkman@fc.ul.pt

J. Lita da Silva et al. (eds.), *Advances in Regression, Survival Analysis, Extreme Values, Markov Processes and Other Statistical Applications*, Studies in Theoretical and Applied Statistics, DOI 10.1007/978-3-642-34904-1_48,

backward elimination or stepwise, with a model evaluation criteria, such as partial F-tests, adjusted R squared, mean squared error, Mallows' C_p statistic, and penalized likelihood criteria [10].

The Bayesian approach to variable selection in regression problems includes several types of methods. Some methods, inspired by those used in the classical approach, combine a search method with a model evaluation criteria, usually a penalized likelihood criteria or a discrepancy function [2]. The informal manual selection [14], and the forward variable selection or the backward variable elimination via DIC [18], are examples of these type of variable selection methods. The Monte Carlo model composition [9] and the "Occam's Window" method [8] are examples of model selection methods based on the search in the model space. However, the intrinsically Bayesian solution to the variable selection problem is often viewed as a problem of parameter estimation, i.e., the estimation of the posterior model probabilities and the estimation of the posterior variable inclusion probabilities. The use of Gibbs sampler and the inclusion of a vector of binary indicator variables in the sampling algorithm allow us to estimate the above-mentioned probabilities. This strategy gave rise to several Bayesian methods for variable selection where the search is over the model and the parameter space jointly [2], namely stochastic search variable selection [5], unconditional priors' method [7], Gibbs variable selection [3], and the reversible jump MCMC [6].

Dellaportas et al. [4] reviewed some Bayesian variable selection methods based on Gibbs sampling and emphasized that their main advantages is that they can be easily implemented in WinBUGS. These authors mentioned that it is impossible to provide guidelines for the most appropriate method for the variable selection problem within generalized linear models framework. O'Hara and Sillanpää [13] also carried out a review of these methods and stated that stochastic search variable selection and reversible jump MCMC methods can all provide good results. However, they claimed that the choice of the better method depends on the choice of the priors and the method of implementation used.

The focus of this chapter will be on the Gibbs variable selection (GVS) method. In this work, a simulation study was performed to investigate the influence of sample dimension and the number of predictors in the GVS performance in the context of linear regression models; besides that, an example was used to illustrate some issues on this subject.

2 Gibbs Variable Selection

Let n represent the number of subjects and p the number of predictors. Denoting by Y_i the response variable, by $x_{i1}, \ldots, x_{ip}$ the observations of p predictors, and by $\boldsymbol{\beta} = (\beta_0, \beta_1, \ldots, \beta_p)$ the vector of unknown parameters, the linear regression model for the ith subject is $Y_i = \beta_0 + \beta_1 x_{i1} + \cdots + \beta_p x_{ip} + \epsilon_i$, $i = 1, \ldots, n$, where the random errors ϵ_i are independent and identically distributed $\mathrm{N}(0, \sigma^2)$, with $\tau = 1/\sigma^2$.

Considering that the uncertainty about the model only refers to the choice of the predictors being included in the linear predictor, assuming that the constant term is included in the model and that possible interactions are not taken into account, the set of all regression models under consideration, represented by $\mathcal{M}$, is such that $\#\mathcal{M} = 2^p$. In the GVS method, each model $m \in \mathcal{M}$ is identified by a vector $\boldsymbol{\gamma} = (\gamma_1, \ldots, \gamma_p)$, with $\gamma_j = 1$ if the jth predictor is included in the model and $\gamma_j = 0$ otherwise, $j = 1, \ldots, p$. The linear predictor of the linear regression model for the ith subject is $\eta_i = \beta_0 + \gamma_1\beta_1 x_{i1} + \cdots + \gamma_p\beta_p x_{ip}$, $i = 1, \ldots, n$. The components of the vector $\boldsymbol{\gamma}$ are included in the modelling process through the linear predictor and through the prior distribution of β_j conditional to γ_j, for $j = 1, \ldots, p$ [4, 11]. As the vector $\boldsymbol{\gamma}$ identifies the model $m \in \mathcal{M}$, the samples $\{\boldsymbol{\gamma}^{(t)}\}_{t=1}^N$, obtained after the convergence of MCMC algorithm, enable us to estimate the posterior probability of each model m:

$$\widehat{\Pr}(m \mid \mathcal{D}) = \frac{\#\{t : \boldsymbol{\gamma}^{(t)} = \boldsymbol{\gamma}\}}{N}, \tag{1}$$

where $\mathcal{D}$ represents the data. These probabilities allow us to identify the maximum *a posteriori* (MAP) model [12], i.e., the model with the highest posterior probability. Estimates of posterior variable inclusion probabilities are also easily obtained:

$$\hat{\gamma}_j = \widehat{\Pr}(\beta_j \neq 0 | \mathcal{D}) = \frac{\#\{t : \gamma_j^{(t)} = 1\}}{N}. \tag{2}$$

With these probabilities we can identify, if it exists, the median probability (MP) model, that is defined as the model that includes the variables whose posterior inclusion probabilities are at least 0.5 [1].

In the absence of information about the importance of each predictor it is usual to consider γ_j independent and identically distributed Bern(1/2), i.e., a prior probability of $1/2^p$ for each model. We assume independent priors for $\beta_0, \beta_1, \ldots, \beta_p$, τ, and $\boldsymbol{\gamma}$. After centering and scaling x_{ij} we consider that (i) the prior distribution of β_0 is normal with zero mean and large variance (say equal to 100000); (ii) the prior distribution of β_j, conditional on γ_j, is normal with zero mean and variance equal to $0.01^{1-\gamma_j} \times 1000$, $j = 1, \ldots, p$. Note that if the jth predictor is to be included in the model then the prior variance should be large, so that the prior distribution of β_j is sufficiently diffuse; if the jth predictor is to be excluded from the model then the prior distribution of β_j has all its mass around zero; (iii) the prior distribution of τ is diffuse gamma with both shape and inverse scale parameter equal to 0.001 [4, 11].

3 Example

The aerobic fitness data [15] is an example of data that have been used by several authors to illustrate variable selection methods. The aerobic fitness, which is measured by the ability to consume oxygen, could be fit by the results of some

Table 1 Estimates of posterior variable inclusion probabilities and posterior model probabilities

		Variables included in the model					
Model	$\widehat{\Pr}(m_{i^{th}} \mid \mathscr{D})$	x_1	x_2	x_3	x_4	x_5	x_6
$m_{1^{st}}$	0.812			•			
$m_{2^{nd}}$	0.044	•		•			
$m_{3^{rd}}$	0.038			•		•	
$m_{4^{rd}}$	0.032			•		•	•
$m_{5^{rd}}$	0.018	•		•		•	
…	…						
$m_{9^{th}}$	0.004	•		•		•	•
…	…						
$\hat{\gamma}_j = \widehat{\Pr}(\beta_j \neq 0 \mid \mathscr{D})$		0.071	0.017	1.000	0.018	0.095	0.055

trivial exercise tests, instead of expensive and cumbersome oxygen consumption measurements. So, the goal is to develop a model to predict aerobic fitness based on the results of some exercise tests. The response variable is oxygen uptake rate (y) measured in ml per kg body weight per minute. The candidate predictors are age in years (x_1), weight in kg (x_2), time in minutes to run 1.5 miles (x_3), heart rate while resting (x_4), heart rate while running (x_5), and maximum heart rate records while running (x_6). The data consist of $n = 31$ observations of these seven variables. The Pearson correlation coefficients between some of the predictors are high, in particular, as expected, the correlation between x_5 and x_6.

The estimates of posterior variable inclusion probabilities and the estimates of posterior model probabilities are presented in Table 1. There is only one variable with posterior inclusion probability greater than 0.5, the variable x_3, and so we can say that the MP model only includes this variable. This model is the one that stands out by presenting the highest posterior probability, i.e., $m_{1^{st}}$ is the MAP model and, in this particular case, it stands out by presenting a posterior probability much higher than the other models.

In this example, the number of possible models is $2^6 = 64$. So, the choice of the model that minimizes the DIC [17] and $\widehat{\text{BIC}}$ [2] measures was made using forward variable selection (FVS) and backward variable elimination (BVE) methods. The FVS and the BVE procedures via DIC favored the model $m_{9^{th}}$, i.e., the model that includes the variables x_1, x_3, x_5, and x_6. When using the $\widehat{\text{BIC}}$ as model evaluation criterion, the two above-mentioned search methods did not lead to the choice of the same model: the FVS favored model $m_{1^{st}}$, which includes the variable x_3, and the BVE favored model $m_{5^{th}}$, which includes variables x_1, x_3, and x_5. When there are high correlations between the predictors, it is usual that these two methods yield different models [10].

Table 2 Simulation results for $p = 2$ and $n = 25$

	Model that gave rise to data			
	m_1	m_2	m_3	m_4
	x_1, x_2	x_1	x_2	const.
$\widehat{\Pr}(m_1 \mid \mathscr{D})$	0.145	0.012	0.005	0.000
$\widehat{\Pr}(m_2 \mid \mathscr{D})$	0.416	0.695	0.016	0.019
$\widehat{\Pr}(m_3 \mid \mathscr{D})$	0.075	0.006	0.277	0.017
$\widehat{\Pr}(m_4 \mid \mathscr{D})$	0.364	0.287	0.702	0.964
$\hat{\gamma}_1 = \widehat{\Pr}(\beta_1 \neq 0 \mid \mathscr{D})$	0.561	0.708	0.021	0.019
$\hat{\gamma}_2 = \widehat{\Pr}(\beta_2 \neq 0 \mid \mathscr{D})$	0.220	0.018	0.282	0.017
$\widehat{\text{BIC}}$ selection	m_1	m_2	m_3	m_4
$\widehat{\text{BIC}}$ value	121.7	117.5	118.3	114.1
DIC selection	m_1	m_2	m_3	m_4
DIC value	113.0	110.9	111.7	109.7

4 Simulation Study

A simulation study was carried out to evaluate the performance of GVS under several combinations of sample sizes and number of predictors. The data generation scheme was adapted from Kuo and Mallick [7]. We considered p predictors, $x_1 \dots, x_p$, obtained through the generation of p independent n-dimensional vectors of a standard normal random variable. A random sample of n observations from $\text{N}(0, 2.5^2)$ was assigned to the random error and for each combination of predictors, a sample of the response variable for the model m_k, $k = 1, \dots, 2^p$, was obtained, giving the same weight to the predictors. Several simulation conditions were considered: $p = 2, 3, 4$ and $n = 25, 50, 75, 100, 125, 150, 175, 200$. Using WinBUGS, v. 1.4.3, GVS was applied to each sample, and estimates of posterior model probabilities and posterior variable inclusion probabilities were obtained. We used the MAP and MP models to evaluate the performance of GVS under the simulated conditions.

In this section, only the most relevant results of the simulation study will be presented and commented.

We start by examining the results of the case $p = 2$ and $n = 25$, presented in Table 2. The models from which the data were generated and the variables included in each one of these models were indicated in the header of the table. The following comments may help in interpreting the table. When the data is generated from the model m_1 (that includes x_1 and x_2), the MAP model is m_2 (that only includes x_1), with a posterior model probability equal to 0.416, followed by m_4 that only includes the constant term. In the case of data generated from the model m_2, the MAP model is actually m_2, with a posterior model probability equal to 0.695. If the data is generated from m_3 (that only includes x_2), the MAP model is m_4. In the case of data generated from m_4, GVS method gives the highest posterior probability to this

Table 3 Simulation results for $p = 2$ and $n = 75$

	Model that gave rise to data			
	m_1	m_2	m_3	m_4
	x_1, x_2	x_1	x_2	const.
$\widehat{\Pr}(m_1\|\mathscr{D})$	0.615	0.014	0.010	0.000
$\widehat{\Pr}(m_2\|\mathscr{D})$	0.003	0.571	0.000	0.010
$\widehat{\Pr}(m_3\|\mathscr{D})$	0.377	0.010	0.987	0.026
$\widehat{\Pr}(m_4\|\mathscr{D})$	0.005	0.405	0.003	0.964
$\hat{\gamma}_1 = \widehat{\Pr}(\beta_1 \neq 0\|\mathscr{D})$	0.618	0.586	0.010	0.010
$\hat{\gamma}_2 = \widehat{\Pr}(\beta_2 \neq 0\|\mathscr{D})$	0.993	0.024	0.997	0.027
$\widehat{\text{BIC}}$ selection	m_1	m_2	m_3	m_4
$\widehat{\text{BIC}}$ value	377.4	374.1	372.3	369.0
DIC selection	m_1	m_2	m_3	m_4
DIC value	364.2	364.2	362.4	362.3

model. In the second part of the table, we present estimates of posterior variable inclusion probabilities. When the data is generated from m_1, the posterior inclusion probabilities for variables x_1 and x_2 are, respectively, 0.561 and 0.220. In this case, the MP model is m_2, which matches with the MAP model. In the third and fourth parts of the table are indicated the models selected by $\widehat{\text{BIC}}$ and DIC criteria, with the corresponding criteria values. With the data generated from each of the models, using the exhaustive search method and the above-mentioned penalized likelihood criteria, the selected model matches the model from which the data was generated. In the case $p = 2$ and $n = 50$ (table of results omitted), the performance of GVS method is quite similar to the previous case. Only the models selected by the $\widehat{\text{BIC}}$ criterion differ from the previous case. When considering $p = 2$ and $n = 75$ (Table 3), the MAP models coincide with the models from which the data were generated, though in some cases the probabilities of the models with the highest posterior probabilities are not much higher than 0.5. In the case $p = 2$ and $n = 100$ (table of results omitted), the performance of GVS increases and the probabilities of the models with the highest posterior probabilities are also a little closer to one.

Let us now consider what happens when we have three possible predictors. In this case we have to monitor eight models and the table to summarize the results is more complex. In the case $p = 3$ and $n = 25$ (table of results omitted), regardless of the model from which the data was generated, the MAP model is the one that only includes the constant. The GVS method is unable to identify the model from which the data was generated. The posterior variable inclusion probabilities do not present any value greater than 0.5, so the MP model is never identified. When considering $p = 3$ and $n = 50$ (table of results omitted), the performance of the GVS method is not very distinct from the previous case. In the case $p = 3$ and $n = 75$ (Table 4), the performance of the GVS method has a little improvement. In some situations (when the models that gave rise to data are m_3, m_5, m_7, and m_8) the MAP model and the MP model were correctly identified; however, the results attained are still poor. For $p = 3$ and $n = 100$ (Table 5), the performance of the GVS method improves

Table 4 Simulation results for $p = 3$ and $n = 75$

	Model that gave rise to data							
	m_1	m_2	m_3	m_4	m_5	m_6	m_7	m_8
	x_1,x_2,x_3	x_1,x_2	x_1,x_3	x_2,x_3	x_1	x_2	x_3	const.
$\widehat{\Pr}(m_1\|\mathscr{D})$	0.156	0.001	0.010	0.002	0.000	0.000	0.000	0.000
$\widehat{\Pr}(m_2\|\mathscr{D})$	0.025	0.189	0.001	0.000	0.010	0.002	0.000	0.000
$\widehat{\Pr}(m_3\|\mathscr{D})$	0.689	0.008	0.854	0.009	0.009	0.000	0.010	0.000
$\widehat{\Pr}(m_4\|\mathscr{D})$	0.001	0.000	0.000	0.175	0.000	0.001	0.009	0.000
$\widehat{\Pr}(m_5\|\mathscr{D})$	0.124	0.797	0.129	0.001	0.976	0.009	0.001	0.011
$\widehat{\Pr}(m_6\|\mathscr{D})$	0.000	0.001	0.000	0.024	0.000	0.209	0.001	0.010
$\widehat{\Pr}(m_7\|\mathscr{D})$	0.004	0.000	0.004	0.682	0.000	0.008	0.865	0.009
$\widehat{\Pr}(m_8\|\mathscr{D})$	0.001	0.004	0.002	0.107	0.005	0.771	0.114	0.970
$\hat{\gamma}_1 = \widehat{\Pr}(\beta_1 \neq 0\|\mathscr{D})$	0.994	0.996	0.994	0.012	0.995	0.012	0.011	0.011
$\hat{\gamma}_2 = \widehat{\Pr}(\beta_2 \neq 0\|\mathscr{D})$	0.182	0.192	0.011	0.202	0.011	0.212	0.010	0.010
$\hat{\gamma}_3 = \widehat{\Pr}(\beta_3 \neq 0\|\mathscr{D})$	0.849	0.009	0.868	0.868	0.009	0.009	0.884	0.009
$\widehat{\text{BIC}}$ selection	m_1	m_2	m_3	m_4	m_5	m_6	m_7	m_8
$\widehat{\text{BIC}}$ value	359.9	354.5	355.1	355.0	349.7	349.7	350.2	344.8
DIC selection	m_1	m_2	m_3	m_4	m_5	m_6	m_7	m_8
DIC value	343.3	341.3	341.9	341.8	339.8	339.7	340.2	338.1

Table 5 Simulation results for $p = 3$ and $n = 100$

	Model that gave rise to data							
	m_1	m_2	m_3	m_4	m_5	m_6	m_7	m_8
	x_1,x_2,x_3	x_1,x_2	x_1,x_3	x_2,x_3	x_1	x_2	x_3	const.
$\widehat{\Pr}(m_1\|\mathscr{D})$	0.438	0.010	0.004	0.009	0.000	0.000	0.000	0.000
$\widehat{\Pr}(m_2\|\mathscr{D})$	0.281	0.817	0.004	0.007	0.008	0.017	0.000	0.000
$\widehat{\Pr}(m_3\|\mathscr{D})$	0.107	0.002	0.627	0.003	0.012	0.000	0.012	0.000
$\widehat{\Pr}(m_4\|\mathscr{D})$	0.001	0.000	0.000	0.492	0.000	0.010	0.005	0.000
$\widehat{\Pr}(m_5\|\mathscr{D})$	0.173	0.169	0.362	0.005	0.978	0.004	0.008	0.020
$\widehat{\Pr}(m_6\|\mathscr{D})$	0.000	0.001	0.000	0.291	0.000	0.858	0.004	0.008
$\widehat{\Pr}(m_7\|\mathscr{D})$	0.000	0.000	0.002	0.074	0.000	0.001	0.626	0.011
$\widehat{\Pr}(m_8\|\mathscr{D})$	0.000	0.001	0.001	0.119	0.002	0.110	0.345	0.961
$\hat{\gamma}_1 = \widehat{\Pr}(\beta_1 \neq 0\|\mathscr{D})$	0.998	0.999	0.998	0.024	0.998	0.021	0.020	0.020
$\hat{\gamma}_2 = \widehat{\Pr}(\beta_2 \neq 0\|\mathscr{D})$	0.720	0.829	0.008	0.800	0.008	0.885	0.009	0.008
$\hat{\gamma}_3 = \widehat{\Pr}(\beta_3 \neq 0\|\mathscr{D})$	0.546	0.012	0.633	0.579	0.012	0.012	0.643	0.011
$\widehat{\text{BIC}}$ selection	m_1	m_2	m_3	m_4	m_5	m_6	m_7	m_8
$\widehat{\text{BIC}}$ value	496.5	491.8	490.9	492.9	486.2	488.1	487.2	482.4
DIC selection	m_1	m_2	m_3	m_4	m_5	m_6	m_7	m_8
DIC value	478.5	477.4	476.5	478.5	475.4	477.3	476.4	475.2

significantly. For the data generated from any of the models, the MAP model is the model that gave rise to data. In this case, when there is a MP model, it coincides with the MAP model.

The tables of results for $p = 4$ (four possible predictors) are not presented here; nevertheless the simulations results lead us to think that we need a minimum sample size of 150 observations to achieve good results.

5 Discussion

The simulation results show that sample size has an important role in the performance of GVS. Smaller sample sizes tend to favor models with less predictors than the ones that gave rise to data. It seems that a balance between sample size and the number of candidate predictors is essential to achieve good results. In the case $p = 2$, a sample of size $n = 50$ is definitely insufficient, but with $n = 75$ the results are satisfactory; for $p = 3$, a sample of size $n = 75$ does not lead to good results, but with $n = 100$ the results suggest that GVS performs well in identifying the correct model; for $p = 4$, it seems that a minimum sample size of $n = 150$ is required to attain good results. The simulation results suggest that, to have a good performance of GVS, we need at least 35 data points for each variable in the model. However, more simulation should be made for $p > 4$. From a classical point of view, Sheskin [16] summarized several rules for the reliability of linear regression results. One of these rules is a minimum of ten observations for each candidate predictor. The simulation study results seem to point out that GVS is more demanding in terms of minimum sample sizes requirements than the classical linear regression techniques.

Acknowledgments Research partially sponsored by national funds through the Fundação Nacional para a Ciência e Tecnologia, Portugal – FCT, under the projects PEst-OE/SAU/UI0447/2011 and PEst-OE/MAT/UI0006/2011.

References

1. Barbieri, M.M., Berger, J.O.: Optimal predictive model selection. Ann. Statist. **32**, 870–897 (2003)
2. Carlin, B.P., Louis, T.A.: Bayes and empirical Bayes methods for data analysis. Chapman and Hall/CRC, New York (2000)
3. Dellaportas, P., Forster, J.J., Ntzoufras I.: On Bayesian model and variable selection using MCMC. Statist. Comput. **12**, 27–36 (2002)
4. Dellaportas, P., Forster, J.J., Ntzoufras, I.: Bayesian variable selection using the Gibbs sampler. In: Dey, D., Ghosh, S., Mallick, B. (eds.) Generalized Linear Models: A Bayesian Perspective, pp. 271–286. Marcel Dekker, New York (2000)
5. George, E.I., McCulloch, R.E.: Variable selection via Gibbs sampling. J. Am. Stat. Assoc. **88**, 881–889 (1993)
6. Green, P.J.: Reversible jump Markov chain Monte Carlo computation and Bayesian model determination. Biometrika **82**(4), 711–732 (1995)
7. Kuo, L., Mallick, B.: Variable selection for regression models. Sankhya B **60**, 65–81 (1998)

8. Madigan, D., Raftery, A.E.: Model selection and accounting for model uncertainty in graphical models using Occam's window. J. Am. Stat. Assoc. **89**, 1535–1546 (1994)
9. Madigan, D., York, J.: Bayesian graphical models for discrete data. Int. Stat. Rev. **63**, 215–232 (1995)
10. Montgomery, D.C., Peck, E.A., Vining, G.G.: Introduction to Linear Regression Analysis, 4th edn. Wiley Series in Probability and Statistics. Hoboken, New Jersey (2006)
11. Ntzoufras, I.: Gibbs variable selection using BUGS. J. Statist. Soft. **7**, 1–19 (2002)
12. Ntzoufras, I.: Bayesian Modeling Using WinBUGS. Wiley Series in Computational Statistics. Hoboken, New Jersey (2009)
13. O'Hara, R.B., Sillanpää, M.J.: A review of Bayesian variable selection methods: what, how and which. Bayesian Anal. **4**(1), 85–118 (2009) doi: 10.1214/09-BA403
14. Paulino, C.D., Amaral Turkman, M.A., Murteira, B.: Estatística Bayesiana. Fundação Calouste Gulbenkian, Lisboa (2003)
15. SAS Institute: SAS User's Guide: Statistics, v. 5. SAS Institute Inc., Cary, NC (1985)
16. Sheskin, D.J.: Handbook of Parametric and Nonparametric Statistical Procedures, 4th edn. Chapman and Hall/CRC, Boca Raton (2007)
17. Spiegelhalter, D.J., Best, N.G., Carlin, B.P., van der Linden, A.: Bayesian measures of model complexity and fit (with discussion). J. Roy. Stat. Soc. B **64**, 583–639 (2002)
18. Teles, J.: Uma Abordagem Bayesiana à Determinação de Modelos. PhD thesis, Faculdade de Motricidade Humana, Universidade Técnica de Lisboa, Portugal (2005)

Dependence of Multivariate Extremes

C. Viseu, L. Pereira, A.P. Martins, and H. Ferreira

Abstract
We give necessary and sufficient conditions for two sub-vectors of a random vector with a multivariate extreme value (MEV) distribution, corresponding to the limit distribution of the maximum of a multidimensional stationary sequence with extremal index, to be independent or totally dependent. Those conditions involve first relations between the multivariate extremal indices of the sequences and secondly a coefficient that measures the strength of dependence between both sub-vectors. The main results are illustrated with an auto-regressive sequence and a 3-dependent sequence.

1 Introduction

Multivariate extreme value (MEV) analysis is frequently applied in the context of modeling environmental data, for which the phenomenon of dependence is often intrinsic. This chapter focuses on the characterization of total dependence and of independence of two MEV distributions.

Let $\mathbf{X} = \{\mathbf{X}_n^{(d)} = (X_{n,1}, \ldots, X_{n,d})\}_{n \geq 1}$ be a d-dimensional stationary sequence with common distribution function (d.f.) $Q(\mathbf{x}^{(d)}) = Q(x_1, \ldots, x_d)$, $\mathbf{x}^{(d)} \in \mathbb{R}^d$, and $\mathbf{M}_n = (M_{n,1}, \ldots, M_{n,d})$ the vector of pointwise maxima, where $M_{n,i}$ is the maximum of ith component of $\mathbf{X}$. Denote $\widehat{\mathbf{M}}_n = (\widehat{M}_{n,1}, \ldots, \widehat{M}_{n,d})$ the corresponding vector of pointwise maxima of the associated d-dimensional sequence,

C. Viseu (✉)
Polytechnic Institute of Coimbra, Coimbra, Portugal
e-mail: cviseu@iscac.pt

L. Pereira · A.P. Martins · H. Ferreira
University of Beira Interior, Covilhã Portugal
e-mail: lpereira@ubi.pt; amartins@ubi.pt; helenaf@ubi.pt

J. Lita da Silva et al. (eds.), *Advances in Regression, Survival Analysis, Extreme Values, Markov Processes and Other Statistical Applications*, Studies in Theoretical and Applied Statistics, DOI 10.1007/978-3-642-34904-1_49,

$\widehat{\mathbf{X}} = \{\widehat{\mathbf{X}}_n^{(d)}\}_{n\geq 1}$, of independent and identically distributed (i.i.d.) random vectors having the same distribution function Q.

In this multivariate setting operations among vectors are defined componentwise, that is, for each $d > 1$ and $\mathbf{a}^{(d)}, \mathbf{b}^{(d)} \in \mathbb{R}^d$, $\mathbf{a}^{(d)} \leq \mathbf{b}^{(d)}$, if and only if $a_j \leq b_j$, for all $j = 1, 2, \ldots, d$.

If there exist sequences $\{(a_{n,1} > 0, \ldots, a_{n,d} > 0)\}_{n\geq 1}$ and $\{(b_{n,1}, \ldots, b_{n,d})\}_{n\geq 1}$, such that for $\mathbf{u}(\mathbf{x}^{(d)}) = \{\mathbf{u}_n(\mathbf{x}^{(d)}) = (a_{n,1}x_1 + b_{n,1}, \ldots, a_{n,d}x_d + b_{n,d})\}_{n\geq 1}$,

$$P\left(\widehat{\mathbf{M}}_n \leq \mathbf{u}_n(\mathbf{x}^{(d)})\right) = P\left(\bigcap_{j=1}^{d}\left\{\widehat{M}_{n,j} \leq a_{n,j}x_j + b_{n,j}\right\}\right) \xrightarrow[n\to\infty]{} G(\mathbf{x}^{(d)}),\ \mathbf{x}^{(d)} \in \mathbb{R}^d,$$

where G is a d.f. with nondegenerate margins, then Q is said to be in the max-domain of attraction of G ($Q \in D(G)$) and G is said to be a MEV distribution function.

We will assume, without loss of generality, that the univariate marginal distributions of G are equal to F.

It is well known that the relationship between the d.f. $G(\mathbf{x}^{(d)})$, $\mathbf{x}^{(d)} \in \mathbb{R}^d$, and its marginal distributions $F(x_j)$, $j = 1, \ldots, d$, can be characterized by its copula function which is a d.f. $D_G : [0,1]^d \to [0,1]$ that satisfies $D_G(F(x_1), \ldots, F(x_d)) = G(\mathbf{x}^{(d)})$, $\mathbf{x}^{(d)} \in \mathbb{R}^d$. The copula function exhibits a number of interesting properties, namely, its stability equation:

$$D_G^t(y_1, \ldots, y_d) = D_G(y_1^t, \ldots, y_d^t),\ \forall t > 0 \text{ and } (y_1, \ldots, y_d) \in [0,1]^d. \quad (1)$$

If the stationary sequence $\mathbf{X}$ satisfies some mixing conditions, $D(\mathbf{u}_n(\mathbf{x}^{(d)}))$ of Hsing [4] or $\Delta(\mathbf{u}_n(\mathbf{x}^{(d)}))$ of Nandagopalan [7], and

$$P\left(\mathbf{M}_n \leq \mathbf{u}_n(\mathbf{x}^{(d)})\right) \xrightarrow[n\to\infty]{} H(\mathbf{x}^{(d)}),\ \mathbf{x}^{(d)} \in \mathbb{R}^d,$$

where H is a d.f. with nondegenerate components, then H is also a MEV d.f.. The MEV d.f. H and G can be related through the multivariate extremal index function, $\theta(\boldsymbol{\tau}^{(d)}) = \theta(\tau_1, \ldots, \tau_d)$ introduced by Nandagopalan [7], which is a measure of clustering among the extreme values of a multivariate stationary sequence.

Definition 1. A d-dimensional stationary sequence $\mathbf{X}$ is said to have multivariate extremal index $\theta^{\mathbf{X}}(\boldsymbol{\tau}^{(d)}) \in [0,1]$, if for each $\boldsymbol{\tau}^{(d)} = (\tau_1, \ldots, \tau_d) \in \mathbb{R}_+^d$ there exists $\mathbf{u}_n^{(\boldsymbol{\tau}^{(d)})} = (u_{n,1}^{(\tau_1)}, \ldots, u_{n,d}^{(\tau_d)})$, $n \geq 1$, satisfying

$$nP(X_{1,j} > u_{n,j}^{(\tau_j)}) \xrightarrow[n\to\infty]{} \tau_j,\ j = 1, 2, \ldots, d,\ P\left(\widehat{\mathbf{M}}_n \leq \mathbf{u}_n^{(\boldsymbol{\tau}^{(d)})}\right) \xrightarrow[n\to\infty]{} \Psi(\boldsymbol{\tau}^{(d)}) \text{ and}$$

$$P\left(\mathbf{M}_n \leq \mathbf{u}_n^{(\boldsymbol{\tau}^{(d)})}\right) \xrightarrow[n\to\infty]{} \Psi^{\theta^{\mathbf{X}}(\boldsymbol{\tau}^{(d)})}(\boldsymbol{\tau}^{(d)}).$$

If $d = 1$ we say that $\mathbf{X}$ has extremal index θ, $0 \leq \theta \leq 1$, when $\forall \tau \in \mathbb{R}_+$, $\exists u_n^{(\tau)}$, $n \geq 1$, satisfying $nP(X_1 > u_n^{(\tau)}) \xrightarrow[n\to\infty]{} \tau$ and $P(M_n \leq u_n^{(\tau)}) \xrightarrow[n\to\infty]{} \exp(-\theta\tau)$ [5].

Similarly to one dimension, the multivariate extremal index is a key parameter that enables one to relate the properties of extreme values of a stationary sequence to those of independent random vectors from the same d-dimensional marginal distribution. However, unlike the one dimensional case, it is not a constant for the whole process, but instead depends on the vector $\boldsymbol{\tau}^{(d)}$.

It is now clear that the existence of $\theta^{\mathbf{X}}(\boldsymbol{\tau}^{(d)})$ allows us to write

$$H(\mathbf{x}^{(d)}) = G^{\theta^{\mathbf{X}}(\boldsymbol{\tau}^{(d)})}(\mathbf{x}^{(d)}) \quad \text{with} \quad \tau_j \equiv \tau_j(x_j) = -\log F(x_j),\ j = 1, 2, \ldots, d.$$

Taking $d = p + q$, it follows, as a consequence of the definition of the multivariate extremal index, that the sequences $\mathbf{X}^{(p)} = \{\mathbf{X}_n^{(p)} = (X_{n,1}, \ldots, X_{n,p})\}_{n\geq 1}$ and $\mathbf{X}^{(q)} = \{\mathbf{X}_n^{(q)} = (X_{n,p+1}, \ldots, X_{n,p+q})\}_{n\geq 1}$ have, respectively, extremal indexes

$$\theta^{\mathbf{X}^{(p)}}(\boldsymbol{\tau}^{(p)}) = \lim_{\substack{\tau_j \to 0+ \\ j=p+1,\ldots,p+q}} \theta^{\mathbf{X}}(\boldsymbol{\tau}^{(p+q)}) \quad \text{and} \quad \theta^{\mathbf{X}^{(q)}}(\boldsymbol{\tau}^{(q)}) = \lim_{\substack{\tau_j \to 0+ \\ j=1,\ldots,p}} \theta^{\mathbf{X}}(\boldsymbol{\tau}^{(p+q)}).$$

The marginal sequence $\{X_{ni}\}_{n\geq 1}$ has extremal index $\theta_i = \lim_{\substack{\tau_j \to 0+ \\ j\neq i}} \theta^{\mathbf{X}}(\boldsymbol{\tau}^{(p+q)})$, $\forall i = 1, \ldots, p + q$.

In the notation of the extremal index we shall omit the sequence, whenever it is clear by the context and the argument of the function.

Hereinafter, let $\mathbf{Y} = (Y_1, \ldots, Y_{p+q})$ and $\widehat{\mathbf{Y}} = (\widehat{Y}_1, \ldots, \widehat{Y}_{p+q})$ be, respectively, two random vectors with distribution functions $G^{\theta^{\mathbf{X}}(\boldsymbol{\tau}^{(p+q)})}$ and G, where $\mathbf{Y}^{(p)} = (Y_1, \ldots, Y_p)$ and $\mathbf{Y}^{(q)} = (Y_{p+1}, \ldots, Y_{p+q})$ denote two sub-vectors of $\mathbf{Y}$ and $\widehat{\mathbf{Y}}^{(p)} = (\widehat{Y}_1, \ldots, \widehat{Y}_p)$ and $\widehat{\mathbf{Y}}^{(q)} = (\widehat{Y}_{p+1}, \ldots, \widehat{Y}_{p+q})$ two sub-vectors of $\widehat{\mathbf{Y}}$.

In Sect. 2 we discuss conditions under which $\mathbf{Y}^{(p)}$ and $\mathbf{Y}^{(q)}$ are independent or totally dependent. These conditions are established first by relations between the extremal indices $\theta^{\mathbf{X}}(\boldsymbol{\tau}^{(p+q)})$, $\theta^{\mathbf{X}^{(p)}}(\boldsymbol{\tau}^{(p)})$, and $\theta^{\mathbf{X}^{(q)}}(\boldsymbol{\tau}^{(q)})$ and secondly by a coefficient that measures the strength of dependence between $\mathbf{Y}^{(p)}$ and $\mathbf{Y}^{(q)}$.

The main results are illustrated in Sect. 3 with an auto-regressive sequence and a 3-dependent sequence.

2 Main Results

If the d.f. Q belongs to the domain of attraction of a MEV distribution, G, and $\mathbf{X}$ has extremal index $\theta(\boldsymbol{\tau}^{(p+q)})$, $\boldsymbol{\tau}^{(p+q)} = (\tau_1, \ldots, \tau_{p+q}) \in \mathbb{R}_+^{p+q}$, then we have

$$G^{\theta(\boldsymbol{\tau}^{(p)})}(\mathbf{x}^{(p)})G^{\theta(\boldsymbol{\tau}^{(q)})}(\mathbf{x}^{(q)}) \leq G^{\theta(\boldsymbol{\tau}^{(p+q)})}(\mathbf{x}^{(p+q)}) \leq \min\{G^{\theta(\boldsymbol{\tau}^{(p)})}(\mathbf{x}^{(p)}), G^{\theta(\boldsymbol{\tau}^{(q)})}(\mathbf{x}^{(q)})\}, \tag{2}$$

for each $\mathbf{x}^{(p+q)} \in \mathbb{R}^{p+q}$ and $\tau_j = -\log F(x_j)$, $j = 1, \ldots, p + q$.

The inequality on the right of Eq. (2) holds true for every multivariate distribution, while the inequality on the left is a property of MEV distributions. The lower and upper bounds are achieved, respectively, when $\mathbf{Y}^{(p)}$ and $\mathbf{Y}^{(q)}$ are independent or totally dependent.

From Eq. (2) we obtain the following bounds for the multivariate extremal index function $\theta(\boldsymbol{\tau}^{(p+q)})$, $\boldsymbol{\tau}^{(p+q)} \in \mathbb{R}_+^{p+q}$.

$$\frac{\max\{\theta(\boldsymbol{\tau}^{(p)})\gamma(\boldsymbol{\tau}^{(p)}), \theta(\boldsymbol{\tau}^{(q)})\gamma(\boldsymbol{\tau}^{(q)})\}}{\gamma(\boldsymbol{\tau}^{(p+q)})} \leq \theta(\boldsymbol{\tau}^{(p+q)}) \leq \frac{\theta(\boldsymbol{\tau}^{(p)})\gamma(\boldsymbol{\tau}^{(p)}) + \theta(\boldsymbol{\tau}^{(q)})\gamma(\boldsymbol{\tau}^{(q)})}{\gamma(\boldsymbol{\tau}^{(p+q)})}, \tag{3}$$

where

$$\gamma(\boldsymbol{\tau}^{(p+q)}) = -\log G(F^{-1}(e^{-\tau_1}), \ldots, F^{-1}(e^{-\tau_{p+q}})) = \lim_{n\to\infty} nP\left(\mathbf{X}_1^{(p+q)} \not\leq u_n^{(\boldsymbol{\tau}^{(p+q)})}\right)$$

$$\gamma(\boldsymbol{\tau}^{(p)}) = \lim_{\substack{\tau_j \to 0+ \\ j=p+1,\ldots,p+q}} \gamma(\boldsymbol{\tau}^{(p+q)}) \text{ and } \gamma(\boldsymbol{\tau}^{(q)}) = \lim_{\substack{\tau_j \to 0+ \\ j=1,\ldots,p}} \gamma(\boldsymbol{\tau}^{(p+q)}).$$

The next result follows immediately from these bounds.

Proposition 1. *Suppose that* $Q \in D(G)$ *and* $\mathbf{X}$ *has extremal index* $\theta(\boldsymbol{\tau}^{(p+q)})$, $\boldsymbol{\tau}^{(p+q)} \in \mathbb{R}_+^{p+q}$.

(i) *If* $\widehat{\mathbf{Y}}^{(p)}$ *and* $\widehat{\mathbf{Y}}^{(q)}$ *are independent, then* $\mathbf{Y}^{(p)}$ *and* $\mathbf{Y}^{(q)}$ *are independent if and only if*

$$\theta(\boldsymbol{\tau}^{(p+q)}) = \frac{\theta(\boldsymbol{\tau}^{(p)})\gamma(\boldsymbol{\tau}^{(p)}) + \theta(\boldsymbol{\tau}^{(q)})\gamma(\boldsymbol{\tau}^{(q)})}{\gamma(\boldsymbol{\tau}^{(p)}) + \gamma(\boldsymbol{\tau}^{(q)})}, \quad \boldsymbol{\tau}^{(p+q)} \in \mathbb{R}_+^{p+q}.$$

(ii) *If* $\widehat{\mathbf{Y}}^{(p)}$ *and* $\widehat{\mathbf{Y}}^{(q)}$ *are totally dependent, then* $\mathbf{Y}^{(p)}$ *and* $\mathbf{Y}^{(q)}$ *are totally dependent if and only if*

$$\theta(\boldsymbol{\tau}^{(p+q)}) = \frac{\max\{\theta(\boldsymbol{\tau}^{(p)})\gamma(\boldsymbol{\tau}^{(p)}), \theta(\boldsymbol{\tau}^{(q)})\gamma(\boldsymbol{\tau}^{(q)})\}}{\max\{\gamma(\boldsymbol{\tau}^{(p)}), \gamma(\boldsymbol{\tau}^{(q)})\}}, \quad \boldsymbol{\tau}^{(p+q)} \in \mathbb{R}_+^{p+q}.$$

The necessary and sufficient conditions for $\mathbf{Y}^{(p)}$ and $\mathbf{Y}^{(q)}$ to be independent or totally dependent given in the previous result demand the evaluation of the extremal index function $\theta(\boldsymbol{\tau}^{(p+q)})$ in each point $\boldsymbol{\tau}^{(p+q)} \in \mathbb{R}_+^{p+q}$. Nevertheless, this task can be simplified with the characterizations, given in Ferreira [3], for independence and total dependence of the multivariate marginals of a MEV distribution. These characterizations are essential to prove the following propositions which guarantee that the independence or total dependence between $\mathbf{Y}^{(p)}$ and $\mathbf{Y}^{(q)}$ only depends on the value of the extremal index in some points.

Proposition 2. *Suppose that $Q \in D(G)$ and the sequence $\mathbf{X} = \{\mathbf{X}_n^{(p+q)}\}_{n\geq 1}$ has extremal index $\theta(\boldsymbol{\tau}^{(p+q)})$, $\boldsymbol{\tau}^{(p+q)} \in \mathbb{R}_+^{p+q}$.*

The sub-vectors $\mathbf{Y}^{(p)}$ and $\mathbf{Y}^{(q)}$ are independent if and only if

$$\theta(\mathbf{1}^{(p+q)}) = \frac{\theta(\mathbf{1}^{(p)})\gamma(\mathbf{1}^{(p)}) + \theta(\mathbf{1}^{(q)})\gamma(\mathbf{1}^{(q)})}{\gamma(\mathbf{1}^{(p+q)})}, \tag{4}$$

where $\mathbf{1}^{(k)} = (1, \ldots, 1)$, $k > 1$, denotes the k-dimensional unitary vector.

Proof. Suppose that $\mathbf{Y}^{(p)}$ and $\mathbf{Y}^{(q)}$ are independent. Since Eq. (3) holds for all $\boldsymbol{\tau}^{(p+q)} \in \mathbb{R}_+^{p+q}$, we have in particular for $\boldsymbol{\tau}^{(p+q)} = (\tau, \ldots, \tau) \in \mathbb{R}_+^{p+q}$, with $\tau \equiv \tau(x) = -\log F(x)$, $x \in \mathbb{R}$,

$$\theta(\boldsymbol{\tau}^{(p+q)}) = \frac{\theta(\boldsymbol{\tau}^{(p)})\gamma(\boldsymbol{\tau}^{(p)}) + \theta(\boldsymbol{\tau}^{(q)})\gamma(\boldsymbol{\tau}^{(q)})}{\gamma(\boldsymbol{\tau}^{(p+q)})}.$$

Now from the fact that $\theta(c\boldsymbol{\tau}^{(k)}) = \theta(\boldsymbol{\tau}^{(k)})$ for each $\boldsymbol{\tau}^{(k)} \in \mathbb{R}_+^k$, $k > 1$ and $c > 0$, we can write

$$\theta(\boldsymbol{\tau}^{(p+q)}) = \theta(\mathbf{1}^{(p+q)}), \quad \theta(\boldsymbol{\tau}^{(p)}) = \theta(\mathbf{1}^{(p)}), \quad \theta(\boldsymbol{\tau}^{(q)}) = \theta(\mathbf{1}^{(q)}),$$

and from Eq. (1), for all $\boldsymbol{\tau}^{(p+q)} = (\tau, \ldots, \tau) \in \mathbb{R}_+^{p+q}$,

$$\begin{aligned}\gamma(\boldsymbol{\tau}^{(p+q)}) &= -\log G(F^{-1}(e^{-\tau}), \ldots, F^{-1}(e^{-\tau})) = -\log D_G(e^{-\tau}, \ldots, e^{-\tau}) \\ &= -\log D_G^{\tau}(e^{-1}, \ldots, e^{-1}) = \tau\gamma(\mathbf{1}^{(p+q)}),\end{aligned} \tag{5}$$

$\gamma(\boldsymbol{\tau}^{(p)}) = \tau\gamma(\mathbf{1}^{(p)})$ and $\gamma(\boldsymbol{\tau}^{(q)}) = \tau\gamma(\mathbf{1}^{(q)})$. Equality (4) is now straightforward.

On the other hand if Eq. (4) is verified, then for $\mathbf{x}^{(p+q)} = (x, \ldots, x)$ we have

$$\begin{aligned}G_{\mathbf{Y}}(\mathbf{x}^{(p+q)}) &= G^{\theta(\mathbf{1}^{(p+q)})}(\mathbf{x}^{(p+q)}) = D_G^{\theta(\mathbf{1}^{(p+q)})}(e^{-\tau}, \ldots, e^{-\tau}) \\ &= D_G^{\theta(\mathbf{1}^{(p+q)})\tau}(e^{-1}, \ldots, e^{-1}) = \exp(-\tau\gamma(\mathbf{1}^{(p+q)})\theta(\mathbf{1}^{(p+q)})) \\ &= \exp(-\tau(\theta(\mathbf{1}^{(p)})\gamma(\mathbf{1}^{(p)}) + \theta(\mathbf{1}^{(q)})\gamma(\mathbf{1}^{(q)}))) = G_{\mathbf{Y}^{(p)}}(\mathbf{x}^{(p)})G_{\mathbf{Y}^{(q)}}(\mathbf{x}^{(q)})\end{aligned}$$

and from Proposition 2.1 [3] we conclude that $\mathbf{Y}^{(p)}$ and $\mathbf{Y}^{(q)}$ are independent. □

Proposition 3. *Suppose that $Q \in D(G)$, $\mathbf{X} = \{\mathbf{X}_n^{(p+q)}\}_{n\geq 1}$ has extremal index $\theta(\boldsymbol{\tau}^{(p+q)})$, $\boldsymbol{\tau}^{(p+q)} \in \mathbb{R}_+^{p+q}$.*

(i) *If $\mathbf{Y}^{(p)}$ and $\mathbf{Y}^{(q)}$ are totally dependent then there exists $\boldsymbol{\tau}^{(p+q)} \in \mathbb{R}_+^{p+q}$ with $\tau_j \equiv \tau_j(x_j) = -\log F(x_j), x_j \in \mathbb{R}$, $j = 1, \ldots, p+q$, such that*

$$\gamma(\boldsymbol{\tau}^{(p)})\theta(\boldsymbol{\tau}^{(p)}) = \gamma(\boldsymbol{\tau}^{(q)})\theta(\boldsymbol{\tau}^{(q)}) = \theta_1\tau_1 \ldots = \theta_{p+q}\tau_{p+q} = d > 0$$

and $\theta(\boldsymbol{\tau}^{(p+q)}) = \left[\gamma\left(\frac{\boldsymbol{\tau}^{(p+q)}}{d}\right)\right]^{-1}$.

(ii) *If there exists* $\boldsymbol{\tau}^{(p+q)} \in \mathbb{R}_+^{p+q}$ *with* $\tau_j \equiv \tau_j(x_j) = -\log F(x_j)$, $x_j \in \mathbb{R}$, $j = 1, \ldots, p+q$, *such that*

$$\gamma(\boldsymbol{\tau}^{(p+q)})\theta(\boldsymbol{\tau}^{(p+q)}) = \theta_1\tau_1 \ldots = \theta_{p+q}\tau_{p+q} = d > 0,$$

then $\mathbf{Y}^{(p)}$ *and* $\mathbf{Y}^{(q)}$ *are totally dependent.*

Proof. **(i)** From Proposition 2.1 [3], if $\mathbf{Y}^{(p)}$ and $\mathbf{Y}^{(q)}$ are totally dependent, then there exists $\boldsymbol{\tau}^{(p+q)} \in \mathbb{R}_+^{p+q}$ such that

$$\theta(\boldsymbol{\tau}^{(p)})\gamma(\boldsymbol{\tau}^{(p)}) = \theta(\boldsymbol{\tau}^{(q)})\gamma(\boldsymbol{\tau}^{(q)}) = \theta(\boldsymbol{\tau}^{(p+q)})\gamma(\boldsymbol{\tau}^{(p+q)}) = d$$
$$= \theta_1\tau_1 = \ldots = \theta_{p+q}\tau_{p+q},$$

with $d \in]0,1[$. Hence

$$\theta(\boldsymbol{\tau}^{(p+q)}) = \frac{d}{\gamma(\boldsymbol{\tau}^{(p+q)})} = \frac{d}{-\log D_G(\exp(-\tau_1), \ldots, \exp(-\tau_{p+q}))}$$
$$= \frac{1}{-\log D_G\left(\exp\left(-\frac{\tau_1}{d}\right), \ldots, \exp\left(-\frac{\tau_{p+q}}{d}\right)\right)} = \frac{1}{\gamma\left(\frac{\boldsymbol{\tau}^{(p+q)}}{d}\right)}.$$

□

Another way to look at issues concerning independence or total dependence is to use parameters that measure the strength of dependence between $\mathbf{Y}^{(p)}$ and $\mathbf{Y}^{(q)}$. We therefore define, in the following result, the dependence structure of $\mathbf{Y}^{(p)}$ and $\mathbf{Y}^{(q)}$ through the coefficient $\epsilon^{(\mathbf{Y}^{(p)},\mathbf{Y}^{(q)})}$ of Ferreira [3]. This coefficient emerged from the extremal coefficient of $\mathbf{Y}$, $\epsilon^{\mathbf{Y}}$, defined in Martins and Ferreira [6] as

$$G^{\theta(\mathbf{1}^{(p+q)})}(\mathbf{x}^{(p+q)}) = F^{\epsilon^{\mathbf{Y}}}(x), \quad x \in \mathbb{R},$$

and the relationship

$$P\left(\mathbf{Y}^{(p)} \le \mathbf{x}^{(p)}, \mathbf{Y}^{(q)} \le \mathbf{x}^{(q)}\right) = \left(G_{\mathbf{Y}}^{(p)}(\mathbf{x}^{(p)})G_{\mathbf{Y}}^{(q)}(\mathbf{x}^{(q)})\right)^{\frac{\epsilon^{\mathbf{Y}}}{\epsilon^{\mathbf{Y}^{(p)}}+\epsilon^{\mathbf{Y}^{(q)}}}}.$$

It is then defined as $\epsilon^{(\mathbf{Y}^{(p)},\mathbf{Y}^{(q)})} = \frac{\epsilon^{\mathbf{Y}}}{\epsilon^{\mathbf{Y}^{(p)}}+\epsilon^{\mathbf{Y}^{(q)}}}$ and has the following properties.

Proposition 4. **(i)** $\epsilon^{(\mathbf{Y}^{(p)},\mathbf{Y}^{(q)})} = \frac{\theta(\mathbf{1}^{(p+q)})\gamma(\mathbf{1}^{(p+q)})}{\theta(\mathbf{1}^{(p)})\gamma(\mathbf{1}^{(p)})+\theta(\mathbf{1}^{(q)})\gamma(\mathbf{1}^{(q)})}$.

(ii) $\epsilon^{(\mathbf{Y}^{(p)},\mathbf{Y}^{(q)})} = 1$ *if and only if* $\mathbf{Y}^{(p)}$ *and* $\mathbf{Y}^{(q)}$ *are independent.*

(iii) *If* $\mathbf{Y}^{(p)}$ *and* $\mathbf{Y}^{(q)}$ *are totally dependent, then* $\epsilon^{(\mathbf{Y}^{(p)},\mathbf{Y}^{(q)})} = \frac{\max\{\epsilon^{\mathbf{Y}^{(p)}},\epsilon^{\mathbf{Y}^{(q)}}\}}{\epsilon^{\mathbf{Y}^{(p)}}+\epsilon^{\mathbf{Y}^{(q)}}}$.

Proof. **(i)** Since

$$\epsilon^{(\mathbf{Y}^{(p)},\mathbf{Y}^{(q)})} = \frac{\theta(\mathbf{1}^{(p+q)}) \log G(\mathbf{x}^{(p+q)})}{\theta(\mathbf{1}^{(p)}) \log G(\mathbf{x}^{(p)}) + \theta(\mathbf{1}^{(q)}) \log G(\mathbf{x}^{(q)}))},$$

the result follows from Eq. (5).

(ii) It is an immediate consequence of (i) and Proposition 2.

(iii) If $\mathbf{Y}^{(p)}$ and $\mathbf{Y}^{(q)}$ are totally dependent, then from Eq. (2), we have

$$\epsilon^{(\mathbf{Y}^{(p)},\mathbf{Y}^{(q)})} = \frac{-\log \min\{G^{\theta(\mathbf{1}^{(p)})}(\mathbf{x}^{(p)}), G^{\theta(\mathbf{1}^{(q)})}(\mathbf{x}^{(q)})\}}{\theta(\mathbf{1}^{(p)}) \log G(\mathbf{x}^{(p)}) + \theta(\mathbf{1}^{(q)}) \log G(\mathbf{x}^{(q)})} = \frac{\max\{\epsilon^{\mathbf{Y}^{(p)}}, \epsilon^{\mathbf{Y}^{(q)}}\}}{\epsilon^{\mathbf{Y}^{(p)}} + \epsilon^{\mathbf{Y}^{(q)}}}. \quad \square$$

3 Examples

Example 1. Let $\{Y_n\}_{n\geq 1}$ be a sequence of i.i.d. random variables with common d.f. $\dot{F}$ and consider an auto-regressive sequence of maxima $\{X_n\}_{n\geq 1}$ defined by

$$X_n = \max\{Y_n, Y_{n+1}\},\ n \geq 1,$$

with marginal distribution function $\dot{F}^2$.

Let $\{u_n^{(\tau_i)}\}_{n\geq 1}$, $i = 1, \ldots, p$, and $\{v_n^{(\tau'_j)}\}_{n\geq 1}, j = p+1, \ldots, p+q$, be sequences of real numbers such that $\lim_{n\to\infty} n(1 - \dot{F}^2(u_n^{(\tau_i)})) = \tau_i$ and $\lim_{n\to\infty} n\dot{F}^2(-v_n^{(\tau'_j)})) = \tau'_j$.

The sequences $\{X_n\}_{n\geq 1}$ and $\{-X_n\}_{n\geq 1}$ have, respectively, extremal indexes $\theta_1 = 1/2$ and $\theta_2 = 1$.

For sequences $\mathbf{X}_n^{(p+q)} = \begin{cases} X_{n,i} = X_n & , i = 1, \ldots p \\ X_{n,i} = -X_n & , i = p+1, \ldots, p+q \end{cases}$, $\mathbf{X}_n^{(p)} = (X_n, \ldots, X_n)$ and $\mathbf{X}_n^{(q)} = (-X_n, \ldots, -X_n)$, we have

$$\lim_{n\to\infty} P(\mathbf{M}_n^{(p)} \leq \mathbf{u}_n^{(\tau^{(p)})}) = \exp\left(-\frac{1}{2} \max_{1\leq j\leq p} \tau_j\right),$$

$$\lim_{n\to\infty} P(\widehat{\mathbf{M}}_n^{(p)} \leq \mathbf{u}_n^{(\tau^{(p)})}) = \exp\left(-\max_{1\leq j\leq p} \tau_j\right),$$

$$\lim_{n\to\infty} P\left(\mathbf{M}_n^{(q)} \leq (v_n^{(\tau'_{p+1})}, \ldots, v_n^{(\tau'_{p+q})})\right) = \exp\left(-\max_{p+1\leq j\leq p+q} \tau'_j\right).$$

Since the order statistics maximum and minimum are asymptotically independent [2, 8] we obtain

$$P(\mathbf{M}_n \leq (u_n^{(\tau_1)}, \ldots, u_n^{(\tau_p)}, v_n^{(\tau'_{p+1})}, \ldots, v_n^{(\tau'_{p+q})})) \xrightarrow[n\to\infty]{} \exp\left(-\frac{1}{2}\max_{1\leq j\leq p} \tau_j - \max_{p+1\leq j\leq p+q} \tau'_j\right)$$

and consequently $\gamma(\boldsymbol{\tau}^{(p+q)}) = \gamma(\boldsymbol{\tau}^{(p)}) + \gamma(\boldsymbol{\tau}^{(q)}) = \max_{1\le j\le p} \tau_j + \max_{p+1\le j\le p+q} \tau'_j$ and $\theta(\boldsymbol{\tau}^{(p+q)})\gamma(\boldsymbol{\tau}^{(p+q)}) = \frac{1}{2} \max_{1\le j\le p} \tau_j + \max_{p+1\le j\le p+q} \tau'_j$. Therefore

$$\theta(\boldsymbol{\tau}^{(p+q)}) = \frac{\theta(\boldsymbol{\tau}^{(p)})\gamma(\boldsymbol{\tau}^{(p)}) + \theta(\boldsymbol{\tau}^{(q)})\gamma(\boldsymbol{\tau}^{(q)})}{\gamma(\boldsymbol{\tau}^{(p)}) + \gamma(\boldsymbol{\tau}^{(q)})}.$$

Example 2. Let $\mathbf{U} = \{U_n\}_{n\ge1}$ be a sequence of i.i.d. random variables with common d.f. H in the domain of attraction of the extreme value distribution F, and independent of the i.i.d. chain $\mathbf{J} = \{J_n\}_{n\ge1}$ such that $P(J_1 = 0) = P(J_1 = 1) = 1/2$.

Let us consider a stationary 1-dependent sequence $\mathbf{Z} = \{Z_n\}_{n\ge1}$, defined as $Z_n = U_n$ if $J_n = 0$ and $Z_n = U_{n+1}$ otherwise, and let $\mathbf{v} = \{v_n\}_{n\ge1}$ be a sequence of normalized levels to $\mathbf{Z}$, and consequently also to $\mathbf{U}$.

We can now define a 3-dependent stationary sequence $\mathbf{X} = \{\mathbf{X}_n = (X_{n,1}, X_{n,2}, X_{n,3})\}$ as

$$(X_{n,1}, X_{n,2}, X_{n,3}) = (Z_n, Z_{n+2}, Z_{n+1}),\ \ n \ge 1,$$

with common distribution function

$$T(x_1,x_2,x_3) = \frac{1}{2}\prod_{i=1}^{3} H(x_i) + \frac{1}{4}H(x_1)H(\min\{x_2,x_3\}) + \frac{1}{4}H(x_2)H(\min\{x_1,x_3\})$$

belonging to the domain of attraction of

$$G(x_1,x_2,x_3) = \begin{cases} F(x_1)F(x_2)F^{\frac{1}{2}}(x_3) & , \ x_1 < x_3 \wedge x_2 < x_3 \\ F(x_1)F^{\frac{3}{4}}(x_2)F^{\frac{3}{4}}(x_3) & , \ x_1 < x_3 \wedge x_3 \le x_2 \\ F^{\frac{3}{4}}(x_1)F(x_2)F^{\frac{3}{4}}(x_3) & , \ x_3 \le x_1 \wedge x_2 < x_3 \\ F^{\frac{3}{4}}(x_1)F^{\frac{3}{4}}(x_2)F(x_3) & , \ x_3 \le x_1 \wedge x_3 \le x_2 \end{cases}$$

Now applying Proposition 2.1 of Smith and Weissman [9] to the sequence $U^{\mathbf{X}} = \{\max\{X_{n,1}, X_{n,2}, X_{n,3}\} = \max\{Z_n, Z_{n+1}, Z_{n+2}\}\}_{n\ge1}$ which verifies the condition $D^{(k)}(v_n),\ k = 2$, of Chernick et al. [1], we easily obtain

$$\theta^{\mathbf{X}}(1,1,1) = \lim_{n\to\infty} \frac{P\left(\max\{Z_1,Z_2,Z_3\} > v_n \ge \max\{Z_2,Z_3,Z_4\}\right)}{P\left(\max\{Z_1,Z_2,Z_3\} > v_n\right)} = \frac{3}{10}.$$

For random vectors $\widehat{\mathbf{Y}} = \left(\widehat{Y}_1, \widehat{Y}_2, \widehat{Y}_3\right)$ and $\mathbf{Y} = (Y_1, Y_2, Y_3)$ with d.f. $G_{\widehat{\mathbf{Y}}} \equiv G$ and $G_{\mathbf{Y}} \equiv G^{\theta(\boldsymbol{\tau}^{(3)})}$, $\mathbf{Y}^{(2)} = (Y_1, Y_2)$ and $\mathbf{Y}^{(1)} = Y_3$ we obtain $\epsilon^{\widehat{\mathbf{Y}}} = \frac{5}{2}$, $\epsilon^{\mathbf{Y}} = \frac{3}{4}$, $\epsilon^{\widehat{\mathbf{Y}}^{(2)}} = 2$, $\epsilon^{\widehat{\mathbf{Y}}^{(1)}} = 1$, $\epsilon^{\mathbf{Y}^{(1)}} = \frac{3}{4} = \epsilon^{\mathbf{Y}^{(2)}}$. Consequently $\epsilon^{(\mathbf{Y}^{(2)},\mathbf{Y}^{(1)})} = \frac{1}{2}$, $\epsilon^{\left(\widehat{\mathbf{Y}}^{(2)},\widehat{\mathbf{Y}}^{(1)}\right)} = \frac{5}{6}$ and

from Proposition 4 we conclude that neither $\mathbf{Y}^{(1)}$ and $\mathbf{Y}^{(2)}$ nor $\widehat{\mathbf{Y}}^{(1)}$ and $\widehat{\mathbf{Y}}^{(2)}$ are independent.
Nevertheless, there exists $\boldsymbol{\tau}^{(3)} = (1,1,1)$ such that $\gamma(\boldsymbol{\tau}^{(3)})\theta(\boldsymbol{\tau}^{(3)}) = \theta_1\tau_1 = \theta_2\tau_2 = \theta_3\tau_3 = \frac{3}{4}$ and from Proposition 3 we can say that $\mathbf{Y}^{(1)}$ and $\mathbf{Y}^{(2)}$ are totally dependent.

Acknowledgments The authors thank the referees for all the helpful remarks. This research was supported by the research unit "Centro de Matemática" of the University of Beira Interior through the Foundation for Science and Technology (FCT).

References

1. Chernick, M., Hsing, T., and McCormick, W.: Calculating the extremal index for a class of stationary sequences. Adv. Appl. Prob. **23**, 835–850 (1991)
2. Davis, R.: Limit laws for the maximum and minimum of stationary sequences. Z. Wahrsch. verw. Gebiete **61**, 31–42 (1982)
3. Ferreira, H.: Dependence between two multivariate extremes. Statist. Prob. Letters **81**(5), 586–591 (2011)
4. Hsing, T.: Extreme value theory for multivariate stationary sequences. J. Mult. Anal. **29**, 274–291 (1989)
5. Leadbetter, M.R.: Extremes and local dependence in stationary sequences. Z. Wahrsch. verw. Gebiete **65**, 291–306 (1983)
6. Martins, A.P., Ferreira, H.: Measuring the extremal dependence. Statist. Prob. Letters **73**, 99–103 (2005)
7. Nandagopalan, S.: Multivariate extremes and estimation of the extremal index. Ph.D. Thesis, Department of Statistics, University of North Carolina, Chapel Hill (1990)
8. Pereira, L.: Valores extremos multidimensionais de variáveis dependentes. Ph.D. Thesis, University of Beira Interior, Portugal (2002)
9. Smith, R.L., Weissman, I.: Characterization and estimation of the multivariate extremal index. Technical report, University of North Carolina at Chapel Hill, NC, USA (1996) In http://www.stat.unc.edu/postscript/rs/extremal.pdf

The manufacturer's authorised representative in the EU is Springer Nature Customer Service Centre GmbH, Europaplatz 3, 69115 Heidelberg, Germany. If you have any concerns regarding our products, please contact ProductSafety@springernature.com

Printed and bound by CPI Group (UK) Ltd, Croydon, CR0 4YY
25/09/2026
02222853-0011